T0189935

Communications
in Computer and Information Science 1148

Commenced Publication in 2007
Founding and Former Series Editors:
Phoebe Chen, Alfredo Cuzzocrea, Xiaoyong Du, Orhun Kara, Ting Liu,
Krishna M. Sivalingam, Dominik Ślęzak, Takashi Washio, Xiaokang Yang,
and Junsong Yuan

More information about this series at http://www.springer.com/series/7899

Neeta Nain · Santosh Kumar Vipparthi ·
Balasubramanian Raman (Eds.)

Computer Vision and Image Processing

4th International Conference, CVIP 2019
Jaipur, India, September 27–29, 2019
Revised Selected Papers, Part II

 Springer

Editors
Neeta Nain
Malaviya National Institute of Technology
Jaipur, Rajasthan, India

Santosh Kumar Vipparthi
Malaviya National Institute of Technology
Jaipur, Rajasthan, India

Balasubramanian Raman
Indian Institute of Technology Roorkee
Roorkee, Uttarakhand, India

ISSN 1865-0929   ISSN 1865-0937 (electronic)
Communications in Computer and Information Science
ISBN 978-981-15-4017-2   ISBN 978-981-15-4018-9 (eBook)
https://doi.org/10.1007/978-981-15-4018-9

This Springer imprint is published by the registered company Springer Nature Singapore Pte Ltd.
The registered company address is: 152 Beach Road, #21-01/04 Gateway East, Singapore 189721, Singapore

Preface

This volume contains the papers from the 4th International Conference on Computer Vision and Image Processing (CVIP 2019). The event was endorsed by the International Association for Pattern Recognition (IAPR) and organized by Malaviya National Institute of Technology, Jaipur, during September 27–29, 2019.

CVIP is a premier conference focused on image, video processing, and computer vision. The conference featured world-renowned speakers, technical workshops, and demonstrations.

CVIP 2019 acted as a major forum for presentation of technological progress and research outcomes in the area of image processing and computer vision, serving as a platform for exchange between academia and industry. The selected papers come from around 202 original submissions by researchers based in several countries including South Korea, Norway, Malaysia, Iceland, Ethiopia, Canada, Bangladesh, India, and the USA. The highly diversified audience gave us the opportunity to achieve a good level of understanding of the mutual needs, requirements, and technical means available in this field of research.

The topics included in this edition of CVIP the following fields connected to computer vision and image processing: data acquisition and modeling, visualization and audio methods, sensors and actuators, data mining, image enhancement and restoration, segmentation, object detection and tracking, video analysis and summarization, biometrics and forensics, deep learning, document image analysis, remote sensing, multi-spectral and hyper-spectral image processing, etc. All the accepted papers were double-blind peer reviewed by three qualified reviewers chosen from our Technical Committee based on their qualifications, areas of interest, and experience. The papers were evaluated on their relevance to CVIP 2019 tracks and topics, scientific correctness, and clarity of presentation. Selection was based on these reviews and on further recommendations by the Program Committee.

The editors of the current proceedings are very grateful and wish to thank the dedicated Technical Committee members and all the other reviewers for their valuable contributions, commitment, and enthusiastic support. We also thank CCIS at Springer for their trust and for publishing the proceedings of CVIP 2019.

September 2019

Neeta Nain
Santosh Kumar Vipparthi
Balasubramanian Raman

Organization

Organizing Committee

Neeta Nain	MNIT Jaipur, India
Santosh Kumar Vipparthi	MNIT Jaipur, India
Partha Pratim Roy	IIT Roorkee, India
Ananda Shankar Chowdhary	Jadavpur University, India

Program Committee

Balasubramanian Raman	IIT Roorkee, India
Sanjeev Kumar	IIT Roorkee, India
Arnav Bhaskar	IIT Mandi, India
Subramanyam Murala	IIT Ropar, India
Abhinav Dhall	IIT Ropar, India

International Advisory Committee

Uday Kumar R. Yaragatti	MNIT Jaipur, India
Anil K. Jain	Michigan State University, USA
Bidyut Baran Chaudhari	ISI Kolkata, India
Mohamed Abdel Mottaleb	University of Miami, USA
Mohan S. Kankanhalli	NUS, Singapore
Ajay Kumar	Hong Kong Poly University, Hong Kong
Ales Prochazka	Czech Technical University, Czech Republic
Andrea Kutics	ICU, Japan
Daniel P. Lopresti	Lehigh University, USA
Gian Luca Foresti	University of Udine, Italy
Jonathan Wu	University of Windsor, Canada
Josep Llados	University of Barcelona, Spain
Kokou Yetongnon	University of Burgundy, France
Koichi Kise	Osaka Prefecture University, Japan
Luigi Gallo	National Research Council, Italy
Slobodan Ribaric	University of Zagreb, Croatia
Umapada Pal	ISI Kolkata, India
Xiaoyi Jiang	University of Münster, Germany

Local Committee

Emmanuel S. Pilli	MNIT Jaipur, India
Dinesh Kumar Tyagi	MNIT Jaipur, India
Vijay Laxmi	MNIT Jaipur, India
Arka Prakash Mazumdar	MNIT Jaipur, India
Mushtaq Ahmed	MNIT Jaipur, India
Yogesh Kumar Meena	MNIT Jaipur, India
Satyendra Singh Chouhan	MNIT Jaipur, India
Mahipal Jadeja	MNIT Jaipur, India
Madhu Agarwal	MNIT Jaipur, India
Kuldeep Kumar	MNIT Jaipur, India
Prakash Choudhary	NIT Hamirpur, India
Maroti Deshmukh	NIT Uttarakhand, India
Subhash Panwar	GEC Bikaner, India
Tapas Badal	Bennett University, India
Sonu Lamba	MNIT Jaipur, India
Riti Kushwaha	MNIT Jaipur, India
Praveen Kumar Chandaliya	MNIT Jaipur, India
Rahul Palliwal	MNIT Jaipur, India
Kapil Mangal	MNIT Jaipur, India
Ravindra Kumar Soni	MNIT Jaipur, India
Gopal Behera	MNIT Jaipur, India
Sushil Kumar	MNIT Jaipur, India

Sponsors

Contents – Part II

Object Detection

Object Recognition

Online Handwriting Recognition

Optical Character Recognition

Security and Privacy

Unsupervised Clustering

Contents – Part I

Image Segmentation

Neural Network

Denoising Images with Varying Noises Using Autoencoders

Snigdha Agarwal, Ayushi Agarwal, and Maroti Deshmukh$^{(\boxtimes)}$

Department of Computer Science and Engineering,
National Institute of Technology, Uttarakhand, Srinagar, India
{snigdha.cse16,ayushi16.cse,marotideshmukh}@nituk.ac.in

Abstract. Image processing techniques are readily used in the field of sciences and computer vision for the enhancement of images and extraction of useful information from them. A key step used in image processing involves the removal of different kinds of noises from the images. Noises can arise in an image during the process of storing, transmitting or acquiring the images. A model qualifies as a satisfactory de-noising model if it satisfies image preservation along with noise removal. There can be various kind of noises in an image such as Gaussian, salt and pepper, Speckle etc. A model which can denoise a different kind of noises is considered to be superior to others. In this paper, we have designed a model using autoencoder which can remove several kinds of noises from images. We have performed a comparative study between the accuracy of each kind using PSNR, SSIM and RMSE values. An increase in the PSNR and SSIM values was seen from the original and noisy image to the original and reconstructed image while a decrease was seen in the value of RMSE.

Keywords: Image processing · Denoising · Autoencoder · CNN · Noise · Cifar-10

1 Introduction

Image processing [1] is the most fundamental part of computer vision. Image processing involves the conversion of an image into its digital equivalent by deploying various operations for the extraction of required features or to get an enhanced version of the original image. A major challenge in the discipline of image processing is the denoising of images [2] which requires the estimation of the original denoised image by the removal or suppression of all kinds of noises from the noisy version of the image. Noise is basically the presence of unwanted signals or disturbances in an image leading to visual distortion. One can define noise in an image as the presence of random variations. Noise can occur due to varying brightness, colors or contrasts in an image. In an image, noise can be in the form of tiny speckles, grains or multi-colored pixels [3].

© Springer Nature Singapore Pte Ltd. 2020
N. Nain et al. (Eds.): CVIP 2019, CCIS 1148, pp. 3–14, 2020.
https://doi.org/10.1007/978-981-15-4018-9_1

Mainly three types of noise present in the image during storing, transmitting or acquiring the images. Gaussian noise [4] in an image is obtained during acquisition of an image. The image containing Gaussian noise is shown in Fig. 1(b). Salt and pepper noise [5] is a kind of data-dropout noise. Also referred to as spikes, this noise is mainly obtained by the presence of errors or sudden disturbances in the input data or the signals. The image containing Salt and Pepper noise is shown in Fig. 1(c). Speckle noise [6] is basically a kind of granular noise which is obtained by randomly fluctuating signals in an image. The image containing Speckle noise is shown in Fig. 1(d).

    (a)   (b)   (c)   (d)

Fig. 1. Effect of noise in girl image: (a) Original girl image, (b) Gaussian noise, (c) Salt and Pepper noise, (d) Speckle noise.

Autoencoders are based on artificial neural networks and make use of unsupervised learning techniques to perform various feature detection and classification techniques. Autoencoders are basically feed forward neural networks in which the input and the output are almost the same using encoder and decoder as shown in Fig. 2. Convolutional autoencoders [9, 25, 26] are modified form of the basic architecture of autoencoders, differing only on the basis of convolutional encoding and decoding layers instead of the layers present in the original architecture. Convolutional autoencoders are much more efficient as compared to the classical autoencoders as they are fully capable of exploiting the structure of the image owing to the utilization of the concept of convolutional neural networks for their structure.

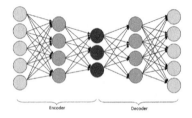

Fig. 2. Structure of autoencoder.

The rest of the paper is structured as follows. Section 2 describes the background work. The proposed model is described in Sect. 3. The experimental setup and results are discussed in Sect. 4. In the end, the work has been concluded in Sect. 5.

2   Background Work

Usage of unsupervised pre-training for the initialization of neural networks in case of supervised tasks has been prevalent from the times when stacked autoencoders [10] and deep belief networks (DBN) [11] were proposed. Unsupervised learning performed layer-by-layer is used for the minimization of the reconstruction error present in the input data. After the pre-training of all the layers present in the model has finished, the network is subjected to a stage of fine-tuning. Later, in order to stop the autoencoder from directly copying the input data to the output, the concept of de-noising autoencoders [12] were introduced which could learn the representations from noisy data. This creates a layer-wise unsupervised pre-training pattern to increase the overall performance of the model. Also, Erhan et al. [13] proved unsupervised pre-training to be a special kind of regularization.

The popularity of unsupervised pre- training for the training of convolutional neural networks has been on a downfall in the recent years owing to the growth and development of newer models and architectures. However, a number of works [14–16] have proved that in spite of the decreasing popularity, unsupervised pre-training can be of great significance to supervised training by the use of extra unlabeled samples for learning representations in the data.

A variety of methods were proposed for the learning of representation followed by the reconstruction of images. Wang et al. [15] made use of the technique of visual tracking for the unsupervised learning of various representations from unlabeled videos. Doersch et al. [14] suggested the use of prediction of the exact position of patches obtained from the image for learning the representations in the image. Context encoder [16], a much more well known concept to the one that is being proposed, learns various representations and features by the process of images using a convolutional network autoencoder along with zero masking corruption.

3   Proposed Model

Image denoising is used for preprocessing of images by removing noise from them. In our model, we have exploited the use of autoencoders as shown in Fig. 3 to achieve the purpose. The encoder compresses the original input data through various layers, extracting the most relevant features until a bottleneck consisting of a latent space representation is reached. The latent space representation [8] is basically a compact and compressed version of the original input. The decoder works on this compressed representation of the code to reconstruct the original image. The final output obtained after the reconstruction is a lossy version of the image. The architecture of our autoencoder consists of various Convolution layers [17] followed by Max pooling layers for encoding and convolutional layers followed by Up sampling layers (deconvolution [18]) for decoding as shown in Fig. 4. Each convolutional layer consists of 32 filters of size 3×3 and padding such that the size of the image remains the same. Each convolution is immediately followed by a batch normalization and an activation (Relu) layer. The Max pooling layers and the up Sampling layers used are of size 2×2.

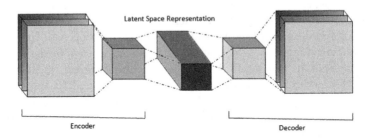

Fig. 3. Autoencoder for image denoising.

Different types of noises were added to images. Gaussian noise [4] was added using Algorithm 1 in which random values for the addition of noise were obtained by the use of a Gaussian function as shown in Eq. 1 where μ is the mean and σ is the variance. Salt and Pepper noise [5] was added using Algorithm 2 where salt and pepper ratio was used as a means to provide the amount of addition of dark and bright spots in colored images. Speckle noise [6] was added using Algorithm 3 in which a random matrix was used to add the noise. These images were passed to our model to minimize the loss function (binary cross entropy) and give the denoised image.

$$P_z = \frac{1}{\sigma\sqrt{2\pi}} e^{-\frac{(z-\mu)^2}{2\sigma^2}} \tag{1}$$

4 Experimental Setup and Results

The aim of our model is to generate a denoised image. The model is trained on Cifar-10 dataset [19]. The Cifar-10 dataset comprises of $60,000$ images of which $50,000$ images were used for training and $10,000$ images were used for testing. Out of these $10,000$ testing images, 7000 were used for validation and 3000 were used for testing. The images were of size $32 \times 32 \times 3$.

The experimental tests were run on Intel(R) Core(TM) $i3 - 5005U$ CPU @ 2.00 GHz and 8.0 GB RAM. We trained our model using binary cross entropy [20] as loss function and adam [21] as optimizer. The batch size was taken as 50 and the total number of iterations were 20. The model was trained with an accuracy of 66.64%.

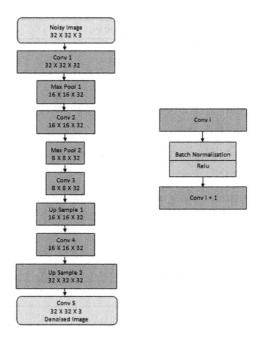

Fig. 4. Autoencoder model architecture with noisy image as input and output parameters of each layer given along with them.

Algorithm 1. Adding gaussian noise.

Input   : Image, mean, variance, noise factor
Output: Noisy image

1 $I \leftarrow image$
2 $h, w, d \leftarrow I.shape$
3 $n \leftarrow$ noise factor
4 $gaussian \leftarrow$ random normal matrix of size $h \times w \times d$ with given mean and variance
5 $I = I + n * gaussian$
6 $clip(I)$
7 **return** I

Each dataset was divided into three parts. In the first part, we added Gaussian noise [4] with mean 0, variance 0.2 and noise factor 0.5. In the second part, we added Salt and Pepper noise [5] with salt pepper ratio 0.5 and amount 0.06. In the third part, we added Speckle noise [6] with a noise factor of 0.5.

In the results we have shown 4 random images from the Cifar-10 dataset along with their corresponding noisy image for each type of noise and the reconstructed denoised image using the autoencoder. The Figs. 5, 6 and 7, shows the results of proposed method on Gaussian, Salt and Pepper and Speckle noise

---

Algorithm 2. Adding salt and pepper noise

Input : Image, salt pepper ratio, Amount
Output: Noisy image

1 $I \leftarrow image$
2 $h, w, d \leftarrow I.shape$
3 $n \leftarrow$ salt pepper ratio
4 $a \leftarrow amount$
5 $num\_salt \leftarrow a * size(I) * n$
6 $random =$ create a random matrix of size num_salt with values of random coordinates
7 $I[random] = 1$
8 $num\_pepper \leftarrow a * size(I) * (1 - n)$
9 $random =$ create a random matrix of size num_pepper with values of random coordinates
10 $I[random] = 0$
11 $clip(I)$
12 **return** I

Algorithm 3. Adding Speckle noise

Input : Image, noise factor
Output: Noisy image

1 $I \leftarrow image$
2 $h, w, d \leftarrow I.shape$
3 $n \leftarrow$ noise factor
4 $speckle \leftarrow$ random matrix of size $h \times w \times d$
5 $I = I + n * speckle$
6 $clip(I)$
7 **return** I

---

respectively. I_1, I_2, I_3, I_4 are original images, N_1, N_2, N_3, N_4 are noised images and D_1, D_2, D_3, D_4 are denoised images.

The quantitative analysis of the results is done on the basis of SSIM, PSNR and RMSE values. Structural Similarity Index (SSIM) [22] is an index or metric for quantitative measure of the losses or degradation of quality of images, resulting from different types of image processing techniques such as image compression or due to the losses caused during the course of transmission. Peak Signal-to-Noise Ratio (PSNR) [23] in another error metric which gives the ratio of the maximum power of the original signal to the power of the noise which causes disturbances in the original signal, affecting the quality of representation. Root Mean Squared Error (RMSE) [24] is a measure which gives the average of the quantity of the error. Thus, the Root Mean Squared Error is obtained by calculating the square root of the mean of the squared difference between the predicted and the original image.

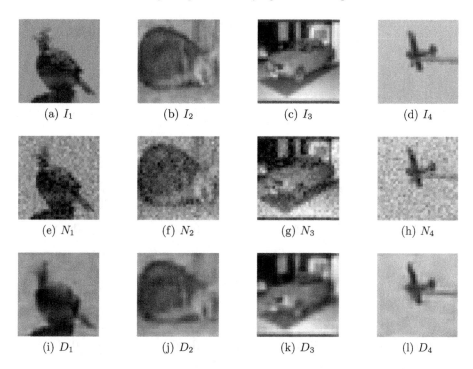

(a) I_1 (b) I_2 (c) I_3 (d) I_4

(e) N_1 (f) N_2 (g) N_3 (h) N_4

(i) D_1 (j) D_2 (k) D_3 (l) D_4

Fig. 5. Experimental result of proposed scheme for Gaussian noise: (a–d) Original image, (e–h) Noised image, (i–l) Denoised image.

The average of the calculated values of SSIM, PSNR and RMSE was obtained between original and noisy image as shown in Table 1 and the average of the calculated values of SSIM, PSNR and RMSE was obtained between original and denoised image as shown in Table 2. The difference in the value of SSIM, PSNR and RMSE for the original and denoised image, and the original and noisy image for each of the three noises was obtained to give us the increase in their corresponding values. Thereafter, the percentage increase in the values was calculated and their average was taken to give us the average percentage increase in accuracy for each of the 3 noises. The percentage increase in accuracy for Gaussian noise was 22.8%, for Salt and Pepper noise 30.6% was and for speckle noise was 16.37%.

Histograms were plotted for the original, noisy and de-noised image for Gaussian noise as shown in Fig. 8, Salt and Pepper noise shown in Fig. 9, and Speckle noise shown in Fig. 10. I_1, I_2, I_3 are original images, N_1, N_2, N_3 are noised images and D_1, D_2, D_3 are denoised images. IH_1, IH_2, IH_3 are histogram of original images I_1, I_2, I_3 respectively. NH_1, NH_2, NH_3 are histogram of noised images N_1, N_2, N_3 respectively. DH_1, DH_2, DH_3 are histogram of denoised images D_1, D_2, D_3 respectively. It was found that the histograms of

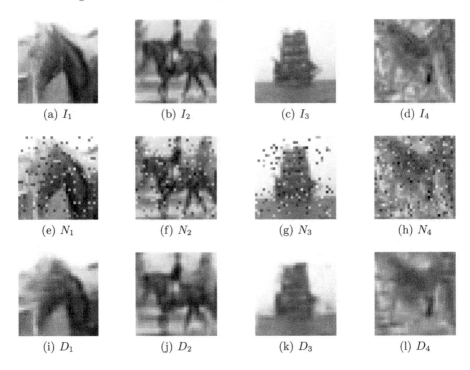

Fig. 6. Experimental result of proposed scheme for Salt and Pepper noise: (a–d) Original image, (e–h) Noised image, (i–l) Denoised image.

Table 1. Quantitative result of proposed method between original and noisy image

Noise	Measure	Result
Gaussian	SSIM	0.767
	PSNR	650.43
	RMSE	0.08
Salt and Pepper	SSIM	0.718
	PSNR	654.1
	RMSE	0.136
Speckle	SSIM	0.84
	PSNR	715.55
	RMSE	0.06

the denoised images were not the exact copies of the original images. On analyzing the histograms, it was found that the shape of the histogram remained intact, just the magnitude changed a bit. Thus, the final image was a lossy version of the original image but the structure remained the same.

(a) I_1	(b) I_2	(c) I_3	(d) I_4
(e) N_1	(f) N_2	(g) N_3	(h) N_4
(i) D_1	(j) D_2	(k) D_3	(l) D_4

Fig. 7. Experimental result of proposed scheme for Speckle noise: (a–d) Original image, (e–h) Noised image, (i–l) Denoised image.

Table 2. Quantitative result of proposed method between original and denoised image

Noise	Measure	Result
Gaussian	SSIM	0.904
	PSNR	736.1
	RMSE	0.05
Salt and Pepper	SSIM	0.899
	PSNR	724.7
	RMSE	0.06
Speckle	SSIM	0.93
	PSNR	751.96
	RMSE	0.04

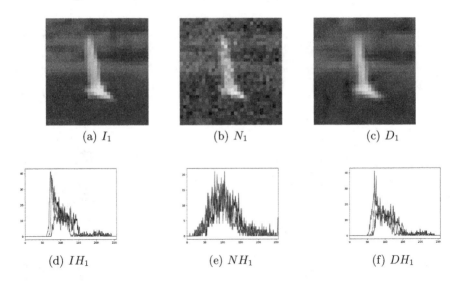

(a) I_1 (b) N_1 (c) D_1

(d) IH_1 (e) NH_1 (f) DH_1

Fig. 8. Histogram result of proposed scheme for Gaussian noise: (a) Original image, (b) Noised image, (c) Denoised image (d) Histogram of original image (e) Histogram of noised image (f) Histogram of denoised image.

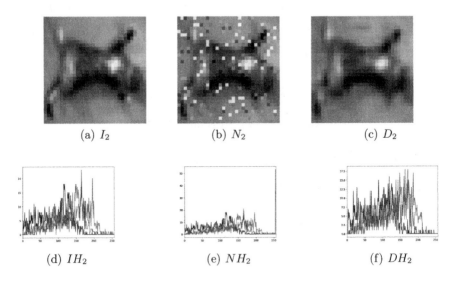

(a) I_2 (b) N_2 (c) D_2

(d) IH_2 (e) NH_2 (f) DH_2

Fig. 9. Histogram result of proposed scheme for Salt and Pepper noise: (a) Original image, (b) Noised image, (c) Denoised image. (d) Histogram of original image (e) Histogram of noised image (f) Histogram of denoised image.

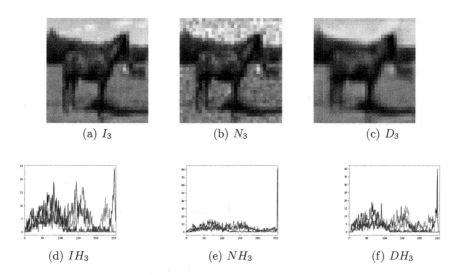

(a) I_3 (b) N_3 (c) D_3

(d) IH_3 (e) NH_3 (f) DH_3

Fig. 10. Histogram result of proposed scheme for Speckle noise: (a) Original image, (b) Noised image, (c) Denoised image. (d) Histogram of original image (e) Histogram of noised image (f) Histogram of denoised image.

5 Conclusion

The key step in image processing involved the removal of different kinds of noises from the images. We designed a model to remove three different kinds (Gaussian, Salt and Pepper, Speckle) of noises from an image. As a result, it was found that the accuracy increased in Gaussian noise by 22.8%, in Salt and Pepper noise by 30.6% and in Speckle noise by 16.37%. Thus, our model worked well for removing various kind of noises from an image.

References

1. Shih, F.Y.: Image Processing and Mathematical Morphology: Fundamentals and Applications. CRC Press, Boca Raton (2017)
2. Katiyar, A., Katiyar, G.: Denoising of images using neural network: a review. In: Advances in System Optimization and Control, pp. 223–227. Springer, Singapore (2019). https://doi.org/10.1007/978-981-13-0665-5_20
3. Boyat, A.K., Joshi, B.K.: A review paper: noise models in digital image processing. arXiv preprint arXiv:1505.03489 (2015)
4. Kamble, V.M., Bhurchandi, K.: Noise estimation and quality assessment of Gaussian noise corrupted images. In: IOP Conference Series: Materials Science and Engineering. vol. 331, no. 1. IOP Publishing (2018)
5. Singh, V., et al.: Adaptive type-2 fuzzy approach for filtering salt and pepper noise in grayscale images. IEEE Trans. Fuzzy Syst. **26**(5), 3170–3176 (2018)
6. Maity, A., et al.: A comparative study on approaches to speckle noise reduction in images. In: 2015 International Conference on Computational Intelligence and Networks. IEEE (2015)

7. Le, Q.V.: A tutorial on deep learning part 2: autoencoders, convolutional neural networks and recurrent neural networks. Google Brain 1–20 (2015)
8. Bojanowski, P., et al.: Optimizing the latent space of generative networks. arXiv preprint arXiv:1707.05776 (2017)
9. Holden, D., et al.: Learning motion manifolds with convolutional autoencoders. In: SIGGRAPH Asia 2015 Technical Briefs. ACM (2015)
10. Bengio, Y.: Learning deep architectures for AI. Found. Trends Mach. Learn. **2**(1), 1–127 (2009)
11. Hinton, G.E., Osindero, S., Teh, Y.W.: A fast learning algorithm for deep belief nets. Neural comput. **18**(7), 1527–1554 (2006)
12. Vincent, P., et al.: Stacked denoising autoencoders: learning useful representations in a deep network with a local denoising criterion. J. Mach. Learn. Res. **11**(Dec), 3371–3408 (2010)
13. Erhan, D., Bengio, Y., Courville, A.C., Manzagol, P., Vincent, P., Bengio, S.: Why does unsupervised pre-training help deep learning? J. Mach. Learn. Res. **11**, 625–660 (2010)
14. Doersch, C., Gupta, A., Efros, A.A.: Unsupervised visual representation learning by context prediction. In: Proceedings of the IEEE International Conference on Computer Vision (2015)
15. Wang, X., Gupta, A.: Unsupervised learning of visual representations using videos. In: Proceedings of the IEEE International Conference on Computer Vision, pp. 2794–2802 (2015)
16. Pathak, D., et al.: Context encoders: feature learning by inpainting. In: Proceedings of the IEEE Conference on Computer Vision and Pattern Recognition (2016). Proceedings of the IEEE International Conference on Computer Vision (2015)
17. Gatys, L.A., Ecker, A.S., Bethge, M.: Image style transfer using convolutional neural networks. In: Proceedings of the IEEE Conference on Computer Vision and Pattern Recognition (2016)
18. Noh, H., Hong, S., Han, B.: Learning deconvolution network for semantic segmentation. In: Proceedings of the IEEE International Conference on Computer Vision (2015)
19. Recht, B., et al.: Do CIFAR-10 classifiers generalize to CIFAR-10?. arXiv preprint arXiv:1806.00451 (2018)
20. Creswell, A., Arulkumaran, K., Bharath, A.A.: On denoising autoencoders trained to minimise binary cross-entropy. arXiv preprint arXiv:1708.08487 (2017)
21. Zhang, Z.: Improved adam optimizer for deep neural networks. In: 2018 IEEE/ACM 26th International Symposium on Quality of Service (IWQoS). IEEE (2018)
22. Cecotti, H., Gardiner, B.: Classification of images using semi-supervised learning and structural similarity measure. In: Irish Machine Vision and Image Processing Conference. Irish Pattern Recognition and Classification Society (2016)
23. Tanabe, Y., Ishida, T.: Quantification of the accuracy limits of image registration using peak signal-to-noise ratio. Radiol. Phys. Technol. **10**(1), 91–94 (2017)
24. Brassington, G.: Mean absolute error and root mean square error: which is the better metric for assessing model performance? In: EGU General Assembly Conference Abstracts, vol. 19 (2017)
25. Dong, L.-F., et al.: Learning deep representations using convolutional autoencoders with symmetric skip connections. In: IEEE International Conference on Acoustics, Speech and Signal Processing (ICASSP), p. 2018. IEEE (2018)
26. Gondara, L.: Medical image denoising using convolutional denoising autoencoders. In: 2016 IEEE 16th International Conference on Data Mining Workshops (ICDMW). IEEE (2016)

Image Aesthetics Assessment Using Multi Channel Convolutional Neural Networks

Nishi Doshi[1]([✉]), Gitam Shikkenawis[2], and Suman K. Mitra[1]

[1] Dhirubhai Ambani Institute of Information and Communication Technology,
Gandhinagar, India
{201601408,suman_mitra}@daiict.ac.in
[2] C R Rao Advanced Institute of Mathematics, Statistics and Computer Science,
Hyderabad, India
gitam365@gmail.com

Abstract. Image Aesthetics Assessment is one of the emerging domains in research. The domain deals with classification of images into categories depending on the basis of how pleasant they are for the users to watch. In this article, the focus is on categorizing the images in high quality and low quality image. Deep convolutional neural networks are used to classify the images. Instead of using just the raw image as input, different crops and saliency maps of the images are also used, as input to the proposed multi channel CNN architecture. The experiments reported on widely used AVA database show improvement in the aesthetic assessment performance over existing approaches.

Keywords: Image Aesthetics Assessment · Convolutional Neural Networks · Deep learning · Multi channel CNNs

1 Introduction

Image aesthetics is one of the emerging domains of research. Image Aesthetics Assessment (IAA) problem deals with giving rating to images on the basis of how pleasant they are for the user to watch. A human is more likely to feel happy looking at a high quality picture rather than low quality images. On the basis of aesthetic value of an image, it helps the user to identify whether he or she is more likely to like the image and view the rest or not. It deals with finding out the aesthetic quality of an image that is classifying an image as into the category of either high quality image or low quality image.

A photo can be clicked from any device. Every device has certain resolution which results in the clarity in the pixels. Thus, use of different devices for clicking the photographs leads to the existence of this problem domain and its classification as well. It also depends on of the photo has been captured by the photographer and the scene covered. To demonstrate this, a few images from the AVA database [1] are shown in Figs. 1 and 2.

The aesthetics assessment problem has been used in many practical applications specially to attract the users by showing visually more appealing images.

© Springer Nature Singapore Pte Ltd. 2020
N. Nain et al. (Eds.): CVIP 2019, CCIS 1148, pp. 15–24, 2020.
https://doi.org/10.1007/978-981-15-4018-9_2

Fig. 1. Low quality images

Fig. 2. High quality images

Various search engines makes use of such a classification. For a given search word, it takes the aesthetic value of an image into consideration while showing any image on top or showing it last. Review applications, which allow the users to upload photographs and write review about the products/place also take aesthetics of the uploaded images into consideration. While displaying the review, photos with high aesthetics quality or clicked from a high quality camera are more likely to be shown first compared to a photo clicked from a low quality camera or having low aesthetic quality.

Being one of the recent problems, a lot of research work is going in the domain of Image Aesthetics. Not only restricting the problem definition to classification; many variants of the problem statements exist and researchers are working on the same. RAPID: Rating Pictorial Aesthetics using Deep Learning was one of the earliest models used for classification [6]. It was the model to propose use of AlexNet architecture as well as double column architecture involving use of global and local views of an image as an input to channels. In Brain Inspired Deep Neural Network model [7], 14 style CNNs are pre-trained, and they are parallelly cascaded and used as the input to a final CNN for rating aesthetic prediction, where the aesthetic quality score of an image is subsequently inferred on AVA dataset [1]. Category Specific CNN having initially classification of images into different classes and then corresponding to every class a different CNN architecture is trained to classify images into different categories: either high quality or low quality [2].

In this article, variants of multi column Convolutional Neural Networks (CNN) are proposed. In particular, the novelty of the approach lies in using informative inputs to the channels in multi column architecture. A pre-trained CNN architecture - VGG19 [9] on ImageNet database [8] is fine tuned for image aesthetics assessment. The images are classified into one of the two classes, i.e. high aesthetic quality or low aesthetic quality. The experiments are performed on the widely used AVA database [3].

The organization of the paper is as follows: Sect. 2 discusses deep CNN architectures and fine tuning the pre-trained deep networks. Variants of multi channel CNN for IAA are discussed in Sect. 3. Experiments on AVA database are reported in Sect. 4 followed by conclusions in 5.

2   Deep Learning for IAA

In this paper, deep convolutional neural networks are used for assessing the aesthetic quality of the images. In particular, widely used deep learning architectures namely Alexnet [4] and VGG19 [9] are used. The architectural details of both networks are discussed in this section along with the procedure of fine tuning an already trained deep neural network.

2.1   CNN Architectures

Mainly two CNN column architectures were involved in the models designed to classify the images: Alexnet and VGG19

AlexNet. AlexNet architecture involves a total of 8 layers: 5 convolutional layers and 3 fully connected layers. After every layer there is a ReLu function calculation. RAPID [6] was the first method that used AlexNet architecture for solving the classification problem for image aesthetics. The architecture can be visualized by the Layer diagram in Fig. 3.

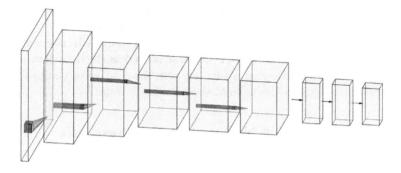

Fig. 3. AlexNet Convolutional Neural Network architecture

VGG19. VGG19 is a convolutional neural network architecture that is already trained on more than hundred million images of ImageNet database [8]. The network is 19 layers deep and classifies images into 1000 classes : some namely mug, mouse, keyboard, pencil. In considering the architecture for training and classifying images into two classes: High quality and Low quality images the initial network architecture is kept the same and the fully connected layers are changed to obtain classification into 2 classes by adding 9 dense layers having ReLu activation after Max Pooling layer of block 5.

The VGG19 architecture is divided into 5 blocks. Each block has certain number of convolutional layers with each block ending with Max Pooling layer. The number of convolutional layers per block are described in the Table 1.

Table 1. Number of convolutional layers in different blocks of CNN

Block number	1	2	3	4	5
Number of convolutional layers	2	2	4	4	4

2.2   Transfer Learning and Fine Tuning

As discussed in CNN architectures, the Deep networks have huge number of parameters and training them requires very large databases. If large enough datasets are not available, transfer learning is shown to be working very well. In transfer learning, a model trained on a large and general dataset is used as the base generic model for new application. Thus, the learned features maps are used without having to start from scratch training a large model on a large dataset. In transfer learning, the weights of the convolution layers are used as it is, only the soft max layers are re-trained to give desired output.

In case of fine tuning, instead of using fixed weights, top convolutional layers are made trainable. Along with fully connected soft max layers, some of the initial weights of convolutional layers (the layers near the fully connected layers that is the last of the layers of the network) are also updated.

In the current proposal, instead of training the deep CNN from scratch, pre-trained VGG19 network on ImageNet database is used as the base network. Top CNN layers and fully connected layers are re-trained to solve the problem of classification of images on the basis of their aesthetic value.

3   Multi Channel CNN Based Image Aesthetic Assessment

The conventional methods of fine tuning a pre-trained deep network involve giving the original image as input to the single channel network and predict the high/low aesthetic quality. However, as suggested in [6], instead of taking just the input image for assessing the quality, considering global and local views of the image in a double column network enhances the classification performance of the IAA technique. In this section, various image pre-processing and feature selection procedures are discussed. Based on the observations made, a double and triple column architecture for IAA is proposed.

3.1   Image Pre-processing and Feature Selection

As already designed deep CNN architectures are used in this article, the sizes of the images to be given as input are fixed, hence the images are required to be resized according to the specifications of respective networks. Both Alexnet and VGG19 support input sizes of 224×224 whereas AVA dataset contains images with different sizes. Hence, all the images were resized to 224×224. In addition to resizing the original image, input images by padding and cropping various portions are generated to negate the effects of scaling and resizing. Also, instead of using the raw image as a input, saliency map are used as input to one of the channels. These pre-processing and feature selection techniques are discussed in the following section.

Original Image. As both VGG19 and AlexNet architectures require images of size 224×224 as input, the images of the dataset were resized to the desired input size. As the pixels become blur after resizing, the aspect ratio of image was taken into consideration while resizing.

Padded Images. To negate the impact of aspect ratio, image was initially padded with zeros to make it a square image. And then the image was resized to 224×224. For example an image of 512×1024 dimensions was first converted to image of size 1024×1024 by padding the image on top and bottom with 256 zeros, and then this 1024×1024 was resized to 512×512 image.

Cropped Portion of Image. Human eyes generally do not look at the entire image as a whole. In computational aspects, crops of image can be used to detect such portions which can be focus of human eye and be responsible for the judgement about the image. In this paper, different cropping techniques are used to capture the essence of the image instead of resizing it to fit the input size.

Center Crop
Center portion of the image is cropped and given as input to channel [6]. The reason to take center crop of 224×224 size is that when a person looks at the picture; it is the center most part of the image where eyes of a human rest first. Human eye tends to see the center position of the image most rather than being focused on corner parts of an image. Hence, considering the center crop of the image would lead to decide relevant features for channel and help towards a better classification into two classes.

Random Crop

Instead of using the fixed center crop, three random cropped patches of size 224×224 were generated for images. A novelty added to this approach was refining the algorithm to generate random crops. While generating these three crops, two things were taken into consideration: (1) it is not the center crop of the image and (2) it does not overlap with other random crops already taken. If the random cropped patches belonged to similar portion of the image, misleading results may be produced as same images are passed in both channels of the network. Hence, the current proposal takes into consideration that when generating random cropped patches of 224×224, the distance between the center of crops has to be more than a threshold, so that two crops belonging to same region or image are not generated. A distance of 100 was kept for x and y coordinate of the generated crop image and previously generated crops of the image.

Saliency Map of Image. While viewing an image, humans do not treat the entire scene equally, mostly the focus is on visually appealing parts. Saliency deals with unique features of image related to visual representation of an image. The saliency map highlights the pixels which have more visual importance in the image. It elaborates the part of image to which our brain gets attracted the most.

Here, we have considered use of static saliency map detection as images are static in nature. Two types of static saliency maps are taken into consideration.

Spectral Residual Map

The algorithm analyzes the log-spectrum of an image, extracts the spectral residual and proposes a fast method to construct spectral residual saliency map which suggests the positions of visually attracted spots of an image.

Fine Grained Map

Human eyes have retina which consists of two types of cells: off center and on center. Specialty of these two types of cells are as follows:

- On center: It responds to bright areas surrounded by dark background.
- Off center: It responds to dark areas surrounded by bright background.

Fine grained saliency map is generated by taking considering the on center and off center differences [10].

Spectral residual and fine grained saliency maps have been shown in Fig. 4.

Thus, various ways of pre-processing and feature extraction from the original input image have been discussed in this Section. Next part discusses using these processed images as input to multi channel CNNs.

3.2 Proposed Multi Channel CNN for IAA

Instead of working on a single channel with the raw image as input, in this article, we have used pre-processed images as discussed before in this Section as input along with the raw image. Thus, building a multi channel convolutional neural

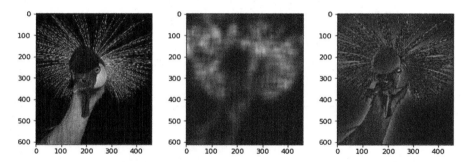

Fig. 4. In order from left: Original high quality image taken from PhotoQuality Dataset [5], corresponding spectral residual saliency map and fine grained saliency map.

network architecture. Experiments have been performed on double and triple column networks. The configuration details of both the networks are discussed below:

Double Column Network. The double column network involves use of two pipelines and concatenation of those channels to generate output classifier. In one part of the network: original form of the image, the padded form of the image and the center cropped form of an image are supplied whereas in the second part of network three variants of random cropped forms of images are given.

In case of AlexNet architecture; initially trained single column network is used as base model for both channels. The parameters after concatenating are trained for 300 epochs and then fine tuning is carried out by making the 4th and 5th convolutional layers trainable.

In case of VGG network; a pre-trained VGG19 network on ImageNet dataset is taken as the base network for both the channels. As in case of AlexNet, 4th and 5th convolutional layers are fine tuned for image aesthetics assessment task.

Triple Column Network. The triple column network involves use of three pipelines and concatenation of those channels to generate output classifier. There is an addition of third pipeline to double column network where two variants of saliency maps that is spectral residual map and fine grained maps are passed.

VGG19 network is only trained for this column network. VGG19 network which had two channels already trained for double column network were used as base model and for third channel; single column network model was used as base model. After concatenating the results of all the three channels the classification of images into high quality and low quality was done. The network design for triple column network is shown in Fig. 5.

4   Experiments

There are many datasets available for testing the validity of the model designed to solve the image aesthetic assessment problem. In this paper, the experimen-

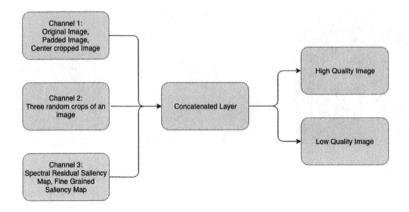

Fig. 5. Triple column network design

tal results are reported on one of the most used database for IAA *i.e.* AVA dataset [3].

AVA Dataset. AVA Dataset [3] consists of images which have votes of users for every rating 1 to 10. As the model developed is for classifying images into two categories: high and low. The ratings were assigned to images on the basis of maximum number of votes corresponding to that image. The number of images corresponding to each rating are as follows:

Table 2. Number of images in AVA dataset corresponding to each and every vote

Rating	1	2	3	4	5	6	7	8	9	10	Total images
Number of images	566	104	1083	24305	147483	74294	6824	743	31	97	255530

As it can be seen from Table 2, out of 255530 images, 147483 that is around 58% images belong to rating 5. Hence, the images belonging to rating 1, 2, 3 and 4 were considered low quality images and images with rating 7, 8, 9 and 10 were considered high quality images.

4.1 Comparison with Various Approaches

Experiments using single column VGG19 and AlexNet architectures were carried out on AVA dataset. In case of AlexNet, the network is trained from scratch for image aesthetic assessment whereas for VGG19, a pre-trained model on ImageNet dataset is used. Here, the pre-trained works as the base model and it is fine tuned for aesthetics assessment on AVA database. The results in terms of accuracy are reported in Table 3. It can be observed that there is significant improvement in testing accuracy in case of VGG19 as compared to AlexNet.

Hence, the double and triple column CNN experiments have been performed by using the VGG19 architecture as the base CNN architecture.

The results obtained after using double and triple channel CNNs are reported in Table 3. It can be observed that more that more 7% enhancement in the testing accuracy is obtained using triple channel CNN over the double channel CNN. It shows that addition of saliency map as feature boost the IAA performance.

For fair comparison, results on AVA dataset using a few existing deep neural network based approaches namely SCNN [6], DCNN [6] and BDN [7] have also been reported in Table 4. It can be observed that the proposed triple channel architecture surpasses all three compared approaches.

Table 3. Different architectures and network results on AVA dataset

Architecture	Network	Train accuracy	Test accuracy
AlexNet	Single Column	0.993	0.6164
VGG19	Single Column Network	0.9987	0.7137
VGG19	Double Column Network	0.8082	0.7444
VGG19	Triple Column Network	0.92	0.823

Table 4. Comparison with existing results on AVA dataset

Network	Accuracy
Single Column Network (SCNN) [6]	71.20
Double Column Network (DCNN) [6]	73.25
Brain Inspired Deep Neural Network (BDN) [7]	78.08
Triple Column Network	**82.3**

5   Conclusions

After conducting various results on AVA dataset for different architectures and observing the results, we came to a conclusion that increasing the number of columns in the architecture did give us better results compared to single column architecture results. We also observed that compared to other proposed architectures like SCNN [6], DCNN [6] and BDN [7], triple column architecture showed the best results that is 82.3% accuracy was achieved. The major reason for achieving such a high accuracy was due to the fact of involving different forms of images such as cropped, padded and saliency maps. Giving balanced and equal weights to these forms of an image helped to train the network more efficiently.

References

1. Deng, Y., Loy, C.C., Tang, X.: Image aesthetic assessment : an experimental survey. IEEE Sig. Process. Mag. **34**, 80–106 (2017)
2. Lihua, G., Fudi, L.: Image aesthetic evaluation using paralleled deep convolution neural network. In: 2016 International Conference on Digital Image Computing: Techniques and Applications (DICTA) (2016)
3. Murray, N., Marchesotti, L., Perronnin, F.: AVA: a large-scale database for aesthetic visual analysis. In: Proceedings of the IEEE Conference on Computer Vision and Pattern Recognition (CVPR), pp. 2408–2415 (2012)
4. Krizhevsky, A., Sutskever, I., Hinton, G.E.: ImageNet classification with deep convolutional neural networks. In: Advances in Neural Information Processing Systems 25. Neural Information Processing Systems Foundation, pp. 1097–1105 (2012)
5. Luo, W., Wang, X., Tang, X.: Content-based photo quality assessment. In: Proceedings of the IEEE International Conference on Computer Vision (ICCV), pp. 2206–2213 (2011)
6. Lu, X., Lin, Z., Jin, H., Yang, J., Wang, J.Z.: RAPID: rating pictorial aesthetics using deep learning. In: Proceedings of the ACM International Conference on Multimedia, pp. 457–466 (2014)
7. Wang, Z., Dolcos, F., Beck, D., Chang, S., Huang, T.S.: Brain-inspired deep networks for image aesthetics assessment. arXiv preprint arXiv:1601.04155 (2016)
8. Russakovsky, O., et al.: ImageNet large scale visual recognition challenge. Int. J. Comput. Vis. **115**(3), 211–252 (2015). https://doi.org/10.1007/s11263-015-0816-y
9. Simonyan, K., Zisserman, A.: Very deep convolutional networks for large-scale image recognition. arXiv:1409.1556 (2014)
10. Wang, B., Dudek, P.: A fast self-tuning background subtraction algorithm. In: Proceedings of the IEEE Workshop on Change Detection (2014)

Profession Identification Using Handwritten Text Images

Parveen Kumar[1,2]([✉]), Manu Gupta[2], Mayank Gupta[2], and Ambalika Sharma[1]

[1] Indian Institute of Technology, Roorkee, Roorkee, India
asharfee@iitr.ac.in
[2] National Institute of Technology Uttarakhand, Srinagar, Garhwal, India
{parveen.cse,gmanu530.cse15,gmayank386.cse15}@nituk.ac.in

Abstract. A writer handwriting depicts various information and it gives the insights into the physical, mental and emotional state of the writer. This art of analyzing and studying handwriting is *graphology*. The prime features of handwriting such as margins, slanted, the baseline can tell the characteristics of a writer. The writer handwriting analysis reveals strokes and patterns through which identification and understanding the personality of a writer is possible. The writing of a person molds into various shapes and styles, starting from school until the struggle for his/her career. If we examine the writings of a person from different stages of his/her life then we will see that there are many differences in the shapes, styles, and sizes of the characters. The proposed work analyze the handwriting data written by the writer's from different professions and classify them based on the top features that characterize their profession. In this paper, the profession of a writer is identified by analyzing the features of writer's offline handwritten images. The previous work mostly includes determining various traits like honesty, emotional stability of a writer. The Proposed work uses the CNN based model for the feature extraction from the writer's offline handwritten images.

Keywords: Handwritten document · Personality prediction · Feature extraction

1 Introduction

In the previous years, handwritten document analysis has remained a demanding research area. Different types of handwriting styles are present which includes cursive handwriting, handwriting without tears, continuous strokes for writer identification. There is high inconsistency in handwriting styles, which requires handwriting tools and techniques to be more robust. Coordination is always there between an individual's brain and handwriting. Mental activities are highly influenced by the job we are engaged, that indirectly influences the handwriting. Handwriting tells about the behavioral characteristics of a human. Every handwriting has some specific characteristics which help in differentiating one writer

© Springer Nature Singapore Pte Ltd. 2020
N. Nain et al. (Eds.): CVIP 2019, CCIS 1148, pp. 25–35, 2020.
https://doi.org/10.1007/978-981-15-4018-9_3

from another. There are many applications of writer identification such as in forgery detection which takes the complex models and models of high accuracy. The main objective of any automated writer identification technique is to identify a writer based on his/her handwritten samples. This can be achieved using two methods, offline and online writer identification. Online writer identification requires the recording of the complete trajectories with the help of special tools, thus, input is a function of time, positions, pressures, angles, and other necessary information. On the other hand, only scanned images of handwritten text are used as input in case of off-line identification. With this little information of only pixels, the identification is somewhat difficult [1].

The grub of the   Meanis very light.

(a)   (b)

Fig. 1. English text line: (a) Engineer writing, (b) Doctor writing.

The off-line writer identification methods can be categorized into two groups: text-dependent and text-independent. In the case of text-dependent methods, there are registered templates for the identification process [29]. These templates are used to compare with the required input image having fixed text contents [2, 3]. In contrast to the text-dependent, the text-independent method requires to extract writer-specific features which exhibit huge variations according to the individuals. It does not make assumptions on the content of the input. Here, we have extended the idea of writer identification task to a more general way of detecting the profession of an individual by his/her writing using a convolution neural network (CNN) based approach. Figure 1 shows handwritten English text samples written by two professionals. Figure 1(a) shows an English text line written by an engineer and Fig. 1(b) shows an English text line written by a doctor. The difference in pattern between two handwritten samples by different writers can be clearly seen [34]. For the identification of a writer, one needs to extract notional written style attributes and details which focus on personal writing habits.

The paper addresses the issue of identification of the writer's profession by leveraging deep convolutional neural network (CNN) as a robust model to produce successful representations for it. Deep CNNs have shown its effectiveness in various problems related to computer vision by improving state-of-the-art results with a considerable edge, including image classification [4,5] and [31], object detection [6,7] and [32], face recognition [8,9], handwriting recognition [10]. Profession identification is a more general idea which is somewhat similar to the writer identification process. Some of the works related to the text-identification that gives the author an idea to extend it to profession identification. Character Recognition and writer identification are two different things, but they include some common process of feature extraction from handwriting.

A slant removal algorithm is proposed [11,12] based on the vertical projection profile that gives the slant angle as one of the features of handwriting. Here, in addition to it, Wigner-Ville distribution [13] is used. Similarly writer verification process [14,15] performs a comparison of one document by another document to check whether or not they are written by the same writer [27]. In case of a writer identification process [14] a large database comprising handwritten samples of known writers is searched in order to find the most probable writer. This states clearly the difference between verification and identification process as the former is a two-class classification whereas the latter is a multi-class classification. Bertolini D et al., uses k-adjacent segment features to perform offline writer identification and achieves a recognition rate of 93.3%. Some of the various classifiers such as hidden markov model (HMM) [16] and [30], support vector machine (SVM) [17] and [33] have been used in previous researches. One such classifier based on the Euclidean distance is used [18]. Deep learning techniques are very effective in automatic learning of features, so they are being applied to many recognition tasks [10]. Prasad et al. used CNN for the writer identification task. They eliminated the last fully connected layer of the model as the above layer have adequately extracted the features in order to identify the writer.

Profession highly influenced the personality traits of a person. The personality traits of an individual from the features extracted are discussed [20–22] and [28]. Joshi et al. proposed a mathematical model to predict the personality traits of the writer. They did not discuss the time complexity and accuracy of the model. A similar approach [21] using Artificial Neural Network is proposed that achieved performance goal with 4500 epochs and 8 hidden layer nodes in a shorter duration of time.

In contrast, this paper address the problem which is more general and difficult that is profession identification of writer. This paper feeds the model with a handwritten image comprising of the text from different fields, and learns successful representation with robustly designed deep CNN model, leading an easier and refined method. The remainder of this paper is organized as follows: the proposed work in Sect. 2. Further parts of Sect. 2.1 present acquisition of dataset, proposed model and discuss the Leaky ReLUs. Section 3 presents the experiments and results of the complete phases starting from the training phase to the testing phase. Finally, Sect. 4 presents the conclusion of the paper.

2 Proposed Work

In the proposed model the convolutional neural network (CNN) is used to identify the profession of a person based on handwriting. The handwriting analysis gives the profession characteristics which are identified by the CNN. This section is further divided into various parts which begin with the dataset acquisition to its use in the training of the proposed model.

2.1 Acquisition of Dataset

We have collected the dataset from the following institutions: Medical, Engineering, Arts and Commerce. We collected the dataset for Doctors, Professors (Teachers), Writers, Engineers, and Art Students. The dataset is produced for each profession and is properly labeled. Each offline dataset image is first preprocessed to get the processed dataset images. On each original image, the median filter is applied to remove the salt and pepper noise with filter size taken is 3×3.

After applying the median filter the image is passed to global thresholding applied on the image with a threshold value 190. The preprocessing result of the original image to the processed image is shown in Fig. 2(a) and (b). Now each line is separated to get the final image for the particular label as shown in Fig. 3.

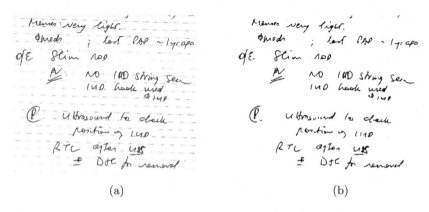

(a)   (b)

Fig. 2. Handwritten text samples: (a) Original, (b) Preprocessed.

Fig. 3. Line segmented from handwritten image.

2.2 Proposed Model

A CNN model is proposed to identify the professions using offline handwriting. The important considerations are taken to extract the features within each profession and features which differentiate each profession from others. This includes the proper incorporation of batch size and proper shuffling of the dataset. For training, 400 images are taken, with each class contains approximately 80 images. For each image that belongs to a class, random patches of 113×113 are cropped and patched together. In this way, each 113×113 patched image contains the handwritten dataset from a different writer that belong to a particular class. This helps the CNN model to extract the features within each profession. Now

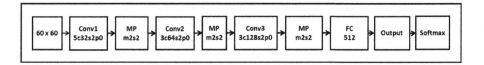

Fig. 4. Network structure of the proposed model.

a shuffled batch of size eight is created which contains different dataset of a particular profession. Now the complete batch is fed to the CNN model to train the weighs. This creates ten batches of each class, of size eight. Total classes are five which makes 50 batches and these batches are randomly shuffled to feed into the CNN. The last step takes care of between class variation.

The proposed model is similar to AlexNet [4]. AlexNet contained eight layers; the first five were convolutional layers, some of them followed by max-pooling (MP) layers, and the last three were fully connected layers (FC) [25]. It used the non-saturating ReLU activation function, which showed improved training performance over tanh and sigmoid [4].

The Proposed model consists of the following layers which are shown in Table 1. The first layer is the lambda_1 (Lambda) which resize the image to 60×60. It is found experimentally that the result does not get affected at that size. The second layer Convolution2D layer. The next layer used is Activation Leaky ReLU. The detailed description of the model is shown in Fig. 4. The convolutional layers are denoted by boxes with ConvN, where N is the index of the layer. The notation like XcYsZpQ depicts that the convolutional layer filters the input size $X \times X$ with Y kernels, a stride of Z pixels and padding of Q pixels. Max-pooling layer is denoted by the boxes with MP. The notation like mXsY specifies that the max-pooling operation is performed in a neighborhood of size $X \times X$ with a stride of Y pixels. The fully-connected layers are denoted by boxes with FC, and the followed number depicts the number of neurons at that layer. All convolutional layers and fully-connected layers are followed by Leaky Rectified Linear Unit layer activation function, described in the next section. FC is followed by dropout layer with ratio $= 0.6$, which is taken experimentally to prevent the over fitting and then it connects to the output layer of five neurons.

Leaky ReLUs allow a small, positive gradient when the unit is not active. It is given as Eq. 1 [26].

$$f(x) = \begin{cases} 0, & x > 0 \\ 0.01x, & \text{otherwise} \end{cases} \tag{1}$$

The next layer used is max pooling *2d* layer. Max pooling is a sample-based discretization process. The objective is to reduce the dimensionality and consider the most important feature in the subregion based on some assumption. After this, the same sequence is repeated and the final output is flattened and fed into a neural network which is connected to a dense layer. Some regularization technique like dropout technique is employed which takes care of the overfitting

Table 1. Layers used in the proposed model.

Layer (type)	Output shape
lambda_1 (Lambda)	(None, 60, 60, 1)
conv1 (Conv2D)	(None, 30, 30, 32)
activation_1 (Activation)	(None, 30, 30, 32)
pool1 (MaxPooling2D)	(None, 15, 15, 32)
conv2 (Conv2D)	(None, 15, 15, 64)
activation_2 (Activation)	(None, 15, 15, 64)
pool2 (MaxPooling2D)	(None, 7, 7, 64)
conv3 (Conv2D)	(None, 7, 7, 128)
activation_3 (Activation)	(None, 7, 7, 128)
pool3 (MaxPooling2D)	(None, 3, 3, 128)
flatten_1 (Flatten)	(None, 1152)
dropout_1 (Dropout)	(None, 1152)
dense1 (Dense)	(None, 512)
activation_4 (Activation)	(None, 512)
dropout_2 (Dropout)	(None, 512)
output (Dense)	(None, 5)
activation_6 (Activation)	(None, 5)

of the training model. The dropout is defined as removing units in a neural network. The summary of the layers used to build the model is given in Table 1.

3   Experiments and Results

The section consists of the complete phases starting from the training phase to the testing phase of the proposed model. A classification report and confusion matrix is also presented to show the correctness of the proposed model.

3.1   Training Phase

The model is trained parallelly with generators that generate the batch_size of images by taking eight different images as input. The details of training are

Table 2. Details of training phase.

Number of epochs	8
Samples per epoch	217
Number of validation samples	56

Table 3. Training accuracy at each epoch.

Epoch no.	Accuracy (%)
1	64.36
2	79.20
3	92.62
4	93.68
5	92.45
6	95.29
7	95.70
8	97.52

shown in Table 2. For each epoch the generator will stop after the limit of Number of samples per epoch is reached. The accuracy during each epoch is shown in Table 3. This is measured using categorical cross-entropy.

Cross-entropy is used for calculating the cost which takes the calculated probability from the activation function and the created one-hot-encoding matrix to calculate the loss. For the right target class, the cost will be lesser, and the cost will be larger for the wrong target class. We define Cross Entropy by $CE(S_i, T_i)$ for i^{th} iteration with output activation vector, S_i and one-hot target vector, T_i as given in Eq. 2.

$$CE(S_i, T_i) = -\sum_{j=1}^{k} T_{ij} log S_{ij} \qquad (2)$$

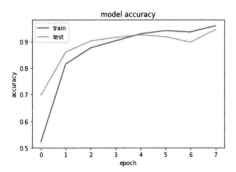

Fig. 5. Training and validation graph during the training of the model.

The training accuracy for the proposed model is 97.50%. Figure 5 depicts the training and validation accuracy graph plot against the iterations performed during the training of the model.

3.2   Testing Phase

The dataset acquired contains 200 images for testing with approximately 40 images of each profession, when applied to the above trained model gives 96.90% accuracy, which is far much better for this novel proposed work on profession identification using offline handwriting, whereas earlier works completely vary from this work and are basically focused on writer identification like Half Deep Writer and Deep Writer given by Xing et al. [19], which gives an accuracy of 97.3% applied on IAM dataset comprising of 657 English writer's stuff.

3.3   Classification Report and Confusion Matrix

To describe the performance of the classification model on a set of test data a confusion matrix is shown in Table 4. The table outlook those classes which are mislabeled. Here, the count of right and wrong predictions are summarized. With the help of the confusion matrix, it can be easily seen that the model works good for the five classes Doctors, Professors (Teachers), Writers, Engineers, and Art Students labeled as 0, 1, 2, 3 and 4, respectively.

Table 4. Confusion matrix.

Label	0	1	2	3	4
0	139	0	2	10	1
1	0	944	27	1	0
2	0	236	541	0	0
3	0	0	3	1627	0
4	4	1	0	16	2500

Table 5. Classification report.

Classes	Precision	Recall	F1-score	Support
Doctors	0.97	0.91	0.94	152
Professors	0.80	0.97	0.88	972
Writers	0.94	0.70	0.80	777
Engineers	0.98	1.00	0.99	1630
Art students	1.00	0.99	1.00	2521

The confusion matrix is used to generate the classification report shown in Table 5. The confusion matrix is used to measure performance parameters. The first parameter is *precision*. To get the value of this parameter we divide the total number of correctly classified test images of a class by the total number of predicted test images of that class. The second parameter is *recall*. To get the value of this parameter we divide the total number of correctly classified test images of a class by the total test images belong to that class. The third parameter is *F1-score* which is the harmonic mean of the precision and recall. It gives a measurement that represents both the recall and precision. The fourth parameter *Support* gives the total test images that belong to the particular class.

4 Conclusion

In this paper, we introduced an approach to identify the profession of a writer by analyzing the writer's off-line handwritten scanned images. A systematic deep convolution neural network is designed to extract the distinguished features from handwritten samples. We present an innovative idea of profession detection which can be used to know about the writer's qualification background and can be very helpful as we can quantify a person to his/her profession based on handwriting. It reveals the deep impact of the profession on the minds of a person that is indirectly reflecting through his/her handwriting.

References

1. Yang, W., Jin, L., Liu, M.: DeepWriterID: an end-to-end online text-independent writer identification system, arXiv preprint arXiv:1508.04945 (2015)
2. Said, H., Tan, T., Baker, K.: Personal identification based on handwriting. Pattern Recogn. **33**(1), 149–160 (2000)
3. Zhu, Y., Tan, T., Wang, Y.: Font recognition based on global texture analysis. IEEE Trans. Pattern Anal. Mach. Intell. **23**(10), 1192–1200 (2001)
4. Krizhevsky, A., Sutskever, I., Hinton, G.E.: Imagenet classification with deep convolutional neural networks. In: Advances in Neural Information Processing Systems, pp. 10971105 (2012)
5. He, K., Zhang, X., Ren, S., et al.: Deep residual learning for image recognition. arXiv preprint arXiv:1512.03385 (2015)
6. Girshick, R.: Fast R-CNN. In: Proceedings of the IEEE International Conference on Computer Vision, pp. 1440–1448 (2015)
7. Ren, S., He, K., Girshick, R., et al.: Faster R-CNN: towards real-time object detection with region proposal networks. In: Advances in Neural Information Processing Systems, pp. 91–99 (2015)
8. Sun, Y., Wang, X., Tang, X.: Deep learning face representation from predicting 10,000 classes. In: Proceedings of the IEEE Conference on Computer Vision and Pattern Recognition, pp. 1891–1898 (2014)
9. Sun, Y., Liang, D., Wang, X., et al.: DeepID3: face recognition with very deep neural networks. arXiv preprint arXiv:1502.00873 (2015)
10. Ciresan, D., Meier, U., Schmidhuber, J.: Multi-column deep neural networks for image classification. In: 2012 IEEE Conference on Computer Vision and Pattern Recognition (CVPR), pp. 3642–3649. IEEE (2012)
11. Kavallieratou, E., Fakotakis, N., Kokkinakis, G.: Slant estimation algorithm for OCR systems. Pattern Recogn. **34**(12), 2515–2522 (2001)
12. Kavallieratou, E., et al.: An integrated system for handwritten document image processing. Int. J. Pattern Recognit. Artif. Intell. **17**(04), 617–636 (2003)

13. Boashash, B., Black, P.: An efficient real-time implementation of the Wigner-Ville distribution. IEEE Trans. Acoust. Speech Signal Process. **35**(11), 1611–1618 (1987)
14. Bulacu, M., Schomaker, L.: Text-independent writer identification and verification using textural and allographic features. IEEE Trans. Pattern Anal. Mach. Intell. **29**(4), 701–717 (2007)
15. Bertolini, D., Oliveira, L.S., Justino, E., et al.: Texture-based descriptors for writer identification and verification. Expert Syst. Appl. **40**(6), 2069–2080 (2013)
16. Schlapbach, A., Bunke, H.: A writer identification and verification system using HMM based recognizers. Pattern Anal. Appl. **10**(1), 33–43 (2007). https://doi.org/10.1007/s10044-006-0047-5
17. Franke, K., Bunnemeyer, O., Sy, T.: Ink texture analysis for writer identification. In: 2002 Proceedings of the Eighth International Workshop on Frontiers in Handwriting Recognition, pp. 268–273. IEEE (2002)
18. Bulacu, M., Schomaker, L.: A comparison of clustering methods for writer identification and verification. In: NULL, pp. 1275–1279. IEEE, August 2005
19. Xing, L., Qiao, Y.: DeepWriter: a multi-stream deep CNN for text-independent writer identification. In: 2016 15th International Conference on Frontiers in Handwriting Recognition (ICFHR), pp. 584–589. IEEE, October 2016
20. Joshi, P., Agarwal, A., Dhavale, A., Suryavanshi, R., Kodolikar, S.: Handwriting analysis for detection of personality traits using machine learning approach. Int. J. Comput. Appl. **130**(15) (2015)
21. Champa, H.N., AnandaKumar, K.R.: Artificial neural network for human behavior prediction through handwriting analysis. Int. J. Comput. Appl. **2**, 36–41 (2010). (09758887)
22. Prasad, S., Singh, V.K., Sapre, A.: Handwriting analysis based on segmentation method for prediction of human personality using support vector machine. Int. J. Comput. Appl. **8**(12), 25–29 (2010)
23. Jain, R., Doermann, D.: Offline writer identification using k-adjacent segments. In: 2011 International Conference on Document Analysis and Recognition (ICDAR), pp. 769–773. IEEE (2011)
24. Li, B., Sun, Z., Tan, T.: Online text-independent writer identification based on Stroke's probability distribution function. In: Lee, S.-W., Li, S.Z. (eds.) ICB 2007. LNCS, vol. 4642, pp. 201–210. Springer, Heidelberg (2007). https://doi.org/10.1007/978-3-540-74549-5_22
25. CS231n Convolutional Neural Networks for Visual Recognition. cs231n.github.io. Accessed 20 Oct 2018
26. Maas, A.L., et al.: Rectifier nonlinearities improve neural network acoustic models. In: Proceedings of ICML, vol. 30, no. 1, p. 3, June 2013
27. Kumar, P., Sharma, A.: DCWI: distribution descriptive curve and cellular automata based writer identification. Expert Syst. Appl. **128**, 187–200 (2019)
28. Meena, Y., Kumar, P., Sharma, A.: Product recommendation system using distance measure of product image features. In: 2018 Second International Conference on Intelligent Computing and Control Systems (ICICCS). IEEE (2018)
29. Kumar, B., Kumar, P., Sharma, A.: RWIL: robust writer identification for Indic language. In: 2018 Second International Conference on Intelligent Computing and Control Systems (ICICCS). IEEE (2018)
30. Kumar, V., Monika, Kumar, P., Sharma, A.: Spam email detection using ID3 algorithm and hidden Markov model. In: 2nd Conference on Information and Communication Technology (CICT 2018), Jabalpur, India (2018)

31. Panwar, P., Monika, Kumar, P., Sharma, A.: CHGR: captcha generation using hand gesture recognition. In: 2nd Conference on Information and Communication Technology (CICT 2018), Jabalpur, India (2018)
32. Bhatt, M., Monika, Kumar, P., Sharma, A.: Facial expression detection and recognition using geometry maps. In: 2nd Conference on Information and Communication Technology (CICT 2018), Jabalpur, India (2018)
33. Katiyar, H., Monika, Kumar, P., Sharma, A.: Twitter sentiment analysis using dynamic vocabulary. In: 2nd Conference on Information and Communication Technology (CICT 2018), Jabalpur, India (2018)
34. Mishra, A., Kumar, K., Kumar, P., Mittal, P.: A novel approach for handwritten character recognition using K-NN classifier. In: 3rd IEEE International Conference on Soft Computing: Theories and Applications (SoCTA 2018), Jalandhar, India (2018)

A Study on Deep Learning for Breast Cancer Detection in Histopathological Images

Oinam Vivek Singh[1]([⊠]), Prakash Choudhary[2], and Khelchandra Thongam[1]

[1] Department of Computer Science and Engineering,
NIT Manipur, Imphal, India
oinamvivek@gmail.com, thongam@gmail.com
[2] Department of Computer Science and Engineering,
NIT Hamipur, Hamirpur, India
choudharyprakash87@gmail.com

Abstract. Pathological examination is the most accurate method for the diagnosis of cancer. Breast cancer histopathology evaluation analyses the chemical and cellular characteristics of the cells of a suspicious breast tumor. A computer-aided automatic classifier with the help of machine learning can improve the diagnosis system in terms of accuracy and time consumption. These types of system can automatically distinguish a benign and malignant pattern in a breast histopathology image. It can reduce the workload of pathologists and can provide a more accurate process. In recent years, like in other areas, deep networks have also attracted for histopathology image analysis. Convolution Neural Network has become a preferred choice for images analysis including breast histopathology. In this paper, we review various deep learning concepts applied to breast cancer histopathology analysis and summarizes contributions to this field. We present a summary of the recent developments and a discussion about the best practices done using deep in breast histopathology analysis and improvements that can be done in future research.

Keywords: Histopathology · Cancer · Deep learning · CNN · Benign · Malignant

1 Introduction

1.1 Breast Cancer

Breast cancer is the most common cancer among women, accounting to 14% of women diagnosed with cancer. In India, it accounts for 27% of all cancers in women. It is the second leading cause of cancer death among women, exceeded only by lung cancer (NICPR 2018). Early detection of breast cancer is very important for treatment. If detected soon, there are more treatment options and a better chance of survival. Late detection reduces survival rate by 3 to 17 times. In India, 2000 new women are diagnosed every day in which 1200 are detected at later stages. This is due to the fact that breast cancer awareness and its treatment is very low.

© Springer Nature Singapore Pte Ltd. 2020
N. Nain et al. (Eds.): CVIP 2019, CCIS 1148, pp. 36–48, 2020.
https://doi.org/10.1007/978-981-15-4018-9_4

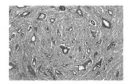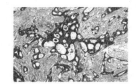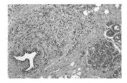

Fig. 1. Hematoxylin and eosin stained histology of breast tissue

Breast cancer occurs when the cells in bosom tissue mutate, keep reproducing. These mutated cells group together to form a tumor. A tumor is said to be cancerous or malignant when they spread to other parts of the breast and successively, through the lymph system and bloodstream, to other body parts. The reason why and how this mutation occurs is not entirely known [1]. The diagnosis of breast cancer follows a three-step model. The first step is to find lesion or changes in the breast structures using appropriate imaging methods such as mammography. Clinical evaluation of the image is done by a medical expert, and where indicated, needle biopsy will be performed (i.e. Hematoxylin and Eosin Stained histology). An example of H&E histological images of breast tissue is shown in Fig. 1.

When an expert suspects the presence of a tumor, a biopsy is performed to obtain the sample of the suspected tissues [2]. A significant amount of women who underwent biopsy were found to have malignant tumors [3]. A biopsy is currently recognized as the only method to validate the presence of cancer [4]. In the traditional clinical setting, this analysis process is manually done by a pathologist. This process depends on the skill and expertise of the pathologist and can result in huge variability in the final estimation. As a result of this possible impact, there is a rise in many researchers working in developing such a computer-aided diagnosis system. A computer-aided system for breast cancer diagnosis in histopathology will extract features from the images and will use these features for diagnosing breast cancer. A CAD for histological image analysis is used for detecting tumor regions, analyzing mitotic activity, nuclear atypia score, the epithelium-stroma, and tubule formation score. They can also be used for classifying the images into various subtypes of breast cancer like Invasive ductal carcinoma or Invasive lobular carcinoma [5, 6]. Such a system will provide a second opinion to the radiologist about the presence of cancer Traditional CAD system for histopathology analysis uses handcrafted features (i.e. morphological, topological and textural). These features will then be fed to different classifiers to classify the images into different classes.

A sensible step towards this challenge is to enable the computers to automatically learn the features according to the problem. This is the main idea behind various deep learning algorithms. Some of the most famous architecture for unsupervised learning are Sparse Auto Encoder (SAE) [7], Deep Belief Network (DBN) and Restricted Boltzmann machine (RBM) [8]. For supervised learning, Convolution Neural Network (CNN) and Recurrent Neural Network (RNN) are the most prominent one. Convolutional neural networks (CNNs) has become the most successful type of models for medical image analysis. CNNs was first introduced in the late seventies [9]. In 1995, Application of CNNs to medical image analysis was done for the first time [10]. The first successful real-world application for hand-written digit recognition using CNN [11]. A turning point in this field came during the ImageNet challenge [12]. A proposed

CNN architecture, called AlexNet, has a large impact on the field of deep-learning by winning that competition with a wide margin. Further progress has been made using deeper architectures in subsequent years that performs much better than AlexNet [13].

Research articles on deep learning in medical image analysis has grown rapidly since 2015 and is now a dominant topic in many conferences. There are also competitions that were held recently to attract researchers and developed new method for improving mammogram classification for breast cancer detection. Some recent breast histopathology competitions include: ICIAR2018 (2018), ICPR2012 (2017), AMIDA (2017), MITOSATYPIA-14 (2016), CAMELYON16 (2016), CAMELYON17 (2017) and TUPAC16 (2016). These competitions have influenced the evaluation of different methods to become more transparent and easier to compare.

Our survey mainly focuses on the application of deep learning in breast histopathological image analysis. Traditional approaches using handcrafted features are excluded. Recently, interest has returned in the topic and gained significant advance. The rest of this paper is organized as: Sect. 2 discusses briefly about the breast histopathology for breast cancer analysis. Section 3, discuss various deep learning architectures and its variations that have been used for breast histopathology image analysis. Section 4 describes various contributions of deep learning in breast histopathology analysis followed by a discussion, conclusion and future works.

2 Breast Histopatholgy

A Breast biopsy performed after a lesion is found in Mammographic examination, in order to get tissues for pathological diagnosis. Tissues or sometimes fluid are removed from the suspicious area and are analyzed under a microscope to check the presence of breast cancer. It is the only method that can surely determine if a lesion is cancerous or not. During the preparation of the histopathology, the different components of the tissues are visualized by staining the elements with different colors. In H&E stain histology, the nuclei parts are stained with blue and the cytoplasm by pink. It is the conventional protocol for staining breast tissues [5]. The glass slide containing the tissue is then coverslip and is digitized at different magnification using a WSI35 scanner.

Nottingham grading system (NGS) is a very popular and preferred grading system for breast histopathology analysis. After the tissues are analyzed, the tissues are graded according to how aggressively a tumor may behave. The pathologist looks at the breast cancer cells under a microscope and derived the scores by assessing three morphological features. They are:

Tubule formation – How many percent of cancer cells are in tubule formation?
Nuclear pleomorphism – How different the tumor cells look from normal cells?
Mitotic activity – How fast cells are growing or dividing?

The three scores are then combined to obtain the histological grade [14]. The analysis of breast tissue components for breast cancer detection on WSI scanned breast tissue images includes nuclei, tubules, epithelium and stroma, and mitotic detection. Deep learning is becoming very popular because of the availability of a big database and powerful system. The recent success of CNNs for natural image analysis has

Table 1. Publicly available database for breast histology analysis.

Database	No. of cases	Magnification	Abnormality	Annotation
ICPR2012 (2017)	5	x40	Mitotic nuclei	- Centroids of around 300 mitosis and Mask in .jpg format
AMIDA13 (2017)	23	x40	Mitotic nuclei	- Centroid of 1157 Mitosis and mask in .TIFF format
MITOS-ATYPIA-14 (2016)	32	x10, x20, x40	Mitosis and Nuclear atypia	- Centroids of mitosis and mask in .jpg format; confidence Degree in .csv file
CAMELYON16 (2016)	400	x40,x10,x1	Metastasis	- Contours of cancer locations in .xml files and WSI masks
TUPAC16 (2016)	500+AXILARY DATASETS	x40	Tumor proliferation	- ROC coordinates with the scores in .csv files
CAMELYON17 (2017)	200	–	Metastasis	- Contours of cancer locations in .xml files and WSI masks
BreakHis	82	x400, x200, x100, x40	Atypia, mitosis and metastasis	–
ICIAR2018 BACH (2018)	162 images	- Microscopy whole slide	Metastasis	- Contours of cancer locations and label in .xml
BreCaHAD (2019)	162 images	x10, x40	mitosis, apoptosis and tumor nuclei	- Centroids of mitosis in.json format and mask in .jpg format

inspired their use on medical images, for example, MRI, CT, histopathology analysis. Some of the publicly available breast histology database and challenges are shown in Table 1. The next sections discussed briefly earning used for histopathology analysis.

3 Deep Neural Network Architecture

Deep architecture is similar to traditional feed-forward ANN in the sense that both are made up of neurons which have biases and learnable weights. Artificial Neural Networks (ANNs) were inspired by how information is processed and the distributed communication nodes of the visual cortex of animal [15]. A deep learning architecture consists of multiple layers that progressively extract lower level to higher level features from raw input. Among the deep architecture, CNNs are the most successful and the most suitable for image-based classification. In, CNNs the network share the weights in such a way that convolution is performed on the images. An Auto Encoder (AE) is another form of ANN most suitable for unsupervised learning models [16]. It is trained to reconstruct its inputs, by forcing the hidden layer to learn good representations of the inputs. It has been extensively employed for segmentation and detection process in breast image analysis. Some of the layers which are commonly used in deep learning networks are:

Input Layer: This layer communicates external environment with the network by presenting a pattern to the network [17].

Convolutional Layer: It is made up of three sub stages. They are:

- *Convolutional filters*: Comprises a set of independent filters that are convolved independently within the image and resulting in feature maps [15, 18].
- *Pooling*: This layer is to progressively reduce the spatial size of the representation so that the number of calculation is reduced [19].
- *Activation function*: It defines the output of a node when an input or set of inputs are fed to it. It is the non-linear element-wise operator which decides the excitation of a neuron [20–22].

Normalization Layer: Normalization layer are used between the convolution layer and the activation function layer to speed up the training process. It aims at acquiring an improved description of the input [22].

Dropout Regularization Layer: This layer helps in minimizing the over-fitting of the network and helps in learning more robust features in the succeeding steps. In this, during the training period, units along with their connections are dropped from the network [23].

Fully Connected Layers: Connected at the end of the network after several convolutional and max pooling layer. Here, all the neurons are connected to all activations in the previous layer. High-level reasoning is performed by this layer. Using a matrix multiplication followed by a bias offset, the activations can be calculated [24].

4   Deep Learning for Breast Histopathology Analysis

In this section, various methods that used deep learning algorithm for analyzing histological components to grade breast cancer on histology data are discussed.

4.1   Nuclei Analysis

The degree of malignancy of a breast cancer and nuclei life cycle affects the form, size, structure and mitotic count of the breast epithelial nuclei. Nucleic pleomorphism can help in predicting the presence and severity of cancer.

An unsupervised two-layer Stacked sparse Auto Encoder framework for nuclei classification was developed by [25]. The first SAE is used to extract the features from the input. The features from the first layer are used to train the second SAE to produce the actual feature that is used for the classifier. In another paper [26], they improve their framework by detecting multiple nuclei automatically. The detection process is performed by calculating locally maximal confidence scores. [27] studied various deep learning approaches for histology image analysis in five breast tissue examination task. They tried to decrease computational cost caused by interrogating all the image pixels [28]. They suggested a solution-adaptive deep hierarchical learning method. In this method, higher levels of magnification were used for analysis only when they are needed. [29] used a CNN model to performed nucleus segmentation while maintaining the shape by generating probability maps. They apply selection-based sparse shape and local repulsive deformable models. [30] also used a CNN model on histology images containing tumor with known nuclei location to calculate the statistics of individual nuclei and surrounding regions. A modified CNN that uses a structures regression layer was proposed by [31] for cell detection. They encode topological information which was not considered in the conventional CNN due to coherency in labeled regions.

4.2   Tubules Analysis

[32] proposed a system that uses a customized CNN to identify tubule nuclei in histopathology images and compute a degree of tubule formation. The degree of the tubule is calculated by the ratio between tubule nuclei to the overall number of nuclei. They concluded that tubule formation indicator is related to the likelihood of cancer occurrence.

4.3   Epithelial and Stromal Region Analysis

[33] proposed a system that uses a patch based DCNN for distinguishing epithelial and stromal components in histology images. Using a superpixel algorithm the images are segmented to many regions. It is also found that the union of DCNN with the Ncut-based algorithm and an SVM classifier gives the optimal results. [34] used two deep CNN for epithelial and stromal analysis. The CNNs are inspired by VGG-Net but they replace the 2 fully connected layers with convolutions. It allows arbitrary input sizes to be fed to the network. The first CNN model was used to classify the histology into the epithelium, stroma, and fat. The second CNN model was trained to analyze stromal areas to detect if the stromal regions is cancerous or not.

4.4   Mitotic Activity Analysis

Mitotic activity rate describes how quickly the cancer cells are reproducing. The mitotic count helps in quantifying the locality and proliferative activity of the tumors. It is measured as the number of mitoses in an area of 2 mm^2. This value provides an evaluation of the aggressiveness of the tumor. [35] used the deep max-pooling CNN architecture worked directly on raw RGB pixels image classification and segmentation. They tried to decrease the deep neural network's variance and bias by averaging the outputs of multiple classifiers with different architectures along with using rotational invariance. This method won the ICPR12 competition with the highest F-score and precision. In 2013, the same team won the AMIDA13 competition by using the same approach plus employing Multi-column CNN [36]. [37] combined a lightweight CNN with hand-crafted features for each candidate region. A cascade of two random forest classifiers was combined and trained. From the result, they concluded that the combination of the two features gives a better performance than the one using only individual features. [38] also used a combination of manually segmentation-based nuclear features such as color, texture, and shape, with the features using a CNN architecture [15]. Such a system have the advantage of handling the appearance varieties in mitotic figures and decreasing sensitivity to the manually crafted features and thresholds. [39] suggested a system that uses a deep cascade neural network with two phases. During the first phase, probable mitosis candidates are found out using a 3-layer CNN. In the second phase, mitotic cells are detected in the mitotic candidates that are obtained in the first phase by using three CaffeNet based CNNs [40]. In other work by [41], a deep regression network along with transferred knowledge is used for automatic mitosis detection. To solve the problem of limited annotated data, [42] proposed a framework for learning from crowds by gathering ground truth from non-expert crowd. In their proposed data aggregation system, a multi-scale CNN model is trained using images annotated by experts. In the next step, using the unlabeled data, aggregation schemes were integrated into CNN layers via an additional crowdsourcing layer (AggNet). It is concluded that deep CNNs can be trained with data collected from crowdsourcing and the new method can improve the performance of the CNN. [48] proposed a novel variant of the Faster-RCNN architecture for faster and accurate detecting mitotic figures in histopathological images. By using a two-stage top down multi-scale region proposal generation, small objects such as the mitotic figures are detected.

4.5   Other Tasks

In [49], a hybrid CNN unit is used to make full use of the local and global features of an image for accurate classification. They introduces a bagging strategies and hierarchy voting tactic to improve the performance of the classifier. [50] proposed a system called the transition module that extracts filters at multiple scales, and then collapsed them using global average pooling to ease network size reduction from convolutional layers to Fully Connected layers. [43] proposed a system for detection of IDC (invasive ductal carcinoma) in histology images for estimating the grade of the tumor. They used a 3 layer CNN to train the system with 162 patients diagnosed with IDC. [44] examined the performance of various CNN architecture such as GoogLeNet, VGG16, etc. in breast cancer metastases detection. They won the Camelyon16 competition for WSIs classification and tumor localization. They concluded that GoogLeNet and VGG16 perform best for patch-based classification of histology images at x40 magnification. [45] also proposed a method that uses CNN to identify slides that has no micro-metastasis or macro-metastasis. They train CNN to obtain per-pixel cancer likelihood maps and performs segmentation in whole-slide images. [27] proposed a method that use a Sparse Auto Encoders to evaluate Stain Normalization under different circumstances like different concentrations of H&E in the same tissue section or the same slides being scanned multiple times on different platforms. Table 2 shows the comparison of various methods that are based on deep learning and their performance in histopathology analysis.

Table 2. Deep learning based methods used histology analysis

DL architecture and reference	Problem	Dataset	Results
[37] Cascade of CNN (2 conv+1fc+RF classifier)	Mitosis detection	- ICPR12 dataset - AMIDA13 dataset	F-score: 0.7345 F-score: 0.319
[42] 3conv+1fc	Mitosis detection	- AMIDA13 dataset	AUC:0.8695
[41] CNN:5conv+3fc+ classifier	Mitosis detection	- ICPR12 dataset	F-score: 0.79
[45] 4conv+2fc	Breast cancer metastasis detection in sentinel lymph nodes	Digitized H&E-stained slides from 271 patients	AUC: 0.88
[43] 2conv+2fc+log softmax classifier	Invasive ductal carcinoma (IDC) detection	169 cases from the Hospital of the University of Pennsylvania and The Cancer Institute of New Jersey	F-score: 0.718
[27] AE: 2layer	Stain normalization	Anonymous - 200training images - 25 testing images	Error: 0.047
[27] AlexNet	- Nuclei segmentation - Epithelium segmentation - Tubule segmentation - Mitosis detection - Invasive ductal carcinoma detection	Anonymous	- F-score: 0.83 - F-score: 0.84 - F-score: 0.83 - F-score: 0.53 - F-score: 0.76
[25] SSAE with 2 hidden layers+classifier	- Nuclei classification	17 patient cases containing 37 H&E images	- F-score: 0.82
[26] SSAE with 2 hidden layers+classifier	- Nuclei detection	537 H&E images	- F-score: 0.8449
[29] CNN:2conv+3fc+ classifier	- Nucleus segmentation	Anonymous	- F-score: 0.78
[31] CNN:2conv+3fc	- Cell detection	32 images from(TCGA) dataset	F-score: 0.913
[32] CNN:3conv+3fc+ classifier	- Tubule detection and classification	174 ER+breast cancer images	F-score: 0.59
[33] 2conv+2fc+ Softmaxclassifier	- Epithelial-Stromal segmentation	- 106 H&E images from NKI dataset - 51H&Eimages General Hospita l (VGH)	- F-score: 0.8521 - F-score: 0.891
[35] DNN1: 5conv+2fc+ Softmax classifier; DNN2: 4conv+2fc+ softmax classifier	- Mitosis detection	ICPR 12 mitosis dataset	- F-score: 0.782
[48] RCNN: VGG-net	- Mitosis detection	ICPR 2012, AMIDA 2013 and MITOS-ATYPIA14	- F-score: 0.955

(continued)

Table 2. (*continued*)

DL architecture and reference	Problem	Dataset	Results
[34] CNN1: VGG-net with 11 Layers	- Classification of tissue into epithelium, stroma, and fat - Stromal regions classification - Breast cancer classification	646 H&E sections (444 cases) from the Breast Radiology Evaluation and Study of Tissues Stamp Project	- ACC: 0.95 - ACC: 0.921 - ROC: 0.92
[44] - GoogleNet - AlexNet - VGG16 - FaceNet	- Breast cancer metastasis detection and localization	Camelyon16 dataset	- ACC: 0.921 - ACC: 0.979 - ACC: 0.9968 - ACC: 0.88

5   Discussion, Conclusions and Future Works

5.1   Discussion

The aim of this survey is to provide insights for researchers, to the application of deep learning architecture in the field of breast cancer histopathology image analysis.

In recent years, as seen in Sect. 4, deep learning architectures have been widely applied in the many areas of histopathology image analysis, in areas such as abnormality detection, abnormality segmentation, and classification. Most of these applications have been tested on different network depths and varying input size to address various issues. It is found that the system using deep learning features outperforms those systems using handcrafted features. Some researchers were able to achieve a better performance by combining CNN based features with hand crafted features. More intelligent combination can be explored for better performance. It is also seen that the use of an SAE can also improve the performance of the system. In some studies [46] comparison of the performance of deep learning methods to the performance of expert pathologists is done. When a histologist examines an image, analysis is done in low magnification followed by a more complex and detailed analysis at higher magnification. In a CAD system, only a few specific magnifications are considered and thus selecting appropriate magnifications for good performance is a challenge. Although some of the system provides good performances, they need a lot of computations. Some system results a decent performance with less computation. In developing such a CAD, a balance of all these limitations must be taken into account. It is also found that the performance of the majority of the methods is directly related to the correctness of the training data.

5.2   Conclusion

Our study showed different applications of deep learning in the field of breast cancer histopathology analysis. Deep learning based approaches have promising results in the field of histopathology analysis but there is still improvement needed to reach clinically acceptable results. This work summarizes the history, different model of deep learning, recent advancements and the current state of art for histopathology analysis. We anticipate that this paper can provide insights for researchers, to the application of deep learning network in the field of breast cancer detection and diagnosis in histopathology.

5.3   Future Works

From some research [47], it was proved that there are a lot of associations between mammogram and histology. Information in a histology have some relation to the occurrence of abnormalities in a mammogram. From the research, it is also known that changes in the cellular and nuclear structure can lead to change in tissues and thus can lead to the formation of micro calcification, masses, and other abnormalities. Most of the existing studies relating these two methods are done through statistical risk analysis and observations. A computer-aided system for associating these two modalities has not developed yet. Considering the biological associations between these two modalities and the development in new deep learning algorithms, a model for better cancer detection can be developed for associating these two modalities. In the future, we will try to develop a model based on deep learning that can automatically associate the mammographic and histologic information for better breast cancer detection

References

1. Breast cancer: prevention and control. http://www.who.int/cancer/detection/breastcancer/en/. Accessed 13 Feb 2018
2. Neal, L., Tortorelli, C.L., Nassar, A.: Clinician's guide to imaging and pathologic findings in benign breast disease. In: Mayo Clinic Proceedings, vol. 85, pp. 274–279 (2010)
3. Kopans, D.B.: The positive predictive value of mammography. Am. J. Roentgenol. **158**(3), 521–526 (1992)
4. Elmore, J.G., et al.: Variability in interpretive performance at screening mammography and radiologists characteristics associated with accuracy. Radiology **253**(3), 641–651 (2009)
5. Veta, M., Pluim, J.P., vanDiest, P.J., Viergever, M.A.: Breast cancer histopathology image analysis: a review. IEEE Trans. Biomed. Eng. **61**(5), 1400–1411 (2014)
6. Gurcan, M.N., Boucheron, L.E., Can, A., Madabhushi, A., Rajpoot, N.M., Yener, B.: Histopathological image analysis: a review. IEEE Rev. Biomed. Eng. **2**, 147–171 (2009)
7. Ng, A.: Sparse autoencoder. In: CS294A LectureNotes, vol. 72, pp. 1–19. Stanford University (2011)
8. Salakhutdinov, R., Hinton, G.E.: Deep Boltzmann machines. In: Proceedings of The Twelfth International Conference on Artificial Intelligence and Statistics (AIS-TATS), vol. 5, pp. 448–455 (2009)
9. Fukushima, K.: Neocognitron: a self-organizing neural network model for a mechanism of pattern recognition unaffected by shift in position. Biol. Cybern. **36**(4), 193–202 (1980). https://doi.org/10.1007/BF00344251
10. Lo, S.-C., Lou, S.-L., Lin, J.-S., Freedman, M.T., Chien, M.V., Mun, S.K.: Artificial convolution neural network techniques and applications for lung nodule detection. IEEE Trans. Med. Imaging **14**, 711–718 (1995)

11. LeCun, Y., Bottou, L., Bengio, Y., Haffner, P.: Gradient-based learning applied to document recognition. Proc. IEEE **86**, 2278–2324 (1998)
12. Krizhevsky, A., Sutskever, I., Hinton, G.: Imagenet classification with deep convolutional neural networks. In: Advances in Neural Information Processing Systems, pp. 1097–1105 (2012)
13. Russakovsky, O., et al.: ImageNet large scale visual recognition challenge. Int. J. Comput. Vis. **115**(3), 1–42 (2014). https://doi.org/10.1007/s11263-015-0816-y
14. Elston, C.W., Ellis, I.: Pathological prognostic factors in breast cancer. I. The value of histological grade in breast cancer: experience from a large study with long-term follow-up. Histopathology **19**(5), 403–410 (1991)
15. LeCun, Y., Kavukcuoglu, K., Farabet, C.: Convolutional networks and applications in vision. In: Proceedings of IEEE International Symposium on Circuits and Systems (ISCAS), pp. 253–256 (2010)
16. Bengio, Y.: Learning deep architectures for AI. Found. Trends® Mach. Learn. **2**(1), 1–127 (2009)
17. Hamidinekoo, A., Suhail, Z., Qaiser, T., Zwiggelaar, R.: Investigating the effect of various augmentations on the input data fed to a convolutional neural network for the task of mammographic mass classification. In: Valdés Hernández, M., González-Castro, V. (eds.) MIUA 2017. CCIS, vol. 723, pp. 398–409. Springer, Cham (2017). https://doi.org/10.1007/978-3-319-60964-5_35
18. Schmidhuber, J.: Deeplearning in neural networks: an overview. Neural Netw. **61**, 85–117 (2015)
19. Krizhevsky, A., Hinton, G.: Learning multiple layers of features from tiny images (2009)
20. Glorot, X., Bordes, A., Bengio, Y.: Deep sparse rectifier neural networks. In: 14th International Conference on Artificial Intelligence and Statistics, vol. 15, pp. 315–323 (2011)
21. Goodfellow, I., Bengio, Y., Courville, A.: Deep Learning. MIT Press, Cambridge (2016)
22. Dahl, G.E., Sainath, T.N., Hinton, G.E.: Improving deep neural networks for LVCSR using rectified linear units and drop out. In: 2013 IEEE International Conference on Acoustics, Speech and Signal Processing, pp. 8609–8613 (2013)
23. Srivastava, N., Hinton, G.E., Krizhevsky, A., Sutskever, I., Salakhutdinov, R.: Drop out: a simple way to prevent neural networks from over fitting. J. Mach. Learn. Res. **15**(1), 1929–1958 (2014)
24. Krizhevsky, A., Sutskever, I., Hinton, G.E.: ImageNet classification with deep convolutional neural networks. In: Advances in Neural Information Processing Systems, pp. 1097–1105 (2012)
25. Xu, J., Xiang, L., Hang, R., Wu, J.: Stacked sparse autoencoder (SSAE) based framework for nuclei patch classification on breast cancer histopathology. In: IEEE 11th International Symposium on Biomedical Imaging (ISBI), pp. 999–1002 (2014)
26. Xu, J., et al.: Stacked sparse auto encoder (SSAE) for nuclei detection of breast cancer histopathology images. IEEE Trans. Med. Imaging **35**(1), 119–130 (2016)
27. Janowczyk, A., Basavanhally, A., Madabhushi, A.: Stain normalization using sparse auto encoders (StaNoSA): application to digital pathology. Comput. Med. Imaging Graph. **57**, 50–61 (2017)
28. Janowczyk, A., Madabhushi, A.: Deep learning for digital pathology image analysis: a comprehensive tutorial with selected use cases. J. Pathol. Inform. (2016). https://doi.org/10.4103/2153-3539.186902
29. Xing, F., Xie, Y., Yang, L.: Anautomatic learning-based framework for robust nucleus segmentation. IEEE Trans. Med. Imaging **35**(2), 550–566 (2016)

30. Veta, M., van Diest, P.J., Pluim, J.P.W.: Cutting out the middleman: measuring nuclear area in histopathology slides without segmentation. In: Ourselin, S., Joskowicz, L., Sabuncu, Mert R., Unal, G., Wells, W. (eds.) MICCAI 2016. LNCS, vol. 9901, pp. 632–639. Springer, Cham (2016). https://doi.org/10.1007/978-3-319-46723-8_73

31. Xie, Y., Xing, F., Kong, X., Su, H., Yang, L.: Beyond classification: structured regression for robust cell detection using convolutional neural network. In: Navab, N., Hornegger, J., Wells, William M., Frangi, Alejandro F. (eds.) MICCAI 2015. LNCS, vol. 9351, pp. 358–365. Springer, Cham (2015). https://doi.org/10.1007/978-3-319-24574-4_43

32. Romo-Bucheli, D., Janowczyk, A., Romero, E., Gilmore, H., Madabhushi, A.:. Automated tubule nuclei quantification and correlation with oncotype DX risk categories in ER+breast cancer whole slide images. In: SPIE Medical Imaging, p. 979106. International Society for Optics and Photonics (2016)

33. Xu, J., Luo, X., Wang, G., Gilmore, H., Madabhushi, A.: A deep convolutional neural network for segmenting and classifying epithelial and stromal regions in histopathological images. Neurocomputing **191**, 214–223 (2016)

34. Bejnordi, B.E., et al.: Deep learning-based assessment of tumor associated stroma for diagnosing breast cancer in histopathology images. arXiv preprint arXiv:1702.05803 (2017)

35. Cireşan, D.C., Giusti, A., Gambardella, L.M., Schmidhuber, J.: Mitosis detection in breast cancer histology images with deep neural networks. In: Mori, K., Sakuma, I., Sato, Y., Barillot, C., Navab, N. (eds.) MICCAI 2013. LNCS, vol. 8150, pp. 411–418. Springer, Heidelberg (2013). https://doi.org/10.1007/978-3-642-40763-5_51

36. Veta, M., et al.: Assessment of algorithms for mitosis detection in breast cancer histopathology images. Med. Image Anal. **20**(1), 237–248 (2015)

37. Wang, H., et al.: Cascaded ensemble of convolutional neural networks and handcrafted features for mitosis detection. In: SPIE Medical Imaging, vol. 9041, p. 90410B. International Society for Optics and Photonics (2014). https://doi.org/10.1117/12.2043902

38. Malon, C.D., Cosatto, E.: Classification of mitotic figures with convolutional neural networks and seeded blob features. Pathol. Inform. **4**(1), 9 (2013). https://doi.org/10.4103/2153-3539.112694

39. Chen, H., Dou, Q., Wang, X., Qin, J., Heng, P.-A.: Mitosis detection in breast cancer histology images via deep cascaded networks. In: Proceedings of the Thirtieth AAAI Conference on Artificial Intelligence, pp. 1160–1166. AAAI Press (2016)

40. Jia, Y., et al.: Caffe: convolutional architecture for fast feature embedding. In: Proceedings of the 22nd ACM International Conference on Multimedia, pp. 675–678. ACM (2014)

41. Chen, H., Wang, X., Heng, P.A.: Automated mitosis detection with deep regression networks. In: 13th IEEE International Symposium on Biomedical Imaging (ISBI), pp. 1204–1207. IEEE (2016)

42. Albarqouni, S., Baur, C., Achilles, F., Belagiannis, V., Demirci, S., Navab, N.: Aggnet: deep learning from crowds for mitosis detection in breast cancer histology images. IEEE Trans. Med. Imaging **35**(5), 1313–1321 (2016)

43. Cruz-Roa, A., et al.: Automatic detection of invasive ductal carcinoma in whole slide images with convolutional neural networks. In: SPIE Medical Imaging, vol. 9041. International Society for Optics and Photonics (2014). https://doi.org/10.1117/12.2043872

44. Wang, D., Khosla, A., Gargeya, R., Irshad, H., Beck, A.H.: Deep learning for identifying metastatic breast cancer. arXiv preprint arXiv:1606.05718 (2016)

45. Litjens, G., et al.: Deep learning as a tool for increased accuracy and efficiency of histopathological diagnosis. Sci. Rep. **6**, 26286 (2016). https://doi.org/10.1038/srep26286

46. Giusti, A., Caccia, C., Cireşari, D.C., Schmidhuber, J., Gambardella, L.M.: A comparison of algorithms and humans for mitosis detection. In: IEEE 11th International Symposium on Biomedical Imaging (ISBI), pp. 1360–1363. IEEE (2014)

47. Boyd, N., Jensen, H.M., Cooke, G., Han, H.L.: Relationship between mammographic and histological risk factors for breast cancer. J. Natl. Cancer Inst. **84**, 1170–1179 (1992)
48. Rao, S.: Mitos-rcnn: a novel approach to mitotic figure detection in breast cancer histopathology images using region based convolutional neural networks. arXiv preprint arXiv:1807.01788 (2018)
49. Guo, Y., Dong, H., Song, F., Zhu, C., Liu, J.: Breast cancer histology image classification based on deep neural networks. In: Campilho, A., Karray, F., ter Haar Romeny, B. (eds.) ICIAR 2018. LNCS, vol. 10882, pp. 827–836. Springer, Cham (2018). https://doi.org/10.1007/978-3-319-93000-8_94
50. Akbar, S., Peikari, M., Salama, S., Nofech-Mozes, S., Martel, A.: The transition module: a method for preventing over fitting in convolutional neural networks. Comput. Methods BioMech. Biomed. Eng. Imaging Vis. **7**, 1–6 (2018)

Face Presentation Attack Detection Using Multi-classifier Fusion of Off-the-Shelf Deep Features

Raghavendra Ramachandra, Jag Mohan Singh[✉], Sushma Venkatesh, Kiran Raja, and Christoph Busch

Norwegian Biometrics Laboratory,
Norwegian University of Science and Technology (NTNU), Trondheim, Norway
{raghavendra.ramachandra,jag.m.singh,susma.venkatesh,kiran.raja,
christoph.busch}@ntnu.no

Abstract. Face recognition systems are vulnerable to the presentation (or spoof or direct) attacks that can be carried out by presenting the face artefact corresponding to the legitimate user. Thus, it is essential to develop a Presentation Attack Detection (PAD) algorithms that can automatically detect the presentation attacks the face recognition systems. In this paper, we present a novel method for face presentation attack detection based on the multi-classifier fusion of deep features that are computed using the off-the-shelf pre-trained deep Convolutional Neural Network (CNN) architecture based on AlexNet. Extracted features are compared using softmax and Spectral Regression Kernel Discriminant Analysis (SRKDA) classifiers to obtain the comparison scores that are combined using a weighted sum rule. Extensive experiments are carried out on the publicly available OULU-NPU database and performance of the proposed method is benchmarked with fifteen different state-of-the-art techniques. Obtained results have indicated the outstanding performance of the proposed method on OULU-NPU database.

Keywords: Face recognition · Spoof detection · Smartphone biometrics · Anti spoofing · Deep learning · Fusion

1 Introduction

The exponential increase in the face recognition applications has raised the security concern of the face recognition devices for the presentation attack (or direct attacks). Further, the extensive deployment of the face recognition applications primarily in the smartphone for authentication has increased the security concern as it provides an ample opportunity for the attacker in an uncontrolled (or with no human monitoring) scenario if the smartphone is stolen or lost. The goal of the presentation attack (or spoof attacks) is to deceive the face recognition system by presenting a face artefact (or Presentation Attack Instrument (PAI))

© Springer Nature Singapore Pte Ltd. 2020
N. Nain et al. (Eds.): CVIP 2019, CCIS 1148, pp. 49–61, 2020.
https://doi.org/10.1007/978-981-15-4018-9_5

corresponding to the legitimate user. The accessible presentation attack instruments include the printed photo, displaying the photo using electronic devices and 3D print masks that have shown the vulnerability on both commercial and academic face recognition. Thus, Presentation Attack Detection (PAD) is a crucial component in designing the biometric applications to improve the security and reliability for the legitimate verification of the data subject.

Face presentation attack detection algorithms are extensively studied in the literatures that has resulted in serval survey articles [7,20,27]. Based on the taxonomy proposed in [20], the available PAD techniques are broadly classified as the hardware-based and software-based approaches. The hardware-based approaches are known to provide robust attack detection accuracy at the cost of additional hardware components and cost. The hardware-based approach explores different ways of capturing the biometrics information together with the meta-information. Examples for hardware-based approaches includes: multi-spectral imaging [24], challenge-response [1], special imaging sensors [25], etc. The software-based approaches based on processing the captured face image and are trained to identify the artefacts. Software-based approaches are low-cost to integrate with the existing face recognition systems but not always generalize to detect different presentation attack instruments. Among the magnitude of software-based approaches, the texture-based features have gained the popularity in detecting face PAD. These texture-based methods include the Local Binary Patterns and its variants that are extensively used for face PAD in both visible and near-infrared imaging. Based on the comparative study of different state-of-the-art face PAD techniques reported in [20] has indicated that the detection of face presentation attacks in the visible spectrum (the majority of the face recognition systems are working) is very challenging. Further, the face presentation attacks on smartphones are still more challenging to detect because it is easy and also highly vulnerability for low coast presentation attacks such as print and electronic display attacks.

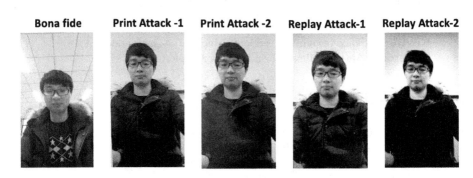

Fig. 1. Example image from OULU-NPU dataset [30]

Face Presentation Attack Detection (PAD) on the smartphone has been an extensively explored problem in the literature [4,6,19,23,30]. Several techniques are explored for detecting the face PAD that includes, Image quality features [6], Moire patterns [19], scale-space features [23], texture features and its variants [23]. Further, face PAD competition on the smartphone-based presentation attack was organized in 2017 on the large-scale database collected with two different types of PAI, including print attack and display attack. More than fifteen algorithms include color texture features, use of color spaces like YCbCr, HSV together with texture features, deep features, image distortion analysis features, Convolution Neural Networks (CNN) and hybrid algorithms that combine more than one algorithm. The extensive evaluation carried out on the OULU-NPU dataset [30] 4950 bona fide, and attack videos indicate the low performance of all thirteen algorithms on four different evaluation protocols. The degraded performance of the algorithms can be attributed to the challenging dataset together with the complex evaluation protocols aimed at generalization. Figure 1 shows the example images from OULU-NPU dataset [30] illustrating the high-quality face artefact images.

In this work, we address the face presentation attack detection in the smartphone environment. To this extent, we propose a new approach based on the multi-classifier fusion of deep features. The deep features are learned using pretrained deep convolution neural network architecture using AlexNet [14]. We then employ two different classifiers, such as Soft-Max and Spectral Regression Kernel Discriminant Analysis (SRKDA) [2]. The comparison scores from these two classifiers are fused using the weighted SUM rule to make the final decision. Extensive experiments are carried out on the publicly available OULU-NPU dataset [30], which is collected using six different smartphones in three different sessions with variation in illumination and background scenes. Further, the performance of the proposed method is compared with the baseline method and also the state-of-the-art methods on all four performance evaluation protocols described by the OULU-NPU dataset. All the results are presented in conformance to the International Standards on PAD (ISO/IEC 30107-1:2016 [11] and ISO/IEC 30107-3 [10]). The main contributions of this paper can be listed as below:

1. Presents a novel approach for face presentation attack detection algorithm based on the multi-classifier fusion of the deep features computed from pretrained AlexNet.
2. Extensive experiments are carried out on the publicly available smartphone-based face presentation attack database OULU-NPU following the same protocol for evaluation.
3. performance of the proposed scheme is compared with fifteen different state-of-the-art face PAD methods, and quantitative results are benchmarked with ISO/IEC 30107-3 [10] metrics.

The rest of the paper is organized as follows: Sect. 2 presents the proposed scheme, Sect. 3 presents the quantitative experiments on the OULU-NPU database and also the comparative performance analysis and Sect. 4 draws the conclusion.

2 Proposed Approach

Figure 2 illustrates the block diagram of the proposed method, which is based on combining the multi-classifier comparison scores computed using the deep features. The proposed method can be structured in three main functional units (1) feature extraction (2) classification (3) fusion. The feature extraction is based on the deep features that are computed from the pre-trained AlexNet [14]. Owing to the availability of small datasets, in this work, we are motivated to use the pre-trained network in the transfer learning analogy. To this extent, we are motivated to use the AlexNet architecture as it has indicated the best performance in the various biometric application, including presentation attack detection [22].

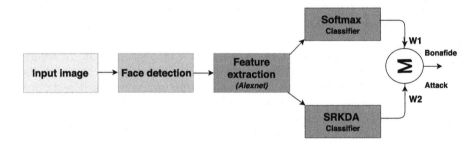

Fig. 2. Block diagram of the proposed method

In this work, we employed the pre-trained AlexNet and fine-tune the Fully Connected (FC8) layer with the training dataset from OULU-NPA database. The fine-tuning is carried out by freezing the weights of the first layers and by boosting the learning rate of the FC8 layer. Thus, we have used the weight learning rate factor as 20 and bias learning rate factor as 20. Further, the data augmentation is carried out using pixel translation, reflection, and rotation to address the over-fitting of the pre-trained network. We then use the fine-tuned AlexNet in which the fully connected layer FC7 is used to representing the features that are further classified using SoftMax and SRKDA classifier.

Figure 3 illustrates the activation of the fine-tuned AlexNet on both bona fide and artefact face samples. For simplicity, we have included the activations corresponding to the first layer conv1 and the late layer relu5 with a maximum response channel. It is interesting to notice the variation of the activations corresponding to bona fide and attack presentations.

In this work, we have employed two different classifiers namely: Softmax and SRKDA. Both of these classifiers are trained using the FC7 features from the fine-tuned AlexNet. The Softmax classifier is trained using the cross-entropy based loss function that can be described as follows:

$$E = \frac{1}{n}\sum_{j=1}^{n}\sum_{i=1}^{k} T_{ij}ln(y_{ij}) + (1 - T_{ij})ln(1 - ln(y_{ij})) \tag{1}$$

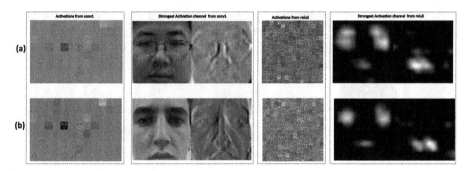

Fig. 3. Illustration of AlexNet features on (a) bona fide (b) attack

where, n is the number of training samples, k is the number of classes, T_{ij} represents the training samples and y_{ij} is the ith output when input vector is x_j.

The SRKDA classifier can perform the discriminant analysis on the data projected in the space induced by a non-linear mapping. Given the training data, the SRKDA solves a set of regularized least-squares problems, and there is no eigenvector computation involved; thus, the use of SRKDA will reduce both time and memory for the computation.

Finally, the comparison score obtained using both Softmax, and SRKDA classifier are combined using a weighted SUM rule to make the final decision. In this work, the weights are computed using a greedy algorithm mentioned in [21] using a development dataset from OULU-NPA database.

3 Experiments and Results

In this section, we present the experimental results of the proposed method on the publicly available face presentation attack dataset (OULU-NPU [30]). Further, the performance of the proposed method is compared with fifteen different state-of-the-art techniques reported in [4]. The performance of the PAD techniques evaluated in this work are reported using ISO/IEC 30107-3 [10] metrics: Attack Presentation Classification Error Rate (APCER) is defined as proportion of attack presentations incorrectly classified as bona fide presentations, and Bona fide Presentation Classification Error Rate (BPCER) is defined as proportion of bona fide presentations incorrectly classified as attacks. In addition, we also present the result in terms of Detection Equal Error Rate (D-EER%).

The OULU-NPU dataset (see Fig. 4) is constructed using six different smartphones in three different sessions with two different Presentation Attack Instrument (PAI) (print attack and display attack). Further database proposes four different protocols for PAD algorithm evaluation to reflect the generalizability of the face PAD techniques to capturing environment, different PAIs, and different smartphone (or sensors). Each protocol has three independent partitions; development, train, and test dataset. The training and development dataset is used to the train and select the parameters of the PAD techniques. The decision threshold corresponding to the BPCER at two different values of APCER, such as 5%

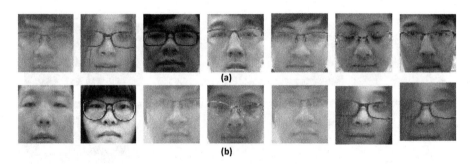

Fig. 4. Examples from OULU-NPU dataset [30] (a) Bona fide samples (b) artefact samples

and 10% is set on the development set. The testing set is used only to evaluate the performance of face PAD techniques. **Protocol 1:** is designed to evaluate the generalization of the face PAD methods for unseen environment conditions like illumination and background scenes. **Protocol 2:** is designed to evaluate the performance of the PAD algorithms on unseen PAIs. **Protocol 3:** is designed to evaluate the sensor interoperability. **Protocol 4:** combines all challenges, as mentioned above (generalization to the sensor (across different smartphone), PAI, and environment conditions). Thus, protocol 4 is very challenging as it evaluates the PAD algorithm generalizing capability across the various real-life scenario. The details about the SOTA in OULU-NPU dataset are shown briefly in Table 1 and more information on OULU-NPU dataset is available in [30]:

Table 1. Overview of existing SOTA on OULU-NPU dataset

Algorithm	Features
Baseline	LBP [5]
CPqD	Inception v3 [28]
Gradiant	LBP [5] from motion, and texture
Gradiant_extra	LBP (additional training) [5] from motion, and texture
HKBU	IDA [29], msLBP [15], Deep feature [12], and AlexNet [14]
Massy_HNU	Guided Image Filtering [8], LBP [16], and Co-occurence matrix [18]
MBLPQ	Multi-Block LPQ [17]
MFT-FAS	BSIF [13]
MixedFASNet	CLAHE [31]
NWPU	LBP from convolutional layers [15]
PML	PML [3] with LPQ
Recod	Squeeze [9] with additional training
SZUCVI	VGG [26] with fine-tuning
VSS	Self-designed deep learning architecture
VSS_extra	Self-designed deep learning architecture

Table 2. Performance of the proposed method: Protocol-1

Algorithm	Development set	Testing set		
	D-EER (%)	D-EER (%)	BPCER @ APCER =	
			5%	10%
Baseline [4]	4.44	9.16	20	10.83
CPqD [4]	0.55	7.39	4.16	1.66
Gradiant [4]	1.11	50	70.83	67.5
Gradiant_extra [4]	0.62	50	64.16	59.16
HKBU [4]	4.37	13.54	17.5	14.16
Massy_HNU [4]	1.11	10	5	5
MBLPQ [4]	2.22	19.47	1.66	0
MFT-FAS [4]	1.87	16.67	12.5	5.83
MixedFASNet [4]	1.18	2.39	9.16	5.83
NWPU [4]	0	44.16	32.5	25
PML [4]	0.55	9.27	0.83	0
Recod [4]	2.22	8.33	9.16	4.16
SZUCVI [4]	35.69	18.54	100	100
VSS [4]	12.22	23.02	55.83	50
VSS_extra [4]	23.95	35.2	95.83	92.5
Proposed method	**0**	**0**	**0**	**0**

Tables 2, 3, 4 and 5 shows the results of the proposed method and fifteen different state-of-the-art methods evaluated using the same protocols from OULU-NPU dataset. Note that, the threshold for BPCER @ APCER = 5% and 10% are set on the development dataset and results are reported only on the testing dataset. Based on the obtained results following are the main observation:

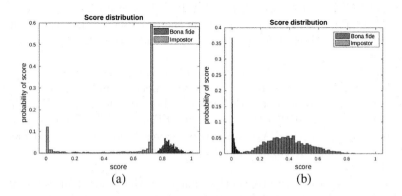

Fig. 5. Distribution of bona fide and artefact scores of proposed method: (a) Protocol 1 (b) Protocol 2

Table 3. Performance of the proposed method: Protocol-2

Algorithm	Development set	Testing set		
	D-EER (%)	D-EER (%)	BPCER @ APCER =	
			5%	10%
Baseline [4]	4.07	11.59	5	3.05
CPqD [4]	2.22	6.95	1.38	0.55
Gradiant [4]	0.83	49.72	59.72	50.27
Gradiant_extra [4]	0.74	49.93	51.38	42.22
HKBU [4]	4.53	9.72	5.27	1.94
Massy_HNU [4]	1.23	7.77	2.22	1.38
MBLPQ [4]	1.85	9.23	4.16	1.94
MFT-FAS [4]	2.22	5.9	1.11	0.83
MixedFASNet [4]	1.2	5	1.11	0.84
NWPU [4]	0	48.12	40.27	35.27
PML [4]	0.83	6.11	0.55	0.27
Recod [4]	3.71	7.52	4.16	2.78
SZUCVI [4]	4.44	6.12	9.17	9.15
VSS [4]	14.81	21.94	36.94	30.83
VSS_extra [4]	23.33	33.61	59.16	51.94
Proposed method	**0**	**0**	**0**	**0**

1. As noted from the qualitative results obtained on the protocol 1 (see Table 2), a majority of the state-of-the-art techniques has indicated a lower D-EER on the development set when compared to that of the testing set. This indicates the limitations of the state-of-the-art techniques on generalizing to the unknown capture environment. The proposed method has indicated the outstanding performance on both development and testing dataset with D-EER = 0%, BPCER = 0% @ APCER = 5% 10%. Figure 5(a) shows the distribution of the PAD comparison scores from the proposed method that indicates the separation of bona fide and attack scores indicating the outstanding performance.

2. Table 3 shows the quantitative results of the proposed and state-of-the-art methods on protocol 2. Here also, it can be observed that the proposed methods have indicated the best performance on both development and testing set with D-EER = 0%, BPCER = 0% @ APCER = 5% 10%. Figure 5(b) shows the distribution of the PAD comparison scores from the proposed method, which is showing the slight separation indicating the robustness of the proposed scheme.

3. Table 4 shows the quantitative performance of both proposed and state-of-the-art techniques corresponding to protocol 3. Similar to the previous experiments, the proposed method has indicated an outstanding performance on

Table 4. Performance of the proposed method: Protocol-3

Algorithm	Development set	Testing set		
	D-EER (%)	D-EER (%)	BPCER @ APCER =	
			5%	10%
Baseline [4]	3.11	5.12	3.33	1.67
CPqD [4]	0.44	3.33	1.67	1.67
Gradiant [4]	0.05	50	75	71.667
Gradiant_extra [4]	0.38	50	75	71.667
HKBU [4]	2.55	3.54	5	0
Massy_HNU [4]	1.33	4.58	1.67	0
MBLPQ [4]	1.38	8.33	1.67	1.67
MFT-FAS [4]	0.44	0.22	0	0
MixedFASNet [4]	0.88	1.45	0	0
NWPU [4]	0	46.67	60	58.33
PML [4]	0.88	1.67	0	0
Recod [4]	1.77	1.66	1.67	1.67
SZUCVI [4]	3.5	5	1.67	1.67
VSS [4]	12.88	18.33	25	18.33
VSS_extra [4]	21.72	26.45	56.67	40.12
Proposed method	**0**	**0**	**0**	**0**

both development and testing set with D-EER = 0%, BPCER = 0% @ APCER = 5% 10%. Figure 6(c) shows the distribution of the PAD comparison scores from the proposed method, which is showing the separation indicating the robustness of the proposed scheme.

4. Table 5 indicates the quantitative performance of the proposed method, together with the state-of-the-art methods. It can be observed that the performance of both state-of-the-art and the proposed method are degraded when compared with the other three different protocols. The degraded performance can be attributed to the design of a protocol that aims to measure the generalizing capability of PAD algorithms for a different environment of capture, sensors, and PAIs. However, based on the obtained results, the proposed method has indicated the best performance on both development and testing dataset with D-EER = 0% and D-EER = 2% respectively. Figure 6(d) shows the distribution of the PAD comparison scores from the proposed method, which is showing the slight overlapping indicating the robustness of the proposed scheme.

Based on the obtained results, the proposed method has indicated the best performance on all four performance evaluation protocols. The best performance of the proposed method can be attributed to the distinctive features extracted using pre-trained and fine-tuned AlexNet, together with the multiple classifiers

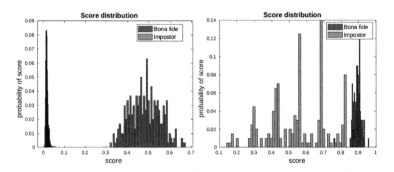

Fig. 6. Distribution of bona fide and artefact scores of proposed method: (a) Protocol 3 (b) Protocol 4

Table 5. Performance of the proposed method: Protocol-4

Algorithm	Development set	Testing set		
	D-EER (%)	D-EER (%)	BPCER @ APCER =	
			5%	10%
Baseline [4]	4.67	15	15	15
CPqD [4]	0.67	5	0	0
Gradiant [4]	0	40	40	30
Gradiant_extra [4]	0.67	45	20	15
HKBU [4]	4.667	10	20	10
Massy_HNU [4]	0.67	6.25	0	0
MBLPQ [4]	2	31.25	0	0
MFT-FAS [4]	10	2	0	0
MixedFASNet [4]	2	6.25	10	5
NWPU [4]	0	45	35	35
PML [4]	0	15	0	0
Recod [4]	2.5	10	5	0
SZUCVI [4]	5.33	6.25	100	65
VSS [4]	10	25	40	35
VSS_extra [4]	14.67	36.25	95	90
Proposed method	0	2	0	0

whose comparison scores are combined using the sum rule. *To facilitate the reproducibility of the proposed method, the comparison scores are as follows (*https:// bit.ly/33CES99*).*

4   Conclusion

In this work, we have proposed a novel technique for face presentation attack detection based on fine-tuning the pre-trained deep CNN AlexNet. Deep features extracted using AlexNet is then classified using multiple classifiers such as Soft-Max and Spectral Regression Kernel Discriminant Analysis (SRKDA). The final decision is computed by combing the comparison scores from multi-classifiers using a weighted SUM rule. Extensive experiments are carried out using the publicly available face presentation attack dataset OULU-NPU. The performance of the proposed method is compared with fifteen different state-of-the-art methods. Obtained results have indicated an outstanding performance of the proposed method in detecting the face presentation attacks.

Acknowledgment. This work is carried out under the partial funding of the Research Council of Norway (Grant No. IKTPLUSS 248030/O70).

References

1. Ali, A., Deravi, F., Hoque, S.: Liveness detection using gaze collinearity. In: 2012 Third International Conference on Emerging Security Technologies, pp. 62–65. IEEE (2012)
2. Baudat, G., Anouar, F.: Generalized discriminant analysis using a kernel approach. Neural Comput. **12**(10), 2385–2404 (2000)
3. Bekhouche, S.E., Ouafi, A., Dornaika, F., Taleb-Ahmed, A., Hadid, A.: Pyramid multi-level features for facial demographic estimation. Expert Syst. Appl. **80**, 297–310 (2017)
4. Boulkenafet, Z., et al.: A competition on generalized software-based face presentation attack detection in mobile scenarios. In: IEEE International Joint Conference on Biometrics (IJCB), pp. 688–696, October 2017. https://doi.org/10.1109/BTAS.2017.8272758
5. Boulkenafet, Z., Komulainen, J., Hadid, A.: Face anti-spoofing based on color texture analysis. In: 2015 IEEE International Conference on Image Processing (ICIP), pp. 2636–2640, September 2015. https://doi.org/10.1109/ICIP.2015.7351280
6. Costa-Pazo, A., Bhattacharjee, S., Vazquez-Fernandez, E., Marcel, S.: The replay-mobile face presentation-attack database. In: International Conference of the Biometrics Special Interest Group (BIOSIG), pp. 1–7, September 2016. https://doi.org/10.1109/BIOSIG.2016.7736936
7. Galbally, J., Marcel, S., Fierrez, J.: Biometric antispoofing methods: a survey in face recognition. IEEE Access **2**, 1530–1552 (2014). https://doi.org/10.1109/ACCESS.2014.2381273
8. He, K., Sun, J., Tang, X.: Guided image filtering. IEEE Trans. Pattern Anal. Mach. Intell. **35**(6), 1397–1409 (2012)
9. Iandola, F.N., Han, S., Moskewicz, M.W., Ashraf, K., Dally, W.J., Keutzer, K.: SqueezeNet: AlexNet-level accuracy with 50x fewer parameters and <0.5 mb model size. arXiv preprint arXiv:1602.07360 (2016)
10. International Organization for Standardization: ISO/IEC DIS 30107–3:2017. Information Technology - Biometric presentation attack detection - Part 3: Testing and Reporting (2016)

11. ISO/IEC JTC1 SC37 Biometrics: ISO/IEC 30107–1:2016. Information Technology - Biometric presentation attack detection - Part 1: Framework. International Organization for Standardization (2016)
12. Jain, A., Vishwanathan, S.V., Varma, M.: SPF-GMKL: generalized multiple kernel learning with a million kernels. In: Proceedings of the 18th ACM SIGKDD International Conference on Knowledge Discovery and Data Mining, pp. 750–758. ACM (2012)
13. Kannala, J., Rahtu, E.: BSIF: binarized statistical image features. In: 21st International Conference on Pattern Recognition (ICPR), pp. 1363–1366 (2012)
14. Krizhevsky, A., Sutskever, I., Hinton, G.E.: ImageNet classification with deep convolutional neural networks. Commun. ACM **60**(6), 84–90 (2017). https://doi.org/10.1145/3065386
15. Maatta, J., Hadid, A., Pietikainen, M.: Face spoofing detection from single images using micro-texture analysis. In: International Joint Conference on Biometrics (IJCB), pp. 1–7, October 2011
16. Ojala, T., Pietikainen, M., Maenpaa, T.: Multiresolution gray-scale and rotation invariant texture classification with local binary patterns. IEEE Trans. Pattern Anal. Mach. Intell. **24**(7), 971–987 (2002). https://doi.org/10.1109/TPAMI.2002.1017623
17. Ojansivu, V., Heikkilä, J.: Blur insensitive texture classification using local phase quantization. In: Elmoataz, A., Lezoray, O., Nouboud, F., Mammass, D. (eds.) ICISP 2008. LNCS, vol. 5099, pp. 236–243. Springer, Heidelberg (2008). https://doi.org/10.1007/978-3-540-69905-7_27
18. Palm, C.: Color texture classification by integrative co-occurrence matrices. Pattern Recogn. **37**(5), 965–976 (2004)
19. Patel, K., Han, H., Jain, A.K.: Secure smartphone unlock: Robust face spoof detection on mobile. Technical report MSU-CSE-15-15, Department of Computer Science, Michigan State University, East Lansing, Michigan, October 2015
20. Raghavendra, R., Busch, C.: Presentation attack detection methods for face recognition systems: a comprehensive survey. ACM Comput. Surv. **50**(1), 8:1–8:37 (2017). https://doi.org/10.1145/3038924
21. Raghavendra, R., Raja, K.B., Venkatesh, S., Busch, C.: Improved ear verification after surgery - an approach based on collaborative representation of locally competitive features. Pattern Recogn. **83**, 416–429 (2018). https://doi.org/10.1016/j.patcog.2018.06.008
22. Raghavendra, R., Venkatesh, S., Raja, K.B., Busch, C.: Transferable deep convolutional neural network features for fingervein presentation attack detection. In: 5th International Workshop on Biometrics and Forensics (IWBF), pp. 1–5, April 2017
23. Raghavendra, R., Venkatesh, S., Raja, K.B., Wasnik, P., Stokkenes, M., Busch, C.: Fusion of multi-scale local phase quantization features for face presentation attack detection. In: 2018 21st International Conference on Information Fusion (FUSION), pp. 2107–2112, July 2018
24. Raghavendra, R., Kiran, R., Sushma, V., Faouzi, C., Busch, C.: On the vulnerability of extended multispectral face recognition systems towards presentation attacks. In: IEEE International Conference on Identity, Security and Behavior Analysis (ISBA), pp. 1–8. IEEE (2017)
25. Raghavendra, R., Raja, K., Busch, C.: Presentation attack detection for face recognition using light field camera. IEEE Trans. Image Process. **24**(3), 1–16 (2015)
26. Simonyan, K., Zisserman, A.: Very deep convolutional networks for large-scale image recognition. arXiv preprint arXiv:1409.1556 (2014)

27. Souza, L., Oliveira, L., Pamplona, M., Papa, J.: How far did we get in face spoofing detection? Eng. Appl. Artif. Intell. **72**(C), 368–381 (2018). https://doi.org/10.1016/j.engappai.2018.04.013

28. Szegedy, C., Vanhoucke, V., Ioffe, S., Shlens, J., Wojna, Z.: Rethinking the inception architecture for computer vision. In: Proceedings of the IEEE Conference on Computer Vision and Pattern Recognition, pp. 2818–2826 (2016)

29. Wen, D., Han, H., Jain, A.K.: Face spoof detection with image distortion analysis. IEEE Trans. Inf. Forensics Secur. **10**(4), 746–761 (2015). https://doi.org/10.1109/TIFS.2015.2400395

30. Zinelabidine, B., Jukka, K., Li, L., Feng, X., Hadid, A.: OULU-NPU: a mobile face presentation attack database with real-world variations. In: IEEE International Conference on Automatic Face and Gesture Recognition (AFGR), pp. 1–7. IEEE (2017)

31. Zuiderveld, K.: Contrast limited adaptive histogram equalization. In: Graphics Gems IV, pp. 474–485. Academic Press Professional Inc., San Diego (1994). http://dl.acm.org/citation.cfm?id=180895.180940

Vision-Based Malware Detection and Classification Using Lightweight Deep Learning Paradigm

S. Abijah Roseline[1], G. Hari[1], S. Geetha[1(✉)], and R. Krishnamurthy[2]

[1] Vellore Institute of Technology - Chennai Campus,
Chennai 600127, Tamilnadu, India
{abijahroseline.s2017,hari.g2017}@vitstudent.ac.in,
geetha.s@vit.ac.in
[2] Sri Vasavi Engineering College, Tadepalligudem 534101,
Andhra Pradesh, India
krishnamurthy.ramanujan@srivasaviengg.ac.in

Abstract. Cyber attackers develop new malicious software to attack their targets every year. Recent sophisticated malware targets financial data and steals the credentials of users. Security analysts design novel methods to defend against malware attacks, but, unfortunately, with the proliferation of newly discovered malware, the methods are inefficient. The need for automated detection of unknown and new malware is still challenging in cybersecurity research. Machine learning approaches are applied for malware detection, however, they require larger feature extraction and feature engineering. The proposed work analyzes and classifies malware based on visualization technique and employs Lightweight Convolutional Neural Networks deep learning model. The model performed better achieving an accuracy of 97% and 95% for the two malware datasets including benign samples. They did not require more hardware resources and model is trained with a low computational cost. The model was evaluated on Malimg dataset and Kaggle's Microsoft Malware Classification Challenge (BIG 2015) dataset.

Keywords: Cybersecurity · Malware classification · Malware visualization · Lightweight approach · Convolutional Neural Networks · Deep learning

1 Introduction

Cyber attacks are an evolving threat to organizations, personnel, and users. Cybersecurity or computer security refers to any approach for defending one's system from malware attacks with the malicious intent of stealing money, accessing personal information, damaging system resources, etc. The attack occurs on the victim's hardware or software, or through social engineering. Malware has to be stopped before it reaches its target or removed immediately once it is detected. Malware authors devise intelligent techniques like code obfuscations to avoid detection by malware detectors. Various modifications or tactics are done to the same malicious code resulting in malware variants and they belong to the same malware family. The security community requires effective and robust detection methods to fight against sophisticated and new malware.

© Springer Nature Singapore Pte Ltd. 2020
N. Nain et al. (Eds.): CVIP 2019, CCIS 1148, pp. 62–73, 2020.
https://doi.org/10.1007/978-981-15-4018-9_6

1.1   Malware Analysis

Malware analysis is usually done using any of the two methods, namely, static analysis [1] and dynamic analysis. In static analysis, the code is analyzed statically without requiring for its execution. Control flow graphs and signatures can be built using static analysis. Though this method is inexpensive, it is not effective when the code is obfuscated or encrypted by malware attackers. Dynamic analysis is analyzing malware's behavior by allowing it to run in a controlled or virtualized environment. This method is effective but it is complex and time-consuming. Another alternative to analyzing malware is visualization method [2], which is based on image processing.

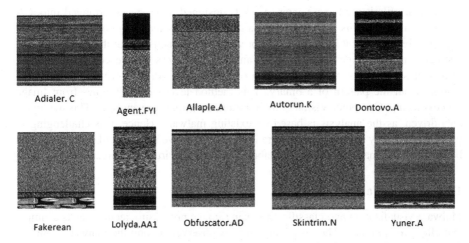

Fig. 1. Malware images of some families in Malimg dataset [2].

Program binaries (malware or benign) are visualized as image patterns, where the various sections of program code are clearly perceived. Malware authors make modifications in some sections in the original code to produce a new variant. Images capture even the small changes, retaining the overall structure. The deviations from the original code are identified by the variations in the patterns. The set of similar pattern images are classified as identical malware samples which belong to the same family. Different malware families include images of dissimilar patterns. Malware images with distinct patterns belonging to different families of Malimg dataset are shown in Fig. 1.

Windows Portable Executable (PE) files are significantly modified by malware attackers. Figure 2 shows the various sections of the PE malware image exhibiting typical image textures. The .text section contains the executable code. The first segment of the .text section contains the code with a fine-grained texture. The remaining segment consists of zeros (black) which indicates zero paddings at the end of the .text section. The .data section comprises uninitialized and initialized data with fine-grained texture leading to a detailed analysis. The .rsrc section contains the resource information for the module generated by the resource compiler.

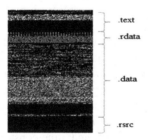

Fig. 2. Sections of a Portable Executable file [2].

A binary file is read as a vector of 8 bits unsigned integers and arranged into a 2-dimensional array. This 2D array is visualized as a grayscale image in the range of [0, 255], where 0 represent black and 255 for white. The dimensions of the image vary for distinct families. Image-based malware analysis takes less feature computation time. This analysis does not require execution or disassembly. Images provide better information about the pattern of a binary. This method leverages techniques from Image Processing and Computer Vision domain for effective malware analysis. This method is data-driven, as the analysis is based on existing malware. Hence, it is challenging to prevent a zero-day attack. The malware images do not give much detail about the actual behavior of the malware except the label given by antivirus software.

1.2   Related Work

Malware visualization and classification using image processing techniques is a simple and effective approach. The visual similarity is observed among the malware families. The images belonging to the corresponding family appear alike in design and texture. Based on this idea, the classification of malware can be done using typical image features, which need no disassembly or code execution. Kancherla et al. [6] proposed a visualization based malware detection approach where the Intensity, Wavelet and Gabor based features were extracted. Their method was robust to code obfuscations without requiring unpacking or decryption.

Han et al. [7] presented the conversion of the opcodes into image matrices (recorded as RGB color pixels) using static as well as dynamic analysis. The similarities among the image matrices were computed. The analysis of obfuscated and packed malware was analyzed by considering the samples from dynamic analysis. A major block selection technique is used for selecting only the blocks with staple behaviors like functions and Application Programming Interface (API) calls, which reduces the computational overhead for analysis. The unknown malware samples were classified in minimal time.

Most of the systems characterize malware based on global features. This leads to a classification model that is inefficient in analyzing advanced malware. Fu et al. [12] presented a fine-grained malware classification method, which is based on a combination of global and local features. Malware is visualized as RGB color images, and texture and color features are taken as global features. Local features are some of the continuous byte sequences observed from the code and data segments of the malicious code. Malware classification was carried out using machine learning techniques like Support Vector Machine, K-Nearest Neighbor, and Random Forest.

The number and diversity of malware attacks are increasing incessantly, making it more difficult to defend against them using traditional methods. Static and dynamic analysis [5] techniques are used to analyze malware binary files. The static analysis leads to the collection of more malware binaries. Labeling malware samples are accomplished using web portals like VirusTotal. Machine learning techniques comply with static analysis methods achieving better performance with the arrival of growing amounts of malware.

Kolter et al. [3] presented the detection and classification of malicious samples using machine learning techniques. A large number of malicious and benign executables were collected and n-grams features of byte codes were extracted leading to many unique n-grams. The best n-gram features were selected for prediction and were evaluated using Naive Bayes, Decision Trees, Support Vector Machines, and Boosting. The boosted J48 decision tree algorithm showed better predictive accuracy than other methods.

Rieck et al. [4] developed an automated incremental method using behavior-based analysis. The new classes of malware with similar behavior were identified using machine learning. The computational overhead is reduced compared to existing analysis methods, with better accuracy and classification of unknown malware.

Traditional machine learning techniques were not able to process raw data like pixel values of an image. The conversion of raw data into a feature vector requires careful engineering and significant domain expertise. The classifier detects or classifies patterns in the input based on this conversion. This ability to take raw data as input and making the machine to learn models automatically is called Representation learning. Deep learning techniques [8] are based on representation learning with multiple levels of representation, where the higher levels show more abstract information about the data.

Conventional methods typically do not allow incremental learning, and dataset with a substantial number of features possibly require expensive hardware resources. Due to significant advances in training models, neural networks are employed as an alternative to the standard machine learning methods. Neural network models are used for malware detection, due to their advantages like incremental feature learning ability, training individual layers whenever required, etc. Deep learning techniques contribute for building automated, generalized and typical models to detect and classify known and unknown malware.

Hardy et al. [13] proposed a Stacked AutoEncoders (SAEs) model for malware detection using the Windows Application Programming Interface (API) calls. Their model uses a layered training process for unsupervised pretraining or feature learning and accomplishes supervised backpropagation. Their system was efficient in terms of accuracy and computational time, thus providing a better solution for malware detection in real-world applications.

Cui et al. [10] proposed an efficient method for detecting malware variants using Convolutional Neural Network (CNN). The imbalanced data problem between different malware families was resolved using an effective data equilibrium technique based on bat algorithm.

Ni et al. [9] presented a malware classification algorithm using SimHash and CNN based on static features. Multi-hash, major block selection, and bilinear interpolation were used to achieve better performance. Their method performed well even with unevenly distributed malware in various families.

Vinayakumar et al. [11] proposed a two-phase scalable malware detection framework. The first phase involves malware classification using static and dynamic analysis methods and detection using deep learning, whereas, in the second phase, malware is classified into their respective families.

This paper presents a visualization approach where the executables are converted into images. The features are extracted and classification of malware and benign files are done using a lightweight Convolutional Neural Network (CNN) architecture. Various malware analysis techniques and the related work in malware visualization and classification using machine learning techniques were discussed. The remaining sections of this paper are organized as follows. The proposed system is elucidated in Sect. 2. Section 3 describes the datasets used for evaluation of the proposed methodology and the experimental results obtained are presented in Sect. 4. The conclusion of the work is discussed in Sect. 5.

2 Proposed Methodology

2.1 Convolutional Neural Networks

Convolutional Neural Networks [8] is a remarkable model of artificial neural networks, proposed by Yann LeCun in 1988. CNN makes use of some of the properties of the visual cortex. The most significant use of this model is image classification. CNN captures a better representation of data and so they do not require feature engineering. When the input image is given to the system, it first converts the images into pixel intensity array values. The images are passed over a sequence of convolutional layers, pooling layers, and fully connected layers, and finally predicts the output. In the convolution layer, an image is read as a smaller matrix called a filter and it produces a convolution. The filter values are multiplied by its original pixel values and the sum of products is obtained. Similarly, the filter moves further by 1 unit and the computations are repeated until the filters are passed across all positions. This results in a smaller matrix.

Likewise, the network consists of several convolutional layers. The result of one convolutional layer is fed as input to the subsequent convolution layer. The nonlinear layer follows each convolution step, where the nonlinear activation function is computed. This layer is a powerful one without which a network would not be modeled. The activation function of this model is Rectified Linear Units (ReLU) [19]. This function sets the threshold value as zero and is represented as $f(x) = \max(0, x)$. If $x > 0$, the size of the array of pixels remains the same, and if $x < 0$, it discards the insignificant information in the network.

Next, the images are passed across a pooling layer where the images are downsampled, resulting in reduced image size. The features that are processed in the convolution layer are not considered here and the images are compressed. Further, the processing results of images across all these layers are given to a fully connected layer. This layer outputs the classes as N-dimensional vectors, from which the correct class is predicted.

A lightweight CNN model is proposed for malware detection and classification. The overall design of the proposed model is shown in Fig. 3. The input images to the proposed lightweight CNN model are reshaped into a size of 32 × 32. The images are converted into 8-bit vectors for further processing. The proposed model includes three convolutional layers for performing classification. The window size of all layers is 3 × 3 convolution filters.

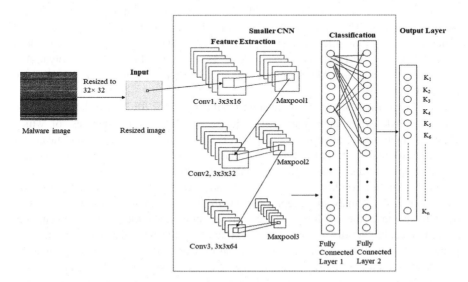

Fig. 3. Proposed system architecture.

The three convolutional layers have an increasing depth of 16, 32 and 64 convolutions each. The convolution stride is set to 1 pixel. The spatial padding (1 pixel) in the convolution layer is done for preserving the spatial resolution after each convolution. Max-pooling is performed over a 2 × 2 pixel window, with stride 2. Three Fully Connected (FC) layers follow the convolutional layers, each having 512 neurons. The last FC has softmax activation with K number of neurons. K neurons is the number of malware families in the datasets. For Malimg dataset, K value is 26 and for BIG 2015 malware dataset, K value is 10.

Adam optimizer and Categorical Cross-entropy loss function are used for training the malware detection model. Adam [16] is an adaptive learning rate optimization algorithm, which updates weights iteratively based on training data. Adam optimizer adopts the advantages of stochastic gradient descent variants such as Adaptive Gradient Algorithm (AdaGrad) and Root Mean Square Propagation (RMSProp). It evaluates adaptive learning rates from calculations of first and second moments of the gradients.

Categorical cross-entropy loss [18] or log loss compares the distribution of the predictions with the actual distribution, where the probability of the actual class is set to 1 and 0 for the other classes. Cross-entropy loss increases as the predicted probability diverges from the actual label.

3 Datasets

The performance of the proposed lightweight CNN model was evaluated on two malware datasets along with benign samples, the Malimg Dataset [2] and Microsoft's BIG 2015 Malware dataset [15]. The experiments are conducted with 1043 benign samples. Benign samples are collected from system files of Windows operating systems. They are tested using the VirusTotal website.

The Malimg dataset consists of 9339 grayscale images belonging to one of the 25 malware families. The families include Adialer.C, Agent.FYI, Allaple.A, Allaple.L, Alueron.gen!J, Autorun.K, Benign, C2LOP.P, C2LOP.gen!g, Dialplatform.B, Dontovo.A, Fakerean, Instantaccess, Lolyda.AA1, Lolyda.AA2, Lolyda.AA3, Lolyda.AT, Malex.gen!J, Obfuscator.AD, Rbot!gen, Skintrim.N, Swizzor.gen!E, VB.AT, Wintrim. BX, and Yuner.A.

The Microsoft Malware Classification Challenge (BIG 2015) [15] dataset (almost half a terabyte when uncompressed) includes 21,741 malware samples, of which 10,868 samples were taken for training and the other 10,873 samples were taken for testing. Each malware file has an Id and class. The Id is a hash value that uniquely identifies the file and class represents one of 9 different malware families, such as Ramnit, Lollipop, Kelihos_ver3, Vundo, Simda, Tracur, Kelihos_ver1, Obfuscator. ACY, Gatak. Each malware is represented with two files, .bytes and .asm files. The . bytes file contains the raw hexadecimal representation of the file's binary content and. asm file contains the disassembled code extracted by the IDA disassembler tool. The proposed system considers the .bytes files and they are converted as malware images.

The distribution of malware samples over classes in the training data is not uniform and the number of malware samples of some families significantly outnumbers the samples of other families. The classes in the two malware datasets are considered for classification, though unbalanced, since they are benchmark datasets. Also, new malware written everyday are the variants of previous malware. Even the classes with less samples (e.g.) Simda, are very important to address new malware.

4 Experimental Results

The malware datasets are split as train and test subsets in 80–20 ratio. The implementation for the proposed malware detection model is performed using Python framework along with packages such as scikit-learn, keras, numpy, matplotlib, imutils, etc. A deep learning library called TensorFlow is used. The experiments were carried out on a system with specifications: Ubuntu 18.06 64-bit OS with Intel® Xeon(R) CPU E3-1226 v3 @ 3.30 GHz × 4 processor, Quadro K2200/PCIe/SSE2 Graphics and 32 GB RAM.

The results for the proposed lightweight CNN model were taken with 1043 benign class samples and each of the two malware datasets. The proposed model was trained on 8306 samples and validated on 2077 samples for Malimg dataset along with benign samples (9339 + 1043). Then, the proposed model was trained on 9529 samples and validated on 2383 samples for BIG 2015 Malware dataset along with benign samples (10868 + 1043).

Table 1. Results of Lightweight CNN model on Malimg and BIG 2015 malware datasets.

S. No	Dataset	No. of epochs	Accuracy (%)	Loss	Precision	Recall	F1-score	Training time (in seconds)	Testing time (in seconds)
1	Malimg dataset	50	97.49	0.08	0.97	0.97	0.97	738.36	0.79
		100	**97.68**	**0.08**	**0.98**	**0.98**	**0.98**	**1484.96**	**0.76**
2	BIG 2015 malware dataset	50	95.67	0.1	0.96	0.96	0.96	809.32	0.75
		100	**96.23**	**0.1**	**0.96**	**0.96**	**0.96**	**1661.65**	**0.90**

Table 1 shows the results of the proposed Lightweight CNN model on Malimg and BIG 2015 malware datasets. Various experiments were conducted on both the malware datasets to select optimal values for parameters. The proposed model is trained for 50 and 100 epochs with a batch size of 32. With a learning rate of 0.0001, the proposed model showed better accuracies and losses for the two datasets.

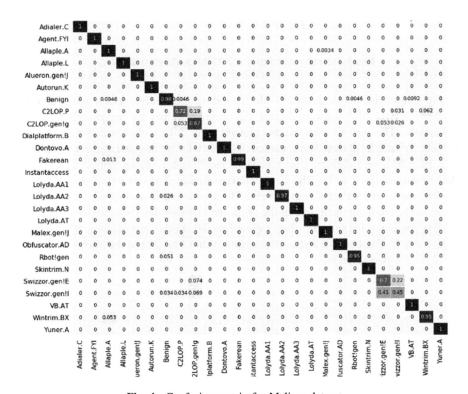

Fig. 4. Confusion matrix for Malimg dataset.

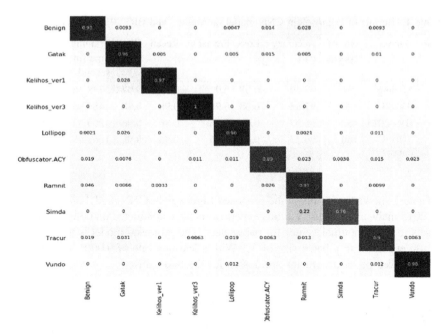

Fig. 5. Confusion matrix for BIG 2015 Malware dataset.

For malimg dataset, the accuracies does not show much improvement for 50 and 100 epochs. For BIG 2015 malware dataset, the accuracy values are comparably better for 100 epochs. The performance results obtained with 100 epochs are better for the two datasets. The model obtained an accuracy of 97.68% for Malimg dataset and 96.23% for BIG 2015 Malware dataset. The precision, recall and f1-score values are 0.98 for Malimg dataset and 0.96 for BIG 2015 Malware dataset. The loss values were 0.08 and 0.1 for two of the datasets respectively. The training and testing time is shown in Table 1. The computational time of the proposed malware detection model is very less.

The confusion matrices for the proposed model on the two datasets are shown in Figs. 4 and 5. The training and testing accuracy and loss graphs are shown in Figs. 6 and 7 for the two datasets.

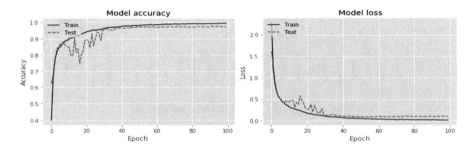

Fig. 6. Plot showing training and validation accuracy and loss with Malimg dataset.

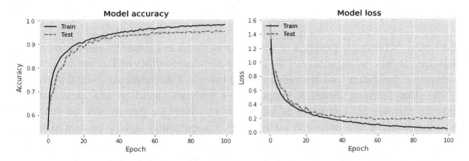

Fig. 7. Plot showing training and validation accuracy and loss with BIG2015 Malware dataset.

The results of the previous works based on machine learning and deep learning techniques in malware detection are shown in Table 2.

Table 2. Results of existing works in malware detection.

S. No	Models	Accuracy (%)
1	Nataraj et al. [2]	98.08
2	Islam et al. [5]	97
3	Agarap et al. [14]	84.92
4	Vinayakumar et al. [11]	93.6
5	Kancherla et al. [6]	95.95
6	Cui et al. [10]	94.5
7	Fu et al. [12]	97.47
8	**Lightweight CNN**	**97.68**

Table 3. Comparison of proposed model with previous works based on trainable parameters.

S. No	Models	Number of trainable parameters
1	CNN 1 [11]	4,197,913
2	CNN 2 [11]	4,222,617
3	CNN A [17]	39,690,313
4	CNN B [17]	20,395,209
5	CNN C [17]	34,519,497
6	Proposed CNN	826,362

The model complexity is studied with the parameters used by a network model. Table 3 shows the model validation of the proposed system by comparing the number of trainable parameters obtained with previous works in malware detection. The total number of parameters of the proposed Lightweight CNN are 828,634, of which the number of trainable parameters is 826,362 and the number of non-trainable parameters is 2,272. The proposed model uses very less number of trainable parameters in comparison with other existing CNN models.

5   Conclusion

Malware is increasingly posing a serious security threat to computers and cyber world. Malware variants belong to any of the existing malware families and image patterns aid to improve the better classification. A Lightweight Convolutional Neural Network (CNN) deep learning model is proposed for malware detection and classification. Malware samples are converted as images, from which the patterns can be analyzed and trained using the proposed architecture. Experimental results on two publicly available benchmark malware classification datasets show the effectiveness of the proposed method, with accuracy of 97.68% and 96.23%. The computational cost is also low. The visualization based approach does not require the execution of malware, so there is no risk of damage to systems and resources. The proposed malware detection system also efficiently identifies the obfuscated and encrypted malware. The proposed Lightweight CNN model outperformed the previous works for malware detection in terms of accuracy, computational cost and model complexity.

References

1. Bergeron, J., Debbabi, M., Desharnais, J., Erhioui, M.M., Lavoie, Y., Tawbi, N.: Static detection of malicious code in executable programs. Int. J. Req. Eng. **2001**(184–189), 79 (2001)
2. Nataraj, L., Karthikeyan, S., Jacob, G., Manjunath, B.S.: Malware images: Visualization and automatic classification. In: Proceedings of the 8th International Symposium on Visualization for Cyber Security, p. 4. ACM (2011)
3. Kolter, J.Z., Maloof, M.A.: Learning to detect and classify malicious executables in the wild. J. Mach. Learn. Res. **7**, 2721–2744 (2006)
4. Rieck, K., Trinius, P., Willems, C., Holz, T.: Automatic analysis of malware behavior using machine learning. J. Comput. Secur. **19**(4), 639–668 (2011). https://doi.org/10.3233/jcs-2010-0410
5. Islam, R., Tian, R., Batten, L., Versteeg, S.: Classification of malware based on integrated static and dynamic features. J. Netw. Comput. Appl. **36**(2), 646–656 (2013). https://doi.org/10.1016/j.jnca.2012.10.004
6. Kancherla, K., Mukkamala, S.: Image visualization based malware detection. In: 2013 IEEE Symposium on Computational Intelligence in Cyber Security (CICS), pp. 40–44. IEEE (2013)
7. Han, K., Kang, B., Im, E.G.: Malware analysis using visualized image matrices. Sci. World J. **2014**, 1–15 (2014). https://doi.org/10.1155/2014/132713
8. LeCun, Y., Bengio, Y., Hinton, G.: Deep learning. Nature **521**(7553), 436–444 (2015)
9. Ni, S., Qian, Q., Zhang, R.: Malware identification using visualization images and deep learning. Comput. Secur. **77**, 871–885 (2018). https://doi.org/10.1016/j.cose.2018.04.005
10. Cui, Z., Xue, F., Cai, X., Cao, Y., Wang, G., Chen, J.: Detection of malicious code variants based on deep learning. IEEE Trans. Industr. Inf. **14**(7), 3187–3196 (2018). https://doi.org/10.1109/tii.2018.2822680
11. Vinayakumar, R., Alazab, M., Soman, K.P., Poornachandran, P., Venkatraman, S.: Robust intelligent malware detection using deep learning. IEEE Access **7**, 46717–46738 (2019). https://doi.org/10.1109/access.2019.2906934

12. Fu, J., Xue, J., Wang, Y., Liu, Z., Shan, C.: Malware visualization for fine-grained classification. IEEE Access **6**, 14510–14523 (2018). https://doi.org/10.1109/access.2018. 2805301
13. Hardy, W., Chen, L., Hou, S., Ye, Y., Li, X.: DL4MD: a deep learning framework for intelligent malware detection. In: Proceedings of the International Conference on Data Mining (DMIN). The Steering Committee of the World Congress in Computer Science, Computer Engineering and Applied Computing (WorldComp), p. 61 (2016)
14. Agarap, A.F., Pepito, F.J.H.: Towards building an intelligent anti-malware system: a deep learning approach using support vector machine (SVM) for malware classification (2017). arXiv preprint arXiv:1801.00318
15. Ronen, R., Radu, M., Feuerstein, C., Yom-Tov, E., Ahmadi, M.: Microsoft malware classification challenge (2018). arXiv preprint arXiv:1802.10135
16. Kingma, D.P., Ba, J.: Adam: A method for stochastic optimization (2014). arXiv preprint arXiv:1412.6980
17. Gibert, D.: Convolutional Neural Networks for Malware Classification. University Rovira i Virgili, Tarragona (2016)
18. Janocha, K., Czarnecki, W.M.: On loss functions for deep neural networks in classification (2017). arXiv preprint arXiv:1702.05659
19. Agarap, A.F.: Deep learning using rectified linear units (ReLU) (2018). arXiv preprint arXiv: 1803.08375

A Deep Neural Network Classifier Based on Belief Theory

Minny George$^{(\boxtimes)}$ and Praveen Sankaran

Department of Electronics and Communication Engineering, NIT Calicut,
Kozhikode, India
minnygeoc@gmail.com, psankaran@nitc.ac.in

Abstract. Classification is a machine learning technique that is used to find the membership of an object to a given set of classes or groups. Neural network (NN) classifier always suffers some issues while examining outliers and data points from nearby classes. In this paper, we present a new hyper-credal neural network classifier based on belief theory. This method is based on credal classification technique introduced in Dempster-Shafer Theory (DST). It allows a data point to belong not only to a specific class but also to a meta-class or an ignorant class based on its mass. The sample which lies in an overlapping region is accurately classified in this method to a meta-class which corresponds to the union of the overlapping classes. Therefore this approach reduces the classification error at the price of precision. But this decrease in precision is acceptable in applications such as medical, defence related applications where a wrong decision would cost more than avoiding some correct decisions. The results and analysis of different databases are given to illustrate the potential of this approach. This idea of hyper-credal classification is extended to convolutional neural network (CNN) classifiers also.

Keywords: Neural network classifier · Outlier · Hyper-credal classification · Specific class · Meta-class · Ignorant class · Medical applications · Convolutional neural network

1 Introduction

In classification problems, the Bayesian classification approach is not acceptable when the complete statistical knowledge about the conditional densities is not available [1]. It led to the development of non-parametric methods such as voting k-nearest neighbour (KNN), distance-based KNN etc. [2,3]. Fix and Hodges developed the popular KNN algorithm or voting KNN in which a data point is allotted the class label with the maximum number of neighbours in its neighbourhood [2]. Later this method is improved by considering the distance between the test point and its neighbours.

In 1995, Denoux proposed a new KNN classifier *i.e.* Evidential KNN (*E-KNN*) [3] based on the concepts of belief function [1]. In this approach, the

© Springer Nature Singapore Pte Ltd. 2020
N. Nain et al. (Eds.): CVIP 2019, CCIS 1148, pp. 74–85, 2020.
https://doi.org/10.1007/978-981-15-4018-9_7

problem of classifying an unseen data point on the basis of its nearest neighbours is done with the help of Dempster-Shafer theory. Each neighbour of a sample to be classified is considered as an item of evidence which supports the class membership and the degree of support is defined as a function of the inter-sample distance. The evidence of the k-nearest neighbours is then combined by Dempster's rule of combination [1]. This approach provides a global treatment of issues such as ambiguity, distance rejection, and imperfect knowledge regarding the class membership of training patterns [3].

Dempster Shafer Theory was used by Chen *et al.* to develop case-based classifier systems [4]. This was done specifically for 3 databases - Wisconsin Breast Cancer Database, Iris plant dataset and the Duke outage dataset. A new data classification method called belief-based k-nearest neighbour method (BK-NN) [5] is introduced by Dezert to overcome the limitations of EK-NN. In BK-NN, the k basic belief assignments (*bba*) are computed based on the distance between the object and its neighbours, the acceptance, rejection thresholds in each class. This new structure of *bba* is able to treat the degree of one object belonging (or not) to a class label of its close neighbour.

The search for the nearest neighbours in the training set is the main drawback of a K-NN classifier when it comes to a large dataset. Therefore Denoux developed a deep neural network classifier in the belief function framework to model uncertain information in an effective manner. He applied this concept and developed an Evidential Theoretic Classifier (ETC) [6] which will classify a test point either to a specific class or to a total ignorant class.

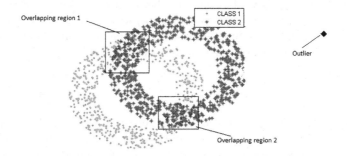

Fig. 1. Pictorial representation of ignorant class in a 2 class problem

Consider a 2 class problem as shown in the Fig. 1. There are 2 overlapping regions and one outlier point as shown in Fig. 1. In a 2 class problem, the output layer of ETC consist of 3 nodes *i.e.* C_1, C_2 and $C_1 \cup C_2$. C_1, C_2 are the masses given to specific classes 1 and 2. The $C_1 \cup C_2$ represents the mass given to outliers and points in the overlapping region. This DST based classifier handles some of the issues that a usual case based classifier can't handle such as the consideration of the distances information, uncertainty, ambiguity and distance rejection.

The concept of meta-class [7] is introduced in DST and it is useful to represent the imprecision present in a dataset. Consider a 3 class problem and 4 overlapping regions will be present in this scenario *i.e.* $C_1 \cup C_2$, $C_1 \cup C_3$, $C_2 \cup C_3$ and $C_1 \cup C_2 \cup C_3$. Among these 4 classes, the first three classes are known as meta-class. Meta-class accounts for the points that lie in the union of several specific classes. But this idea is not incorporated in the existing ETC. Therefore we propose a new classifier based on belief theory to classify a point to either a specific class or meta-class or ignorant class. The meta-class and outlier classes are taken into account for giving importance to the uncertainty and imprecision present in a dataset in the proposed method.

1.1 Evidence Theory

Shafer introduced evidence theory had in 1976 [1]. It is also known as Dempster-Shafer theory or belief theory as it uses Dempster's fusion rule for combining the basic belief assignments [8,9]. The lack of belief about a proposition is represented effectively in this method which is not possible in the probabilistic approach. DST consists of many models such as the theory of hints, transferable belief model etc. But in this work, we concentrate on the non-probabilistic approach introduced in DST.

Construction of a frame of discernment (Ω) which is a finite set of mutually exclusive and exhaustive hypotheses is the starting step in DST [10]. In an n classification problem, the frame of discernment is the set of all possible classes *i.e.* $\Omega = \{c_1, c_2, ..c_n\}$ then the power set 2^Ω is the set of all possible sub-sets of Ω including the empty set ϕ. For example if $n = 3$ then $\Omega = \{c_1, c_2, c_3\}$ and $2^\Omega = \{\phi, c_1, c_2, c_3, \{c_1, c_2\}, \{c_1, c_3\}, \{c_2, c_3\}, \Omega\}$. The union $\{c_i, c_j\}$ represents the possibility of a test point belongs to either c_i or c_j. According to the theory of evidence, a mass value m between 0 and 1 is given to each subset of the power set. *i.e.* $m : 2^\Omega \rightarrow [0, 1]$. The function m is called the mass function (or the basic probability assignment) whenever it verifies two axioms [6]:

$$m(\phi) = 0 \quad and \quad \sum_{A \subseteq 2^\Omega} m(A) = 1 \tag{1}$$

The core of a belief function is defined as the set of elements $A \in 2^\Omega$ having a positive basic belief assignment (*bba*) and these positive *bba*'s are known as focal elements [10]. If A corresponds to a specific class, then $m(A)$ can be considered as the exact or specific belief committed to class A. $m(A \cup B)$ denotes the unpredictability in deciding the correct class between A and B.

In DST, the belief function $Bel(.)$ and the plausibility $Pl(.)$ are known as the lower and upper bounds of the probability of A [4]. Before the final decision making step in a classifier, it is better to transform the belief function to a probability function in a scale of 0 to 1. Pignistic probability transformation function $BetP(.)$ is used for the above purpose and it is defined as follows [6].

$$BetP(a) = \sum_{B \epsilon 2^\theta, A \subseteq B} \frac{|A \cap B|}{|B|} m(B) \tag{2}$$

where $|A|$ represents the cardinality of element A.

1.2   Evidence Theoretic Classifier (ETC)

Denoux introduced evidence theoretic classifier and it is an extension of - supervised neural network models and evidence theoretic K-NN (*E-KNN*) method [6]. The computational complexity of the search for the nearest neighbours in *E-KNN* is alleviated by synthesizing the learning set in the form of prototypes in ETC [6]. Here the assignment of a pattern to a class is made by computing its distances to q prototypes or class representatives: $p^1, p^2, ...p^q$. Each prototype is assigned with a degree of membership u_q with each class c_i such that $\sum_{q=1}^{M} u_q = 1$.

Consider a dataset $X = \{x_1, x_2, ...x_k\}$ to be classified into n distinct groups $\Omega = \{c_1, c_2, ...c_n\}$. If d_i is the distance between a point x and prototype p_i then the information provided by each prototype is represented by a *bba* [6] and it is

$$m^i(c_q) = \alpha^i u_q^i \phi(d^i) \tag{3}$$

$$m^i(\Omega) = 1 - \alpha^i \phi(d^i) \tag{4}$$

where $q = 1, 2, ...n$, $\phi(d^i) = \exp(-\gamma^i (d^i)^2)$ and $m^i(c_q)$, $m^i(\Omega)$ are the bbas associated with specific and total ignorant class respectively. Then the n *bbas* are combined to get a belief function associated with each class using Dempster's Rule or with the conjunctive rule. Finally, the belief associated with each class is normalized and converted using the pignistic probabilistic distribution function. Then the data point will be allocated to the class with maximum pignistic probability [6]. If there is an n class problem, then the ETC will give $n + 1$ output probabilities. *i.e.* n specific class probability values and one total ignorant class probability. The mass given to the frame of discernment Ω can be used for rejection of that particular point if the associated uncertainty is too high. But this method lacks the representation of points that lie in overlapping regions (meta-class). Therefore we propose a new method which gives equal importance to meta-class $(C_i \cup C_j)$ along with specific classes (C_i) and total ignorant class (Ω).

2   Belief Theory Based Hyper-Credal N.N. Classifier

A new data classifier named belief based neural network classifier is proposed to overcome the limitations of ETC and the fusion of *bbas* in the final stage of this proposed classifier is inspired from the BCKN method [7]. The novelty in the proposed classifier is obtained by incorporating the idea of hyper-credal classification to a neural network structure. *i.e.* - the proposed classifier can predict whether a point belongs to a specific class, meta-class and ignorant (outlier) class. A specific class represents the group of points that lie close to any of the given classes. If an object is classified to a meta-class, then it shows that the point belongs to one among the specific classes given in the meta-class. And this can reduce the misclassification or error rate. The mass given to Ω is known as ignorant class mass and it denotes the possibility of a data point not to be in any of the given classes. The allocation of a point to a meta-class sometimes decrease the precision in classification. But it is not an issue in medical, defence related applications where the key point is to avoid taking wrong decisions [14,15]. And these decisions can be made precise by using a secondary resource which will be always available in medical, military related applications.

The proposed classifier is constructed with 3 hidden layers between input and output layers. The first layer is responsible for the selection of prototypes and calculation of distance metric. Second layer calculates the basic belief assignments for each point and the combination of these *bbas* are done at the third layer. The functionality of each layer is explained in the next section.

2.1   The Framework of Basic Belief Assignment

Consider the dataset $X = \{x_1, ...x_k\}$ and let $\Omega = \{c_1, c_2, ...c_n\}$. Here the assumption is that there is no unknown class present in the given training dataset of the classifier. Then q number of data points namely $p_1, p_2, ...p_q$ are selected as the representation of the dataset from each class. The selection of these prototypes can be done either by clustering or clustering followed by k-NN procedure. In the former, each class is clustered to form the class representatives. In latter, each class in the dataset is clustered and those cluster points that are near to a test point are taken for calculating the mass function. Euclidean distance between each point in the dataset and the selected class representative points are calculated. Let this distance be represented by d_{ij} and

$$d_{ij} = ||x_i - p_j|| \tag{5}$$

where $i = 1, 2, ..k$ and $j = 1, 2, ...q * n$. The smaller distance d_{ij} shows that x_i more likely belongs to a specific class and if the distances between a sample point and prototypes from different classes are equal, then there is a chance for that test point to be in a meta-class. Therefore we should allocate the highest mass to the *bba* of a point with the lowest distance and vice versa. The *bba* of a test point x associated with the class c_g is given as

$$m_{ij}(c_g) = \alpha * u_g * e^{-\gamma * (d_{ij})^2} \tag{6}$$

$$m_{ij}(\Omega) = 1 - \alpha * u_g * e^{-\gamma * (d_{ij})^2} \tag{7}$$

where $0 < \alpha < 1$, u_g is the measure of membership of each selected p_j to the respective c_g and γ is a positive constant associated to class c_g.

2.2   Combining the Basic Belief Assignments

There are total $n * q$ *bbas* associated with the hyper-credal classification of a test point. Out of this $n * q$ *bbas*, the q *bbas* will belong to the same class and the rest will contribute to the conflicting masses. Therefore the fusion process will consist of:

Step 1: Combining all the q *bbas* belong to the correct class in the case of a single test point.

Step 2: Fusion of the combined masses obtained in the previous step to get the conflicting and specific masses.

There are many combination rules to perform the step 1 such as Dempster's rule, Smet's rule, Murphy's rule etc. We have adopted Murphy's rule - averaging of the masses.

$$m(c_i) = \frac{1}{Q} \sum_{j=1}^{Q} m_j(c_i) \tag{8}$$

$$m(\Omega) = \frac{1}{Q} \sum_{j=1}^{Q} m_j(\Omega) \tag{9}$$

Then step 2 is the fusion of all the combined masses in the previous step. The fusion rule used in this step is inspired by the DS rule and DP rule. And it is given by [7]

$$m_{1,s}(A) = \begin{cases} \sum_{B_1 \cap B_2 = A | A \neq \phi} m_{1,s-1}(B_1) m_s^{w_s}(B_2) & A \notin \psi \\ \sum_{B_1 \cap B_2 = \phi | B1 \cup B2 = A} m_{1,s}(B_1) m_s^{w_s}(B_2) & A \epsilon \psi \end{cases} \tag{10}$$

where ψ represents the meta-class. Then the masses are normalized by dividing the mass of specific class, meta-class and ignorant class by the sum of it with respect to a point.

3   Belief Theory Based Hyper-Credal CNN Classifier

The idea of hyper-credal classification can be extended to a convolutional neural network classifier also. CNNs are the neural networks that are used for image classification, object identification etc. The idea of belief theory can be applied to CNNs for reducing the error rate in classification. The proposed classifier is constructed by cascading a CNN for feature extraction and the belief based NN classifier structure. The block diagram of the proposed classifier is shown in Fig. 2.

The idea is to use first a CNN structure for extracting the important features present the dataset. After extracting the important features and characteristics of the data, we can apply the classifier described in Sect. 2. Therefore we will get specific classes, meta classes and ignorant class at the output.

Fig. 2. Block diagram of belief theory based CNN classifier

4   Results

The proposed method has applied to several distinct datasets such as Iris dataset, Wine dataset and Arrhythmia dataset. These are some of the benchmark datasets from the UCI machine learning repository. A 3 class problem is the simplest one in the case of a hyper-credal classification. Therefore the analysis is done first on 3 class problems. Iris dataset consists of 150 samples which belong to 3 classes - Iris virginica, iris versi colour, iris setosa with 4 features - sepal length & width, petal length & width. In the training phase, different prototypes from each class are selected and it is done by the k-means clustering algorithm.

Table 1. ANN classification using 4 layer neural network

Hidden layers	Training accuracy (in %)	Testing accuracy (in %)	Parameters
$10, 5, 3, 3$	89.63	56.8	135
$15, 10, 5, 5, 3$	84.45	73.5	308
$20, 15, 10, 3$	83.7	86.6	608
$25, 20, 15, 3$	88.15	93.3	1008

Table 1 shows the number of nodes of an artificial neural network to classify the Iris dataset at different training accuracy levels. As the number of nodes in a hidden layer increases, the number of parameters in an ANN will also increases and it led to the decrease of overall accuracy due to overfitting.

Table 2. Iris dataset

Prototypes	Iris Dataset-ETC [6]		Iris Dataset-Proposed	
	Error rate	Imprecision	Error rate	Imprecision
5	0.0435	0	0	0.5
10	0.0429	0	0	0.4
15	0.0421	0	0	0.4
20	0.0418	0	0	0.3
25	0.0410	0	0	0.2

Table 2 represents the error rate and imprecision rate of the Iris dataset for different trials. In each trial, different combinations of data is taken for training and testing. Each trial is repeated 500 times and average is taken as imprecision and error rate. The error rate and imprecision rates are used as the parameters used to compare the efficiency of ETC and the proposed method. A classified

Table 3. Wine dataset

Prototypes	Wine Dataset-ETC [6]		Wine Dataset-Proposed	
	Error rate	Imprecision	Error rate	Imprecision
5	0.3421	0	0.1579	0.5263
10	0.2764	0	0.0789	0.5526
15	0.2587	0	0.0526	0.3158
20	0.2561	0	0.0263	0.6579
25	0.2156	0	0.0032	0.9737

Table 4. Arrhythmia dataset

Prototypes	Arrhythmia-ETC [6]		Arrhythmia-Proposed	
	Error rate	Imprecision	Error rate	Imprecision
5	0.2719	0	0.1842	0.6053
10	0.2177	0	0.1053	0.5789
15	0.2145	0	0.0789	0.7632
20	0.2084	0	0.0522	0.8684
25	0.2054	0	0.0323	0.7554

test sample is said to be incorrectly classified if it is assigned into class N_i such that $N_i \cap N_j = \phi$ where N_j denotes the actual class label.

$$Error\ rate = \frac{J_{err}}{J_s} \tag{11}$$

where J_{err} is the number of samples incorrectly classified and J_s denotes the total number of test samples.

A classified sample is said to be imprecisely assigned if it is classified into class Ni such that $N_i \cap N_j \neq \phi$ and N_j denotes the actual class. Imprecision rate (I.R.) is

$$I.R = \frac{J_{imp}}{J_s} \tag{12}$$

where J_{imp} is the number of imprecisely classified samples.

In ETC, the imprecision rate always equals to zero. Because in ETC, the mass given to total ignorant class is distributed back to the rest of all specific classes equally. Therefore there won't be any point that is classified in the overlapping region or outlier portion as per ETC. The ETC consists of total 4 classes-3 specific classes, 1 total ignorant class in its output (in the case of a 3 class problem) whereas the proposed classifier is having 7 classes in total for a 3 class problem in its output. The 7 classes that are present in the proposed classifier are 3 specific classes, 3 meta-classes, 1 ignorant class.

Table 3 denotes the analysis of classification of ETC and belief based hyper-credal classifier when wine dataset is applied. Wine dataset consists of total

178 data points with 13 features. Analysis of arrhythmia [14] dataset is done in Table 4. The arrhythmia dataset consists of values from 12 lead ECG of 452 patients. The arrhythmia dataset contains 16 classes:- 1: normal, 2–15: different heart disease conditions, 16: others [16,17]. Out of these 16 classes, the main 3 classes are selected. The proposed method outperforms ETC in all these dataset cases.

Table 5. Artificial dataset

Prototypes	Artificial Dataset-ETC [6]		Artificial Dataset-Proposed	
	Error rate	Imprecision	Error rate	Imprecision
5	0.6682	0	0.3894	0.2895
10	0.6676	0	0.3697	0.2368
15	0.6666	0	0.3508	0.5789
20	0.6663	0	0.2266	0.6579
25	0.6645	0	0.0861	0.8684

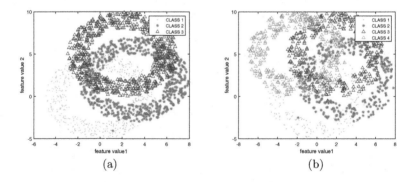

(a)   (b)

Fig. 3. (a) Artificial dataset - 3 classes (b) Artificial dataset - 4 class problem

Finally, a dataset is made artificially to analyse the performance when there is enough number of data points in the overlapping region of many classes.

Figure 3(a) shows the allocation of data points of the artificial dataset on a 2D plane. Table 5 shows that the proposed method only gives the best result in this case also.

The Table 6 shows the performance of the variants of the proposed classifier by changing the different parameters such as distance measure between test point and prototypes, applying clustering followed by k-NN to find class representatives etc. The classifiers are applied to a 4 class problem where the dataset is shown in Fig. 3(b). Here each class in this artificial dataset contains 10000 points.

Table 6. Comparison of variants of the proposed classifiers

Classifier	Error	Imprecision
ETC [6]	**0.6698**	0
Proposed-Euclidean distance and clustering	0.3833	0.2531
Proposed-Mahalanobis distance and clustering	0.3711	0.3697
Proposed-Euclidean distance and clustering followed by kNN	0.2441	0.2170

The last classifier in Table 6 shows the best performance among all the variants. In this classifier, first each class is clustered and q prototypes are selected by considering only the k-NN to each test point.

The belief based hyper-credal CNN is applied to an Invasive Ductal Carcinoma (IDC) dataset of breast cancer available in Kaggle [18]. The original dataset consisted of 162 whole mount slide images of Breast Cancer (BCa) specimens scanned at 40x. From that, 277,524 patches of size 50×50 were extracted (198,738 IDC negative and 78,786 IDC positive). Here a new dataset of 5521 images of 2 classes (non-IDC and IDC) is created and used to evaluate the performance of the proposed classifier. 16 sample images from the dataset are shown in Fig. 4. The dataset is applied to a pretrained 72 layer ResNet CNN for extracting the features from the image dataset. The output from ResNet *i.e.* the feature vectors are then applied to the proposed belief based classifier.

Table 7 shows the accuracy of the CNN classifier and the proposed classifier by varying the number of class representatives. The architecture of the CNN classifier is ResNet with 72 layers in total including convolutional, pooling layers *etc.* and AlexNet with 25 layers. The proposed method shows the best results when compared to the CNN classifier. In the proposed classifier the feature extracted from ResNet only used to construct the belief assignments. This clearly shows that we can improvise any CNN classifier by replacing CNN layers after feature extraction with the proposed classifier structure. The proposed method outperforms both CNN classifiers.

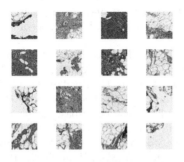

Fig. 4. Sample images from IDC dataset

Table 7. Comparison of different CNN classifiers with the proposed belief based CNN classifier

CNN classifier			Belief based CNN classifier	
			Prototypes	Test accuracy
Pretrained network	ResNet [19]	AlexNet [19]	5	**94.69**
Layers	72	25	10	**94.07**
Test accuracy	**91.94**	**93.54**	15	**95.78**
			20	**96.18**
			25	**96.87**

5 Conclusion

A new hyper-credal classifier based on belief theory has been developed to deal with imprecise and uncertain data. This proposed neural network classifier is designed based on credal classification idea introduced in DST. The main advantage of this method is the overall reduction in error rate and this method reduce over-fitting. The error rate was either always less than ETC proposed by T. Denoux. The other important feature of this classifier is its ability to deal with the uncertain and imprecise points by incorporating the idea of meta-class and ignorant class. Meta-class will contain all the points that will lie in the overlapping region of different classes and ignorant class will accommodate the outliers. The advantage of these two extra classes is that one can take a precise decision about the points that belong to meta or ignorant class than making an erroneous decision. This is very important in medical, defence related applications and the proposed classifier can serve the purpose. And this idea is extended and analysed in CNNs also.

References

1. Shafer, G.: A Mathematical Theory of Evidence, vol. 1. Princeton University Press, Princeton (1976)
2. Fix, E., Hodges, J.L.: Discriminatory analysis, non parametric discrimination: consistency properties. Technical report 4, USAF, School of Aviation Medicine, Randolph Field, TX (1951)
3. Denoeux, T.: A k-nearest neighbor classification rule based on Dempster-Shafer theory. IEEE Trans. Syst. Man Cybern. **25**(5), 804–813 (1995)
4. Chen, Q., Whitbrook, A., Aickelin, U., Roadknight, C.: Data classification using the Dempster-Shafer method. J. Exp. Theor. Artif. Intell. **26**(4), 493–517 (2014)
5. Liu, Z., Pan, Q., Dezert, J.: A new belief-based K-nearest neighbor classification method. Pattern Recognit. Detect. **46**, 834–844 (2013)
6. Denoeux, T.: A neural network classifier based on Dempster-Shafer theory. IEEE Trans. Syst. Man Cybern. Part A: Syst. Hum. **30**(2), 131–150 (2000)
7. Liu, Z., Pan, Q., Dezert, J.: Classification of uncertain and imprecise data based on evidence theory. Neurocomputing **133**, 459–470 (2014)

8. Le Hegarat-Mascle, S., Bloch, I., Vidal-Madjar, D.: Application of Dempster-Shafer evidence theory to unsupervised classification in multisource remote sensing. Trans. Geosci. Remote Sens. **35**(4), 1018–1031 (1997)

9. Denoeux, T.: Logistic regression, neural networks and Dempster-Shafer theory: a new perspective. Knowl. Based Syst. **176**, 54–67 (2019). HAL Id: hal-01830389

10. Smarandache, F., Dezert, J.: Advances and Applications of DST for Information Fusion: Collected Works. Infinite Study, vol. 1. American Research Press, Rehoboth (2004)

11. Basir, O., Karray, F., Zhu, H.: Connectionist-based Dempster-Shafer evidential reasoning for data fusion. IEEE Trans. Neural Netw. **16**(6), 1513–1530 (2005)

12. Al-Ani, A., Deriche, M.: A Dempster-Shafer theory of evidence approach for combining trained neural networks. In: The 2001 IEEE International Symposium on Circuits and Systems, ISCAS, vol. 3 (2001)

13. Gong, B.: An algorithm of data fusion using artificial neural network and Dempster-Shafer evidence theory. In: IITA International Conference on Control, Automation and Systems Engineering (2009)

14. Goldberger, A.L.: Clinical Electrocardiography: A Simplified Approach, 9th edn. Elsevier, Philadelphia (2018)

15. Hu, Y., Fan, X., Zhao, H., Hu, B.: The research of target identification based on neural network and D-S evidence theory. In: International Asia Conference on Informatics in Control, Automation and Robotics (2009)

16. Guidi, G., Karandikar, M.: Classification of arrhythmia using ECG data (2014). http://cs229.stanford.edu

17. Mitra, M., Samanta, R.K.: Cardiac arrhythmia classification using neural networks with selected features. In: International Conference on Computational Intelligence: Modeling Techniques and Applications (2013)

18. https://www.kaggle.com/paultimothymooney/breast-histopathology-images

19. Aloysius, N., Geetha, M.: A review on deep convolutional neural networks. In: International Conference on Communication and Signal Processing (ICCSP), 6 April 2017, pp. 0588–0592 (2017)

Real-Time Driver Drowsiness Detection Using Deep Learning and Heterogeneous Computing on Embedded System

Shivam Khare[1], Sandeep Palakkal[1(✉)], T. V. Hari Krishnan[1], Chanwon Seo[2], Yehoon Kim[2], Sojung Yun[2], and Sankaranarayanan Parameswaran[3]

[1] Samsung R&D Institute India Bangalore, Bangalore, India
{shivam.khare,sandeep.pal,hari.tv}@samsung.com
[2] Samsung Research, Seoul, Republic of Korea
{chanwon.seo,yehoon.kim,sojung15.yun}@samsung.com
[3] Uncanny Vision Solutions Pvt. Ltd., Bangalore, India
sankar.p@uncannyvision.com

Abstract. Timely detection of driver drowsiness is a crucial problem in advanced driver assistance systems, because around 22–24% of road accidents are caused by driver being sleepy. A drowsiness detection solution should be very accurate and run in real-time. All the existing deep learning solutions for drowsiness detection are computationally intensive and cannot be easily implemented on embedded devices. In this paper, we propose a real-time driver drowsiness detection solution implemented on a smartphone. The proposed solution makes use of a computationally light-weight convolutional neural network (CNN), which requires only around 4 million multiply-and-accumulate operations per image and has a test accuracy of 94.4%. The computational requirement is 650x less than that of the state of the art solution. We implemented the proposed CNN, along with a face detector CNN, on a smartphone using ARM-NEON and MALI GPU in a heterogeneous computing design. This implementation achieves a real-time performance of 60 frames-per-second.

Keywords: Drowsiness detection · Deep learning · Heterogeneous computing · CNN model compression

1 Introduction

Driver drowsiness has been reported to be a major cause of 22–24% of crashes and approximately 3% of fatalities [11]. These numbers motivates the need for developing an accurate and real-time drowsiness detection (DD) solution. Although multiple input modalities such as steering wheel pattern, EEG, EOG, computer vision techniques etc. [15] have been used for detecting drowsiness, we restrict the discussion to deep learning based computer vision techniques due to their superior accuracy [2,4,5,7,9,19].

© Springer Nature Singapore Pte Ltd. 2020
N. Nain et al. (Eds.): CVIP 2019, CCIS 1148, pp. 86–97, 2020.
https://doi.org/10.1007/978-981-15-4018-9_8

In [2], a 4-layer custom CNN was proposed that achieved 78% accuracy on a custom created drowsiness detection dataset. In [12], drowsiness is detected by feature level as well as decision level fusions using different CNNs such as AlexNet [8] and VGG-FaceNet [13]. This approach achieved an accuracy of 73.06% on NTHU dataset [17]. A slightly modified version of ResNet-50 [3] is used to detect a closed eye versus an open eye in [5]. In [7], VGG-19 [16] was used for feature extraction, without fine-tuning, to detect multiple categories of driver distractions and achieves 80% accuracy on a custom dataset. In a more recent work [9], a combination of CNN and Long Short-Term Memory (LSTM) blocks were explored to exploit spatio-temporal features for detecting drowsy state. This approach has been shown to achieve 90% accuracy on NTHU dataset. The approaches reviewed so far use well-known CNNs such as AlexNet [8], VGG-Net [16] and ResNet [3], or custom defined CNNs, which are computationally intensive and unsuitable for embedded implementation. Thus, there is a need to develop computationally light CNNs, along with efficient implementation techniques for embedded systems.

To efficiently implement a deep learning solution on an embedded device such as a smartphone, there are multiple levels of optimization possible. There are methods to prune and optimize deep learning models. On the other hand, specialized hardware accelerators such as DSPs and GPUs can be used to accelerate CNNs. Multiple processing units can also be run in parallel in a heterogeneous computing fashion.

In this paper, we present a light-weight CNN based solution for DD and detail its implementation on a smartphone using a heterogeneous computing framework. The solution presented in this paper achieves real-time performance of 60 fps on a smartphone. We propose a light-weight CNN for DD that requires only 4.1 M MAC operations and has 94.4% classification accuracy on a custom dataset collected by the authors in a simulated driving set-up. High accuracy and speed make the proposed solution attractive for implementation on a low resource hardware in the car itself, without requiring to run CNNs on a remote server. This solution has potential to make high impact in ADAS and autonomous cars.

The proposed solution makes use of two CNNs. First, the driver's face, eyes, nose and mouth are located in the input RGB color image using MTCNN [20]. Second, the proposed light-weight drowsiness detector CNN (Fig. 1) is used to classify the face into drowsy or non-drowsy classes from the face, eyes and mouth regions localized by MTCNN. The proposed CNN is largely inspired by a computationally heavy prototype proposed in [14] for DD. In the paper, the prototype [14] is analyzed with deep visualization techniques [18] and rank analysis [10] to gain insights into how many convolutional (conv) layers and conv neurons are sufficient to solve the DD problem. Based on the analysis, we design a lighter CNN, which is 650x lighter in terms of computational requirements yet achieving higher classification accuracy, when compared to the prototype. In a heterogeneous fashion, we run MTCNN on ARM-NEON processor and the proposed drowsiness detector CNN on MALI GPU on a smartphone.

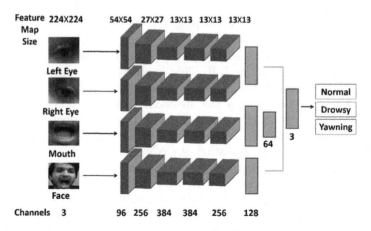

Fig. 1. DDDN-B4 architecture [14]

Organisation of the Paper: Sect. 2 discusses the prototype CNN proposed by [14]. In Sect. 3, we perform deep visualization and rank analysis of the prototype CNN, following which we derive the light-weight CNN. Section 4 discusses the experimental results as well as the heterogeneous implementation of the overall solution on a smartphone. Finally, Sect. 5 concludes the paper.

2   The Prototype Network: DDDN

In this section, we describe DDDN (driver drowsiness detection network), the CNN proposed in [14] for driver drowsiness detection. The authors had presented two versions of DDDN, namely, Baseline-4 (DDDN-B4) and Baseline-2 (DDDN-B2) models. Referring to Fig. 1, we note that DDDN-B4 has four input branches, each receiving a 224 × 224 sized crop of the input image: the face, right eye, left eye or mouth crop. Each input branch consists of five conv layers followed by one or two fully connected (fc) layers. We note that the structure of each input branch is similar to AlexNet architecture [8] and uses the same number of conv kernels in every branch. The outputs of the final fc layers of the four branches are concatenated into a single vector, which is further processed by an fc layer. The final 3-way fc layer with softmax operation classifies the driver state into one of *Normal, Yawning* or *Drowsy*. Drowsy state is defined when the driver has closed her eyes and face is possibly in a nodded state. The DDDN-B2 model [14] uses only two input branches of DDDN-B4: the left eye and mouth branches.

In [14], DDDN-B2 was further compressed by knowledge distillation, which uses a teacher-student framework to train a lighter, student neural network with the help of a heavier, teacher neural network. In [14], DDDN-B2 model was used as teacher network to train DDDN-C2. When compared to DDDN-B2, DDDN-C2 has less number of neurons in each layer, achieves 2.8x model size reduction, and results in 27% increase in fps. However, the test accuracy of DDDN-C2 is less than that of DDDN-B2 model by 4.3%.

3   The Proposed Light-Weight DDDN

In this paper, the primary goal is to design a light-weight CNN for drowsiness detection that can run in real time on embedded devices. Obvious steps towards this goal are to reduce the number of layers and number of conv kernels per layer. In this section, starting with DDDN-B4 (Sect. 2), we systematically derive a light-weight DDDN, namely, DDDN-L, that has fewer layers and kernels. Instead of teacher-student training method used in [14], we follow a different approach to obtain a light-weight model, as discussed in detail in Sects. 3.1 and 3.2.

3.1   Deep Visualization and Pruning of DDDN-B4

Several researchers have proposed visualization of deep features learned by a trained CNN for understanding their effectiveness in solving a given problem (refer [18] and references therein). We performed deep visualization of DDDN-B4 using the method proposed in [18] to understand the scope for possible reduction in network size. In this analysis, several drowsy face images were passed through the network and the feature maps generated by conv layers were examined. Figure 2 shows the visualization results for conv layers of left eye branch for a certain input image. The zoomed versions of the input image, activation of a selected kernel and deconvolution output [18] are shown in the left side of each figure. From Fig. 2, we make the following observations.

1. As only a few kernels are activated in each layer, there is a scope for huge reduction in number of neurons.
2. The activations in the initial layers (first layer is shown) are identical to input image, implying that these kernels have not learnt any meaningful features. Hence, the initial layers can be possibly removed without any adverse effects.

Although we show only first and final conv layers of one branch (left-eye) due to space limitations, the visualization results in other branches also follow a similar pattern. Based on the above observations, the depth of the network is

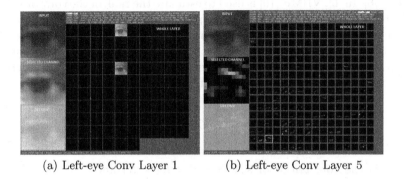

(a) Left-eye Conv Layer 1   (b) Left-eye Conv Layer 5

Fig. 2. DDDN-B4 model and visualization

first reduced in every branch of DDDN-B4 by retaining only the first 3 layers. This pruned model (DDDN-P4) is shown in Fig. 3. The conv filter sizes and number of filters in each layer and branch remain the same as before, but the input image size is reduced from 224×224 to 112×112. DDDN-P4 has reduced the computational complexity, yet shows higher classification accuracy when compared to DDDN-B4 (See Table 3).

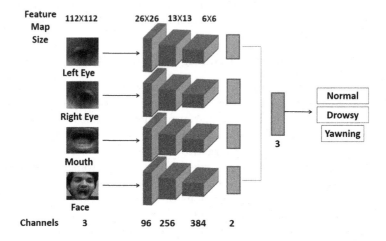

Fig. 3. DDDN-P4 architecture

3.2   Rank Analysis Using VBMF

In this section, we further optimize DDDN-P4 model derived in Sect. 3.1 by systematically reducing the number of conv filters in each layer. The problem of optimizing a deep CNN has been formulated as a low-rank approximation problem of a matrix or tensor, which represents the kernel weights of a conv or fc layer [1,6]. In [6], VBMF [10] is used to estimate the rank of the conv or fc layer weight tensor with the objective of computing a Tucker tensor decomposition. Suppose $\mathcal{K}$ is the conv kernel weight tensor of size $S \times S \times D \times N$ for a given layer, where S and D are the spatial size (width and height) and depth (number of channels), respectively, of the conv kernel and N is the number of kernels. By Tucker decomposition [6], $\mathcal{K}$ can be written as

$$\mathcal{K}_{i,j,d,n} = \sum_{r_3=1}^{R_3} \sum_{r_4=1}^{R_4} \mathcal{C}_{i,j,r_3,r_4} U_{d,r_3}^{(3)} U_{n,r_4}^{(4)}, \qquad (1)$$

where $\mathcal{C}$ is the core tensor of size $S \times S \times R_3 \times R_4$ and $U^{(3)}$ and $U^{(4)}$ are factor matrices of size, respectively, $D \times R_3$ and $N \times R_4$. This is equivalent to decomposing conv layer into three separate layers comprising of R_3, R_4 and N

kernels, respectively, of spatial size 1×1, $S \times S$ and 1×1 and of depth D, R_3 and R_4. If R_3 and R_4 are less than D and N respectively, the computational demand of resulting network will be less than that of the original network [6].

As described in [6], R3 and R4 for the DDDN-P4 model were estimated using VBMF. The results are tabulated in Table 1. We note that the ranks of conv layer tensor are very low (≤ 4) and hence DDDN-P4 can be replaced by a very light neural network having very few number of conv neurons.

Table 1. VBMF rank estimation of DDDN-P4

Layer name	Eye branch		Mouth branch		Face branch	
	R3	R4	R3	R4	R3	R4
conv1	1	1	2	2	1	3
conv2	1	0	1	1	2	0
conv3	2	2	2	2	4	4

3.3 The Proposed Light DDDN Model (DDDN-L)

Based on the VBMF rank analysis of DDDN-P4 presented in Sect. 3.2, the number of conv filters were reduced. Since the ranks are very low (≤ 4), conv layers of DDDN-P4 are not decomposed into 3 conv layers as was done in [6]. Instead, the number of filters are reduced to 4 in each conv layer of DDDN-P4. The resulting configuration, which we call DDDN-L, is shown in Fig. 4.

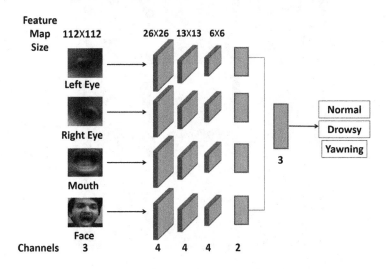

Fig. 4. The proposed DDDN-L architecture

4   Experiment Discussions and Implementation Details

In this section, we discuss the details of the dataset, network training and implementation of the solution on smartphone, while also presenting the experimental results. Based on the experimental results, we show that DDDN-L achieves better accuracy when compared to DDDN-B4, while substantially reducing the computational requirement.

4.1   Dataset

We expanded the custom dataset presented in [14] by adding more participants and used for all our experiments. The participants acted 3 types of drowsy states: normal, drowsy (eyes closed or face nodding) and yawning. Each state lasted for 30–40 s. The dataset originally had 30 subjects and we added an additional 32 subjects resulting in a total 62 subjects. The dataset consists of 145k images. The newly collected data increases ethnic and gender variation in the data, thereby making the model more robust. Also, the data was collected from multiple angles and distances to reduce overfitting to angle, pose and camera-to-user distance. To avoid person-based bias in the experiments, we used images from different subjects in training, validation and test datasets. The training, validation and test datasets respectively consist of 40, 12 and 10 subjects. A few samples from the dataset are shown in Fig. 5.

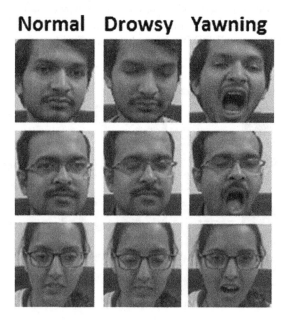

Fig. 5. Samples from the dataset

4.2   CNN Training

In order to force each input branch of DDDN to learn to perform specific tasks such as detection of closed eye, yawn mouth or nodded face, we train the branches independently. That is, we train the eye branch to classify between a fully closed eye and open eye. The eye branch is trained for both left eye and right eye, since both eyes have almost the same structure. Similarly, the mouth branch is trained to classify mouth into yawn or normal (closed or talking) state. The face branch is trained to discriminate between nodded and normal faces (frontal as well as slightly turned faces). Each branch is trained for 10k iterations, while reducing the learning rate every 3k iterations. Similar to transfer learning, the softmax layer of each branch is then removed and the outputs of branches are combined through a concatenation layer, followed by the final fc layers as shown in Fig. 4. The entire network is then trained by keeping 100 times lower learning rate for the conv layers compared to fc layers, since the conv layers have been pre-trained. This training method ensures that the branches are effectively trained for the subtasks and overall network performs the DD classification effectively. The optimal values of hyperparameters, such as the learning rate and step decay rate, were found using grid search and the values that resulted in best accuracy are tabulated in Table 2.

Table 2. DDDN-L training hyperparameters and accuracy

Branch	Learning rate	Step decay rate	Weight decay	Accuracy (%)
Eye	0.001	0.8	0.0001	94.62
Mouth	0.0005	0.1	0.0005	94.58
Face	0.0005	0.5	0.0001	91.19
Overall	0.0005	0.1	0.0001	94.39

4.3   Heterogeneous Computing Based Implementation

As in [14], we use MTCNN [20] for face and face landmark detection before localizing and passing the face, mouth and eye crops through the proposed DDDN-L for drowsiness detection. We implemented the overall driver drowsiness solution on a smartphone in a heterogeneous computing design. In this design, MTCNN and DDDN-L networks are run in parallel on ARM-NEON and MALI GPU, respectively, in a pipelined fashion. The details of this setup are captured in Fig. 6. In this setup, camera pipeline generates a frame which is sent to MTCNN. MTCNN is executed on NEON using a custom NEON accelerated deep learning library. On the other hand, DDDN-L is executed on the GPU using custom created, highly optimized GPU kernels. This implementation helped us achieve real-time performance. The pipeline diagram with time taken for execution of each CNN is shown in Fig. 7.

Table 3. Summary of results for different drowsiness detection CNNs

	ResNet-50 [5]	VGG-FaceNet [5]	DDDN-B4 [14]	DDDN-P4 (Pruned)	DDDN-L (Proposed)
Network size (MB)	17.92	502.00	53.83	19.16	**0.036**
# Operations (MACs)	4.12 G	15.48 G	2.65 G	430 M	**4.1 M**
Test accuracy (%)	74.7	85.2	91.3	**94.77**	94.39
Time: MALI GPU (ms)	413.00	824.00	256.00	97.00	**8.00**

Table 4. Confusion matrix of DDDN-L model

Actual class	Predicted class			Recall (%)
	Normal	Drowsy	Yawn	
Normal	3986	196	130	92.44
Drowsy	420	8126	44	94.6
Yawn	20	68	2728	96.87
Precision (%)	90.05	96.85	94.03	

4.4   Results and Discussion

All experimental results are summarized in Table 3. We compare the proposed networks (DDDN-P4 and DDDN-L) in terms of network size, number of Multiply and Accumulate operations (MACs), test accuracy and run-time with DDDN-B4 proposed in [14]. In addition, we also compare two CNN configurations used for drowsiness detection in [5]: ResNet-50 and VGG-FaceNet. As done in [5], we used the pre-trained weights of ResNet-50 and VGG-FaceNet, and fine-tuned the final fc layers on our custom dataset described in Sect. 4.1.

In terms of model size and MAC operations, the proposed DDDN-L network is the lightest network. When compared to DDDN-B4, DDDN-L is smaller by a factor of 1500 in model size and 650 in MAC operations. In terms of accuracy, DDDN-P4 achieves the best accuracy of 94.77%. DDDN-L achieves a slightly lower accuracy of 94.39%. However, it is important to note that the former is 12x faster than the later. DDDN-L is the fastest network and is 30x faster than the original prototype, DDDN-B4.

We note that when we compare the results achieved by the proposed DDDN-L with that of standard CNNs such as ResNet-50 and VGG-FaceNet, the reduction in the amount of MACs and improvement in speed are orders of magnitude apart. Moreover, ResNet-50 and VGG-FaceNet achieve less accuracy. VGG-FaceNet shows better accuracy than ResNet-50 probably because the former was pre-trained on faces and hence able to extract facial features more effectively. We believe that we can improve the accuracy of ResNet-50 and VGG-FaceNet with further fine-tuning, given their representational capacities. However, the reduction in computational complexity achieved by the proposed DDDN-L makes it

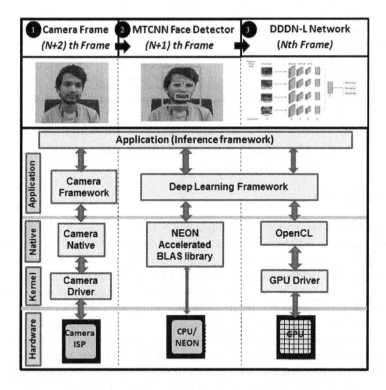

Fig. 6. Heterogeneous computing implementation

irrelevant. In fact, the results prove that for solving drowsiness detection, we do not need large set of features learned by huge networks such as ResNet. While large networks are suitable for solving large scale visual understanding problems, a very limited set of features, sufficient to detect a closed eye, yawning mouth or nodded face, solves drowsiness detection.

In Table 4, the confidence matrix and class-wise precision and recall values of DDDN-L are tabulated. We note that precision and recall of DDDN-L are above 90% for all classes.

As shown in Fig. 7, DDDN-L runs in 8 ms on MALI GPU while MTCNN takes 16 ms on NEON. Hence, the latency of the solution is 24 ms. Without pipelining, the throughput of the solution would be 41 fps. However, the use of multiple accelerators in a heterogeneous setup makes pipelining possible, leading to around 50% increase in throughput. Consequently, the solution achieves 60 fps speed on device.

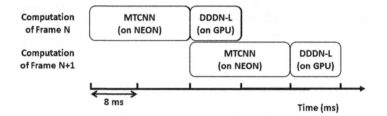

Fig. 7. Pipelined heterogeneous DD implementation

5   Conclusion

This paper presented a very light-weight CNN based solution for driver drowsiness detection for real-time execution on embedded devices. Starting from a state-of-the-art, computationally demanding CNN, we arrived at a lighter CNN using analysis techniques such as deep feature visualization and rank analysis maintaining a 94.39% accuracy. Proper implementation on a smartphone exploiting heterogeneous computing and pipelining helped to achieve 60 fps on the device. Thus, the paper explored model level optimization as well as hardware level acceleration to implement a deep learning solution on a embedded device. The accuracy and speed performance of the proposed light-weight CNN show that driver drowsiness detection requires very limited set of deep features, when compared to more complex classification tasks such as ImageNet challenge. In future, we will focus on fusion of MTCNN with drowsiness detection network, better load balancing between NEON and GPU and further improvement of classification accuracy on real life datasets.

References

1. Denton, E.L., Zaremba, W., Bruna, J., LeCun, Y., Fergus, R.: Exploiting linear structure within convolutional networks for efficient evaluation. In: NIPS, pp. 1269–1277 (2014)
2. Dwivedi, K., Biswaranjan, K., Sethi, A.: Drowsy driver detection using representation learning. In: IEEE IACC, pp. 995–999 (2014)
3. He, K., Zhang, X., Ren, S., Sun, J.: Deep residual learning for image recognition. In: CVPR, pp. 770–778 (2016)
4. Huynh, X.-P., Park, S.-M., Kim, Y.-G.: Detection of driver drowsiness using 3D deep neural network and semi-supervised gradient boosting machine. In: Chen, C.-S., Lu, J., Ma, K.-K. (eds.) ACCV 2016. LNCS, vol. 10118, pp. 134–145. Springer, Cham (2017). https://doi.org/10.1007/978-3-319-54526-4_10
5. Kim, K.W., Hong, H.G., Nam, G.P., Park, K.R.: A study of deep CNN-based classification of open and closed eyes using a visible light camera sensor. Sensors **17**(7), 1534 (2017)
6. Kim, Y.D., Park, E., Yoo, S., Choi, T., Yang, L., Shin, D.: Compression of deep convolutional neural networks for fast and low power mobile applications. In: ICLR (2016)

7. Koesdwiady, A., Bedawi, S.M., Ou, C., Karray, F.: End-to-end deep learning for driver distraction recognition. In: Karray, F., Campilho, A., Cheriet, F. (eds.) ICIAR 2017. LNCS, vol. 10317, pp. 11–18. Springer, Cham (2017). https://doi.org/10.1007/978-3-319-59876-5_2

8. Krizhevsky, A., Sutskever, I., Hinton, G.E.: Imagenet classification with deep convolutional neural networks. In: NIPS, pp. 1097–1105 (2012)

9. Lyu, J., Yuan, Z., Chen, D.: Long-term multi-granularity deep framework for driver drowsiness detection. arXiv preprint arXiv:1801.02325 (2018)

10. Nakajima, S., Sugiyama, M., Babacan, S.D., Tomioka, R.: Global analytic solution of fully-observed variational Bayesian matrix factorization. JMLR 14(1), 1–37 (2013)

11. NHTSA: Traffic safety facts: drowsy driving. Technical report, National Highway Traffic Safety Administration, Washington D.C. (2011). https://crashstats.nhtsa.dot.gov/Api/Public/ViewPublication/811449.pdf

12. Park, S., Pan, F., Kang, S., Yoo, C.D.: Driver drowsiness detection system based on feature representation learning using various deep networks. In: Chen, C.-S., Lu, J., Ma, K.-K. (eds.) ACCV 2016. LNCS, vol. 10118, pp. 154–164. Springer, Cham (2017). https://doi.org/10.1007/978-3-319-54526-4_12

13. Parkhi, O.M., Vedaldi, A., Zisserman, A.: Deep face recognition. In: BMVC, vol. 1, p. 6 (2015)

14. Reddy, B., Kim, Y.H., Yun, S., Seo, C., Jang, J.: Real-time driver drowsiness detection for embedded system using model compression of deep neural networks. In: CVPR Workshops, pp. 438–445 (2017). https://doi.org/10.1109/CVPRW.2017.59

15. Sahayadhas, A., Sundaraj, K., Murugappan, M.: Detecting driver drowsiness based on sensors: a review. Sensors 12(12), 16937–16953 (2012)

16. Simonyan, K., Zisserman, A.: Very deep convolutional networks for large-scale image recognition. arXiv preprint arXiv:1409.1556 (2014)

17. Weng, C.-H., Lai, Y.-H., Lai, S.-H.: Driver drowsiness detection via a hierarchical temporal deep belief network. In: Chen, C.-S., Lu, J., Ma, K.-K. (eds.) ACCV 2016. LNCS, vol. 10118, pp. 117–133. Springer, Cham (2017). https://doi.org/10.1007/978-3-319-54526-4_9

18. Yosinski, J., Clune, J., Fuchs, T., Lipson, H.: Understanding neural networks through deep visualization. In: ICML Workshop on Deep Learning (2015)

19. Yu, J., Park, S., Lee, S., Jeon, M.: Representation learning, scene understanding, and feature fusion for drowsiness detection. In: Chen, C.-S., Lu, J., Ma, K.-K. (eds.) ACCV 2016. LNCS, vol. 10118, pp. 165–177. Springer, Cham (2017). https://doi.org/10.1007/978-3-319-54526-4_13

20. Zhang, K., Zhang, Z., Li, Z., Qiao, Y.: Joint face detection and alignment using multitask cascaded convolutional networks. IEEE Signal Process. Lett. 23(10), 1499–1503 (2016)

A Comparative Analysis for Various Stroke Prediction Techniques

M. Sheetal Singh[1]([✉]), Prakash Choudhary[2],
and Khelchandra Thongam[1]

[1] Department of Computer Science and Engineering,
National Institute of Technology Manipur, Imphal, India
msheetalsingh@live.com
[2] Department of Computer Science and Engineering,
National Institute of Technology Hamirpur, Hamirpur, HP, India
pc@nith.ac.in

Abstract. Stroke is a major life-threatening disease mostly occurs to a person of age 65 years and above but nowadays also happen in younger age due to unhealthy diet. If we can predict a stroke in its early stage, then it can be prevented. In this paper, we evaluate five different machine learning techniques to predict stroke on Cardiovascular Health Study (CHS) dataset. We use Decision Tree (DT) with the C4.5 algorithm for feature selection, Principal Component Analysis (PCA) is used for dimension reduction and, Artificial Neural Network (ANN) and Support Vector Machine (SVM) are used for classification. The predictive methods discussed in this paper are tested on different data samples based on different machine learning techniques. From the different methods applied, the composite method of DT, PCA and ANN gives the optimal result.

Keywords: CHS dataset · Support Vector Machine · Artificial Neural Network · Decision Tree · C4.5 · PCA

1 Introduction

Paradigm shifts of Artificial intelligence (AI) in medical domain will transform the future of health care technology. AI has potential to assist the current demand of computer added diagnosis (CAD) for researchers for accurate prediction of disease. A stroke is one of the major cause of death for a person above 65 years. If we can predict whether a person will experience a stroke or not, then he/she can be saved from that life-threatening disease. Early detection and treatment can save one's life and money. In our study, Cardiovascular Health Study (CHS) dataset is used for prediction of stroke. CHS data set is a complex dataset with lots of inconsistent and unwanted data. Therefore, understanding the CHS data set is very challenging. The main problem is to understand the dataset and extract the hidden knowledge. A highly effective predictive method is desired to increase the efficiency and precision.

A powerful Machine learning (ML) techniques are required which is capable of predicting the outcome from data without stringent statistical assumptions. The most

N. Nain et al. (Eds.): CVIP 2019, CCIS 1148, pp. 98–106, 2020.
https://doi.org/10.1007/978-981-15-4018-9_9

common ML techniques that used to induce predictive model from the dataset are Support Vector Machine (SVM), Decision Tree (DT) and Artificial Neural Network (ANN). These three techniques are widely used for AI models for predicting the outcomes. SVM is a powerful supervised machine learning technique used for classification [1]. A decision tree is one of the simplest yet a fairly accurate predictive technique. DT is commonly used for deriving a strategy to reach a particular goal. ANN has widely used ML technique, and we use feed forward back propagation neural network for stroke prediction. In this paper, C4.5 algorithm is used in DT for feature selection, and PCA is used for dimension reduction.

The rest of the paper is organized as: Sect. 2 describes the related works on prediction of events. Section 3 describes the methods and techniques used in our model. And Sect. 4 presents the obtained outcomes of different method used for predictions followed by conclusion.

2 Related Work

2.1 Support Vector Machine (SVM)

In [2] authors used three methods, MLR, RBFNN, and SVM for the prediction of toxicity activity of two different datasets. The first Dataset includes 76 compounds and their corresponding toxicity values. Similarly, the second dataset includes 146 compounds. And both datasets were divided into two dataset 80% for training and the remaining 20% for testing. After applying MLR, RBFNN, and SVM, the results were compared based on RMS error. It shows that SVM performed better classification and generalization ability than the other two methods.

In [3], an SVM based system is applied to the International Stroke trial (IST) Database to predict the risk factor of stroke. The dataset includes information about patient history, risk factors, hospital details and symptoms of stroke. The author apply different kernel functions on 300 training samples and tested the trained system with 50 samples. From all the different kernel functions linear kernel function comes out with the greater accuracy of 91.0%.

2.2 Decision Tree (DT)

In [4], to predict prognosis in severe traumatic brain injury the decision tree with the C4.5 algorithm is used. The author used Waikato Environment for Knowledge Analysis (Weka) tool to implement C4.5 algorithm on the Traumatic Brain Injury (TBI) dataset. TBI dataset consists of 748 patient's records with 18 attributes each. After implementation of the generated model, 87% accuracy is obtained.

In [5], the decision tree with the C4.5 algorithm is used to extract features from the pre-processed data set. Where authors uses Gain Ratio (a constituent function) from the whole C4.5 algorithm to select the best feature for better classification.

Table 1. Comparative analysis of related works.

	[6]	[4]	[3]
Dataset	Demographic Data Non stroke = 6,7647 Stroke = 250	TBI dataset. 748 Records	IST Database (19,435 patients)
After Pre-Processing	Non Stroke = 500 Stroke = 250	728 Records 18 Attributes	300 Training 50 Testing
Classification Algorithm	ANN	Decision Tree (C 4.5 Algorithm)	SVM (Linear Kernel)
Accuracy	74.0%	87.0%	91.0%

2.3 Artificial Neural Network (ANN)

In [6], the author use demographic data from Mahidol University, Thailand as a dataset and compared Nave Bayes, DT and Neural Network classifiers for stroke prediction. The dataset consist of 68,147 patient records from which only 250 were stroke patients. For better classification the size of the non-stroke records were reduced to 500 and randomly created nine small datasets. The all nine datasets were normalized and the one with maximum similarity to the actual dataset was selected for further classification. After classification confusion matrix was used to calculate the accuracy of the classifiers. The result shows Neural Network gives the optimal result with minimum false negative value.

In [7], the author use ANN classification to predict Thrombo-Embolic Stroke. The dataset is collected from 50 stroke patients and, after eliminating insignificant inputs 20 parameters were achieved which were used as input for the model. ANN with back-propagation algorithm is used for training and testing the inputs. The proposed model achieved an accuracy of 0.89.

In [8], the author compared three classification algorithms Nave Bayes, Decision Tree and Neural Networks for the prediction of stroke. A medical institute provide a dataset of patients with the symptoms of stroke disease which consists of patient information related to stroke disease. The provided information consists of many duplicate, inconsistent and noisy records. So, all the unwanted records were eliminated and PCA is applied to reduce the dimension of the obtained pre-processed dataset. After the dimension were reduced the dataset were used as an input for the selected three classifier. After comparing the results Neural Networks gave better accuracy than both Nave Bayes and DT.

After analyzing all the related works Table 1 shows the comparative analysis of the related works.

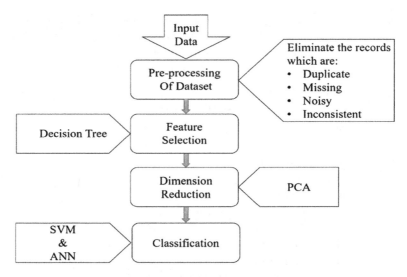

Fig. 1. A schematic diagram of the methodology implemented for stroke prediction.

3   Methodology

This section present a methodology to improve models for stroke diseases prediction. Figure 1 presents an overview of the steps implemented for prediction of stroke.

3.1   Dataset

The dataset we use in our work is Cardiovascular Health Study (CHS) dataset. It is a population-based longitudinal study of coronary heart disease and stroke in adults aged 65 years and older [9]. Available at the National Heart, Lung and Blood Institute (NHLBI) official website. The CHS dataset includes more than 600 attributes for each 5,888 samples. More than 50% of the information were not related to stroke. Table 2 shows the disease type, keyword and class present in the CHS dataset.

3.2   Pre Processing of Dataset

In CHS dataset, missing value and large number of other attribute beside the stroke makes it very challenging for direct use. In the whole dataset about 60% of baseline attributes are missing and having some features which are not directly related to the stroke (i.e. entry in some attribute are blank due to forbidden to answer or unknown). Therefore, to make the CHS dataset appropriate for experimental study a pre-processing is applied and eliminates the missing data, duplicate records, noisy and inconsistent data. Furthermore, six data selection was made with the composition of two disease and stroke is common in each data. And, remaining six disease selected which are nearly close to stroke. In total the dataset having 1,824 examples each with 357 feature and 212 cases of stroke. Table 3 shows the pre-processed dataset with the amount of stroke and non-stroke disease present in it and a keyword is provided to each dataset.

Table 2. Disease types.

Sl. No.	Disease type	Keyword	Class
i	No-event	NO	0
ii	Myocardial infarction	MI	1
iii	Agina	AG	2
iv	Stroke	ST	3
v	Congestive heart failure	CHF	4
vi	Claudication	CL	5
vii	Transient ischemic attack	TIA	6
viii	Angioplasty	AN	7
ix	Coronary artery bypass surgery	CABS	8
x	Other death	Non-CHD	9
xi	ECG MI (silent)	EMS	10
xii	Other CHD deaths	OCHD	11

Table 3. Selected dataset after pre-processing.

Dataset keyword	Composition of disease	Stroke	Non-stroke
MIST	MI and Stroke	212	184
AGST	Agina and Stroke	212	249
CHFST	CHF and Stroke	212	297
CLST	Claudication and Stroke	212	52
TIAST	TIA and Stroke	212	69
ANST	Angioplasty and Stroke	212	79

3.3   Feature Selection Using DT

For feature selection process we use DT method with the C4.5 algorithm as it is an improved version of ID3 and can handle continuous and mix-valued data sets. With the help of the C4.5 algorithm, the best features with higher impact are selected as implemented. The steps for feature selection using C4.5 algorithm are as follows:

- Calculate frequencies of all the attributes in the dataset.
- Calculate entropies of all the attributes in the dataset.
- Calculate the Information Gain of all the Attributes in the dataset.

- Calculate Split Info value of all the attributes in the dataset.
- Calculate Gain Ratio value using Information Gain value and Split Info value of all the attributes in the dataset.
- Select the attributes with higher gain ratio value for the classification process.

3.4 Dimension Reduction Using PCA

Once the pre-processing is done, the obtained dataset contains 1000+ samples which is a large amount of data to be used as an input for the prediction of stroke. A tool is required to identify the hidden knowledge on the dataset. Hence we used a dimension reduction tool called Principal Component Analysis (PCA) [10, 11], which reduces a large dataset into a small set which contains the principal components or uncorrelated variables. The reduction in dimension increases the accuracy and reduces the run time.

3.5 Classification Models

In this work, we are applying five methods. The methods are as follows:

- SVM
- ANN
- PCA + ANN
- DT + ANN
- DT + PCA + ANN

In the above-mentioned methods, DT is used for feature selection purpose only.

Support Vector Machine (SVM). Support Vector Machine (also known as support-vector network) is a machine learning technique for two-group classification problems. These machines are supervised models with associated learning algorithms that analyse data used for classification and regression analysis. In this work first, we normalize the dataset (−1 to 1) then we are applying linear kernel SVM using LIBSVM tool [2].

Artificial Neural Network (ANN). Artificial Neural Networks (NN) are important data mining tool which attempts to mimic brain activity to be able to learn by examples. ANN is mainly used for classification and clustering [12]. In this work, a multilayer neural network is used with back propagation as a training method [13].

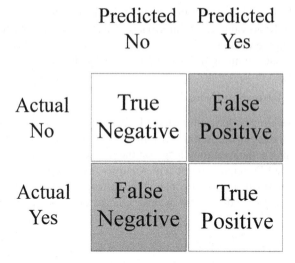

Fig. 2. Confusion matrix.

4   Result Analysis

We use Confusion Matrix (Fig. 2) for outline the performance of the classification models used, and the rule for calculating sensitivity, specificity and accuracy is given below.

Sensitivity = TP/(TP + FN)
Specificity = TN/(FP + TN)
Accuracy = (TP + TN)/(TP + TN + FN + FP)

Where,

TP = true positive
TN = true negative
FP = false positive and
FN = false negative

The result of the experiments are shown in Tables 4 and 5.

Table 4. Comparative experimental results.

Dataset	Methods and accuracy (%)				
	SVM	ANN	PCA + ANN	DT + ANN	DT + PCA + ANN
MIST	64.2	52.5	55.9	62.7	66.4
AGST	68.5	43.5	65.2	65.2	72.5
CHFST	65.7	51.3	64.5	64.5	67.1
CLST	93.68	82.5	92.5	92.5	95.0
TIAST	94.0	78.6	92.9	92.9	95.2
ANST	96.6	90.9	95.5	95.5	97.7

Table 5. Experimental results.

DataSet	Confusion Matrix	Specificity	Sensitivity	Accuracy
MIST	20 7 / 14 18	74.04%	56.25%	64.4%
AGST	26 10 / 9 29	72.22%	76.31%	72.5%
CHFST	44 18 / 7 7	7.96%	50.0%	67.1%
CLST	5 1 / 1 33	83.33%	97.05%	95.0%
TIAST	6 0 / 2 34	94.44%	100%	95.2%
ANST	8 1 / 0 35	100%	88.89%	97.7%

5   Conclusion

The proposed paper presented an extensive comparative study of the different classification methods for stroke prediction. As we compared five methods with different combinations observed that the combination of the Decision tree, PCA and ANN gives the best result than other four methods. This work shows the predictive capacity of the machine learning algorithms with a small set of input parameters.

References

1. Cortes, C., Vapnik, V.: Support-vector networks. Mach. Learn. **20**(3), 273–297 (1995). https://doi.org/10.1023/A:1022627411411
2. Zhao, C., Zhang, H., Zhang, X., Liu, M., Hu, Z., Fan, B.: Application of support vector machine (SVM) for prediction of toxic activity of different data sets. Toxicology **217**(2), 105–119 (2006). http://www.sciencedirect.com/science/article/pii/S0300483X05004270
3. Jeena, R.S., Kumar, S.: Stroke prediction using SVM. In: 2016 International Conference on Control, Instrumentation, Communication and Computational Technologies (ICCICCT), Kumaracoil, pp. 600–602 (2016). https://doi.org/10.1109/iccicct.2016.7988020
4. Hssina, B., Merbouha, A., Ezzikouri, H., Erritali, M.: A comparative study of decision tree ID3 and C4.5. Int. J. Adv. Comput. Sci. Appl. (IJACSA) (2014). https://doi.org/10.14569/SpecialIssue.2014.040203. Special Issue on Advances in Vehicular Ad Hoc Networking and Applications
5. Singh, M.S., Choudhary, P.: Stroke prediction using artificial intelligence. In: 2017 8th Annual Industrial Automation and Electromechanical Engineering Conference (IEMECON), August 2017, pp. 158–161 (2017)
6. Kansadub, T., Thammaboosadee, S., Kiattisin, S., Jalayondeja, C.: Stroke risk prediction model based on demographic data. In: 2015 8th Biomedical Engineering International Conference (BMEiCON), November 2015, pp. 1–3 (2015)
7. Shanthi, D., Sahoo, D.G., Saravanan, D.N.: Designing an artificial neural network model for the prediction of thrombo-embolic stroke (2004)
8. Gayathri, P.: Effective analysis and predictive model of stroke disease using classification methods (2012)
9. Dataset: Cardiovascular Health Study (CHS). https://biolincc.nhlbi.nih.gov/studies/chs/. Accessed 08 May 2016
10. Jolliffe, I.T., Cadima, J.: Principal component analysis: a review and recent developments. Philos. Trans. Roy. Soc. Lond. A Math. Phys. Eng. Sci. **374**(2065) (2016). http://rsta.royalsocietypublishing.org/content/374/2065/20150202
11. Freire, V.A., de Arruda, L.V.R.: Identification of residential load patterns based on neural networks and PCA. In: 2016 12th IEEE International Conference on Industry Applications (INDUSCON), November 2016, pp. 1–6 (2016)
12. Cilimkovic, M.: Neural networks and back propagation algorithm. Institute of Technology Blanchardstown, Dublin 15, Ireland (2010)
13. Rojas, R.: Neural Networks - A Systematic Introduction. Springer, Berlin (1996). https://doi.org/10.1007/978-3-642-61068-4

A Convolutional Fuzzy Min-Max Neural Network for Image Classification

Trupti R. Chavan$^{(\boxtimes)}$ and Abhijeet V. Nandedkar

SGGS Institute of Engineering and Technology, Nanded 431606, Maharashtra, India
chavantrupti89@gmail.com, avnandedkar@sggs.ac.in

Abstract. Convolutional neural network (CNN) is a well established practice for image classification. In order to learn new classes without forgetting learned ones, CNN models are trained in offline manner which involves re-training of a network considering seen as well as unseen data samples. However, such training takes too much time. This problem is addressed using proposed convolutional fuzzy min-max neural network (CFMNN) avoiding the re-training process. In CFMNN, the online learning ability is added to network by introducing the idea of hyperbox fuzzy sets for CNNs. To evaluate the performance of CFMNN, benchmark datasets such as MNIST, Caltech-101 and CIFAR-100 are used. The experimental results show that drastic reduction in the training time is achieved for online learning of CFMNN. Moreover, compared to existing methods, the proposed CFMNN has compatible or better accuracy.

Keywords: Convolutional neural network · Image classification · Online learning · Hyperbox fuzzy set · CFMNN

1 Introduction

Artificial intelligence has influenced a wide range of applications in various fields. Image classification is one of the most fundamental issues in such applications. In literature, image classification techniques are broadly divided into shallow methods and deep neural networks (DNNs). Convolutional neural network (CNN) is the most famous DNN for image classification. It is preferred over shallow approaches due to its advantages like more generic features, scalability to large datasets, etc. It is an usual exercise that the new concepts are made available to the classification network. Offline learning is utilized for learning of such freshly added class samples. It involves scratch training of the CNN using earlier and recently supplied data. But, it takes longer training time and requires already learnt samples which may not be always accessible. Hence, a network must has the facility to train in online mode [10].

The online learning approaches in the literature are mainly dependent on complete, partial or no usage of old class samples while training unseen data. Xiao et al. [20] achieved online learning using complete set of earlier samples. In this method, the classes are divided into superclasses and partitioning of new

© Springer Nature Singapore Pte Ltd. 2020
N. Nain et al. (Eds.): CVIP 2019, CCIS 1148, pp. 107–116, 2020.
https://doi.org/10.1007/978-981-15-4018-9_10

classes is done into component models. This leads to more complex, memory expensive and time consuming system. Käding et al. [4] utilized AlexNet [6] for online learning and the approach is influenced by fine tuning. Due to the fine tuning of all layers, more training time is needed. In [3], an expected model is employed for online learning. The drawback of this method is that it emphasis on the active selection of relevant batches of unlabeled data. Lomonaco and Maltoni proposed CORe50 dataset [9] for online learning and used the fine tuning of pre-trained VGGNET model. In this work, the temporally adjacent frames are fused rather than considering single frame during classification. If more frames are fused, it adversely affects the classification performance. To tackle the problem of online learning, hybrid VGGNET [1] is proposed which is based on the transfer learning approach. The parameters of convolutional layers of base model are shared during online learning mode and classification layers are updated.

A generative adversarial network (GAN) is applied to generate the exemplars as partial data for training [19]. In this work, the class imbalance issue is handled by combining cross entropy and distillation loss. Rebuffi et al. [11] considered partial learnt samples for online learning and the work is based on the exemplar approach. Such methods using some part of earlier data have limitations such as requirement of seen samples and memory consuming. Also, these techniques need more training time which increases almost linearly with the earlier tasks.

An online learning approach based on use of only new data is presented in [18]. To decide whether to predict a label or ask for actual label, recurrent neural network is applied. This work employs the reinforcement learning and one shot learning. Another approach [8] is motivated by transfer learning and implemented knowledge distillation loss function. In this work, the knowledge from a large network is transferred to a smaller network. The drawback of such learning is it gives poor performance for dissimilar new tasks and addition of large number of tasks. A tree-like network is presented in Tree-CNN [13]; it grows hierarchically during online learning. It has attempted to reduce the training efforts in terms of weight updations per epoch.

Ren et al. [12] introduced dynamic combination model for online learning which uses ensemble method. Such model has multiple sub-classifiers which are not dependent on the base model and are aggregated using decision tree. An attempt of online learning is done in [21] using dynamically expandable network. This network is inspired by selective re-training concept and dynamically choses the network capacity whenever provided with new samples. The existing online learning methods are incompetent due to need of earlier data, more training duration, insufficient loss functions, unfair proportion of earlier and new classes, complexity and memory consumption.

The proposed convolutional fuzzy min-max neural network (CFMNN) handles the online learning issue of image classification using only new class samples. The motivation behind the work is hybridization of a convolutional and fuzzy neural network. It learns the new data in online mode without re-training a network. The novelty of the proposed work is that the hyperbox fuzzy set is introduced for CNNs to add the online learning capability. The proposed network has feature extraction network (FEN) and classifier. The FEN is similar to the CNN and help to extract features from the image. The classifier uses hyperbox fuzzy sets to classify given image. The experimental results indicate that the performance of network is compatible with existing methods.

The paper is arranged as: Sect. 2 introduces the proposed online learning method. The experimental results are presented in Sect. 3 and the work is concluded with brief remarks in Sect. 4.

2   Online Image Classification Using CFMNN

The details of convolutional fuzzy min-max neural network are explained as follows:

2.1   Architecture of CFMNN

The proposed CFMNN as shown in Fig. 1 is divided into two parts such as feature extraction network (FEN) and classifier. The FEN is motivated by CNN and extracts the significant attributes from the input images. It mainly has convolutional, pooling and fully connected (FC) layers. The network is provided with RGB image of size $(S \times S \times 3)$. The first convolutional layer has (D_1) filters of dimensions $(F_1 \times F_1)$. It is followed by the pooling layer which decreases the size of activations for next layer. After the cascaded structure of convolutional and pooling layers, FC layers are applied. The first two FC layers (FC1 and FC2) have l_1 and l_2 nodes, respectively.

The classifier of CFMNN consists of three layers, namely input feature vector (F_A), fuzzy hyperbox sets (F_B) and classification nodes (F_C). It computes the fuzzy membership of input pattern for different hyperboxes and determines class label of that pattern. The input layer, F_A is provided with the normalized feature vector (Z_i) which are obtained from FEN. The nodes (b_j) in fuzzy hyperbox set layer (F_B) evaluate the fuzzy membership function for hyperboxes. The weights of connections between layers F_A and F_B are denoted by V and W which are set of min and max points of hyperboxes, respectively. These parameters are updated using the expansion process of FMNN [16]. The F_C layer has T nodes representing classes to be learnt and the connections between layers F_B and F_C are indicated by U. This layer finds the fuzzy union of membership values and takes the classification decision.

The training process of CFMNN includes training of FEN and classifier. It is briefed in the following subsections.

2.2   Training of FEN

In order to train FEN, the F_B layer is removed and classification nodes are directly connected as fully connected softmax layer. The parameters of FEN and softmax layer are indicated as θ_{FEN} and θ_{soft}, respectively. The complete set of parameters of a network is $\theta = (\theta_{FEN}, \theta_{soft})$. Let X and Y are successively the

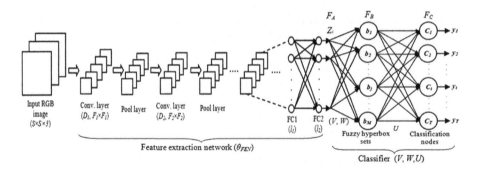

Fig. 1. Architecture of CFMNN (conv: convolutional layer, pool: pooling layer, FC: fully connected layer)

set of N training images and labels. $(X_i, Y_i) \in (X, Y)$ represents the ith pair of training sample and label. The output of network, is computed as the activation function $(f(.))$ of dot product between parameters and input. It is mentioned in (1)

$$\hat{Y}_i = f(X_i, \theta) = f(\theta X_i) \tag{1}$$

The training loss which is evaluated using (2) has data loss $(L_i(.))$ and regularization loss $(R(.))$.

$$L(\theta, X) = \frac{1}{N} \sum_i L_i(\hat{Y}_i, Y_i) + \lambda R(\theta) \tag{2}$$

where λ controls the regularization penalty, $R(.)$. The data loss is found utilizing softmax function while L_2 regularization is used for regularization loss. The stochastic gradient descent (SGD) optimization is applied to update the parameters of network [15].

The feature vector (Z_i) obtained from the FEN using (3) is considered as input to classifier.

$$Z_i = f(\theta_{FEN} X_i) \tag{3}$$

2.3 Training of Classifier

This subsection describes the training of classifier. The set of feature vectors obtained for training samples (X) is denoted by letter Z. The ith pair of feature vector and its label is (Z_i, Y_i), where $Z_i = (Z_{i1}, Z_{i2}, ..., Z_{in})$ and $Z_i \in (0, 1)$. During the training of classifier the hyperboxes are generated and hence the number of nodes in F_B layer keep extending. This expansion process allows to revise the existing hyperboxes along with creating new hyperboxes. Let j^{th} hyperbox is denoted as $H_j = \{V_j, W_j, L_j\}$, where V_j, W_j, and L_j successively specify the min point, max point and label of the hyperbox. To train the classifier, an input pattern (Z_i) is chosen and hyperboxes with same class label are obtained. If there does not exist any hyperbox for that category, a new hyperbox is created.

The closest hyperbox is found provided hyperboxes matching the input pattern label are already present. For such situation, the expansion condition presented in (4) is verified [16].

$$\min_{j} \left(\sum_{k=1}^{n} \left(\max(w_{jk}, z_{ik}) - \min(v_{jk}, z_{ik}) \right) \right) \leq H_{\theta} \tag{4}$$

where H_{θ} is the maximum permitted hyperbox size and $0 \leq H_{\theta} \leq 1; \forall k = 1, 2, \ldots\ldots, n$.

In case of no possibility of hyperbox expansion, a new hyperbox is formed. Otherwise, the min and max points of hyperbox are updated by applying (5) and (6).

$$v_{jk}^{new} = \min(v_{jk}^{old}, z_{ik}) \tag{5}$$

$$w_{jk}^{new} = \max(w_{jk}^{old}, z_{ik}) \tag{6}$$

The nodes in F_B layer implement the fuzzy membership function (b_j) [16] which is given in (7).

$$b_j(Z_i, V_j, W_j) = \frac{1}{n} \sum_{k=1}^{n} \left[1 - f(g_1, \gamma) - f(g_2, \gamma) \right] \tag{7}$$

where $g_1 = (z_{ik} - w_{jk})$ and $g_2 = (v_{jk} - z_{ik})$. $f(.)$ calculates a ramp threshold function using (8).

$$f(g, \gamma) = \begin{cases} 1, & \text{if } g\gamma > 1 \\ g\gamma, & \text{if } 0 \leq g\gamma \leq 1 \\ 0, & \text{if } g\gamma < 0 \end{cases} \tag{8}$$

where γ denotes sensitivity parameter which decides the slope of the ramp function.

The F_C layer predicts the class label depending on the membership values and assigns label from maximum membership hyperbox. The elements, u_{jt}, of matrix U are assigned according to (9).

$$u_{jt} = \begin{cases} 1, & \text{if } H_j \in \text{class } t \\ 0, & \text{otherwise} \end{cases} \tag{9}$$

To compute the output of F_C layer (C_t), fuzzy union operation mentioned in (10) is used.

$$C_t = \bigcup_{j \in T} b_j = \max_{j=1}^{M} b_j u_{jt} \tag{10}$$

where M indicates the total number of hyperboxes created in training process of classifier.

The following section presents the experimental results in detail.

3   Experimental Results

The aim of the experiments is to validate the competency of online learning for CFMNN. The experimentation is done on Intel Core i7 processor (8 GB RAM) with NVIDIA GTx 1050 Ti GPU (4 GB RAM). The coding platform considered for implementation of the proposed method is Python. Accuracy and training time are used to measure the performance. The experimental setup and results are discussed as follows:

3.1   Experimental Setup

The experiments are conducted on three datasets such as MNIST [7], Caltech-101 [2] and CIFAR-100 [5] to verify online learning ability of CFMNN. MNIST dataset consists of 60,000 training and 10,000 testing images of 0–9 handwritten digits. The size of images in the dataset is (28×28). In Caltech 101 dataset, 101 object classes and 1 background class are available with varying image size. The dataset is splitted in two sets: train and test set. There are 30 images/class in train set and up to 50 images/class in test set. The CIFAR-100 dataset contains 100 categories with 50,000 training and 10,000 testing images of dimensions (32×32).

In this work, three different network architectures [7,13,15] are regarded as a FEN. The input image size is varying for each of these networks. Thus, rescaling of image to the respective input dimensions of networks is performed. From each pixel of image, (127) is subtracted for normalization of data. Some of the parameters such as initial learning rate, weight decay, momentum and batch size are set to 0.001, 5×10^{-4}, 0.9 and 64, respectively. The training is carried out for 50 epochs. For MNIST dataset, 5 classes are considered for training of FEN; while for Caltech-101 and CIFAR-100 datasets, 50 classes are utilized. The left amount of classes from each dataset are further used as new classes during the online learning process. The parameters of classifier, namely expansion criterion (H_θ) and sensitivity (γ) are kept as 0.1 and 4, respectively [16].

3.2   Performance of CFMNN

The details of online learning results obtained for MNIST dataset are presented in Table 1. The implementation for MNIST is done on CPU and hence compared to other datasets, the training time is more. For experimentation conducted on Caltech 101 and CIFAR-100 dataset, GPU is used. To train 5 classes with accuracy of 99.53%, LeNet [7] needs 741.76s (5th column, 2nd row, Table 1). The proposed CFMNN generates the hyperboxes in 24.03 s. The scratch (LeNet) and online (CFMNN) training results of remaining (6–10) classes are mentioned from 3rd row onwards. To learn entire dataset, LeNet attains 99.65% (3rd column, 7th row, Table 1) accuracy, which is increased to 99.71% in case of CFMNN (4th column, 7th row, Table 1). The training time taken by LeNet and CFMNN is 1378.6 s and 87.64 s, respectively.

Table 1. Online learning results of CFMNN on MNIST dataset

No. of classes to be learnt	No. of newly added classes	Average test accuracy (%)		Average training time (s)	
		LeNet [7]	CFM-NN	LeNet [7]	CFM-NN
5	–	99.53	–	741.76	–
6	1	99.40	**99.52**	852.45	**35.22**
7	2	99.44	**99.59**	955.26	**47.49**
8	3	99.31	**99.64**	999.12	**61.24**
9	4	99.40	**99.68**	1251	**75.18**
10	5	99.65	**99.71**	1378.6	**87.64**

Table 2. Comparison of CFMNN with existing methods

Method	% Test accuracy	Training time (s)
MNIST dataset [7]		
LeNet [7]	99.65	1378.60
Dropout [17]	99.21	–
ECC [14]	99.14	–
CFMNN	**99.71**	**87.64**
Caltech 101 dataset [2]		
VGGNET [15]	90.63	12736.00
Hybrid VGGNET [1]	89.95	4659.10
CFMNN	**89.36**	**17.61**
CIFAR-100 dataset [5]		
iCaRL [11]	61.32	–
LwF [8]	52.50	–
CFMNN	**63.79**	**2.03**

The performance of CFMNN on MNIST, Caltech-101 and CIFAR-100 datasets is also compared with existing methods [1,7,8,11,14,15,17] and the results are mentioned in Table 2. The comparison of CFMNN with [7,14,17] for MNIST dataset demonstrates that CFMNN has relatively improved performance. Although the information of training time for approaches [14,17] is not available, it can be stated that these methods could require more training time due to comparatively larger architectures. Likewise, VGGNET [15], hybrid VGGNET [1] and CFMNN learns Caltech-101 dataset with 90.63%, 89.95% and 89.36% accuracy, respectively. The training time needed for VGGNET, hybrid VGGNET and CFMNN is 12736 s, 4659.1s and 17.61s, respectively. Moreover, for CIFAR-100 dataset, iCaRL [11], LwF [8] and CFMNN successively achieve 61.32%, 52.5% and 63.79% accuracy. The performance of CFMNN is improved compared to iCaRL and LwF. Among these methods, iCaRL is based on the partial use of earlier samples for training recently added data, whereas LwF and CFMNN utilize only new samples. Thus it can be inferred that the training time for proposed method is less than that of iCaRL as it conserves the exemplar generation time during the training of network.

The comparison of training time for existing and CFMNN is also illustrated with the help of Fig. 2 in which the training time for Caltech-101 dataset is plotted against the number of classes. Different y-axis scales are considered for existing and proposed method due to huge difference in time requirement. The scale marked on left side is used for existing methods [1,15], whereas scale on right side is utilized for CFMNN. Figure 2 clearly suggests that CFMNN takes very less training time compared to the existing approaches.

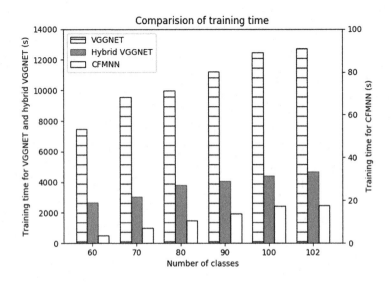

Fig. 2. Comparison of training time for Caltech 101 dataset

4   Conclusion

This work attempts to tackle issue of online learning for image classification without re-training a network. The proposed convolutional fuzzy min-max neural network is inspired from the concept of hyperbox fuzzy sets. It has feature extraction network and classifier. The FEN serves the purpose of feature extraction and classifier helps to categorize the input pattern. The hyperbox fuzzy sets are employed in the classifier. The proposed CFMNN is validated using standard datasets, viz. MNIST, Caltech-101 and CIFAR-100. For these datasets, CFMNN successively attains 99.71%, 89.36% and 63.79% accuracy and needs training time of 87.64s, 17.61s and 2.03s. The results show that compared to existing methods, a compatible or better accuracy is obtained using CFMNN. Also, the training time required for CFMNN is tremendously reduced.

References

1. Chavan, T., Nandedkar, A.: A hybrid deep neural network for online learning. In: Ninth International Conference on Advances in Pattern Recognition (ICAPR), pp. 1–6 (2017). https://doi.org/10.1109/ICAPR.2017.8592942
2. Li, F.F., Fergus, R., Perona, P.: Learning generative visual models from few training examples: an incremental bayesian approach tested on 101 object categories. In: 2004 Conference on Computer Vision and Pattern Recognition Workshop, pp. 178–178 (2004). https://doi.org/10.1109/CVPR.2004.383
3. Käding, C., Rodner, E., Freytag, A., Denzler, J.: Active and continuous exploration with deep neural networks and expected model output changes. CoRR, vol. abs/1612.0 (2016). http://arxiv.org/abs/1612.06129
4. Käding, C., Rodner, E., Freytag, A., Denzler, J.: Fine-tuning deep neural networks in continuous learning scenarios. In: Chen, C.-S., Lu, J., Ma, K.-K. (eds.) ACCV 2016. LNCS, vol. 10118, pp. 588–605. Springer, Cham (2017). https://doi.org/10.1007/978-3-319-54526-4_43
5. Krizhevsky, A.: Learning multiple layers of features from tiny images. Masters thesis, University of Toronto (2009)
6. Krizhevsky, A., Sutskever, I., Hinton, G.: ImageNet classification with deep convolutional neural networks. In: Advances in Neural Information Processing Systems, vol. 25, pp. 1097–1105 (2012)
7. Lecun, Y., Bottou, L., Bengio, Y., Haffner, P.: Gradient-based learning applied to document recognition. Proc. IEEE $86(11)$, 2278–2324 (1998)
8. Li, Z., Hoiem, D.: Learning without forgetting. IEEE Trans. Pattern Anal. Mach. Intell. $40(12)$, 2935–2947 (2018)
9. Lomonaco, V., Maltoni, D.: CORe50: a new dataset and benchmark for continuous object recognition. CoRR, vol. abs/1705.0 (2017). http://arxiv.org/abs/1705.03550
10. Parisi, G., Kemker, R., Part, J., Kanan, C., Wermter, S.: Continual lifelong learning with neural networks: a review. Neural Netw. 113, 54–71 (2019). https://doi.org/10.1016/j.neunet.2019.01.012
11. Rebuffi, S., Kolesnikov, A., Sperl, G., Lampert, C.: iCaRL: incremental classifier and representation learning. In: 2017 IEEE Conference on Computer Vision and Pattern Recognition (CVPR), pp. 5533–5542 (2017)
12. Ren, B., Wang, H., Li, J., Gao, H.: Life-long learning based on dynamic combination model. Appl. Soft Comput. 56, 398–404 (2017). https://doi.org/10.1016/j.asoc.2017.03.005
13. Roy, D., Panda, P., Roy, K.: Tree-CNN: a deep convolutional neural network for lifelong learning. CoRR, vol. abs/1802.0 (2018). http://arxiv.org/abs/1802.05800
14. Simonovsky, M., Komodakis, N.: Dynamic edge-conditioned filters in convolutional neural networks on graphs. In: 2017 IEEE Conference on Computer Vision and Pattern Recognition (CVPR), pp. 29–38 (2017)
15. Simonyan, K., Zisserman, A.: Very deep convolutional networks for large-scale image recognition. CoRR, vol. abs/1409.1 (2014)
16. Simpson, P.: Fuzzy min-max neural networks - part 2: clustering. IEEE Trans. Fuzzy Syst. $1(1)$, 32–45 (1993)
17. Srivastava, N., Hinton, G., Krizhevsky, A., Sutskever, I., Salakhutdinov, R.: Dropout: a simple way to prevent neural networks from overfitting. J. Mach. Learn. Res. 15, 1929–1958 (2014)

18. Woodward, M., Finn, C.: Active one-shot learning. CoRR, vol. abs/1702.0 (2017). http://arxiv.org/abs/1702.06559
19. Wu, Y., et al.: Incremental classifier learning with generative adversarial networks. CoRR, vol. abs/1802.0 (2018). http://arxiv.org/abs/1802.00853
20. Xiao, T., Zhang, J., Yang, K., Peng, Y., Zhang, Z.: Error-driven incremental learning in deep convolutional neural network for large-scale image classification. In: Proceedings of 22nd ACM International Conference on Multimedia, pp. 177–186 (2014). https://doi.org/10.1145/2647868.2654926
21. Yoon, J., Yang, E., Lee, J., Hwang, S.J.: Lifelong learning with dynamically expandable networks. In: International Conference on Learning Representations (2018). http://arxiv.org/abs/1708.01547

Anomalous Event Detection and Localization Using Stacked Autoencoder

Suprit D. Bansod$^{(\boxtimes)}$ (ID) and Abhijeet V. Nandedkar

Shri Guru Gobind Singhji Institute of Engineering and Technology, Nanded, India
{bansodsuprit,avnandedkar}@sggs.ac.in

Abstract. Anomalous event detection and localization from the crowd is a challenging problem to the computer vision community. It is an important aspect of intelligent video surveillance. Surveillance cameras are set up to monitor anomalous or unusual events. But, the majority of video data, related to normal or usual events, is accessible. Thus, analysis and recognition of anomalous events from huge data are very difficult. In this work, an automated system is proposed to identify and localize anomalies at local level. The proposed work is divided into four steps, namely preprocessing, feature extraction, training of stacked autoencoder and anomaly detection and localization. Preprocessing step removes background from video frames. To capture the dynamic nature of foreground objects, magnitude of optical flow is computed. Deep feature representation is obtained over the raw magnitude of optical flow using stacked autoencoder. Autoencoder extracts high-level structural information from motion magnitudes to distinguish between normal and anomalous behaviors. The performance of proposed approach is experimentally evaluated on standard UCSD and UMN dataset developed for anomaly detection. Result of the proposed system demonstrate its usefulness in anomaly detection and localization compared to existing methods.

Keywords: Anomalous event detection · Surveillance · Magnitude of optical flow · Stacked autoencoder

1 Introduction

Surveillance of crowded places is essential for public security. Behavior detection [18] of people in the crowd is a critical issue in surveillance task which helps to detect anomalous or abnormal activities. It is observed in the past decade that abnormal activities have increased in crowded places. Anomaly is defined as any event which is deviating from regular or usual. Such an irregular event is rare, or chances of its occurrence are less. At any given moment information about all possible anomalies cannot be available, hence anomaly detection is very difficult. Anomaly detection is mostly context-dependent, i.e. a normal activity in

© Springer Nature Singapore Pte Ltd. 2020
N. Nain et al. (Eds.): CVIP 2019, CCIS 1148, pp. 117–129, 2020.
https://doi.org/10.1007/978-981-15-4018-9_11

an environment may be anomalous in another and vice-versa. Anomaly detection problem is therefore categorized as unsupervised where normal activities in a typical environment are learned beforehand. An activity which is unknown during testing is treated as anomalous. Many times, it is also termed as one class classification.

Motion is an important attribute of behavior detection. Appearance and motion cues of objects are pursued to decide about abnormality. To represent appearance and motion of objects, hand-crafted optical flow features [1], spatiotemporal features [6] and trajectory features [19] were commonly adopted by researchers. These features constitute spatial and temporal information of video but were inadequate to capture the greater detail of objects. In the last three to four years, newly developed deep learning technique is applied to detect anomalies. Deep learning involves computation of deep features from an image with the help of the convolutional neural network (CNN) [5]. CNN can capture the fine detail of image through convolution, pooling, and fully connected layers. CNN is a supervised approach as it classifies the input image into one of the predefined classes. Deep learning also consists of an unsupervised technique like autoencoders [10]. An autoencoder is a neural network that applies backpropagation, setting target values to be equal to the input. Autoencoder accepts a 1D data input as against CNN which requires 2D input image. So, the image is converted to a vector and then passed to autoencoder for training.

In the literature, different approaches were proposed to detect and locate anomalies using deep learning approaches. Zhou et al. [22] designed a spatiotemporal 3D CNN model from small video patches to detect anomalies. Spatial cues captured appearance details and dynamicity of objects were extracted through temporal convolutions over certain frames. Some abnormal patches were employed for training to understand both normal and abnormal behaviors. Xu et al. [20] proposed appearance and motion DeepNet (AMDN) to detect anomalous activities. Low-level appearance features such as image pixels and optical flow motion features were computed at each patch. Advanced features were calculated by stacked denoising autoencoders (SDAEs) from these features. Anomalies were detected by one class SVM (OC-SVM). Bao et al. [2] designed an unsupervised system to detect and locate anomalies. Optical flow features were used as low-level features and high-level features were procured from PCANet. Sun et al. [15] developed a deep one class model (DOC) for abnormal event detection using optical flow magnitude features. DOC model consisting of CNN to extract deep features and OC-SVM was used to identify anomalies.

Narasimhan et al. [9] computed structural similarity index measure (SSIM) at the local level. High-level features were extracted by SDAEs. The small patches were combined using mean pooling technique and treated as global features. Both local and global features were presented to Gaussian classifiers and abnormalities were detected on the basis of distance metric. Yu et al. [21] proposed a joint representation learning of appearance and motion using 3D deep CNN. A small patch of image intensities acts as appearance features and motion information was captured by optical flow at the patch level. Softmax classifier was used to detect normal and anomalous events.

It is observed from previous approaches for anomaly detection and localization that both supervised and unsupervised approaches were proposed. Unsupervised deep approaches mainly depend on understanding of normal scene behaviors. To detect anomalies, OC-SVM and reconstruction error measures were used. The efficiency of these classifiers relied on feature extraction capability of unsupervised techniques. Supervised approaches make use of abnormal scene behaviors during training. It allowed classifiers to perceive the characteristics of both abnormal and normal behaviors. Also, it reduced the chance of false positive detection. To improve the localization of anomalies, it was noticed that input frames were divided into small patches and representative features were excavated at the local level.

Considering the above findings, in this work, optical flow magnitude features are computed at the local level by dividing the frame into patches of fixed size where the objects are present in the frame. Motion magnitude in consecutive three frames is monitored to confirm about objects in the frame are in motion. Patches having enough motion magnitude are passed to stacked autoencoder for training. Both normal and abnormal patches are trained in an unsupervised fashion. The patch is classified as normal or anomalous by a softmax classifier. The proposed method primarily focuses on:

- detection and localization of anomalies at the local level.
- analysis of significant motion of foreground region using patch selection strategy.
- enhancement of discriminating ability of low-level features to machine transformed high-level features using stacked autoencoder.
- classification of the event by a hybrid learning approach which involves learning of normal and abnormal behaviors using unsupervised approach and classification in a supervised manner.

The remainder of the paper is organized as follows, Sect. 2 elaborates the proposed method, Sect. 3 discusses experimental results along with a comparison of contemporary approaches. Section 4 summarizes and concludes the work.

2   Anomalous Event Detection and Localization Using Stacked Autoencoder

In this section, anomaly detection and localization system using stacked autoencoder (ADLSAE) as shown in Fig. 1 is described in detail. Similar to local level approaches [9,22], ADLSAE analyzes foreground objects by dividing the input frame into small patches. The proposed approach is divided into four steps: (i) preprocessing, (ii) feature extraction, (iii) training of SAE, and (iv) anomaly detection and localization.

120 S. D. Bansod and A. V. Nandedkar

2.1 Preprocessing

Background removal is a necessary preprocessing step in this method. Gaussian filter is applied to every frame to eliminate the noise introduced while capturing the video. In this work, the background removal technique, Visual Background Extractor (ViBe) proposed by Barnich et al. [3] is used to separate foreground objects from the background. ViBe has two advantages: (i) it learns background from the first frame only and (ii) it is free from ghost effects. The background removal of an input frame is as shown in Fig. 1.

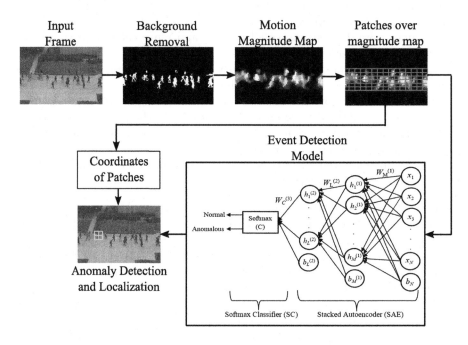

Fig. 1. Framework of proposed system

2.2 Feature Extraction

To obtain the dynamicity of objects in motion, optical flow proposed by Sun et al. [14] is employed. Optical flow is computed for each pixel and provides velocity along horizontal and vertical direction for a pixel. Magnitude of optical flow derived from these velocities is the raw feature in this work, as given in Eq. (1).

$$mag = \sqrt{u_x^2 + v_y^2} \qquad (1)$$

where u_x, v_y — velocity along horizontal and vertical direction. This improved optical flow implementation preserves structural properties and enhances flow estimation. The flow magnitude maintain object boundaries and magnitude values are considerably high for objects moving faster than other objects. The magnitude map of every frame is divided into small patches of fixed size according to background removal output. When a frame is divided into patches, number of moving pixels in the patch varies, so motion magnitude also varies. To find out patches having significant motion, two-step patch selection strategy is implemented. In the first step, an empirical value of the threshold is chosen and magnitude values above the threshold are conserved. In the second step, patches from consecutive three frames are monitored for the same location. If all the three patches are in motion with a specific number of pixels-in-motion then the patch in the third frame is retained, otherwise excluded from future processing.

2.3   Training of Stacked Autoencoder

The consequential patches which follow the patch selection strategy possess structural information. Autoencoders [10] are designed to capture structural content in the data while training. An autoencoder is a neural network that uses backpropagation training algorithm by setting target values to be equal to input values. It is an unsupervised type of learning. If the hidden layer units are less than input, it learns the compressed representation of the input. An autoencoder consists of an encoder-decoder structure. Figure 1 shows the architecture of stacked autoencoder (SAE) with two hidden layer. $W_M^{(1)}, W_K^{(2)}$ are the hidden layer parameters and $W_C^{(3)}$ are the output softmax layer parameters. Moreover, $b_N, b_M^{(1)}$ and $b_K^{(2)}$ are the bias inputs of input and hidden layers, respectively. In the representation of SAE, decoder part is avoided since features extracted by encoder are provided to the encoder of second layer. Thus, SAE hierarchically extracts compressed high-level features in an unsupervised fashion. Training of SAE involves finding the network parameters $\theta = (W, b_h, b_x)$ by minimizing the error between input and its reconstruction. Sparsity can be included in an autoencoder by adding a regularizer to cost function. The regularizer is a function of the average activation of neuron (ρ_i). Also, adding a sparsity regularizer term to cost function, autoencoder learns representation where each neuron in the hidden layer fires for a small number of training samples. Sparsity regularizer is calculated using Kullback-Leibler divergence between sparsity parameter (ρ) and the average activation of neuron. The cost function for training a stacked autoencoder is given in Eq. (2).

$$E = \frac{1}{N} \sum_{i=1}^{N} (x_i - \widehat{x_i})^2 + \lambda * \sum_{i=1}^{N} \sum_{j=1}^{M} \sum_{k=1}^{K} W_{ijk}^2 + \beta * \sum_{j=1}^{M} \sum_{k=1}^{K} KL\left(\rho||\widehat{\rho_{jk}}\right) \qquad (2)$$

where W_{ijk} − L2-regularization term added on the weights to avoid overfitting situations, λ − coefficient of L2-regularization, β − coefficient of sparsity regularization.

Training parameter θ produces a representation $X \to Z_k^{(2)}$ which is a new and advanced feature representation of motion magnitudes of both normal and abnormal patches. The high-level feature and its label after second hidden-layer can be written as $(Z_k^{(2)}, y_k)$, where $k \in 1, 2, \ldots, K$. Learning of features in both the hidden-layers is achieved through the unsupervised scheme. Hence, the label information is not used during training of SAE. As the training of SAE is complete, learned high-level feature representation of motion magnitudes are fed to the output layer.

Anomaly detection is a two-class problem, consists of normal and abnormal classes. In this work, a softmax classifier (SC), a supervised model is used as an output layer which categorizes the input patch in its correct type. Softmax layer is stacked to SAE to fine-tune the whole network in a supervised manner. Softmax classifier utilizes a sigmoid function to predict the output, as mentioned in Eq. (3).

$$f(z_k) = \frac{1}{1 + exp(-W_C^{(3)T} * z_k)} \tag{3}$$

where $W_C^{(3)}$ – parameters of softmax layer. The loss between predicted output and the actual label is governed by cross-entropy loss. The cross-entropy loss is considered while training the network to update the network parameter $W_C^{(3)}$ using stochastic conjugate gradient descent (SCGD) [8] optimization.

2.4   Event Detection Model

A trained SAE with network parameter θ is ready for anomaly detection. Consider a nth frame from a set of I test frames provided as input to the system. Motion magnitude map of the frame is divided into D, $(d = 1, 2, \ldots, D)$ number of small patches of size $(p \times p)$ and its locations; initial coordinates, height and width (x, y, h, w) are recorded. For each patch, high-level feature representation is obtained by SAE and finally, it is passed through the output softmax layer to identify the abnormality. SC has two nodes for two classes. Output loss value associated with the predicted class is lower than the other. It means that if the cross-entropy loss value is less for anomalous node (L_a) than normal (L_n) then the current patch is treated as anomalous, marked as 1 and vice-versa, as given in Eq. (4). If a single patch from the test frame is found to be anomalous then the whole frame is considered as anomalous, as mentioned in Eq. (5).

$$P(d) = \begin{cases} 1 \text{ for } L_a < L_n \\ 0 \text{ for } L_n < L_a \end{cases} \tag{4}$$

$$I(n) = \begin{cases} abnormal \text{ if } \sum_{d=1}^{D} P(d) > 0 \\ normal \quad \text{otherwise} \end{cases} \tag{5}$$

Localization of anomalies depends on decision provided by Eq. (4), i.e the patch detected as abnormal by the classifier with low loss. Also, $P(d) = 1$ means at least one patch is anomalous in the frame. The patches identified as abnormal are localized with the help of its locations accumulated initially and thus anomaly localization is achieved.

3   Experimental Results

The aim of experiments is to detect and locate anomalies at the local level from input video. To validate the performance of the proposed method, datasets developed by UCSD (University of California and San Diego) [7] and University of Minnesota (UMN)[17] are used. Performance of proposed ADLSAE is compared with existing methods such as [2,4,9,12,13,15,16,20–22]. UCSD and UMN datasets used for implementation and analysis of the proposed method are described in brief as follows:

- **UCSD Dataset:** UCSD dataset has two subsets, UCSD Ped1, and UCSD Ped2. UCSD Ped1 dataset consists of 34 train and 36 test sequences with 200 frames each. The frame dimension is fixed to (238×158). UCSD Ped2 dataset has 16 train and 12 test sequences with varying (120–180) frames. The dimension of each frame is fixed to (320×240). UCSD dataset is primarily composed of the cycle, skater, truck, baby cart, car, etc. type of anomalies. Walking on the pavement is the normal activity in both datasets. Pixel level ground truths are also provided to verify results of anomaly localization.
- **UMN Dataset:** UMN dataset is divided into three scenes, namely ground, museum, and court. Each scene has train and test sequences with a varying number of frames. Frame dimension is fixed to (360×240). Running suddenly is the abnormal activity and walking is a normal activity.

3.1   Implementation Details

This subsection gives a brief idea about the implementation details of the proposed method. Motion magnitude maps obtained from optical flow are divided into small patches of size (16×16) over the foreground region. Patch size of (16×16) is found to be appropriate to accommodate objects aptly. To train SAE, both normal and abnormal patches satisfying patch selection strategy, as described in Sect. 2.2, are utilized. The patches are selected from randomly chosen normal and abnormal frames from train and test set sequences. Remaining frames and their patches from test set other than training are used during testing. The procedure to train SAE is mentioned in Sect. 2.3. Some of the hyperparameters selected for the training of SAE are: number of nodes, 256, 128 and 2 for layer 1, layer 2 and output layer, respectively. Encoder transfer function is chosen as saturated linear and for decoder it is pure linear. The maximum epochs selected are 3000, $\lambda = 0.0001, \beta = 2$, and optimizer is SCGD. The experimentation work is carried out on the system with CPU specifications of Intel i7 processor 3.40 GHz and 8 GB RAM. The software platform used is MATLAB, the Deep Learning Toolbox and class of autoencoders available in it.

3.2   Performance Evaluation Protocol

Li et al. [6] introduced a strategy to evaluate the performance of anomaly detection and localization systems. To validate the performance, two types of analysis, namely frame level and pixel level are conducted. In frame level analysis, if at least a single pixel is detected as anomalous then the whole frame is considered as anomalous. For performance comparison at the frame level, two parameters such as Area under Curve (AUC) and Equal Error Rate (EER) are used. AUC and EER are obtained from receiver operating characteristics (ROC) curve. ROC curve is the graph between true positive rate (TPR) and false positive rate (FPR) detection of the frame. AUC is defined as a region under the ROC curve, whereas, EER is the value for which $FPR = 1 - TPR$. AUC and Rate Detection (RD) are the two measures used for comparison at the pixel level. RD is given by $RD = 1 - EER$. In case of pixel level, a frame is detected as anomalous if it contains an anomaly and at least 40% pixels of the anomalous region are identified as anomalous. For better performance of the system, frame level factors AUC should be high and EER should be low; while the pixel level values AUC and RD should both be high.

3.3 Results and Discussion

Figure 2(a), (b) and Table 1 shows the comparison of ROC curves and frame level measures with existing methods for UCSD Ped1 and UCSD Ped2 datasets. It is clear from ROC curve and Table 1 that proposed ADLSAE achieved 91.10% of AUC, 10.96% of EER and 93.30% of AUC, 11.42% of EER for USCD Ped1 and USCD Ped2 datasets, respectively. The results for both datasets are comparable to previous proposed approaches. The proposed approach mainly depends on the raw magnitude of optical flow values. If the objects are moving very close to each other, motion magnitude value increases in the patch. In a normal frame, patches whose motion magnitude value is higher than other patches with comparatively less magnitude lead to false detection as abnormal.

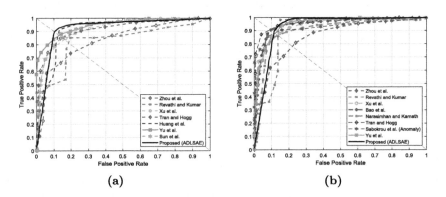

Fig. 2. Frame level ROC curves for (a) UCSD Ped1 dataset and (b) UCSD Ped2 dataset

Figure 3(a), (b) and Table 1 indicates the ROC curve and comparison of AUC and RD at the pixel level for UCSD Ped1 and UCSD Ped2 datasets. It is observed

Table 1. Performance comparison for UCSD dataset at frame and pixel level

Sr. no.	Method	Frame level		Pixel level	
		UCSD Ped1 AUC/EER (%)	UCSD Ped2 AUC/EER (%)	UCSD Ped1 AUC/RD (%)	UCSD Ped2 AUC/RD (%)
1	Zhou et al. [22]	85/24	86/24.4	87/81.3	88/81.9
2	Revathi and Kumar [12]	82.19/18.26	72.64/18	55.45/40	82/75
3	Xu et al. [20]	92.1/16	90.8/17	67.2/52.14	–
4	Bao et al. [2]	–	94.49/10	–	86.89/82.04
5	Tran and Hogg [16]	91.6/14.8	95/9.5	66.1/64.16	83.9/82.53
6	Narasimhan and Kamath [9]	–	90.09/16	–	81.40/78.20
7	Huang et al. [4]	92.6/11.2	–	69.71/61.3	–
8	Sabokrou et al. (Anomaly) [13]	–	92.88/11	–	86.99/83.51
9	Yu et al. [21]	94.4/12.8	94.8/11.1	76.2/68.46	–
10	Sun et al. [15]	91.4/15.6	–	69.1/61.7	–
11	**Proposed (ADLSAE)**	**91.10/10.96**	**93.30/11.42**	**84.43/82.48**	**86.35/84.67**

from Fig. 3 and Table 1 that the proposed ADLSAE secured 84.43% of AUC and 82.48% of RD for UCSD Ped1 dataset, whereas 86.35% of AUC and 84.67% of RD for UCSD Ped2 dataset. Pixel level results for both datasets achieved the second highest result than other approaches for AUC. Also, both datasets have the highest result for RD which is desirable for anomaly localization performance.

The proposed method can detect pixel level anomalies quite well because it learns motion magnitudes at the local level. Thus, it can distinguish between normal and abnormal object patches more correctly. It proves the efficacy of combination of motion features and autoencoder in anomaly detection and localization. Figure 4 shows the anomaly localization results from UCSD datasets. It is seen that the proposed method can localize distant anomalies accurately. The anomalies like skater, which moves slowly than cycle and truck anomalies is localized more precisely. The challenging cycle anomaly from the crowd is identified meticulously. Due to high crowd density, sometime the patch of normal object gets detected as anomalous. But, it will not detect in future frames as objects move away from each other, motion magnitude decreases compared to anomalous objects.

Figure 5(a) and Table 2 shows the frame level ROC and comparison of previous methods for UMN dataset. It is explicit from ROC curve and Table 2 that Scene 1 of UMN dataset achieved 98.86% of AUC and EER of 2.2%, Scene 2 of UMN dataset has 98% of AUC and 3.84% of EER and Scene 3 attained 98.71% of AUC and 2.25% of EER. The results for UMN dataset are comparable to other methods scene-wise. The proposed ADLSAE has less EER values for Scene 1 and Scene 3. UMN dataset consists of running of people randomly as the

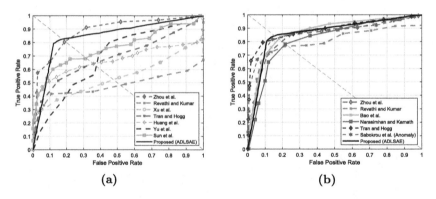

Fig. 3. Pixel level ROC curves for (a) UCSD Ped1 dataset and (b) UCSD Ped2 dataset

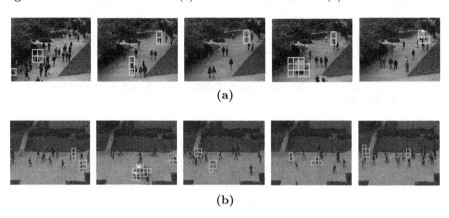

Fig. 4. Anomaly localization results from (a) UCSD Ped1 dataset and (b) UCSD Ped2 dataset

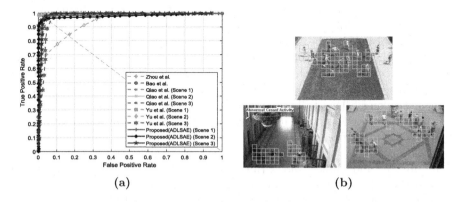

Fig. 5. Result from UMN dataset (a) ROC curves at frame level and (b) anomaly localization results

Table 2. Performance comparison for UMN dataset at frame level

Sr. no.	Method	AUC/EER (%)
1	Zhou et al. [22]	99.63/16.27
2	Bao et al. [2]	99.04/2.6
3	Qiao et al. (Scene1) [11]	98.33/2.6
	Qiao et al. (Scene2) [11]	99.56/4
	Qiao et al. (Scene3) [11]	98.95/1.8
4	Yu et al. (Scene1) [21]	99.4/3.3
	Yu et al. (Scene2) [21]	99.8/0.9
	Yu et al. (Scene3) [21]	97.8/6.4
5	**Proposed (ADLSAE) (Scene1)**	**98.86/2.2**
	Proposed (ADLSAE) (Scene2)	**98/3.84**
	Proposed (ADLSAE) (Scene3)	**98.71/2.25**

only anomaly. The walking activity does not possess motion magnitudes more than running activity. Hence, it becomes easier to recognize the anomalies from frames. Though pixel level ground truths are not available for UMN dataset it is possible to localize anomalies by local level approach, as shown in Fig. 5(b).

4   Conclusion

In this work, a local level anomaly detection and localization method is implemented. Magnitude of optical flow is chosen as the primary motion feature. The motion feature is computed over the foreground region with the help of background removal. The motion of foreground objects is captured by dividing objects into small patches. High-level meaningful representations of motion magnitude are obtained using stacked autoencoder. The proposed ADLSAE is a hybrid approach, i.e. high-level feature learning takes place through unsupervised and supervised techniques. Normal and anomalous behavior of objects is learned in an unsupervised manner by autoencoder and a softmax classifier, a supervised approach is utilized to distinguish between normal and anomalous behaviors. To avoid false positive detection, the patch selection strategy is realized which determine patches having significant motion magnitude. Anomaly localization is achieved with the help of decision provided by softmax classifier and locations of patches. For the pixel level, the proposed method accomplished the highest RD value of 82.48% for UCSD Ped1 and 84.67% for UCSD Ped2 dataset. It shows the effectiveness of ADLSAE to identify anomalies and localize them. The scene-wise analysis is performed for UMN dataset which has the least EER of 2.2% for Scene 1. In future Generative autoencoders will be employed to further improve the system performance.

References

1. Bansod, S.D., Nandedkar, A.V.: Crowd anomaly detection and localization using histogram of magnitude and momentum. Vis. Comput. **36**, 309–320 (2020). https://doi.org/10.1007/s00371-019-01647-0
2. Bao, T., Karmoshi, S., Ding, C., Zhu, M.: Abnormal event detection and localization in crowded scenes based on PCANet. Multimed. Tools Appl. **76**(22), 23213–23224 (2016). https://doi.org/10.1007/s11042-016-4100-0
3. Barnich, O., Droogenbroeck, M.: ViBe: a universal background subtraction algorithm for video sequences. IEEE Trans. Image Process. **20**(6), 1709–1724 (2011)
4. Huang, S., Huang, D., Zhou, X.: Learning multimodal deep representations for crowd anomaly event detection. Math. Prob. Eng. **2018**, 1–13 (2018)
5. Krizhevsky, A., Sulskever, I., Hinton, G.E.: ImageNet classification with deep convolutional neural networks. In: Advances in Neural Information and Processing Systems (NIPS), vol. 60, no. 6, pp. 84–90 (2012)
6. Li, W., Mahadevan, V., Vasconcelos, N.: Anomaly detection and localization in crowded scenes. IEEE Trans. Pattern Anal. Mach. Intell. **36**(1), 18–32 (2014)
7. Mahadevan, V., Li, W., Bhalodia, V., Vasconcelos, N.: Anomaly detection in crowded scenes. In: IEEE Conference on Computer Vision and Pattern Recognition (CVPR), pp. 1975–1981 (2010)
8. Møller, M.F.: A scaled conjugate gradient algorithm for fast supervised learning. Neural Netw. **6**(4), 525–533 (1993)
9. Narasimhan, M.G., Kamath, S.: Dynamic video anomaly detection and localization using sparse denoising autoencoders. Multimed. Tools Appl. **77**(11), 13173–13195 (2017). https://doi.org/10.1007/s11042-017-4940-2
10. Ng, A.: Sparse autoencoder. CS294A Lecture Notes, vol. 72, pp. 1–19 (2011)
11. Qiao, M., Wang, T., Li, J., Li, C., Lin, Z., Snoussi, H.: Abnormal event detection based on deep autoencoder fusing optical flow. In: Chinese Control Conference (CCC), pp. 11098–11103 (2017)
12. Revathi, A.R., Kumar, D.: An efficient system for anomaly detection using deep learning classifier. SIViP **11**(2), 291–299 (2016). https://doi.org/10.1007/s11760-016-0935-0
13. Sabokrou, M., Fayyaz, M., Fathy, M., Moayed, Z., Klette, R.: Deep-anomaly: fully convolutional neural network for fast anomaly detection in crowded scenes. Comput. Vis. Image Underst. **172**, 88–97 (2018)
14. Sun, D., Roth, S., Black, M.J.: Secrets of optical flow estimation and their principles. In: IEEE Conference on Computer Vision and Pattern Recognition (CVPR), pp. 2432–2439 (2010)
15. Sun, J., Shao, J., He, C.: Abnormal event detection for video surveillance using deep one-class learning. Multimed. Tools Appl. **78**(3), 3633–3647 (2017). https://doi.org/10.1007/s11042-017-5244-2
16. Tran, H.T.M., Hogg, D.C.: Anomaly detection using a convolutional winner-take-all autoencoder. In: Proceedings of the British Machine Vision Conference (BMVC), pp. 1–13 (2017)
17. Unusual Crowd Activity Dataset. http://mha.cs.umn.edu/movies/crowdactivity-all.avi/
18. Vishwakarma, S., Agrawal, A.: A survey on activity recognition and behavior understanding in video surveillance. Vis. Comput. **29**(10), 983–1009 (2013). https://doi.org/10.1007/s00371-012-0752-6

19. Wu, S., Moore, B., Shah, M.: Chaotic invariants of Lagrangian particle trajectories for anomaly detection in crowded scenes. In: IEEE Conference on Computer Vision and Pattern Recognition (CVPR), pp. 2054–2060 (2010)

20. Xu, D., Yan, Y., Ricci, E., Sebe, N.: Detecting anomalous events in videos by learning deep representations of appearance and motion. Comput. Vis. Image Underst. **156**, 117–127 (2017)

21. Yu, J., Yow, K.C., Jeon, M.: Joint representation learning of appearance and motion for abnormal event detection. Mach. Vis. Appl. **29**(7), 1157–1170 (2018). https://doi.org/10.1007/s00138-018-0961-8

22. Zhou, S., Shen, W., Zeng, D., Fang, M., Wei, Y., Zhang, Z.: Spatial-temporal convolutional neural networks for anomaly detection and localization in crowded scenes. Sig. Process. Image Commun. **47**, 358–368 (2016)

Kernel Variants of Extended Locality Preserving Projection

Pranjal Bhatt$^{(\boxtimes)}$, Sujata, and Suman K. Mitra

Dhirubhai Ambani Institute of Information and Communication Technology,
Gandhinagar, India
pranjal.daiict@gmail.com, sujata16k@gmail.com, suman_mitra@daiict.ac.in

Abstract. In recent years, non-linear dimensionality reduction methods are getting popular for the handling image data due to non-linearity present in data. For the image recognition task, non-linear dimensionality reduction methods are not useful as it is unable to find the out-of-sample data representation in the reduced subspace. To handle non-linearity of the data, the kernel method is used, which find the feature space from higher dimensional space. One can find the reduce subspace representation by applying the linear dimensionality reduction techniques in the feature space. Extended Locality Preserving Projection (ELPP) tries to capture non-linearity by maintaining neighborhood information in the reduce subspace but fails to capture complex-nonlinear changes. So kernel variants of ELPP are proposed to handle non-linearity present in the data. This article addressed kernel variants of the ELPP which explored the complex non-linear changes of the facial expression recognition. The proposed kernel variants of the ELPP is applied for face recognition on some benchmark databases.

Keywords: ELPP · IGO-ELPP · EULER-ELPP · Facial expression recognition · Face reconstruction

1 Introduction

Image data usually represent in the higher dimensional space, but contains lot of redundancy so it is possible to represent image data in the lower dimensional space. Dimensionality reduction techniques are used to find subspace from higher dimensional space by reserving the as much information as possible. PCA (Principal Component Analysis) [1], LDA (Linear Discriminant Analysis) [2], ICA (Independent Component Analysis) [3] are some of the dimensionality reduction techniques which preserve the global information of data from higher dimensional space to reduce subspace. Usually Image data representation in higher dimensional space is non-linear. While PCA, LDA, ICA assumed that data representation is linear in the higher dimensional space. Locality Preserving Projection (LPP) [4], Neighborhood Preserving Projection (NPE) [5], Locality Preserving Discriminant Projection (LPDP) [6], Orthogonal Neighborhood Preserving Projections (ONPP) [6] and their variants preserve the local structure information in

© Springer Nature Singapore Pte Ltd. 2020
N. Nain et al. (Eds.): CVIP 2019, CCIS 1148, pp. 130–142, 2020.
https://doi.org/10.1007/978-981-15-4018-9_12

the reduce subspace. This techniques give good results when data-points are linearly separable in the higher dimensional space. LPP uses the nearest-neighbor approach for finding the neighborhood for the data-point. But it will fail when there is overlapping region present in the dimensional space [7]. So extension of the LPP, Extended Locality Preserving Projection (ELPP) [7] is proposed to overcome the shortcoming of the LPP.

ELPP captures the non-linearity present in the data by maintaining the neighborhood information. But Image data contains complex non-linear representation in the higher dimensional space due to illumination variations and occlusions. So ELPP fails to capture complex non-linear changes present in the data. For handling the non-linearity of the data, kernel function is used which maps the higher dimensional data to feature space using the kernel function. In the feature space data becomes linearly separable so one can apply the linear dimensionality reduction techniques to the feature space. Kernel Variants of PCA (K-PCA), LDA (K-LDA) and LPP (K-LPP) have already been proposed. In this article Kernel Variants of the ELPP, in particular Euler-ELPP and IGO-ELPP are proposed, which are able to handle complex non-linear changes present in the data and also remove the outlier from the Image data.

This article is organized in the following manner. Section 2 gives some basic idea about Extended Locality Preserving Projection (ELPP). Section 3 contains brief introduction of Kernel methods for linear dimensionality reduction methods. Section 4 contained the Proposed Kernel Variants of ELPP. Face recognition experiments on some benchmark dataset using proposed approach and various other linear dimensionality reduction techniques are reported in Sect. 5 followed by conclusion.

2 Extended Locality Preserving Projection

Extended Locality Preserving Projection (ELPP) [7] is an extension of the Locality Preserving Projection (LPP) [4] which specifically focused on improving energy preservation compared to LPP and resolving ambiguity in case of overlapping class regions.

In the adjacency graph construction, the decision of two data-points as a neighbor is depending upon the K-means algorithm. K-means algorithm gives the natural grouping of the data-set.

This can be achieved using a modified similarity matrix. Similarity matrix of ELPP is defined as Eq. 1.

$$\mathbf{S}_{ij} = \begin{cases} 1 & \text{if } x_{ij} \leq a \\ 1 - 2\left(\frac{x_{ij}-a}{b-a}\right)^2 & \text{if } a \leq x_{ij} \leq \frac{a+b}{2} \\ 2\left(\frac{x_{ij}-b}{b-a}\right)^2 & \text{if } \frac{a+b}{2} \leq x_{ij} \leq b \\ 0 & \text{otherwise} \end{cases} \tag{1}$$

Here, a and b specify the range of values along with the function changes its values and can be controlled. Generally, the value of b is taken as the maximum pairwise distance between two data-points in the same cluster. x_{ij} indicates the Euclidean distance between data points i and j. From the above similarity matrix, diagonal matrix D is constructed, Which can be calculated as D $=\sum_i S_{ij}$. Laplacian matrix is computed as L = D − S. Objective function of the ELPP is:

$$argmax \sum_{ij} (y_i - y_j)S_{ij} \qquad (2)$$

By solving the Eq. 2 we will get in the form of $W^T X L X^T W$ subject to the constraint $W^T X D X^T W = 1$, Which can be solved using generalized eigenvalue problem $X L X^T W = \lambda X D X^T W$. The detail of it can be obtained from [7].

3   Kernel-Methods for Linear Dimensionality Reduction Methods

The linear dimensionality reduction techniques such as PCA [1], ICA [3], LDA [2] may fail i.e not able to produce the good result, if the underlined structure of the data is non-linear. Other dimensionality reduction techniques such as LPP [4], ONPP [6], LPDP [6] try to capture the non-linearity present in the data based on the neighborhood information. If the underlying data structure contains complex non-linearity then the above techniques will not give good results.

One approach to handle non-linearity in the data is by projecting the data in high dimensional feature space. After transforming data in the feature space, data can be linearly separable. So linear dimensionality reduction techniques is applied in the feature space F to find reduced subspace.

A function $\Phi : R^n \rightarrow F$ is used to map the data from original non-linear n-dimensional space to Feature space F. It has been found out that finding the feature space according to function Φ is same as choosing the kernel K. where kernel K is defined by :

$$K(x_i, x_j) = \langle \phi(x_i), \phi(x_j) \rangle = \phi(x_i)^\top \phi(x_j) \qquad (3)$$

Kernel function [8] is finding the dot product between two data-points. The kernel function is used for finding the similarity between the two data-points.

There are various kernel function proposed in the literature, some of the popular kernel functions are:

– Polynomial Kernel: $K(x_i, x_j) = (1 + x_i * x_j)^n$
– Gaussian Kernel: $K(x_i, x_j) = \exp\left(-\frac{||x_i - x_i||}{2\sigma^2}\right)^2$

Here, we are proposing two newly developed kernels best suited for handling complex non-linear image data.

4   Proposed Kernel Variants for ELPP

In this paper, we proposed some kernel variants of the ELPP. ELPP is an extension of the LPP which helps to reduce the overlapping region ambiguity [7]. ELPP gives the best result when the underlying data is in linear separable form. In the case of image data due to high expression, noise, and varying illumination, the underlying data is non-linear and contains the complex non-linear changes, thus ELPP will not give the best result for the complex non-linear data distribution in the original space.

One approach to handle the non-linearity in the data is to apply kernel-based methods to the data-points and project the data into high dimensional feature space after that apply the dimensionality reduction method in feature space.

4.1   Proposed Euler-ELPP

Euler-ELPP is the Euler-kernel [9] version of the ELPP. ELPP try to capture the neighborhood information from original space to reduced space. But may fail when the underlying structure contains complex non-linear changes. For capturing the non-linear changes in the data, Euler kernel is used. Euler kernel is applied to data-point for transforming the original data to feature space F. After the data transformation to the feature space, the dimensionality reduction techniques are applied in feature space.

Euler kernel utilizes the robust dissimilarity by using the cosine-dissimilarity function instead of the l_2 norm. Cosine-dissimilarity is robust to outliers [9]. Euler kernel is trying to capture dissimilarity measure between the pixel intensities and project into the feature space.

Let, we have set of, n images, $I_j \in \mathbb{R}^{m \times n}, (j = 1, ...n)$ of size $m \times n$. Each image I_j is transformed in the vector format $x_j \in \mathbb{R}^{p \times 1}$, where $p = m \times n$. If dataset contains n samples then, $X \in \mathbb{R}^{p \times n}$. For Euler-ELPP, goal is to transform high dimensional datapoint to feature space, first step is to normalise the data-point in range [0,1]. After the data-point X is normalise into range [0,1], X is transformed into feature space F using Euler kernel.

Each pixel intensities of X is transformed into complex representation Z using,

$$\mathbf{z}_j = \frac{1}{\sqrt{2}} \begin{bmatrix} e^{i\alpha\pi\mathbf{x}_j(1)} \\ \vdots \\ e^{i\alpha\pi\mathbf{x}_j(p)} \end{bmatrix} = \frac{1}{\sqrt{2}} e^{i\alpha\pi\mathbf{x}_j} \tag{4}$$

Compute z_j using Eq. 4 and $\mathbf{Z} = [\mathbf{z}_1 \cdots \mathbf{z}_n] \in \mathbb{C}^{p \times n}$ matrix is formed. Where, $\mathbf{z}_j$ points in the feature space.

Objective function of the Euler-ELPP is turned out according to Eq. 5. y_i and y_j are transformed data-points in the reduce space corresponding to x_i and x_j.

$$\min \sum_{ij} (\mathbf{y_i} - \mathbf{y_j})^2 \mathbf{S}_{ij} \tag{5}$$

For computing the similarity matrix, Z-shaped function is used [7].

$$\mathbf{S}_{ij} = \begin{cases} 1 & \text{if } z_{ij} \leq a \\ 1 - 2\left(\frac{z_{ij}-a}{b-a}\right)^2 & \text{if } a \leq z_{ij} \leq \frac{a+b}{2} \\ 2\left(\frac{z_{ij}-b}{b-a}\right)^2 & \text{if } \frac{a+b}{2} \leq z_{ij} \leq b \\ 0 & \text{otherwise} \end{cases} \tag{6}$$

For finding the similarity matrix the first step is to build the clusters from the data-point using the K-means approach. For building the clusters, class label information is used, data-point belongs to the same cluster have the same class label.

Value of z_{ij} indicates the Euclidean distance between two data-point z_i and z_j. For finding the value of a and b the cluster information is used. Value of a is taken relatively small while, the value of b is the same as cluster diameter. The S_{ij} matrix formed using the Z-shaped function, a and b are the controlling parameter of the Z shaped function.

ELPP maintains the neighborhood information in the Laplacian graph. Laplacian matrix formed with the help of the diagonal matrix and similarity matrix. A diagonal matrix is calculated using Eq. 7.

$$\mathbf{D} = \sum_i \mathbf{S}_{ij} \tag{7}$$

Laplacian matrix is calculated as follows:

$$\mathbf{L} = \mathbf{D} - \mathbf{S}$$

If U is the transformation matrix, the objective function of the ELPP is turned out to be:

$$\min \mathbf{U}^T \mathbf{ZLZ}^T \mathbf{U}$$

The constrain is applied to the objective function is:

$$\mathbf{U}^T \mathbf{ZDZ}^T \mathbf{U} = 1$$

Transformation matrix U is obtained by solving the generalized eigenvalue problem as follows.

$$\mathbf{ZLZ}^T \mathbf{U} = \lambda \mathbf{ZDZ}^T \mathbf{U} \tag{8}$$

Each data-point from the feature space is transformed into reduced subspace using the Eq. 9.

$$\mathbf{Y} = \mathbf{U}^T \mathbf{Z} \tag{9}$$

Now, let us define another kernel, which is Image Gradient Orientation (IGO).

4.2 Proposed IGO-ELPP

Image Gradient Orientation (IGO) kernel [10] is proposed to obtain feature space from the higher dimension space. IGO kernel uses the cosine dissimilarity instead of the l_2 norm. IGO kernel is robust to outlier present in the data. Euler kernel used the pixel intensity values for transforming datapoint from higher dimensional space to feature space. While Image Gradient Orientation(IGO) uses gradient orientation for obtaining feature space from higher dimensional space. Cosine dissimilarity between two images is obtained by the Eq. 10:

$$d^2\left(\Phi_i, \Phi_j\right) \triangleq \sum_{k \in \mathcal{P}} \{1 - \cos\left[\Delta\Phi_{ij}(k)\right]\} \tag{10}$$

$\Delta\Phi_{ij}$ is defined as a gradient orientation difference of the image point i and j. Cosine dissimilarity is robust to an outlier. IGO kernel is able to remove outlier present in the data and match the similarity based on the cosine similarity instead of the l_2 norm.

For linear dimensionality reduction techniques, the goal is to find out the transformation matrix which maps the data-point of the original space to reduce dimensional subspace. Suppose there are n images, $\mathbf{I}_i$ where $i = 1, \ldots, n$ of size $m_1 \times m_2$, $\mathbf{I}_i \in \Re^{m_1 \times m_2}$. In case of IGO-ELPP first step is to transform the data-point in the feature space using IGO kernel.

For applying the IGO kernel image gradient orientation is found out. Gradient Orientation is defined by Φ_i for image data-point $\mathbf{I}_i$. $\Phi_i \in [0, 2\pi)^{m_1 \times m_2}$, Φ_i can be calculated as follows.

$$\Phi_i = \arctan \mathbf{G}_{i,y}/\mathbf{G}_{i,x} \tag{11}$$

$$\mathbf{G}_{i,x} = h_x \star \mathbf{I}_i, \quad \mathbf{G}_{i,y} = h_y \star \mathbf{I}_i \tag{12}$$

$\mathbf{G}_{i,x}$ and $\mathbf{G}_{i,y}$ represent the result of the convolution operation with Image $\mathbf{I}_i$, using the h_x and h_y filter, which approximates variation along with the horizontal and vertical direction. Sobel operator [11] is used as a filter for calculating the gradient orientation.

$$\mathbf{G}_{i,x} = \begin{bmatrix} -1 & 0 & +1 \\ -2 & 0 & +2 \\ -1 & 0 & +1 \end{bmatrix} \star \mathbf{I}_i \quad \mathbf{G}_{i,y} = \begin{bmatrix} -1 & -2 & -1 \\ 0 & 0 & 0 \\ +1 & +2 & +1 \end{bmatrix} \star \mathbf{I}_i \tag{13}$$

Computed $\mathbf{z_i}$ corresponding to Φ_i, made a matrix $\mathbf{Z} = [\mathbf{z}_1 \cdots \mathbf{z}_n] \in \mathbb{C}^{p \times n}$. Where $\mathbf{z_i}$ represent the image data-point in the feature space.

$$\mathbf{z}_i\left(\Phi_i\right) = e^{j\Phi_i} \tag{14}$$

Objective Function of the IGO-ELPP is:

$$\min \sum_{ij} \left(\mathbf{y_i} - \mathbf{y_j}\right)^2 \mathbf{S}_{ij} \tag{15}$$

For computing the similarity matrix $\mathbf{S}_{ij}$, Z-shaped function is used.

$$\mathbf{S}_{ij} = \left\{ \begin{array}{ll} 1 & \text{if } z_{ij} \leq a \\ 1 - 2\left(\frac{z_{ij}-a}{b-a}\right)^2 & \text{if } a \leq z_{ij} \leq \frac{a+b}{2} \\ 2\left(\frac{z_{ij}-b}{b-a}\right)^2 & \text{if } \frac{a+b}{2} \leq z_{ij} \leq b \\ 0 & \text{otherwise} \end{array} \right\} \tag{16}$$

For finding the value of the similarity matrix, the first step is to build the clusters of the data-point using the K-means approach. For building the clusters from the data-point, class label information is used. The same class label indicates data-point belongs to the same cluster.

Value of z_{ij} indicates the Euclidean distance between two data-point $\mathbf{z}_i\ (\varPhi_i)$ and $\mathbf{z}_j\ (\varPhi_j)$. Value of a is taken as relatively small while value of b is the same as cluster diameter. The s_{ij} matrix formed the Z-shaped function, a and b are the controlling parameter of the Z shaped function. Value of a and b is dependent on the cluster formation so procedure for building the similarity matrix is adaptive to the data-point distribution in the feature space.

IGO-ELPP maintains the locality information in the Laplacian matrix. Laplacian matrix formed with the help of the diagonal matrix and similarity matrix. A diagonal matrix is calculated as follows:

$$\mathbf{D} = \sum_i \mathbf{S}_{ij} \tag{17}$$

Lapalcian matrix is obtained as follows:

$$\mathbf{L} = \mathbf{D} - \mathbf{S}$$

If U is the transformation matrix, the objective function of the ELPP is turned out to be:

$$\min \mathbf{U}^T \mathbf{Z} \mathbf{L} \mathbf{Z}^T \mathbf{U}$$

The constrain is applied to the objective function is:

$$\mathbf{U}^T \mathbf{Z} \mathbf{D} \mathbf{Z}^T \mathbf{U} = 1$$

Transformation matrix U is obtained by solving the generalized eigenvalue problem as follows.

$$\mathbf{Z} \mathbf{L} \mathbf{Z}^T \mathbf{U} = \lambda \mathbf{Z} \mathbf{D} \mathbf{Z}^T \mathbf{U} \tag{18}$$

Each data-point from the feature space is transformed into reduced subspace using Eq. 19.

$$\mathbf{Y} = \mathbf{U}^T \mathbf{Z} \tag{19}$$

5   Experiments

This section contains the experimental results to show the effectiveness of the proposed Euler-ELPP and IGO-ELPP approach. The high dimensional image is projected to low dimensional learned kernel-subspace with very few dimensions. Performance of the proposed dimensionality reduction technique has been tested on various benchmark data sets.

Results of proposed Euler-ELPP and IGO-ELPP are compared with the ELPP and other linear dimensionality reduction techniques. Four datasets i.e. JAFFE dataset, VIDEO dataset, Oulu-CASIA dataset, CK+ dataset are used for testing of the variants of the ELPP. The dataset is divided into testing and training samples. Testing data are projected in the lower dimensional subspace using the transformation matrix. Nearest Neighbour Classifier(K-NN) is used for measuring the accuracy of the Proposed approach in face recognition task. Note that our aim is to show the efficiency of kernel variants of ELPP. More sophisticated classifier such as SVM could have been used for recognition. But a simple classification such as K-NN is used instead.

JAFFE Dataset

The Japanese Female Facial Expression (JAFFE) Dataset [12] contains 213 images of 7 facial expressions i.e happy, angry, surprise, fear, sadness, disgust, neutral of 10 Japanese female models. Figure 1 shows the sample images from JAFFE dataset. The dataset containing images of size 256×256. If the images are represent in the vectorized format, then size of the data set is 65536×213. Where each column represent the single image and each row represent the feature of the image.

Fig. 1. Examples of facial expressions from JAFFE dataset

CK+ Dataset

The Cohn-Kanade(CK+) [13] Facial Expression Dataset consist of 123 subjects of age varying between 18 to 50 years giving 6 facial expressions. Figure 2 shows the sample images from CK+ dataset. The CK+ database contains images of size 640×490. There are 921 samples considered from CK+ dataset. If the images represented in the vectorized format then size of the dataset is 313600×921.

Fig. 2. Examples of facial expressions from CK+ dataset

VIDEO Dataset

VIDEO [7] dataset consists of videos of 11 subjects. Each video contains four different expressions i.e Angry, Normal, Smiling and Open mouth. Figure 3 shows the sample images from VIDEO dataset. 11290 images are extracted from the video of size 220×165. If images are represented in the vectorized format then the size of the dataset is 36300×11290.

Fig. 3. Examples of facial expressions from VIDEO dataset

Oulu-CASIA Dataset

Oulu-CASIA [14] facial expression dataset, has 480 sequences of images taken from 80 subjects. All the sequences in the dataset start with Neutral expression and end with a peak expression. Instead of taking the whole sequence, to reduce computation complexity, 7 peak expression images from each sequence are considered from different expression categories. Dataset has 6 different expressions i.e. Anger, Disgust, Fear, Happy, Sad, Surprise. Figure 4 shows the sample images from Oulu-CASIA dataset. 3360 image of size 96×79 extracted. If images are represented in the vectorized format then the size of the dataset 7584×3360.

Fig. 4. Examples of facial expressions from Oulu-CASIA dataset

Table 1. Accuracy of Various Techniques with best-reduced Dimension(r), which is dependent on the actual dimension.

Techniques	JAFFE (r = 50)	CK+ (r = 60)	VIDEO (r = 70)	Oulu-CASIA (r = 50)
PCA	78.25	83.05	89.25	89.5
LPP	84.72	87.62	91.63	92.3
ELPP	86.5	91.5	95	95.2
Gassian-KPCA	85.23	87.62	92.4	92.3
Euler-PCA	86.78	88.9	93.67	94.5
IGO-PCA	85	90.3	94.7	95.3
Euler-ELPP	**92.2**	**98.9**	**98.7**	**98.4**
IGO-ELPP	**94.5**	**98.2**	**99.2**	**98.7**

5.1   Face Recoginition Results

Face Recognition results using PCA, LPP, ELPP, Gaussian-KPCA, Euler-PCA, IGO-PCA and proposed approaches Euler-ELPP and IGO-ELPP are reported in Table 1. It can be observed that proposed kenerlized versions of ELPP provide good results compared to earlier approaches. K-NN classifier is used for measuring the recognition rate. Recognition Results in graphical format for JAFFE and CK+ dataset are shown in Figs. 5 and 6.

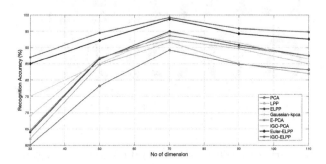

Fig. 5. Recognition accuracy % with varying number of dimensions (r) for CK+ dataset

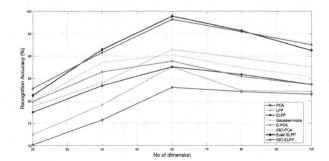

Fig. 6. Recognition accuracy % with varying number of dimensions (r) for JAFFE dataset

5.2   Face Reconstruction Results

Along with the face recognition, many of the dimensionality reduction techniques check for reconstruction error. Though our aim is not t propose the Reconstruction, yet we are furnishing reconstruction results for the completeness.

Reconstruction of the images are done for checking the quality of the image after applying the proposed reduction approach. JAFFE dataset is considered for obtaining the reconstruction results. For achieving the computational efficiency, images are resized into 100×100. After resizing the image, size of the dataset is turned out to be 213×10000. Out of 10000 dimensions 213 dimensions are considered for obtaining the reconstruction results. Figure 7 shows the reconstruction of the image after applying the ELPP, Euler-ELPP, IGO-ELPP. From the result, it is clear that IGO is not a suitable kernel for reconstruction.

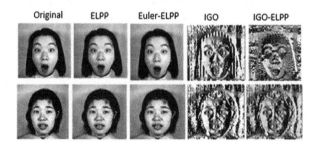

Fig. 7. Reconstruction result of sample images

As it is stated earlier that, Euler-kernel and IGO-kernel are robust to the outlier. Proposed approaches EULER-ELPP and IGO-ELPP are also tested for the same property. Outlier images are reconstructed after applying the Euler-ELPP and IGO-ELPP reduction techniques. Figure 9 demonstrate that the proposed approaches are also robust to the outlier. Euler-ELPP and IGO-ELPP are successfully able to remove outlier from the image. For reconstruction experiments, VIDEO dataset is considered. Figure 8 shows the outlier image formation, while Fig. 9 shows the result of outlier removal for kernel (Euler and IGO) variants.

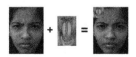

Fig. 8. Outlier image formation

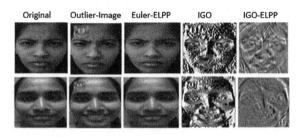

Fig. 9. Outlier-reconstruction results for sample images

6   Conclusion

Extended Locality Preserving Projection tries to capture non-linearity present in the data by maintaining the neighborhood information in the reduced subspace. Kernel variants of ELPP is proposed with the aim of capturing the complex non-linear changes present in the face image data and thereby improve the face recognition performance. Apart from improvement in the recognition task, the Euler-ELPP and IGO-ELPP are able to remove outlier present in the data. Notable improvement is observed in the recognition accuracy using kernel variants of ELPP over conventional non-kernelized ELPP. Higher recognition accuracy is achieved on all four dataset using proposed kernel variants EULER-ELPP and IGO-ELPP. The reconstruction results are also presented to show the removal of outlier.

References

1. Turk, M., Pentland, A.: Eigenfaces for recognition. J. Cogn. Neurosci. **3**(1), 71–86 (1991)
2. Belhumeur, P.N., Hespanha, J.P., Kriegman, D.J.: Eigenfaces vs. fisherfaces: recognition using class specific linear projection. IEEE Trans. Pattern Anal. Mach. Intell. **7**, 711–720 (1997)
3. Hyvarinen, A.: Survey on independent component analysis. Neural Comput. Surv. **2**(4), 94–128 (1999)
4. He, X., Niyogi, P.: Locality preserving projections. In: Advances in Neural Information Processing Systems, pp. 153–160 (2004)
5. He, X., Cai, D., Yan, S., Zhang, H.-J.: Neighborhood preserving embedding. In: Tenth IEEE International Conference on Computer Vision (ICCV'05) vol. 1, vol. 2, pp. 1208–1213. IEEE (2005)
6. Shikkenawis, G., Mitra, S.K.: Locality preserving discriminant projection. In: IEEE International Conference on Identity, Security and Behavior Analysis (ISBA 2015), pp. 1–6. IEEE (2015)
7. Shikkenawis, G., Mitra, S.K.: On some variants of locality preserving projection. Neurocomputing **173**, 196–211 (2016)
8. Schölkopf, B., Smola, A., Müller, K.-R.: Kernel principal component analysis. In: Gerstner, W., Germond, A., Hasler, M., Nicoud, J.-D. (eds.) ICANN 1997. LNCS, vol. 1327, pp. 583–588. Springer, Heidelberg (1997). https://doi.org/10.1007/BFb0020217
9. Liwicki, S., Tzimiropoulos, G., Zafeiriou, S., Pantic, M.: Euler principal component analysis. Int. J. Comput. Vision **101**(3), 498–518 (2013)
10. Tzimiropoulos, G., Zafeiriou, S., Pantic, M.: Subspace learning from image gradient orientations. IEEE Trans. Pattern Anal. Mach. Intell. **34**(12), 2454–2466 (2012)
11. Kanopoulos, N., Vasanthavada, N., Baker, R.L.: Design of an image edge detection filter using the sobel operator. IEEE J. Solid-State Circuits **23**(2), 358–367 (1988)
12. Lyons, M.J., Akamatsu, S., Kamachi, M., Gyoba, J., Budynek, J.: The Japanese female facial expression (JAFFE) database. In: Proceedings of Third International Conference on Automatic Face and Gesture Recognition, pp.14–16 (1998)

13. Kanade, T., Cohn, J.F., Tian, Y.: Comprehensive database for facial expression analysis. In: Proceedings Fourth IEEE International Conference on Automatic Face and Gesture Recognition (Cat. No. PR00580), pp. 46–53. IEEE (2000)
14. Yu-Feng, Y., Dai, D.-Q., Ren, C.-X., Huang, K.-K.: Discriminative multi-layer illumination-robust feature extraction for face recognition. Pattern Recogn. **67**, 201–212 (2017)

DNN Based Adaptive Video Streaming Using Combination of Supervised Learning and Reinforcement Learning

Karan Rakesh, Luckraj Shrawan Kumar[✉], Rishabh Mittar,
Prasenjit Chakraborty, P. A. Ankush, and Sai Krishna Gairuboina

Samsung Research Institute, Bengaluru, India
karanrakesh1@gmail.com, {shrawan.lr, r.mittar,
prasenjit.c}@samsung.com, ankush.preddy@gmail.com,
saikrishnag.vit@gmail.com

Abstract. Video streaming has emerged as a major form of entertainment and is more ubiquitous than ever before. However, as per the recent surveys, poor video quality and buffering continue to remain major concerns causing users to abandon streaming video. This is due to the conditional rule-based logic used by state-of-the-art algorithms, which cannot adapt to all the network conditions. In this paper, a Deep Neural Network (DNN) based adaptive streaming system is proposed, which is trained using a combination of supervised learning and reinforcement learning that can adapt to all the network conditions. This method aims to pre-train the model using supervised learning with a labelled data set generated using state-of-the-art rule based algorithm. This pre-trained model will be used as the base model and is trained with reinforcement learning, which aims to maximize quality, minimize buffering and maintain smooth playback. Training can happen on Personal Computer (PC) based server or edge server setup as well as On-Device, which can even be beneficial in providing user personalization based on network throughput collected on the device. It has been shown that this method will give users a superior video streaming experience, and achieve performance improvement of around 30% on QoE over the existing commercial solutions.

Keywords: Deep Learning · Adaptive bitrate streaming · On device training

1 Introduction

A 2019 report [1] from MRF Forecast shows that video streaming has grown rapidly over the past decade and is projected to grow at 20.8% through 2023. Users' major complaints with video playback usually are buffering and low quality playback. Presently, Adaptive Bit-Rate (ABR) algorithms are used during video streaming playback to decide on which video bitrate quality the player should play depending on the past and current network conditions. However, users' problems continue to persist despite the numerous approaches designed to tackle them. Figure 1 from MUX Report on video streaming [2] released in 2017 shows that 85% of viewers stop watching because of stalling and buffering.

© Springer Nature Singapore Pte Ltd. 2020
N. Nain et al. (Eds.): CVIP 2019, CCIS 1148, pp. 143–154, 2020.
https://doi.org/10.1007/978-981-15-4018-9_13

This is primarily due to the inability of algorithms that are manually tuned to accommodate the vast variety of network conditions. This is where the ability of DNNs to observe network patterns and learn to handle complex network fluctuations gives them a distinct advantage over traditional handcrafted algorithms. In recent years, DNNs have been able to successfully tackle many such previously hard-to-solve problems [3]. Simultaneously, the advent of Reinforcement Learning (RL) has made it possible to focus on high-level goals, e.g., the agent can be told to learn to win rather than manually tweaking parameters, and allows for the model to develop a policy to achieve the objective. This frees the developers from low level tweaking and allows them to develop solutions that can tackle real-world problems.

Fig. 1. Percentage of users who stop watching videos due to issues during playback.

The current proposed solution aims to leverage the power of DNNs, the stability and speed of Supervised Learning (SL) and the flexibility of Reinforcement Learning to develop a mobile-friendly solution. It is focused on improving the users' playback experience by improving quality while maintaining or reducing buffering. MIT has created a Neural Net called Pensieve [4] which aims at applying artificial intelligence into video streaming. They use reinforcement learning for training their model on real network traces. Their approach suffers certain drawbacks which make the commercial use of the algorithm difficult. This proposed solution uses a combination of learning techniques and both real and synthetic data, to achieve a similar result while being able to converge at an optimal solution earlier. In this paper, it is shown despite being less reliant on data collection and taking less training time, the proposed solution is able to achieve performance equal to or slightly more than Pensieve [4].

2   Related Work

Past approaches to ABR streaming such as BOLA [5], buffer-based [6] and rate-based algorithms [7] have revolved around rule-based logics, working based on either video buffer or network bandwidth available to player. However, the dynamic nature of network behavior makes it impossible for pre-defined set of rules to be truly able to adapt to give the best possible performance. Additionally, having a rule-based setup removes any scope for performance improvement by incorporating user's personal streaming pattern. There have been attempts at making bitrate decisions by optimizing directly for a handcrafted equation [8]. Though it yields better results than traditional rule based algorithms, it still fails to generalize well as it is sensitive to errors in throughput predictions.

In [9], a different approach is taken to solve the rule based algorithm pitfalls. Here the parameters of any algorithm are left configurable by the algorithm. The algorithm uses a combination of mean throughput and standard deviation to determine the network state. It accordingly sets the ABR parameters to best suit the model for those type of network conditions. The issue with such an approach is the quantization of network states, which will map a variety of network conditions into a single label classification. Also, tuning such parameters at run time make the model sensitive to network change and also will result in poor performance in unseen network conditions due to the lack of a learnt configuration for such a scenario. A DNN based solution to avoid rule-based adaptive bitrate algorithms called Pensieve [4] was proposed by MIT. They used pure reinforcement learning to train the model from scratch on a set of network traces in a small range (0 to 6 Mbps). However, some of the drawbacks of their approach are discussed below. Firstly, there is an element of randomness associated with the state spaces which the model did not encounter during training i.e. we cannot be certain of the performance of the model in new network scenarios. Secondly, another major drawback is the viability of their input space in a real world deployment scenario – their model is trained considering knowledge of chunk sizes of all future chunks for a video, which is not possible in a real-world streaming scenario. It is also trained over only a small subset of fixed bitrates. Finally, developing a mobile-friendly, readily deployable model using pure RL approach will require large amount of training data and time, which is not very easy to obtain.

In [10], the combination of Supervised Learning followed by Reinforcement Learning has been successfully implemented and tested for playing the game of Go. This allows for faster and more relevant exploration followed by an extensive exploitation phase to maximize the objective. Even though this architecture is designed for local game playback scenario, streaming playback resembles with it in terms of the large number of state spaces. Hence we take motivation from the similar approaches and try to implement for our streaming playback use-case.

3   Design

In this section, the design and implementation of the proposed solution is described. It consists of a system that uses supervised learning to build a base model over which reinforcement learning is carried out. The neural network architecture remains similar to the one used by Pensieve [4]. The supervised learning is carried out on synthetically generated traces to ensure maximum coverage while reinforcement learning ensures good performance. The synthetic dataset generation for the supervised learning step and actual supervised training will be discussed in Sects. 3.1 and 3.2 respectively. This is then followed by a discussion of the reinforcement-learning step in Sect. 3.3.

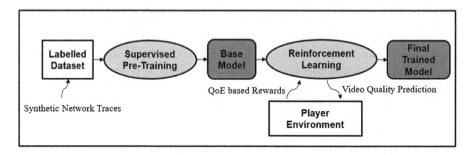

Fig. 2. Proposed training methodology.

Finally, the proposed hybrid inference approach as well as on-Device training are detailed in Sects. 3.4 and 3.5. A pictorial representation of the design is outlined in Fig. 2.

3.1 Synthetic Trace Generation

The first step of the proposed solution is to generate dataset for the Supervised Learning (SL). It can be used to model existing behavior of an algorithm by tweaking the dataset to contain input-output pairs generated from running the particular algorithm. Since the performance of the model is reliant on its generalization, there is a possibility of sub-optimal performances in state spaces which were unexplored during training. Hence, one of the main objectives of supervised learning in the proposed solution is to model a commercial algorithm like buffer based and to help explore all the possible states to minimize the possibility of any unexplored state. Hence, there is a requirement of generating sufficient number of state spaces such that they capture the real world variation, so that this dependency can be eliminated. This is where synthetic trace generation plays a crucial part since it allows us to model real-world traces but tweak the bandwidth range, intensity and frequency of network fluctuations as well as cover all the possible spectrum of range. This ability to generate traces allows for exploration of network scenarios that are hard to capture and ensures they are taken care of in a suitable manner. Using this synthetically generated dataset also allows for modelling scenarios that might not yet be available like 5G as we can tweak the ranges suitably to model its use cases. This helps in future proofing the proposed solution.

The data synthesis is done by creating a corpus of network traces ranging from 128 Kbps to up to 80 Mbps with varying granularities. The lower end of the scale have a very small gap between consecutive bitrates and it increases as higher bitrates are involved. The data is generated using a Markovian model, where each state represents the bandwidth at that point in time. State transition probabilities are varied to help model both stable (stationary use case) and unstable (moving use case) network conditions. Even within a particular state, it is nearly impossible to expect the throughput to be the same across any two points in time. Hence, the value is picked from a Gaussian distribution centered on the bandwidth for the current state. This acts as noise over the current state bandwidth and helps model the real world situation appropriately. The variance falls under the 0.05 to 0.5 range.

3.2 Supervised Learning

The next step of the proposed solution is to construct a base model, which models an existing commercial ABR algorithm. Since the model is trained to mimic the algorithm, this step allows us to guarantee performance on the level of an existing algorithm. In the absence of this step, the model is exposed to the risk of failing to generalize to particular scenarios. Additionally, SL is significantly faster than Reinforcement Learning (RL), hence the model is able to skip the unnecessary time spent exploring the various state spaces and making suboptimal decisions.

As shown in Fig. 2, SL pre-training is carried out by utilizing a simulated video player, which is modified to accommodate the chosen ABR algorithm. The proposed setup uses buffer-based [5] ABR algorithm as it provides stable performance and is relatively easy to prototype due to absence of highly complex relationships. The simulation is run over the generated synthetic traces, described in Sect. 3.1, and tuples of the input states and the mapped output states are created. The input states consist of buffer, throughput, bitrate and average chunk size parameters. This is joined with the output bitrate predicted by the commercial ABR algorithm. This dataset is used to train the model.

Model training is carried using standard supervised learning techniques like RMSProp optimizers with categorical cross-entropy loss calculation. The categories are predefined bitrates chosen so as to be as close to actual video manifest file bitrates. The output bitrates are converted into one-hot encoding to work seamlessly. Since manifest file bitrate values and total bitrates available vary from one video to another, a masking logic takes care of the required mapping Training a model using the supervised learning method with the dataset generated, allows for creation of a stable base model which can be trained further using reinforcement learning.

3.3 Reinforcement Learning

The base model created using the mentioned procedure is then used as a starting point to run reinforcement learning. The base model provides a platform that allows reinforcement learning to focus on rewards without having to intensively explore in the start. This significantly shortens the training time required to achieve a trained model and RL based rewards allow improvement over the ABR algorithm used for training in Sect. 3.2.

The model is trained using the state-of-the-art Asynchronous Advantage Actor Critic algorithm (A3C) [11]. The algorithm uses two neural networks, namely the actor and the critic. The actor controls the behavior, bitrate prediction in our case. The critic measures how good the action taken by the actor is. The base model from Sect. 3.2 is used as the actor in this setup and the critic is trained from scratch, but with a higher learning rate. Physically running the video and logging the results will slow down the process of learning since it is coupled to the video length. Hence, a simulated player is used as the environment, where the actor acts as the ABR algorithm. The player simulates the video playback considering the input network conditions, and the bitrate selected by the actor. Since the player can simulate video playback without having to actually run the video and wait for it to finish, the approach reduces the time required to complete a video from the order of minutes to milliseconds.

Multiple actor critic agents are used, each of which receives random traces to run the video on. This also helps speed up the time required to obtain a finished model. Every reinforcement-learning problem must have a goal towards which it converges. In this case, it is to maximize the Quality of Experience (QoE) achieved by the model for any video being played.

Quality of Experience. There is no recognized metric to quantitatively measure the performance of a video playback experience. In order to give the model a goal to move towards, a metric was developed by Pensieve [4] to train and benchmark the model, called Quality of Experience (QoE) [4].

QoE is defined as,

$$QoE\_n = b\_n - R * r\_n - S * | b\_n - b\_(n - 1) | \qquad (1)$$

where "b" and "r" designate the magnitude of content bitrate and rebuffer count, respectively, for the nth chunk, and "R" and "S" designate the penalty multiplier for rebuffering and content playback smoothness, respectively. The goal of the model is to maximize the QoE.

Different metrics will yield a different overall outcome, hence to have a benchmark to compare against; the QoE metric of Pensieve [4] has been used, as it is a state-of-the-art neural net implementation.

3.4   Smart Strategy Module

A smart strategy module is used to optimize user device performance along with playback experience. This is done by placing a copy of the model on both the server as well as on-device as shown in Fig. 3. Decision is made on the go about which model will suit the situation better. The various parameters taken into consideration include device parameters like battery, CPU usage and memory as well as network parameters like throughput and round trip time. These parameters are gathered by a smart module, which makes a decision on the optimal choice of model to be used for inferencing. For instance, in a low battery scenario it will make sense to make predictions on the server whereas in a low network scenario it will make more sense to make the predictions on-Device itself. Additionally, the version of the model currently present on the device also impacts the decision-making. For example, in cases such as when the user does not update on-device model device software. The smart module is capable of analyzing all these given scenarios and making optimal runtime decisions.

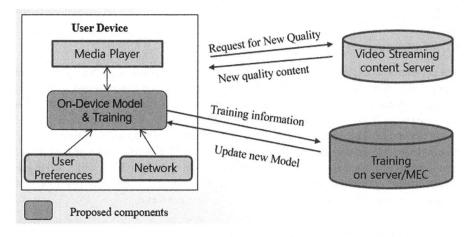

Fig. 3. Hybrid deployment with smart strategy module

3.5 On-Device Training

Machine Learning and Neural Network Frameworks like TensorFlow and Caffe provide libraries solely for on-device inferencing. For on-device training, DL4J [13] was chosen as the Deep Learning framework. This is done for mainly two reasons. Firstly, it provides basic APIs necessary for training a neural network on Android devices. Secondly it is written in Java and therefore can take advantage of the Java Virtual Machine.

The trained model from 3.3 is retrained using Advantage Actor Critic Algorithm [11]. The device is initially loaded with this trained model for inferencing. As user streams videos, data needed for training is simultaneously collected and stored locally. After collecting enough amount of data from the user, the model is further trained and is updated with the newer version for further inferencing. Training is done at strategic times, such as during charging, to decrease power consumption. A diagrammatic representation of the system for on-device training is depicted in Fig. 4.

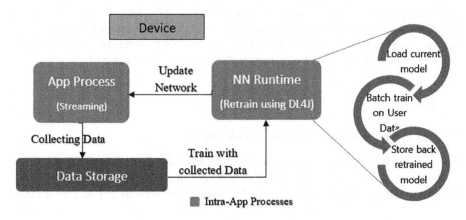

Fig. 4. On-device training methodology

4   Evaluation

4.1   Test Setup

It is necessary to evaluate the model's performance in all possible scenarios to ensure strong performance irrespective of the network scenarios. The initial testing and model validation occurs concurrently with the training on the simulated video player. This allows execution of preliminary trails to test the model's performance over a large number of traces. This helps by allowing rapid validation of the models. Once that step is complete, tests are carried out through running tests in a controlled environment and then later performance is tested by carrying out actual video playback while driving in various traffic conditions and geographical areas. The performance of our model is compared against Google's Dash player and against Pensieve.

4.2   Lab Testing

After preliminary testing on the simulated environment, the model is ported onto an Android device and used for actual video playback. This process involves freezing the model in Tensor-flow, i.e., freezing the model's weights and biases to prevent changes and removal of all training related variables. This frozen graph is then converted to a mobile compatible format using in-built APIs. Finally Tensorflow Mobile APIs are leveraged to load, run and fetch the results. Tests are conducted in a controlled environment using a basic network shaper. Traces are gathered from local networks as well as from publically available datasets like FCC.

4.3   Drive Testing

After validating the model through purely simulated validation and then semi-simulated lab testing, several real world tests are conducted through periodic drive testing. Devices are loaded with these trained models and comparisons against other solutions are made. This elaborate process helps arrive at a model, which can be confidently deployed as a final product. It involves a side-by-side comparison of multiple devices containing the trained model versus the alternative algorithm to be compared against. Both the devices are connected to the same network provider, have the same device configuration and are kept side by side to minimize any major fluctuations in overall throughput received by the devices being compared.

5   Results

5.1   QoE Comparision vs Google's Dash Player

Google's Dash Player, which is developed using their Exoplayer framework [12], is an open-source video player being used widely by developers in the Android ecosystem. Exoplayer uses a rate-based algorithm to determine the next chunk bitrate to be fetched.

Since the network fluctuates in an unpredictable manner, there is safety factor, which is multiplied to the bandwidth estimate to ensure safer playback void of buffering versus best possible playback considering other factors as well. As shown in Fig. 5, proposed solution helps in achieving 40% less buffering and QoE improvement by 30% when compared to Exoplayer. This is due to the model being aware of more parameters like buffer and history of network conditions and being able to take decisions based on its earlier training.

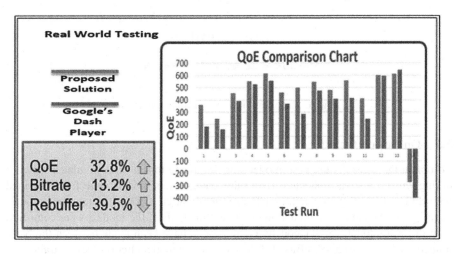

Fig. 5. Snapshot view of QoE improvements

5.2   QoE Comparision vs. Pensieve

Pensieve [4] uses pure reinforcement learning to train the model from scratch on a set of network traces in a small range (0 to 6 Mbps). The proposed solution uses a combination of learning techniques and both, real and synthetic data, to achieve a similar result while being able to converge at an optimal solution much earlier. Hence, as shown in Fig. 6, despite being less reliant on real network data collection and taking 50% the training time, the proposed solution is able to achieve performance equal to or slightly more (up to 10%) than Pensieve [4].

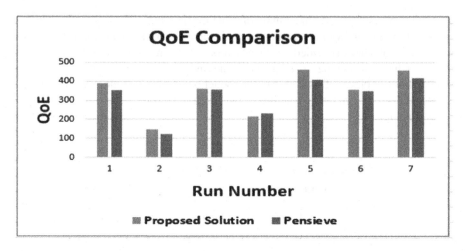

Fig. 6. Snapshot view of QoE improvements

5.3   On-Device Training

To obtain these results the base model was first made to run on a simulated network during which data was collected. Using this data various models were further trained with different hyper parameters (batch size, iteration, etc.). Then the retrained models were run on the similar simulated network and the QoE of the models were compared with the base model simultaneously. Comparisons have been made between the basic trained model, referred to as Base ISF Model and the model after on-Device training is completed (as mentioned in Fig. 7). The results are collected over multiple tests and plotted. An improvement of approximately 10% was observed with just a small amount of iterations (~ 100 iterations) which reinforces the viability of on-Device personalization with locally collected network traces for this use case.

Another series of test runs were conducted keeping every hyper parameter constant except batch size whose value ranged from 32 to 100. The trend with respect to batch size are shown in Fig. 8(a). Next, a similar test was conducted by keeping all parameters constant and varying the epochs (9 h of information constitutes the dataset). These results are presented in Fig. 8(b).

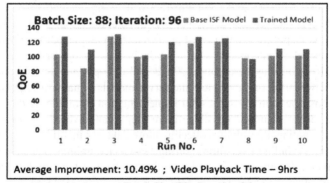

Batch Size:88 , Iteration 96

Fig. 7. Snapshot view of QoE comparison of base ISF model Vs on-device trained model with different batch size and iterations of training

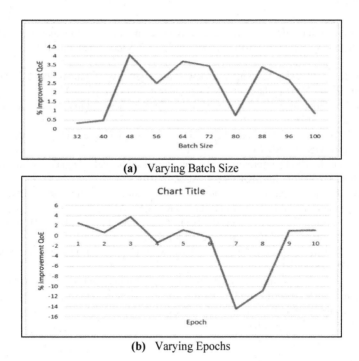

(a) Varying Batch Size

(b) Varying Epochs

Fig. 8. Trend of QoE change with varying hyper parameters

6 Conclusion

The proposed solution aims at providing optimal video streaming experience to mobile device users by maximizing video quality being chosen during video playback while simultaneously minimizing the cases of buffering. This has been made possible through the use of Neural Network and a combination of learning techniques. The use of SL ensures stable performance across any network fluctuation scenario. The RL step helps the model tweak its policy to ensure best performance in all the major scenarios that the model is expected to encounter. The solution is able to outperform commercially available ABR algorithms by around 30%. It will also provide smooth and seamless inferencing between the on-device and the server models. Finally, user personalization will also be possible with the successful implementation of offline on-device training.

To enhance the solution, possible future steps can involve exploring alternate reinforcement learning algorithms, collecting more data and modifying the inputs to the neural network. Avoiding over-fitting during On-Device training also remains a challenging problem. In future, the model could be deployed on mobile edge server (MEC) and localized models on each MEC server could be explored, which are fine-tuned to the network conditions in that particular region.

References

1. MRF Forecast. https://www.marketresearchfuture.com/reports/video-streaming-market-3150
2. Mux Report. https://static.mux.com/downloads/2017-Video-Streaming-Perceptions-Report.pdf
3. Vargas, R., et al.: Deep learning: a review. Adv. Intell. Syst. Comput. **29**, 232–244 (2017)
4. Mao, H., et al.: Neural adaptive video streaming with pensieve (2017)
5. Spiteri, K., Urgaonkar, R., Sitaraman, R.K.: BOLA: near-optimal bitrate adaptation for online videos (2016)
6. Huang, T.Y., et al.: A buffer-based approach to rate adaptation: evidence from a large video streaming service. In: SIGCOMM. ACM (2014)
7. Sun, Y., et al.: CS2P: improving video bitrate selection and adaptation with data-driven throughput prediction. In: SIGCOMM. ACM (2016)
8. Yin, X., Jindal, A., Sekar, V., Sinopoli, B.: A control-theoretic approach for dynamic adaptive video streaming over HTTP. In: SIGCOMM. ACM (2015)
9. Akhtar, Z., Nam, Y.S., Govindan, R., et al.: Oboe: auto-tuning video ABR algorithms to network conditions. In: SIGCOMM. ACM (2018)
10. Silver, D., et al.: Mastering the game of Go with deep neural networks and tree search. Nature **529**, 484 (2016)
11. Mnih, V., et al.: Asynchronous methods for deep reinforcement learning. In: International Conference on Machine Learning (2016)
12. Exoplayer. https://github.com/google/ExoPlayer
13. DL4J. https://deeplearning4j.org/

A Deep Convolutional Neural Network Based Approach to Extract and Apply Photographic Transformations

Mrinmoy Sen[(⊠)] and Prasenjit Chakraborty

Samsung R&D Institute India, Bangalore, India
{mrinmoy.sen,prasenjit.c}@samsung.com

Abstract. Sophisticated image editing techniques like colour and tone adjustments are used to enhance the perceived visual quality of images and are used in a broad variety of applications from professional grade image post-processing to sharing in social media platforms. Given a visually appealing reference image that has some photographic filter or effects applied, it is often desired to apply the same effects on a different target image to provide it the same look and feel. Interpreting the effects applied on such images is not a trivial task and requires knowledge and expertise on advanced image editing techniques, which is not easy. Existing deep learning based techniques fail to directly address this problem and offer partial solutions in the form of Neural Style Transfer, which can be used for texture transfer between images. In this paper, a novel method using a convolutional neural network (CNN) is introduced that can transfer the photographic filter and effects from a given reference image to a desired target image via adaptively predicting the parameters of the transformations that were applied on the reference image. These predicted parameters are then applied to the target image to get the same transformations as that of the reference image. In contrast to the existing stylization methods, the predicted parameters are independent of the semantics of the reference image and is well generalized to transfer complex filters from the reference image to any target image.

Keywords: Convolutional neural networks · Photographic filters · Image stylization

1 Introduction

Professional photographers and social media influencers generally use sophisticated image editing tools to stylize and enhance the perceived visual quality of images. These editing techniques include but are not limited to transformations of the color distribution or tone curve, adjusting brightness and contrast, adjusting pixel level hue and saturation and so on. Operating these sophisticated tools is time consuming and requires considerable skill and expertise in digital photography domain. Although popular social media platforms like Instagram and Snapchat provide users a set of preset filters, they are quite limited and may not be applicable to all kinds of images.

In this paper, a novel method that can extract the photographic effects and transformations applied on a given reference image is proposed. It is demonstrated that the

© Springer Nature Singapore Pte Ltd. 2020
N. Nain et al. (Eds.): CVIP 2019, CCIS 1148, pp. 155–162, 2020.
https://doi.org/10.1007/978-981-15-4018-9_14

extracted transformations can then be applied to any target image so that it looks similar to the reference image. It is shown that a deep convolutional neural network can be trained with image exemplars such that it learns a wide range of complex transformations that constitute traditional photographic filters. Through the experiments conducted, it is demonstrated that using the trained CNN model any target image can be stylized with the same photographic filter applied on a given reference image.

2   Background and Related Work

Early work on image stylization [1, 2] relied on texture synthesis and color transfer through matching the statistics of the color distributions. Recent advances in deep learning and convolutional neural networks has led to remarkable success to address wide array of complex problems in various domains. A number of deep learning based automatic stylization techniques already exist. Lee et al. in [3] propose a method for learning content specific style ranking from a large image dataset and select best exemplars for color and tone enhancement. Deng et al. in [4] propose a method using GANs that can perform image enhancement image in a weakly supervised setup. Hu et al. [5] use reinforcement learning that learns a meaningful sequence of operations to retouch the raw image. These techniques rely on automatic stylization and cannot stylize an image based on a given reference image.

More closely related to this work are the works of Yan et al. [6] and Zhu et al. [7] that demonstrates that it is possible to train a deep learning model that can learn specific photographic styles and transfer the same style to a novel input image. However, their methods needs complex feature engineering, restricted to learn a single style and is inefficient in terms of running time and complexity.

Target   Reference   WCT [8]   Photo-WCT [9]   Proposed Method
(Artistic Style Transfer) (Photo-Realistic Style Transfer)

Fig. 1. Comparison with Neural Style Transfer.

2.1   Comparison with Style Transfer

Neural Style Transfer (NST) has proven to be very effective in image stylization by transferring color, texture and patterns from a style image to a target image. However,

NST is basically a form of texture synthesis and cannot be used for photographic filter transfer. NST extracts the style of an image using correlation between features extracted in deep feature space and fails to capture the transformations represented by traditional photographic filters. In addition, NST will fail to retain the spatially varying effects like vignette that are present in many photographic filters. This is demonstrated in Fig. 1, that both state-of-the-art Artistic [8] and Photo-Realistic Style Transfer [9] is not effective for transferring photographic filters.

3 Proposed Method

Polynomial transformations has been used effectively to address many image-processing tasks [10, 11]. Inspired from [11], it is proposed that application of photographic filter on an image can be formulated as applying a parametric transformation on the image. Specifically, channel wise polynomial transformations that are essentially a set of three functions, each applied separately to the RGB channels is used to address this task. It is also shown that applying these parametric transformations preserves the content of the images without introducing any artifacts. CNNs has been proven to be powerful feature extractors and have been used in numerous recent works to great success. In this work, it is demonstrated that a CNN can be trained to efficiently predict the coefficients of the polynomial transformation that are applied on an image.

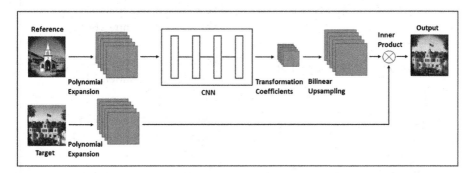

Fig. 2. Filter applied on the reference image is extracted by the CNN as a set of polynomial transformations, which is then multiplied with the target image to produce the output image.

Given a reference image R in which some photographic effect or transformations have been applied, the goal is to extract those transformations and apply them to a given novel target image I. As a pre-processing step, polynomial expansion is applied on both the reference image and the target image. The polynomial expansion of the reference image is then passed through a CNN to estimate a set of intermediate transformation coefficients for each channel. The post-processing step involves performing bilinear upsampling of the intermediate coefficients to produce per-pixel polynomial coefficients for each channel. Lastly, the polynomial expansion of the target image is multiplied by the predicted coefficients of the reference image to produce the output image I', that has the same transformations as that of R. This method is expressed by the following expression:

$$I'_c = \sum_{i+j+k \leq d} x^c_{ijk} \tilde{I}^i_R \tilde{I}^j_G \tilde{I}^k_B \qquad (1)$$

where, I'_c is the pixel value for channel c of the transformed target image I; $\left\{x^c_{ijk}\right\}$, $c \in \{R, G, B\}$ is the bilinearly upsampled per-pixel coefficients predicted from the CNN model from the reference image R and $\tilde{I}$ is the polynomial expansion of degree d of the input image I. The overall pipeline of the proposed method is shown in Fig. 2.

The degree of the polynomial transformations used is an important aspect of the proposed method. As the degree increases the capability to extract more complex transformations increases. As in [11], experiments are performed with polynomial expansion up to degree 3 and the results show that it is sufficient to capture a wide variety of complex filters and effects.

Layer	Output Size	Operations
conv 1	112 x 112 x 64	kernel 7 x 7, channels 64, stride 2
conv2_x	56 x 56 x 64	maxpool 3 x 3, stride 2 $\begin{pmatrix}\text{kernel 3 x 3, channels 64}\\\text{kernel 3 x 3, channels 64}\end{pmatrix}$ ×2
conv3_x	28 x 28 x 128	$\begin{pmatrix}\text{kernel 3 x 3, channels 128}\\\text{kernel 3 x 3, channels 128}\end{pmatrix}$ ×2
conv4_x	14 x 14 x 256	$\begin{pmatrix}\text{kernel 3 x 3, channels 256}\\\text{kernel 3 x 3, channels 256}\end{pmatrix}$ ×2
conv5_x	7 x 7 x 512	$\begin{pmatrix}\text{kernel 3 x 3, channels 512}\\\text{kernel 3 x 3, channels 512}\end{pmatrix}$ ×2
avg pool	1 x 1 x 512	average pool 7 x 7
flatten	512	flatten
fc	47040	fully connected layer

Fig. 3. CNN architecture (for degree 3 polynomial expansion).

A Resnet-18 [12] based CNN architecture is used in this work as the base network. Images of resolution 224 × 224 is used for both the reference and the target image.

Considering degree d for the polynomial expansion, results in images of $\binom{d+3}{d}$ channels. The first convolutional layer is modified to adapt to the $\binom{d+3}{d}$ channel image as the input instead of the usual 3 channel RGB input. This is followed by four residual blocks that produces a feature map of $7 \times 7 \times 512$ dimension. An average pooling and flattening operation results in a feature vector of dimension 512.

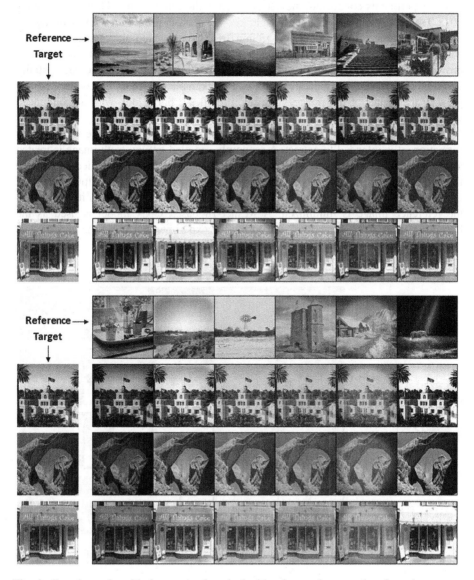

Fig. 4. Sample results with the proposed method with reference images taken from the test set.

Finally a fully connected layer produces the per channel transformation coefficients in a patch-wise manner. It is then interpolated to produce a per-pixel transformation, which is then multiplied with the polynomial expansion of the target image to produce the output image. The patch size is a tunable parameter and can be varied to choose between speed and accuracy. The network architecture is shown in Fig. 3.

4   Results

In the experiments conducted, the dataset used is the same as in [11], which contains 20,000 randomly sampled images from Places-205 dataset processed with 22 Instagram filters to generate 440,000 filtered-ground truth image exemplar pairs. The filters chosen are quite diverse and varying in terms of their overall effect. Some filters provide global adjustments to contrast, tone, saturation while others include spatially varying adjustments like vignette, borders. For example, the filter 'Poprocket' adds a glow to the center of the image, 'Toaster' provides a burnt look with a pronounced vignette, and 'Willow' is a monochromatic filter with white borders.

During training, each filtered image is treated as the reference mage whose transformations it is intended to extract. A randomly sampled ground truth image is treated as the input image in which the extracted transformations is to be applied to produce the output image. Finally, the mean squared error (MSE) between the output image and the corresponding filtered version of the input image having the same filter as that of the reference image is minimized. A Resnet-18 based CNN model is used as the backbone of the network architecture during the experiments. During the experiments conducted, images of resolution 224×224 has been used.

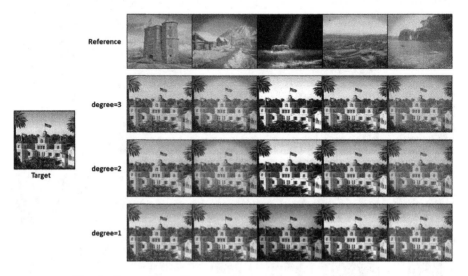

Fig. 5. Comparison with varying degree of polynomial expansion.

The model is trained for 10 epochs using batch size of 25 and a learning rate of 0.0001. Training approximately takes around 24 h in an Nvidia P40 GPU system. The average computation time to generate the output image using the trained model is approximately 100 ms, which demonstrates the efficiency of the proposed method. As demonstrated by the sample results in Fig. 4 the proposed method is able to extract and apply complex filters and transformations. The final output images are visually coherent with no visible artifacts. It is evident from the results that the trained model is able to capture the various effects represented by the different filters and generalizes well to any kind of target image. Experiments are conducted with the degree of the polynomial expansion used to estimate the transformations. Results with different degrees of the polynomial expansion show that both degree 2 and degree 3 can suitably capture all the varying transformations represented by the filters. In contrast, as shown in Fig. 5 polynomial expansion of degree 1 fails at extracting and applying certain effects.

4.1   Error Metrics

To evaluate the quality of the generated images MSE (mean squared error) and PSNR (peak signal-to-noise ratio) are used as the metrics. These metrics are calculated on a test set of 92000 images, which is different from that of the training set. The formula for calculating MSE, given a ground truth image I and its approximation K of size $m \times n$ is given by:

$$MSE = \frac{1}{mn} \sum_{i=0}^{m-1} \sum_{j=0}^{n-1} [I(i,j) - K(i,j)]^2 \tag{2}$$

PSNR is defined using the MSE as:

$$PSNR = 10.log_{10}\left(\frac{MAX_I^2}{MSE}\right) \tag{3}$$

where, MAX_I is the maximum possible pixel value of the image.

In the experiments conducted, these metrics are evaluated with varying degree of the polynomial expansion used. The experimental results are demonstrated in Table 1. As expected, the best results are observed with third degree polynomial expansion in both the metrics.

Table 1. Error metrics calculation. For MSE lower values are better. For PSNR higher values are better.

Polynomial degree (d)	MSE	PSNR
d = 3	0.48×10^{-3}	33.63
d = 2	1.04×10^{-3}	30.09
d = 1	5.93×10^{-3}	22.37

5 Conclusion

In this paper, a novel method that uses deep CNNs to efficiently extract photographic transformations from a reference image and apply the same transformations on a given input image is introduced. It is demonstrated through the experiments conducted that the model is capable of learning a wide range of complex effects and transformations and can produce visually coherent output images as results. At present, the proposed method can only extract and apply the filters that it is trained upon, which is a limitation of the current method. A generalization of this method to unseen filters and effects is intended to be addressed in a future work. The proposed method can be used as part of any photo editing application allowing users to extract photographic filters or effects from any image and apply them on their own images.

References

1. Heeger, D.J., et al.: Pyramid-based texture analysis/synthesis. In: International Conference on Image Processing, Washington, DC, USA, vol. 3, pp. 648–651 (1995)
2. Reinhard, E., et al.: Color transfer between images. In: IEEE Computer Graphics and Applications, vol. 21, no. 5, pp. 34–41, July-August 2001
3. Lee, J.-Y., et al.: Automatic content-aware color and tone stylization. In: IEEE Conference on Computer Vision and Pattern Recognition (CVPR) (2016)
4. Deng, Y., et al.: Aesthetic-driven image enhancement by adversarial learning. In: ACM Multimedia (2018)
5. Hu, Y., et al.: Exposure: a white-box photo post-processing framework. ACM Trans. Graph. **37**, 1–17 (2018)
6. Yan, Z., et al.: Automatic photo adjustment using deep neural networks. ACM Trans. Graph. **35**, 1–15 (2016)
7. Zhu, F., et al.: Exemplar-based image and video stylization using fully convolutional semantic features. IEEE Trans. Image Process. **26**, 3542–3555 (2017)
8. Li, Y., et al.: Universal style transfer via feature transforms. In: International Conference on Neural Information Processing Systems (NIPS) (2017)
9. Li, Y., et al.: A closed-form solution to photorealistic image stylization. In: ECCV (2018)
10. Ilie, A., et al.: Ensuring color consistency across multiple cameras. In: Tenth IEEE International Conference on Computer Vision (2005)
11. Bianco, S., et al.: Artistic photo filter removal using convolutional neural networks. J. Electron. Imaging **27**(1), 011004 (2017). https://doi.org/10.1117/1.jei.27.1.011004
12. He, K., et al.: Deep residual learning for image recognition. In: IEEE Conference on Computer Vision and Pattern Recognition (CVPR) (2016)

Video Based Deception Detection Using Deep Recurrent Convolutional Neural Network

Sushma Venkatesh, Raghavendra Ramachandra[⊠], and Patrick Bours

Norwegian University of Sceince and Technology, Gjøvik, Norway
{sushma.venkatesh,raghavendra.ramachandra,patrick.bours}@ntnu.no

Abstract. Automatic deception detection has been extensively studied considering their applicability in various real-life applications. Since humans will express the deception through non-verbal behavior that can be recorded in a non-intrusive manner, the deception detection from video using automatic techniques can be devised. In this paper, we present a novel technique for the video-based deception technique using Deep Recurrent Convolutional Neural Network. The proposed method uses the sequential input that can capture the spatiotemporal information to capture the non-verbal behavior from the video. The deep features are extracted from the sequence of frames using a pre-trained GoogleNet CNN. To effectively learn the extended sequence, the bi-directional LSTMs are connected to the GoogleNet and can be jointly trained to learn the perceptual representation. Extensive experiments are carried out on a publicly available dataset [5] with 121 deceptive and truthful video clips reflecting a real-life scenario. Obtained results demonstrate the outstanding performance of the proposed method when compared with the four different state-of-the-art techniques.

Keywords: Deception detection · Deep learning · Multimodal deception

1 Introduction

Deception is an act of concealing the truth to mislead a person or to hide the information and it is observed with the existence of mankind. A person tends to perform deception in various circumstances in their lifetime. Deception performed may be of low-stake or high-stake. Depending on various situations, certain deception performed may not have serious consequences. Whereas some of the deceptions performed, for instance: in high-security places like airport security checks or in situations where a person trying to mislead the immigration officials by producing the wrong ID may have serious consequences. Though an immigration officer is trained to identify people performing deception, usually the ability of a person to detect deception is limited, as suggested by Depaulo et al. [1].

© Springer Nature Singapore Pte Ltd. 2020
N. Nain et al. (Eds.): CVIP 2019, CCIS 1148, pp. 163–169, 2020.
https://doi.org/10.1007/978-981-15-4018-9_15

In ancient times, verbal and non-verbal deception detection cues were identified to distinguish between variety and truth. As deception is challenging to identify as some of the deception detection cues goes unnoticed [3], in the later times various deception detection techniques including polygraph, functional magnetic resonance (fMRI) came into existence. Polygraph is one of the popular method used to identify a deceptive person by recording the physiological changes occurring in the body by physically attaching various sensors. In some situations, it is inconvenient to use the polygraph method since it requires the physical attachment of sensors to the body as well as human expertise is required for the controlled setting of a questionnaire to decide the deceptive behavior of a person. In some cases, polygraph method is error-prone and gives biased results.

Earlier work on deception detection by [5] have explored automatic deception detection using verbal and non-verbal features together with a new publicly available dataset. Verbal features are extracted using bag-of-words and non-verbal features are extracted using MUMIN coding scheme. Analysis performed on the three different features provides a classification accuracy of 82%. In [9] presented a new automatic deception detection technique using multiple modalities. Motion features that include the micro-expressions in the videos are extracted using IDT (Improved Dense Trajectory), audio features are extracted using MFCC and text features are extracted using Glove. The classification accuracy on the multimodal dataset gives an improved performance when compared to that of a unimodal technique. In [2], deep learning based multimodal deception detection technique using 3D-CNN for video features extraction was proposed. In addition to this, convolutional neural networks (CNN) is used to extract the textual data and for the audio features with high dimensional features are extracted using the openSMILE toolkit. Finally, all features including the micro expression features are fused to obtain a recognition accuracy of 96.14%. In [6], have presented a 2D appearance based methodology to distinguish the three-dimensional facial features that include the micro facial expressions like mouth motion, eye blink, wrinkle appearance and eyebrow motion. They have achieved recognition accuracy of 76.92% on their private database. Recent work on multimodal deception detection by [8] uses the features of the micro-expression. In addition to this, audio features are extracted using Mel filtering cepstral co-efficient by windowing the audio signals and textual features are extracted using bag-of-N-grams. Finally, all three modalities are fused at the decision level to make the final decision that shows a recognition performance of 97%. Thus, based on the available works on the deception detection it can be noted that (1) the use of multimodal information will help to reach the robust deception detection performance (2) the use of automatic deception detection techniques using non-verbal behavior has indicated a better performance when compared to that of human annotated deception detection methods.

In this work, we propose a novel framework for video-based deception detection by combining the transfer learning using pre-trained image sequence classification and a Bi-directional LSTM. We term this forming as Deep Recurrent Convolutional Neural Networks. Thus, the proposed method directly connect a visual Convolutional model to Bidirectional LSTM network that can capture the temporal state dependencies. The main objective of the proposed method is to use the uni-modal information from video based on non-verbal cues to detect the deception. Thus, the proposed method can over come the need of multi-modal information and thus more reliable for various real-life applications where only video of the interrogation is available. Extensive experiments are carried out on a publicly available dataset [5] with 121 deceptive and truthful video clips reflecting a real-life scenario. The following are the main contributions of this work:

- Novel approach for video-based deception detection by exploring temporal information based on the deep visual features and bi-directional LSTM. We show that the proposed method can significantly improve the deception detection accuracy.
- Extensive experiments are carried out on the publicly available dataset [5] on multimodal deception detection.
- The performance of the proposed method is compared with four different state-of-the-art methods.

The rest of the paper is organised as follows: Sect. 2 presents the proposed method for robust deception detection, Sect. 3 illustrates the quantitative results of the proposed method together with the state-of-the-art multi-modal techniques and Sect. 4 draws the conclusion.

2   Proposed Method

Figure 1 shows the block diagram of the proposed method for reliable deception detection from video. The crucial factor revealing deception detection is based on the non-verbal behavioral cues that include the head movements, hand gestures, expressions, etc. Thus, it is essential to quantify the non-verbal behavior cues that are evolving with time. Hence, we treat this problem as sequential inputs in which a video with arbitrary length is given as the input to get the fixed output as either truth or deceit. The proposed Deep Recurrent Convolutional Neural Network is based on the pre-trained deep CNN network to extract the features which are then classified using a Bi-directional LSTM.

Given the video $V = V_1, V_2, \ldots, V_N$ with N frames, we extract the corresponding features $V_f = V_{f1}, V_{f2}, \ldots, V_{fN}$ using deep pre-trained network. In this work, we have used pre-trained CNN GoogleNet [7] to extract the features by considering its robust performance in object classification and action detection [7]. We then connected a bi-directional LSTM before the last fully connected layer (or after the dropout layer) of the GoogleNet that can perform the classification of the video to either deceit or truth. When compared to the traditional Recurrent Neural network (RNN) that shows limitation to learn the long-term dynamics due to the vanishing gradients, the use of LSTMs can overcome these limitations by incorporating the memory units that allows the network to learn and remember the long sequences. In this work, we have used the single layer

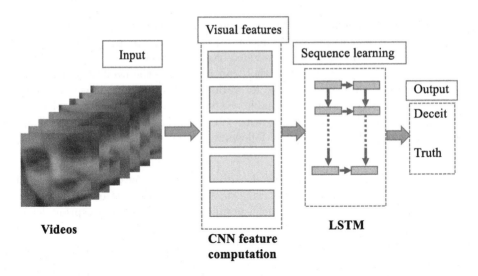

Fig. 1. Block diagram of the proposed method

bi-directional LSTMs with 2000 hidden units. Given the training video set, the proposed Deep Recurrent Convolutional Neural Network is trained in the end-to-end fashion which is then used to classify the test video.

3   Experiments and Results

In this section, we present the experimental results of the proposed method together with the performance comparison with the four different state-of-the-art techniques. Experiments are carried out on the publicly available multimodal deception dataset [5] that is collected using the real-life scenarios. The multimodal deception dataset consists of 121 video recordings that are collected from the public multimedia sources that host the video sequences from the real-life court trials and television interviews. Out of 121 video sequences, 60 corresponds to truthful and 61 corresponds to the deceit and the average length of the video corresponds to 28 s. Further, this dataset also provides the multimodal information like text transcripts, voice data and manually annotated non-verbal cues. The performance of the proposed method and the state-of-the-art is evaluated using a leave-one-out cross-validation with 25 different trials as mentioned in [9]. The performance of the deception detection method is reported using classification accuracy or correct classification rate (CCR %) (Fig. 2).

Table 1 indicates the quantitative performance of the proposed method, together with the four different state-of-the-art methods. Note that, the state-of-the-art techniques used in this work are based on the multimodal information (voice, face & video) while the proposed method is based only on the video information. However, to provide an overview of the performance corresponding

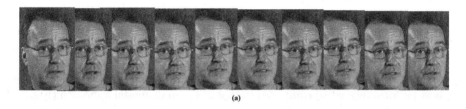

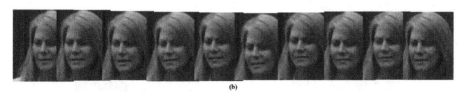

Fig. 2. Example of video frames from multimodal deception dataset [5] (a) Truth (b) Deceit

Table 1. Quantitative performance of the proposed method

Type	Modality	Algorithm	CCR (%)
Unimodal	Audio	Venkatesh et al. [8]	46
		Venkatesh et al. [8]	76
	Text	Wu et al. [9]	24
		Venkatesh et al. [8]	84
	Video (micro exression)	Venkatesh et al. [8]	88
		Wu et al. [9]	77.31
		Pérez-Rosas et al. [4]	73.55
		Proposed method	**100**
Multi-modal	Fusion of audio, text & video	Venkatesh et al. [8]	97.00
		Wu et al. [9]	87.73
		Perez et al. [5]	82.00
		Krishnamurthy et al. [2]	96.14

to the state-of-the-art on unimodal deception detection, we have also included the recent techniques for the comparison. Thus, the following are the main observations:

– Among different unimodal characteristics, the use of video-based information has indicated the best performance with the state-of-the-art techniques. Pérez-Rosas et al. [4] has proposed the human annotated non-verbal cues (39 gestures) that are classified using a Random forest classifier shows the CCR = 73.55%. However, the automatic extraction of non-verbal cues proposed in Wu et al. [9] shows the performance of CCR = 77.31%.
– Multimodal approaches are widely used for automatic deception detection. The performance results shows the better performance when compared to that of the uni-modal approach.

– Based on the obtained results, the proposed method has indicated the outstanding performance with CCR = 100%. The outstanding performance of
the proposed method can be attributed to the deep features together with the
bi-directional LSTM can provide robust features to detect the deception from
a video. Further, it can also be noted that the proposed method indicates the
best performance when compared to that of the multimodal approach (Fig. 3).

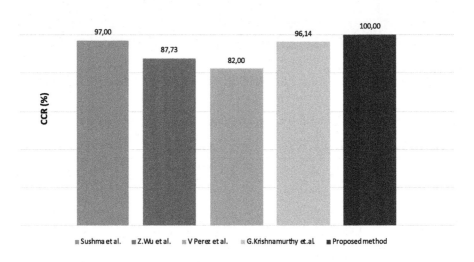

Fig. 3. Comparative performance of the proposed method

4   Conclusion

In this work, we have presented a novel technique for automatic deception detection using Deep Recurrent Convolutional Neural Network. The proposed method
is based on the sequential input of video frames to the deep pre-trained network
and the bi-directional LSTMs to detect the deception based on the non-verbal
cues reliably. The proposed method is based on using the GoogleNet pre-trained
CNN that is used to extract the features from a sequence of video frames. Extensive experiments are carried out on the publicly available dataset [5] with 121
deceptive and truthful video clips reflecting a real-life scenario. Experimental
results have demonstrated the outstanding performance of the proposed method
when compared with the four different state-of-the-art techniques.

References

1. DePaulo, B.M., Kashy, D.A., Kirkendol, S.E., Wyer, M.M., Epstein, J.A.: Lying in everyday life. J. Pers. Soc. Psychol. **70**(5), 979 (1996)
2. Krishnamurthy, G., Majumder, N., Poria, S., Cambria, E.: A deep learning approach for multimodal deception detection. arXiv preprint arXiv:1803.00344 (2018)
3. Levine, T.R., et al.: Sender demeanor: individual differences in sender believability have a powerful impact on deception detection judgments. Hum. Commun. Res. **37**(3), 377–403 (2011). https://doi.org/10.1111/j.1468-2958.2011.01407.x
4. Pérez-Rosas, V., Abouelenien, M., Mihalcea, R., Burzo, M.: Deception detection using real-life trial data. In: Proceedings of ACM on International Conference on Multimodal Interaction, pp. 59–66 (2015)
5. Pérez-Rosas, V., Abouelenien, M., Mihalcea, R., Xiao, Y., Linton, C., Burzo, M.: Verbal and nonverbal clues for real-life deception detection. In: Proceedings of Conference on Empirical Methods in Natural Language Processing (2015)
6. Su, L., Levine, M.: Does lie to me lie to you? An evaluation of facial clues to high-stakes deception. Comput. Vis. Image Underst. **147**, 52–68 (2016)
7. Szegedy, C., et al.: Going deeper with convolutions. In: Proceedings of the IEEE Conference on Computer Vision and Pattern Recognition, pp. 1–9 (2015)
8. Venkatesh, S., Ramachandra, R., Bours, P.: Robust algorithm for multimodal deception detection. In: 2019 IEEE Conference on Multimedia Information Processing and Retrieval (MIPR), pp. 534–537, March 2019
9. Wu, Z., Singh, B., Davis, L.S., Subrahmanian, V.: Deception detection in videos. arXiv preprint arXiv:1712.04415 (2017)

Deep Demosaicing Using ResNet-Bottleneck Architecture

Divakar Verma$^{(\boxtimes)}$, Manish Kumar, and Srinivas Eregala

Samsung R&D Institute Bengaluru, Bengaluru, India
`divakarverma96@gmail.com`, {`man.kumar,srinivas.e`}`@samsung.com`

Abstract. Demosaicing is a fundamental step in a camera pipeline to construct a full RGB image from the bayer data captured by a camera sensor. The conventional signal processing algorithms fail to perform well on complex-pattern images giving rise to several artefacts like Moire, color and Zipper artefacts. The proposed deep learning based model removes such artefacts and generates visually superior quality images. The model performs well on both the sRGB (standard RGB color space) and the linear datasets without any need of retraining. It is based on Convolutional Neural Networks (CNNs) and uses a residual architecture with multiple 'Residual Bottleneck Blocks' each having 3 CNN layers. The use of 1×1 kernels allowed to increase the number of filters (width) of the model and hence, learned the inter-channel dependencies in a better way. The proposed network outperforms the state-of-the-art demosaicing methods on both sRGB and linear datasets.

Keywords: Demosaicing · RGB · Bayer · Moire artefacts · CNN · Residual Bottleneck architecture

1 Introduction

De-mosaicing is the first and the foremost step of any camera ISP (Image Signal Processing) pipeline. Color image sensor can only capture one color at any pixel location in a fixed bayer pattern forming a mosaic/bayer image. An interpolation method is needed to fill the missing colors at each pixel location in the mosaiced image and this process is known as De-mosaicing. A common challenge faced for demosaicing is the unavailability of the actual ground truth images where each pixel contains the actual R (red), G (green) and B (blue) components. It is not feasible to capture all the color components at any given pixel location. So, the common approach is to take high quality images and treat them as the ground truth. These images are then mosaiced into bayer images which goes as an input to the demosaicing algorithm.

Traditional interpolation algorithms take advantage of correlation between R, G and B components of bayer image. Since G component has double sampling frequency, interpolation of G is done first, followed by R and B. Interpolation is done along both horizontal and vertical direction and combined using various

© Springer Nature Singapore Pte Ltd. 2020
N. Nain et al. (Eds.): CVIP 2019, CCIS 1148, pp. 170–179, 2020.
https://doi.org/10.1007/978-981-15-4018-9_16

metrics. In MSG [1] algorithm, authors improved the interpolation accuracy by using Multi-Scale color Gradients to adaptively combine color-difference-estimates from different directions. In ARI (Adaptive Residual Interpolation) [2], authors used R as a guided filter to interpolate G at R&B (guided upsampling) and vice versa to interpolate R&B at G. Due to inherent sensor noise, interpolation based algorithm sometimes fails to demosaic the complicated patterns near the edges, leading to moire, zippering and other color artefacts. To remove the moire artefact from images, camera uses low pass filter but that reduces the sharpness of the image. To address these challenges, deep learning algorithms have been proposed which show significant improvement over traditional interpolation based methods.

1.1   Related Work

Numerous deep learning architectures have been proposed for demosaicing and with the advancements in the processing power, the networks are becoming deeper and deeper. The authors of 'A Multilayer Neural Network for Image Demosaicing' [3] had proposed a 3 layered deep network which achieved a PSNR (Peak signal-to-noise ratio) of 36.71 on 19 Kodak images and showed initial promise that deep learning network could prove to be useful for demosaicing. Gharbi et al. [4] uses a 15 layered network with a residual learning approach. It was able to outperform all the interpolation based demosaicing methods and deep learning based networks by achieving 41.2 PSNR on Kodak dataset [5]. Tan et al. [6] uses a two stage network which is similar to interpolation algorithms such as MSG and AHD [7]. The Green channel is used as a guide for interpolation of Red and Blue channels. First, the demosaicing kernels are learned using the L2 loss [8] on Green channel and then in the second stage, the loss is calculated on all channels. Thus, the Green channel guides the interpolation of the final RGB channels. On Kodak-24 image dataset, it achieved a PSNR of 42.04 and on McMaster (McM) [9] dataset, it achieved a PSNR of 39.98. The network proposed in DMCNN-VD [10] is even deeper and consists of 20 convolutional layers. It also uses a residual learning approach and achieved a PSNR of 42.27 on the Kodak-24 dataset.

However, the above mentioned deep learning networks do not generalize well on all kind of images and hence, will require a re-training for the specific kind of images. The proposed deep learning architecture addresses these issues and outperforms the state-of-the-art deep learning based demosaicing network on both linear and sRGB datasets. For the first time, a bottleneck residual network for demosaicing has been proposed which can generalize across different types of datasets. The proposed network is a fully convolutional neural network and uses multiple residual blocks.

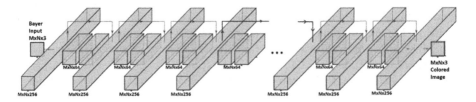

Fig. 1. Proposed Deep Learning model for demosaicing with 10 residual blocks

2   Proposed Deep CNN Architecture

The proposed bottleneck residual network architecture for demosaicing generalizes well and generates superior quality images with minimal artefacts. The
network is inspired from Residual Network (ResNet) architecture [11]. The proposed network is able to handle the complicated patterns in the image and gives
much better visual quality.

The proposed network is based on CNNs and uses a residual architecture
with each residual block having a bottleneck structure [12]. The network has
10 such residual blocks each having 3 convolutional layers. The network has a
varying width of 256 and 64 channels. The bottleneck structure allows faster
learning and at the same time learns more number of features.

The input to the network is a bayer image which is split into 3 channels
- Red(R), Green(G) and Blue(B), with each channel having interleaved zeros.
The starting convolutional layer in the network uses 3 × 3 filters and converts
the dimensions of the 3-channeled bayer input to 256 channels that goes as an
input to the first residual block. Each residual block has 3 CNN layers. The first
layer uses 1 × 1 filters to change the dimension of 256-channeled input from the
previous residual block and convert it to a 64-channeled output. This output is
then passed through a ReLU activation layer. The second CNN layer operates
on a reduced dimensional output of 64 channels from the previous layer. This
layer uses a filter size of 3 × 3 which helps the model to learn important features
and interchannel relationships. The output from this layer is 64-channeled and is
passed through a ReLU activation layer. The third CNN layer uses 1 × 1 filters
to restore the dimensions from 64 to 256 channels. Using a skip connection, the
output from the third CNN layer is added with the original input (256-channeled)
of the given residual block. This output now goes as an input to the next residual
block. After the 10th (last) residual block, the final convolutional layer of the
network uses 3 × 3 filters and converts the output having 256 channels into a 3-
channeled color image. This is the final output of the network and has the same
dimensions as of the input bayer image. Figure 1 shows the proposed network
architecture. The network uses an L2 loss function between the ground truth
and the output of the model.

Figure 2 shows few possibilities of different input bayer images possible for the
network. The input is generated from the ground truth RGB image by mosaicing
it in a bayer fashion. The basic form is shown in (a) which is a single channeled

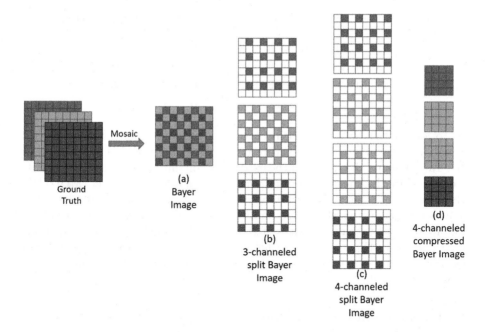

Fig. 2. Comparison of different possible forms of the input image to the network

image with all the three color components interleaved in the same plane. This form is generally not preferred as an input to the network because it adds an additional burden on the network to learn the relationship between the interleaved color components. For example, the network needs to learn that two alternate pixels belong to the same color channel. So, a common approach is to split the color components into different channels. The proposed architecture uses a 3-channeled bayer image as shown in (b). The interleaved white blocks in the channels are the places where no color component is present and have been initialized with zeroes. The Green channel contains 50% of the color components whereas the Red and Blue contains 25% each. For uniformity, the Green channel can further be split into two channels, as shown in (c), so that each channel contains 25% of the color components. This approach was not adopted because it would have increased the training parameters and made the model more complex. The four channels shown in (c), can be compressed by packing the color pixels together, as shown in (d). This would lead to the loss of spatial information of the pixels and hence make it difficult for the network to learn some important information, like edges, which is of utmost priority for demosaicing. Hence, form (b) was chosen as the input for the proposed network.

The proposed model was trained solely on sRGB dataset and still it is able to generalize well across linear dataset. Due to the limited availability of linear datasets, the model was not trained on the linear dataset. So, to test the model on linear datasets, the images were transformed to sRGB domain and demosaiced using the network already trained on the sRGB dataset. During the experiment, it was found that the model performed equally well for the linear dataset.

Table 1. Comparison of BottleNeck architectures with different widths

	Kodak12	McM	Kodak24	Panasonic	Canon
128-64	43.8	39.28	42.24	42.34	44.41
256-64	**43.86**	**39.29**	**42.3**	**42.42**	**44.43**

To confirm the role of the width of the architecture, the model was tested with a modified version of the architecture having a smaller width of 128 instead of 256. Table 1 shows the results of the experiment on different datasets. Panasonic and Canon are the linear datasets of Microsoft Demosaicing Dataset (MDD) [13] while the rest of them are sRGB datasets. The first row shows the results of the architecture having widths of 128 and 64. The second row shows the results of the proposed model having widths of 256 and 64. It can thus be confirmed that, more number of channels (width) helps the network to learn more number of features required for demosaicing. Hence, increasing the width of the network improves the quality of demosaiced image.

3   Experiments and Results

In all the mentioned experiments, Bayer color filter array was used, as it is the most commonly and widely used color filter array in cameras. The network was trained on Waterloo Exploration Dataset (WED) [14] dataset which contains 4,744 colored images of roughly 600 × 400 resolution. The dataset was augmented by shifting 1 pixel along horizontal and vertical direction, all four rotations and flipping. Shifting an image by 1 pixel helps to capture all the color components at any given pixel location when mosaicing the ground truth image into bayer image. Rotations and flipping helps to generate different orientations of the same image and helps the network to learn a wide variety of patterns and orientations. Finally, image patches of size 128 × 128 was cropped from this augmented dataset for training. Total number of training images generated was 735,920.

Table 2. PSNR comparison for sRGB dataset

	Kodak-12	McM	Kodak-24
MSG	NA	NA	41.00
ARI	41.47	37.60	NA
DMCNN-VD	43.45	39.54	42.27
Gharbi	41.2	39.5	NA
Tan	NA	38.98	42.04
Kokkinos [15]	41.5	39.7	NA
MMNet [16]	42.0	**39.7**	NA
Proposed	**43.86**	39.29	**42.30**

Table 2 shows the quantitative comparison on sRGB datasets. Kodak-12 and Kodak-24 are the sets of 12 and 24 Kodak images respectively. Different authors have used different Kodak sets to measure the performance. We have compared our results on both the Kodak datasets. The proposed model outperforms other algorithms on Kodak sets. In case of McMaster (McM) dataset, the results are not far behind. Table 3 shows the quantitative comparison on MDD. The proposed method outperforms other demosaicing algorithms and is the state-of-the-art. Note that the PSNR 42.86 achieved is the weighted average of 200 Panasonic and 57 Canon images.

Table 3. PSNR comparison for linear (MDD) dataset

ARI	RTF [17]	DMCNN-VD	Kokkinos	Gharbi	MMNet	Proposed
39.94	39.39	41.35	42.6	42.7	42.8	**42.86**

Table 4 shows the comparison for two networks with widths of 256 and 128 for linear dataset. In the table, the first row (128-64) refers to the bottleneck architecture with widths 128 and 64. Similarly, 256-64 refers to the bottleneck architecture with widths 256 and 64. The prefix 'lin_sRGB' refers to the method where the testing linear images were first converted to sRGB domain, then demosaiced and finally converted back to linear domain to find the PSNR values. The data clearly shows that the network with 256-width outperforms the 128-width network in both linear and sRGB domain demosaicing.

Table 4. PSNR Comparison of bottleneck architecture on linear datasets

	Panasonic(200)	Canon(57)
128-64	41.92	44.05
128-64 lin_sRGB	42.14	44.41
256-64	41.94	44.07
256-64 lin_sRGB	**42.42**	**44.43**

Figures 3 and 4 shows the qualitative comparison on sRGB datasets. In Fig. 3 top row image (green star), it can be observed that DMCNN-VD fails to produce sharp edges inside the marked region. In Fig. 4, a blue-colored artefact can be observed in the marked region when looked closely which is absent in the proposed image. Figure 5 shows the qualitative results on linear MDD dataset. First two images (a, b) are Ground Truth and the proposed method's output. Next three images (c, d, e) are snapshots taken directly from the DMCNN-VD paper. The authors have increased the saturation and brightness for these images to highlight the chroma artefacts. The proposed model is not fine-tuned

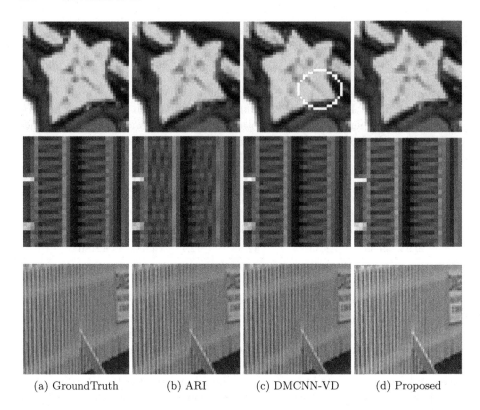

(a) GroundTruth   (b) ARI   (c) DMCNN-VD   (d) Proposed

Fig. 3. Visual comparison with ARI and DMCNN-VD (Color figure online)

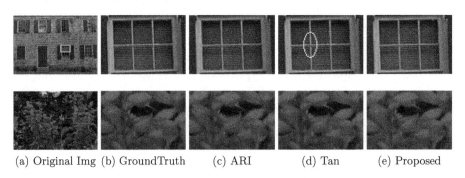

(a) Original Img (b) GroundTruth   (c) ARI   (d) Tan   (e) Proposed

Fig. 4. Visual comparison with ARI and Tan on Kodak (top row) and McM (bottom row) datasets (Color figure online)

using any linear dataset and even then, it is able to match the visual quality of DMCNN-VD-Tr, which is a fine tuned version of DMCNN-VD on MDD dataset using transfer learning. In the top row, the DMCNN-VD-Tr output appears to have lost the chroma information for the monument but the proposed model

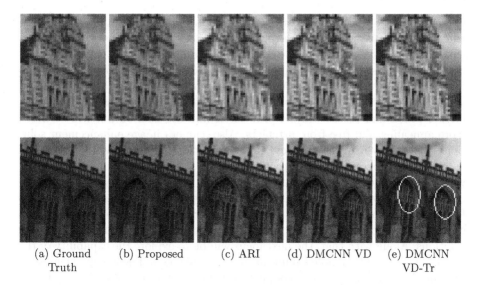

(a) Ground   (b) Proposed   (c) ARI   (d) DMCNN VD   (e) DMCNN
    Truth   VD-Tr

Fig. 5. Visual comparison on linear images (MDD dataset)

Fig. 6. Demosaicing on raw data captured from a smartphone

preserves the color. The proposed model also outperforms DMCNN model in terms of PSNR metric, as shown in Table 3.

The proposed method was also tested on a real-life image dataset. Figure 6 shows demosaicing algorithms applied on the raw images captured at 12 MP by a smartphone and it can be seen that the proposed model has generalized well. Random noise and zipper artefacts can be clearly seen on MSG demosaiced images. The proposed model minimizes all such artefacts.

4   Conclusion and Future Work

In this paper, a novel approach for demosaicing has been proposed. The proposed method is the state-of-the-art and confirms the ability to generalize well, across different types of datasets. Most of the computational photography techniques and computer vision algorithms rely on edge detection. Images with artefacts on the edges such as zippering and chroma are likely to give poor segmentation results, thus, further affecting the processed image. Therefore, it is crucial to solve such issues at the very start of the Image Processing Pipeline. With a superior quality at the initial steps of the camera pipeline, it is expected that further processing blocks will perform better and the final output will be much more appealing and free from artefacts. Also, camera image enhancement solutions such as low-light imaging and super resolution, rely heavily upon per pixel quality. It is expected that the proposed method, which has minimal artefacts, will directly benefit these solutions.

The future work involves exploring the effects of demosaicing algorithms on the computational photography solutions like HDR and Super-Resolution and evaluate the extent to which the proposed demosaicing algorithm improves these solutions. Along with that, the next focus will be to explore the capability of the proposed network to handle simultaneous demosaicing and denoising. Demosaicing and denoising is a tightly coupled problem, solving one greatly affects the other. A wide research is going on to address both of them simultaneously and many deep learning architectures have been proposed. Additionally, it will be explored if such a network can be compressed and optimized for an on-device ISP pipeline without significant loss in performance.

References

1. Pekkucuksen, I., Altunbasak, Y.: Multiscale gradients-based color filter array interpolation. IEEE Trans. Image Process. **22**, 157–165 (2013)
2. Monno, Y., Kiku, D., Tanaka, M., Okutomi, M.: Adaptive residual interpolation for color image demosaicking. In: Proceedings of IEEE ICIP 2015, pp. 3861–3865 (2015)
3. Wang, Y.Q.: A multilayer neural network for image demosaicking. In: 2014 IEEE International Conference on Image Processing (ICIP), pp. 1852–1856. IEEE, October 2014
4. Gharbi, M., Chaurasia, G., Paris, S., Durand, F.: Deep joint demosaicking and denoising. ACM Trans. Graph. (TOG) **35**(6), 191 (2016)
5. Kodak Dataset. http://r0k.us/graphics/kodak
6. Tan, R., Zhang, K., Zuo, W., Zhang, L.: Color image demosaicking via deep residual learning. In: IEEE International Conference on Multimedia and Expo (ICME) (2017)
7. Hirakawa, K., Parks, T.W.: Adaptive homogeneity-directed demosaicing algorithm. IEEE Trans. Image Process. **14**(3), 360–369 (2005)
8. Janocha, K., Czarnecki, W.M.: On loss functions for deep neural networks in classification. arXiv preprint arXiv:1702.05659 (2017)
9. Zhang, L., Wu, X., Buades, A., Li, X.: Color demosaicking by local directional interpolation and nonlocal adaptive thresholding. J. Electron. Imaging **20**(2), 023016 (2011)
10. Syu, N.-S., Chen, Y.-S., Chuang, Y.-Y.: Learning deep convolutional networks for demosaicing. arXiv preprint arXiv:1802.03769 (2018)

11. He, K., Zhang, X., Ren, S., Sun, J.: Deep residual learning for image recognition. arXiv preprint arXiv:1512.03385 (2015)
12. He, K., Zhang, X., Ren, S., Sun, J.: Identity mappings in deep residual networks. In: Leibe, B., Matas, J., Sebe, N., Welling, M. (eds.) ECCV 2016. LNCS, vol. 9908, pp. 630–645. Springer, Cham (2016). https://doi.org/10.1007/978-3-319-46493-0_38
13. Syu, N.S., Chen, Y.S., Chuang, Y.Y.: Learning deep convolutional networks for demosaicing. arXiv preprint arXiv:1802.03769 (2018)
14. Ma, K., et al.: Waterloo exploration database: new challenges for image quality assessment models. IEEE Trans. Image Process. **26**(2), 1004–1016 (2016)
15. Kokkinos, F., Lefkimmiatis, S.: Deep image demosaicking using a cascade of convolutional residual denoising networks. In: Proceedings of the European Conference on Computer Vision (ECCV), pp. 303–319 (2018)
16. Kokkinos, F., Lefkimmiatis, S.: Iterative joint image demosaicking and denoising using a residual denoising network. IEEE Trans. Image Process. **28**, 4177–4188 (2019)
17. Khashabi, D., Nowozin, S., Jancsary, J., Fitzgibbon, A.W.: Joint demosaicing and denoising via learned nonparametric random fields. IEEE Trans. Image Process. **23**(12), 4968–4981 (2014)

Psychological Stress Detection Using Deep Convolutional Neural Networks

Kaushik Sardeshpande[(⊠)] and Vijaya R. Thool

Department of Instrumentation Engineering, SGGSIE&T,
Nanded 431606, Maharashtra, India
kds.sirdeshpande@gmail.com, vrthool@sggs.ac.in

Abstract. Many psychological motives and life incidences are answerable for inflicting psychological stress. It's the primary reason for inflicting many cardiovascular diseases. This paper presents a study on psychological stress detection with the aid of processing the Electrocardiogram (ECG) recordings using Convolutional Neural Networks (CNN) as a classification approach. The main purpose of this study was to trace students under stress during their oral exam. A dataset of ECG recordings of 130 students was taken during the oral exam. A customized CNN is designed for stress recognition, and it has achieved 97.22% and 93.10% stress detection accuracy for filtered and noisy datasets, respectively.

Keywords: Electrocardiogram (ECG) · Deep learning · Convolutional Neural Networks · Stress detection · Scalogram technique

1 Introduction

Physiological stress is a common latest-life disease. Human work-life is changing day by day and has become too hectic. This contributes and creates stress to the unstable mental situation. There are various parameters based mainly on which stress can be categorized, but one in each of which is its effect on the human body. It is therefore labeled as acute and chronic based on the effect of pressure on our body. Acute stress is a temporary kind of stress and not always lasting for a long time. For our performance index, this type of stress is constantly essential. While chronic stress can also result in several continuous changes in our body parameters such as blood pressure, ECG, body temperature, etc. [9]. Stress may be a cause of permanent illnesses such as high blood pressure, heart-related illnesses, etc. Therefore, it is essential to recognize stress in its earliest phase, to keep away from the other issues that may arise from physiological stress.

Deep Learning is an sustainable technique, suitable for any kind of classification or regression problem [8]. A neural network with greater than two hidden layers can be stated as a deep neural network and it's an emerging topic within the era of machine learning. As per as the literature survey for the stress recognition is referred, there are numerous machine learning strategies were used for

N. Nain et al. (Eds.): CVIP 2019, CCIS 1148, pp. 180–189, 2020.
https://doi.org/10.1007/978-981-15-4018-9_17

the recognition of stress. Many barriers of the classical classification strategies are been overcame in deep learning, such as Vanishing gradient, over-fitting and Computational Load [11]. This is why deep learning is gaining recognition in recent days.

While surveying the literature, several studies were found regarding stress recognition [1–3, 5, 9, 10, 12, 13]. Wang and Lin [1] have used k-nearest neighbor (KNN) classifier for the stress recognition. They have used the physionet driver stress detection database for their study. They have applied distinct feature techniques using Principal Component Analysis (PCA) and Linear Discriminant Analysis (LDA), based totally upon which they've carried out some experiments with special function selection criteria.

Liew and Seera [3] have used their own database for stress recognition, which includes Heart Rate Variability (HRV) and Salivary samples of the subjects. They have additionally accomplished Trier Social stress test (TSST) on the subjects and defined the stress levels based on content of enzymes, Cortisol and Alpha-amylase inside the saliva samples. For the stress recognition they have used Fuzzy ARTMAP neural network.

Ample amount of literature is available for the different machine learning techniques. The intelligent deep learning techniques were used by Acharya and Fujita [6], Tsinalis and Paul [21], Li and Dan [13] and by Romaszko and Lukasz [20] for their respective applications on signal or image processing. Dudhane [14] have also used feed forward neural network for Interstitial Lung disease classification. While machine learning techniques was used by Keshan and Parimi [16], Zhang and Wen [15] and Boonnithi and Sansanee [17] for the detection of stress. Yin and Zhang [4] have worked on stress detection. Their focus was to detect the workload levels by analyzing the EEC signals of the subjects using Deep Learning approach. Whereas Hambarde and Talbar [7] have worked on prostate cancer detection using IR Imaginary with deep learning techniques.

The problem statement was to identify the psychological stress, with a better and powerful classification model and that too directly from the ECG signals. This article contributes to advancing standard classification methods by using smart deep learning methods for stress recognition. Here we used a twenty-three-layered Convolutional Neural Network (CNN) to recognize psychological stress using the scalogram method, which is a time-frequency representation of the ECG signals. In the following sections, the dataset, methodology, network architecture and pre-processing are explained.

2   Dataset

We have created our own dataset for this work. The dataset consists of the ECG recordings of students of Department of Instrumentation Engineering from SGGSIE&T, Nanded (M.S), India. The ECG's of about 130 students were taken during oral examination and during normal curriculum. During the oral examination ECG were taken for multiple times, such as before the oral exam and after the oral exam. There are of about 240 three-lead 2-min short-term ECG recordings of students taken during resting and stressed condition using the body surface electrodes in the entire dataset. The ECG's were recorded with the help of BIOPAC-MP150 ECG Acquisition System from the departmental laboratory.

3   Methodology

3.1   Pre-processing

The database was taken with the help of body surface electrodes, during which some noise has occurred because of the few obvious reasons. Minimizing this noise was the difficult task of making the database much less noisy. To eliminate this noise a bandstop IIR filter has been designed to take away the notch occurred at 50 Hz and 150 Hz i.e. at the odd harmonics. After elimination of this notch, the ECG turns into noise free and is further processable, Fig. 1 shows the pre-processed ECG signals. For noise removal we find lot of literature on filtering techniques for ECG signals, but the filter we designed did the desired job of filtering. For this, the literature concerning about these studies was referred.

After de-noising the dataset, it is needed to categorise it as stressed and non-stressed subjects. For this, we have chosen the traditional way, i.e feature extraction followed by classification. Among available feature extraction techniques we simulated the time domain features mentioned by Task force [18]. The statistical features of the dataset have been analysed, which includes time domain feature of the HRV signals which were extracted from the respective ECG signals, including QRS detection using Pan-Tompkins algorithm [19]. The subject were categorised by analysing their experimental condition and according to the feature limits set by the Task force of the European Society of Cardiology [18]. After the dataset got labelled i.e categorised, it was applied to CNN for the classification.

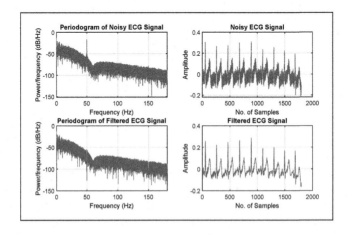

Fig. 1. Pre-processing

3.2   Scalogram Technique

To use CNN for this unique classification problem, the scalogram technique was used. The scalogram is basically a visible time-frequency illustration of a signal using wavelet transform. This is analogous to spectrogram concept which represents the spectrum of a signal. Scalogram is also known as the absolute value of the continuous wavelet transform (CWT) coefficients of a signal. The dataset, after de-noising and labelling, have been undergone through scalogram technique. Sacalograms converted the entire signal dataset into its equivalent image dataset, Fig. 2 shows the equivalant scalogram image of an ECG signal. To create the scalograms, the CWT filter banks of a signal are computed, as it is the suitable method of creating the scalograms. The filter bank is used to obtain the CWT of the ECG signals and generates the scalogram image database from the respective wavelet coefficients.

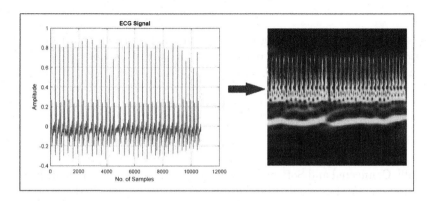

Fig. 2. ECG signal and its scalogram image

3.3   Network Architecture

Twenty three layered Deep Neural Network is used for the recognition of stress. The network consisted of several layers, listed as follows. There were four stages of the network, 1. Convolution 2. Pooling 3. Activation and 4. Classification. There are 4 convolution layers along with 3 other layer such as Batch-Normalization layer, Max-Pooling layer and the RELU Layer. This group of 4 layers is repeated for 4 instances. The detailed network architecture is shown in Fig. 3.

As we know, Convolution Layer does the convolution operation on the inputs and the filters, to obtain feature maps. Here we have used four convolution layers of different filter sizes and feature maps as per given in Fig. 3.

Max-Pooling Layer pulls out the feature data, whereas REctified Linear Unit (RELU) Layer is the activation function used in CNN to activate the neurons.

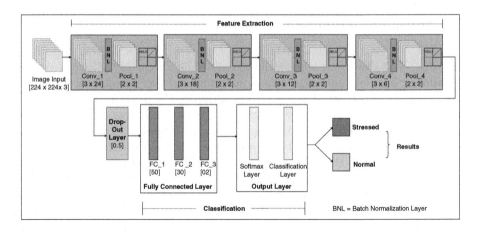

Fig. 3. CNN architecture

We have used these layers along with batch normalization layer with each convolution layer together form a one convolution unit. The Eq. 1 represents the activation in CNN i.e the RELU layer.

$$\varphi(x) = \begin{cases} x & x > 0 \\ 0 & x \le 0 \end{cases}$$
$$= max(0, x) \tag{1}$$

Fully Connected and Soft-max Layer are the classification stages of the CNN. The drop-out layer is applied before this, to cut down unwanted features from the feature maps. In softmax layer, the softmax function is applied to an output, which is used for the multi-class classification tasks using the concepts of logistic regression, while the classification layer classifies the output data.

3.4   Experimental Setup

All the experiments were done on computer with Intel core i7-6700 CPU @ 3.4 GHz processor with 8 GB RAM & 64-bit operating system. Software used was MATLAB, version 2018a. The dataset was split in 70 : 30 ratio for training and testing of the model. The CNN used, was tuned with back propagation algorithm with Stochastic Gradient descent with momentum with customized layer architecture of the CNN. The network was trained with 25 batch size and 0.001 learning rate.

The similar kind of work is done by Acharya and Fujita [6] and by Tsinalis and Paul [21]. Their work was focused on designing a CNN for the signal processing problem. Acharya and Fujita [6] designed a CNN for detection of myocardial infarction using ECG signals, while Tsinalis and Paul [21] designed a CNN to detect sleep stage using EEG signals.

The training parameters for the network were kept simple. At first the HR analysis of the dataset has been done and its has been categorised using classical classifiers. For HRV analysis, eight time-domain features of HRV signal were extracted. The classical classifiers like of SVM, KNN, LDA and Decision Tree with fivefold cross validation were applied to this feature-set. After this, the designed CNN was trained with equivalent image dataset. Our dataset has also been applied to the pre-trained CNN models Alex-Net and Google-Net. The features maps extracted by the CNN model were taken out and also been applied to the classical classifiers SVM, KNN and Decision Tree to obtain comparative results.

4   Experimental Results

Equivalent image dataset was obtained using the scalogram technique for the training on CNN. The results of the proposed work are divided in two sections, results of classical classifiers on the HRV analysis of the dataset and 2nd is results on CNN. Among classical classifiers SVM & KNN gave accuracy of 73.30% and 71.40% respectively for the HRV analysis in Table 1.

Table 1. Results of classifiers applied to HRV analysis

Classifier	Sens.	Spec.	Acc.	PPV	NPV
SVM	76.19%	69.04%	73.3%	78.68%	65.90%
Decision Tree	82.53%	52.38%	70.50%	72.22%	66.66%
KNN	68.25%	76.19%	71.40%	81.13%	61.53%
LDA	77.77%	50%	65.7%	70%	60%

After HRV analysis, the scalograms were applied to the designed CNN. The proposed network has given accuracy of 97.22% on the filtered dataset, with Positive Predictive Value (PPV) of 100% and Negative Predictive Value (NPV) of 93.50%. While the same network has given accuracy of 93.10% accuracy for the noisy dataset, (SP) with PPV of 100% and NPV of 83.90%. The confusion matrix of the network is in Fig. 4. After the network gets trained, the features extracted by CNN, i.e. the outputs of last convolution layer were taken as a feature set and been applied to the classical classifiers. The size of feature set was $(168 \times 488, 358)$ and $(72 \times 488, 358)$ for training and testing, respectively. Due to the large size of feature map we couldn't apply it to the LDA classifier. The results of these classifiers are compared with a proposed CNN model in Table 2. The pre-trained CNN models, Alex-net and Google-Net has been applied to our dataset. Their comparative results with proposed CNN model are tabulated in Table 3. It was observed that out of entire dataset, 56% subjects were stressed.

Table 2. Results of classifiers on features extracted by CNN

Dataset	Classifier	Sens.	Spec.	Acc.	PPV	NPV
Filtered	SVM	89.10%	100%	93.10%	100%	83.90%
	Decision Tree	95.10%	93.50%	94.44%	95.10%	93.50%
	KNN	94.90%	87.90%	91.70%	90.20%	93.50%
	Proposed Network	100%	95.30%	**97.22%**	100%	93.50%
Noisy	SVM	90.20%	87.10%	88.90%	90.20%	87.10%
	Decision Tree	81.40%	79.30%	80.60%	85.40%	74.20%
	KNN	89.70%	81.80%	86.10%	85.40%	87.10%
	Proposed Network	89.10%	100%	**93.10%**	100%	83.90%

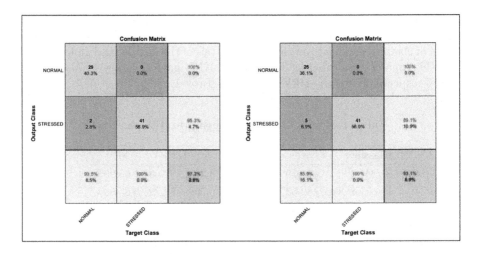

Fig. 4. Confusion matrix for filtered and noisy dataset

Table 3. Comparison of proposed network with pre-trained CNN models

Dataset	Network	Accuracy
Filtered	**Proposed Network**	**97.22%**
	Google Net	84.72%
	Alex Net	83.33%
Noisy	**Proposed Network**	**93.10%**
	Google Net	83.33%
	Alex Net	77.78%

5   Discussion

Table 4 focuses the various techniques and the work that has been set up for the stress recognition and their comparison with our work. The majority of the researchers have executed the studies using machine learning techniques [1–3,5,9,10,12,13]. Out of these, a few have designed the algorithms for the standard databases while some has designed for their own databases.

Table 4. Comparison of the studies for stress recognition

Sr. No.	Authors Year	Features	Dataset used	Classifier used	Performance
1	Wang and Lin (2012) [1]	Trend and parameter based feature generation	Physionet dataset	k-nearest neighbour	Highest Accuracy of 97.78%
2	Liew and Seera (2016) [3]	HRV features, Salivary samples and Trier Social Stress test	Their own dataset	Fuzzy ARTMAP (FAM)	Highest Accuracy for classification 80.75%
3	Sets and Arnrich (2010) [2]	Statistical features of EDA signals	Their own dataset while at work, taken by EDA sensor	SVM & LDA	Accuracy using LDA - 82.8% using SVM - 81.3%
4	Melillo and Bracale (2011) [5]	Time and frequency domain analysis of HRV with different sets of features	Their own dataset, while university examination	LDA classifier	Acc. 90%, with Sens 86% and Spec 95%
5	Tanav and Saadi (2014) [10]	Time and frequency domain analysis of HRV	Their own dataset	Naive Bayes Classifier	Highest Accuracy of 90%
6	Zhai and Barreta (2006) [12]	Statistical Feature extraction	Their own dataset	SVM classifier	Highest Accuracy of 90.1%
7	**Proposed Method**	**Convolutional Neural Network**	Our own dataset, while oral examination	**Deep Learning, CNN**	**The noisy dataset Acc - 93.10% filtered dataset Acc - 97.22% And** better results on machine learning techniques using features extracted by CNN

The overall performance of the proposed network architecture was quite excellent in both feature extraction and classification aspects. Tables 1 and 2 shows the results in figures. Following are a few highlights of our studies.

- There was no separate technique for Feature Extraction in addition to Classification, the proposed network did both the tasks.
- This study turned into an effective implementation of scalogram strategies for the stress recognition using ECG Signals.
- Even classical classifiers gave nice results on features extracted by CNN, this is the beauty of CNN in the field of feature extraction.
- The proposed network gave better results, than that of the pre-trained CNN models and the classical classification strategies.
- The database we have taken, contains clinical records of people from the Indian Sub-Continent, the physiology of these people is different from others.

Despite of all these strong points enlisted above, we feel some areas where, there is a scope for improvement in our study.

- With use of some advanced software and hardware combination this proposed method can be adopted for the real time stress recognition.
- With some more detailed study, three or four classes of stress could be identified.

We achieved the results using CPU system, so using a GPU system the computation time for the training process could be reduced.

6   Conclusion

This paper provides a look at on psychological stress detection using ECG signals. In the era of hectic work-lifestyles, there are bunch of problems, that can be caused because of the psychological stress, so its detection in its earlier stage is very crucial. The proposed approach is an effective implementation of CNN for signal processing by way of the use of scalogram method. For experimental cause psychological stress in students have been detected for the duration of their oral exam.

The classical classifiers gave good results when they applied to the features extracted by CNN than that of the results on HRV analysis. This is a completely sturdy factor about CNN and make us conclude that CNN is a good classifier and a feature extractor than the classical techniques for the identical.

References

1. Wang, J.-S., Lin, C.-W., Yang, Y.-T.C.: A k-nearest-neighbor classifier with heart rate variability feature-based transformation algorithm for driving stress recognition. Neurocomputing **116**, 136–143 (2013)
2. Setz, C., et al.: Discriminating stress from cognitive load using a wearable EDA device. IEEE Trans. Inf. Technol. Biomed. **14:2**, 410–417 (2010)
3. Liew, W.S., et al.: Classifying stress from heart rate variability using salivary biomarkers as reference. IEEE Trans. Neural Netw. Learn. Syst. **27.10**, 2035–2046 (2016)
4. Yin, Z., Zhang, J.: Cross-session classification of mental workload levels using EEG and an adaptive deep learning model. Biomed. Signal Process. Control **33**, 30–47 (2017)
5. Melillo, P., Bracale, M., Pecchia, L.: Nonlinear Heart Rate Variability features for real-life stress detection. Case study: students under stress due to university examination. Biomed. Eng. Online **10.1**, 96 (2011)
6. Acharya, U.R., et al.: Application of deep convolutional neural network for automated detection of myocardial infarction using ECG signals. Inf. Sci. **415**, 190–198 (2017)
7. Hambarde, P., Talbar, S.N., Sable, N., Mahajan, A., Chavan, S.S., Thakur, M.: Radiomics for peripheral zone and intra-prostatic urethra segmentation in MR imaging. Biomed. Signal Process. Control **51**, 19–29 (2019)

8. Zhang, J., Zong, C.: Deep neural networks in machine translation: an overview. IEEE Intell. Syst. **30**(5), 16–25 (2015)

9. Hjortskov, N., et al.: The effect of mental stress on heart rate variability and blood pressure during computer work. Eur. J. Appl. Physiol. **92(1–2)**, 84–89 (2004). https://doi.org/10.1007/s00421-004-1055-z

10. Tanev, G., et al.: Classification of acute stress using linear and non-linear heart rate variability analysis derived from sternal ECG. In: 2014 36th Annual International Conference of the IEEE, Engineering in Medicine and Biology Society (EMBC). IEEE (2014)

11. Kim, P.: MATLAB Deep Learning: With Machine Learning, Neural Networks and Artificial Intelligence (2017)

12. Zhai, J., Barreto, A.: Stress detection in computer users based on digital signal processing of noninvasive physiological variables. In: 28th Annual International Conference of the IEEE, Engineering in Medicine and Biology Society, 2006. EMBS 2006. IEEE (2006)

13. Li, D., et al.: Classification of ECG signals based on 1D convolution neural network. In: 2017 IEEE 19th International Conference on e-Health Networking, Applications and Services (Healthcom). IEEE (2017)

14. Dudhane, A., et al.: Interstitial lung disease classification using feed forward neural networks. In: International Conference on Communication and Signal Processing 2016 (ICCASP 2016). Atlantis Press (2016)

15. Zhang, J., et al.: Recognition of real-scene stress in examination with heart rate features. In: 2017 9th International Conference on Intelligent Human-Machine Systems and Cybernetics (IHMSC), vol. 1. IEEE (2017)

16. Keshan, N., Parimi, P.V., Bichindaritz, I.: Machine learning for stress detection from ECG signals in automobile drivers. In: 2015 IEEE International Conference on Big Data (Big Data). IEEE (2015)

17. Boonnithi, S., Phongsuphap, S.: Comparison of heart rate variability measures for mental stress detection. In: Computing in Cardiology, 2011. IEEE (2011)

18. Heart rate variability, standards of measurement, physiological interpretation, and clinical use. Task Force of the European Society of Cardiology. Circulation **93**, 1043–1065 (1996)

19. Pan, J., Tompkins, W.J.: A real-time QRS detection algorithm. IEEE Trans. Biomed. Eng. **3**, 230–236 (1985)

20. Romaszko, L.: Signal correlation prediction using convolutional neural networks. In: Neural Connectomics Workshop (2015)

21. Tsinalis, O., et al.: Automatic sleep stage scoring with single-channel EEG using convolutional neural networks. arXiv preprint arXiv:1610.01683 (2016)

Video Colorization Using CNNs and Keyframes Extraction: An Application in Saving Bandwidth

Ankur Singh[(✉)], Anurag Chanani, and Harish Karnick

Indian Institute of Technology Kanpur, Kanpur, India
{ankuriit,achanani,hk}@iitk.ac.in

Abstract. A raw colored video takes up around three times more memory size than it's grayscale version. We can exploit this fact and send the grayscale version of a colored video along with a colorization model instead of the colored video to save bandwidth usage while transmission. In this paper, we tackle the problem of colorization of grayscale videos to reduce bandwidth usage. For this task, we use some colored keyframes as reference images from the colored version of the grayscale video. We propose a model that extracts keyframes from a colored video and trains a Convolutional Network from scratch on these colored frames. Through the extracted keyframes we get a good knowledge of the colors that have been used in the video which helps us in colorizing the grayscale version of the video efficiently.

Keywords: Image colorization · Convolution/deconvolution · Bandwidth · Mean shift clustering · Histograms · Keyframes

1 Introduction

Learning based colorization algorithms for grayscale videos and images have been the subject of extensive research in the areas of computer vision and machine learning. Apart from being alluring from an artificial intelligence point of view, such potential has vast practical implementations starting from video restoration to image improvement for enhanced understanding. Colorizing a grayscale image can be hugely beneficial, since grayscale images contain very less information thus adding color can add a lot of information about the semantics.

Another motivation for video colorization that we propose, is it's capacity to save data while transmitting a video. A raw colored video takes upto three times more memory than it's grayscale version. Hence sending a grayscale video instead of a colored one while streaming and then colorizing it on the receiver's end can help save data and in turn the bandwidth. In this paper, we propose a convolutional neural network model that is trained on the keyframes of a raw colored video. This model is transmitted along with the grayscale version of the colored video and on the receiver's end this model colorizes the grayscale video.

© Springer Nature Singapore Pte Ltd. 2020
N. Nain et al. (Eds.): CVIP 2019, CCIS 1148, pp. 190–198, 2020.
https://doi.org/10.1007/978-981-15-4018-9_18

Apart from our convolutional neural network model we also propose a keyframe extraction method that extracts keyframes from a video by comparing colored histograms of all the frames in that video.

Also, in general the colorization problem is mulimodal since in image and video colorization a given grayscale image can have varying colored outputs when tested with different colorization models. For eg. a grayscale image of a ball can have different colored outputs from different colorizing models. Some models may output a green colored ball while some may output a blue colored ball. This might differ from the actual color of the ball. Hence, in this paper we also tackle this problem by using few colored keyframes of the video to colorize the grayscale video. Having a sense of the colors that have been used in the video will help a great deal in predicting the actual colors of the rest of the frames of the video.

Hence our work serves two purposes:

– Saving bandwidth while transmitting a video by sending grayscale version of a raw colored video along with a CNN model trained on the keyframes of the

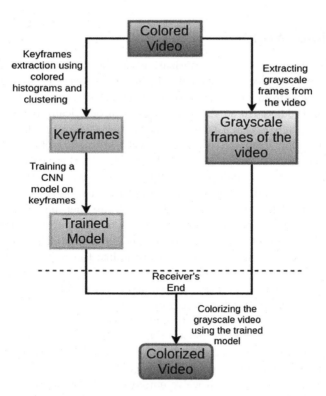

Fig. 1. Sending grayscale version of a raw colored video alongwith a convolutional neural network model trained on keyframes of the video and then colorizing the grayscale video on the receiver's end. (Color figure online)

video and then colorizing the grayscale video on the receiver's end as shown in Fig. 1.

– Tackling the problem of multimodality in image and video colorization (shown in Fig. 2) by colorizing a video using few keyframes of the video.

Fig. 2. A grayscale image can have multiple colored output image from Alexei Afros homepage

2 Previous Work

The start of automated image colorization can be dated back to 2002, when Welsh et al. [8] presented an approach which could colorize a grayscale image by transferring colors from a related reference image. Our work on video colorization is inspired by Baldassarre et al. [1] system on automatically colorizing images. Apart from the convolutional neural network that we have used, they have also employed Inception Resnet [7] as a high level feature extractor which provides information about the image contents that helps in their colorization. Their network consists of four main components: an encoder, a feature extractor, a fusion layer and a decoder. The encoding and the feature extraction parts obtain mid and high-level features, respectively, which are then merged in the fusion layer. Finally, the decoder uses these features to estimate the output. Iizuka et al. [3] and Larsson et al. [5] have developed similar models. Zhang et al. [9] use a classification loss in their architecture unlike the regression loss that we have used.

The work on Keyframes extraction is inspired from Zhuang et al. [10] work on color histograms. In a color histogram, a 1D array contains the total pixels that belong to a particular color in the image. All the images are resized to the same shape before their histograms are taken so that they have equal number of pixels. To discretize the space, images are represented in RGB colorspace using some important bits for every color component.

The main purpose for which we've employed color histograms in keyframes extraction is that they are very easy to compute and show striking properties despite their simplicity. They are often used for content based image retrieval. They are also highly invariant to the translation and rotation of objects in the image since they do not relate spatial information with the pixels of the colors.

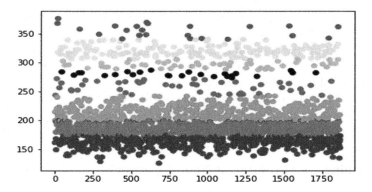

Fig. 3. Result of Meanshift algorithm applied on the frames of the video, Clusters have been represented in different colors. X axis represents indices of the frames. Y axis denotes the Hellinger distance of a frame from the sample image. (Color figure online)

3   Proposed Method

We introduce a two step process for our approach of colorizing grayscale videos using keyframes extraction.

The first step deals with the extraction of keyframes of the video. The second step involves training a Convolutional Neural Network on these keyframes and colorizing the rest of the video using the trained model.

3.1   Keyframes Extraction

We extract keyframes of a video by comparing colored histograms of all frames with a sample image. In our experiments we have taken the sample image to be a black image(all pixels equal to zero).

We extract a 3D RGB color histogram with 8 bins per channel for all the frames. This yields a 512-dimensional feature vector for a frame once flattened. For comparing two histograms we use the Hellinger distance which is used to measure the "overlap" between the two histograms.

Formally, let H be the 512 dimensional colored histogram of our sample image. Let h_i be the 512 dimensional colored histogram of the ith frame. We calculate the Hellinger distance $d(H, h_i)$ between H and h_i by:

$$d(H, h_i)\sqrt{1 - \frac{1}{\sqrt{\bar{H}\,\bar{h_i}N^2}}\sum_{j=0}^{511}\sqrt{H[j]h_i[j]}}$$

N = total number of bins of the histogram, $\bar{x} = \frac{1}{N}\sum_j x[j]$

Additionally, we multiply the hellinger distance by a factor of 10,000 to ease out calculations that follow this step.

Once, we have the distances for all the frames against our sample image we use mean shift clustering [2] to cluster frames whose distances from the random image are close to each other. The mean shift algorithm is a non parametric clustering technique that does not need initial information about the number of clusters. This property is essential in our problem since we don't have any prior knowledge about the number of clusters present in a particular video. Result of clustering on a 1 min video is shown in Fig. 3.

After we have the clusters we can choose every x^{th} frame from the cluster depending upon the number of frames we want. We have found emperically that $\times$ equal to 30 does a good job.

3.2   Training a Convolutional Neural Network

For the training part, we consider images in the CIELab color space. Here L stands for lightness, a stands for the green red color spectra and b stands for the blue yellow color spectra. A CIELab encoded image has one layer for grayscale, and it packs three color layers into two. This means that the original grayscale image can be used in our final prediction. Also, we only have two channels to predict. Starting from the L component X_L, the purpose of our model is to estimate the remaining two components X_a and X_b.

Preprocessing
The pixel values of all three image components namely L, a and b are centered and scaled to get values within the $[-1, 1]$ range. All images are converted from RGB color space to CIELab color space to feed them into our model.

Architecture
The architecture of our model is inspired from [1]. Given the L component of an image, our model estimates it's a and b components and combines them with the L component to get the final colored image. We have used 12 convolutional layers with 3×3 kernels and 3 upsampling layers as shown in Fig. 4 and Table 1. In the second, fourth and the sixth convolutional layer, a stride of two is applied which halves the dimension of their output, resulting in less number of computations [6]. We have made use of padding to preserve the layer's input dimension. Upsampling has been performed so that the height and width of the output are twice that of the input. This model applies a number of convolutional and upsampling layers in order to output a final image with dimensions H x W x 2. The 2 output channels are a and b. These are merged with the L component to get the colored image.

Training
We obtain the optimal parameters of the model by minimizing a function which is defined over the predicted output of our network and the target output. In order to quantify the model loss, we employ the mean squared loss between the estimated pixel colors in a, b space and their real value. While training, we back propagate this loss to update the model parameters using Adam Optimizer [4] with a learning rate of 0.001. During training, we impose a fixed input image size to allow for batch processing.

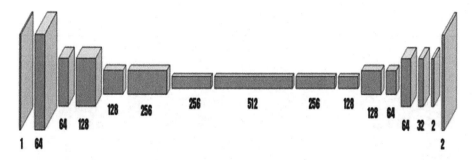

Fig. 4. Architecture of the network: 12 convolutional and 3 upsampling layers have been used. In the 2nd 4th and the 6th convolutional layer a stride of 2 has been applied. A final image with dimensions H × W × 2 is obtained. The 2 output channels are merged with the L component to get the final colored image. (Color figure online)

Table 1. Architecture of the network

Layer	Kernels	Stride
Convolution	(64, 3, 3)	(1, 1)
Convolution	(64, 3, 3)	(2, 2)
Convolution	(128, 3, 3)	(1, 1)
Convolution	(128, 3, 3)	(2, 2)
Convolution	(256, 3, 3)	(1, 1)
Convolution	(256, 3, 3)	(2, 2)
Convolution	(512, 3, 3)	(1, 1)
Convolution	(256, 3, 3)	(1, 1)
Convolution	(128, 3, 3)	(1, 1)
Upsampling	–	–
Convolution	(64, 3, 3)	(1, 1)
Upsampling	–	–
Convolution	(32, 3, 3)	(1, 1)
Convolution	(2, 3, 3)	(1, 1)
Upsampling	–	–

4   Experiments and Results

We tested our model on the popular video compressing benchmarking dataset: the Xiph HD[1] library of 24 1080 p videos. We used the popular SSIM (Structural Similarity) metric to measure the quality of the output video against the original video. The results for the quality of the reconstructed video have been mentioned in Table 3.

---

[1] https://media.xiph.org/video/derf/.

Table 2. Results

Grayscale	Ground truth	Zhang et al	Ours

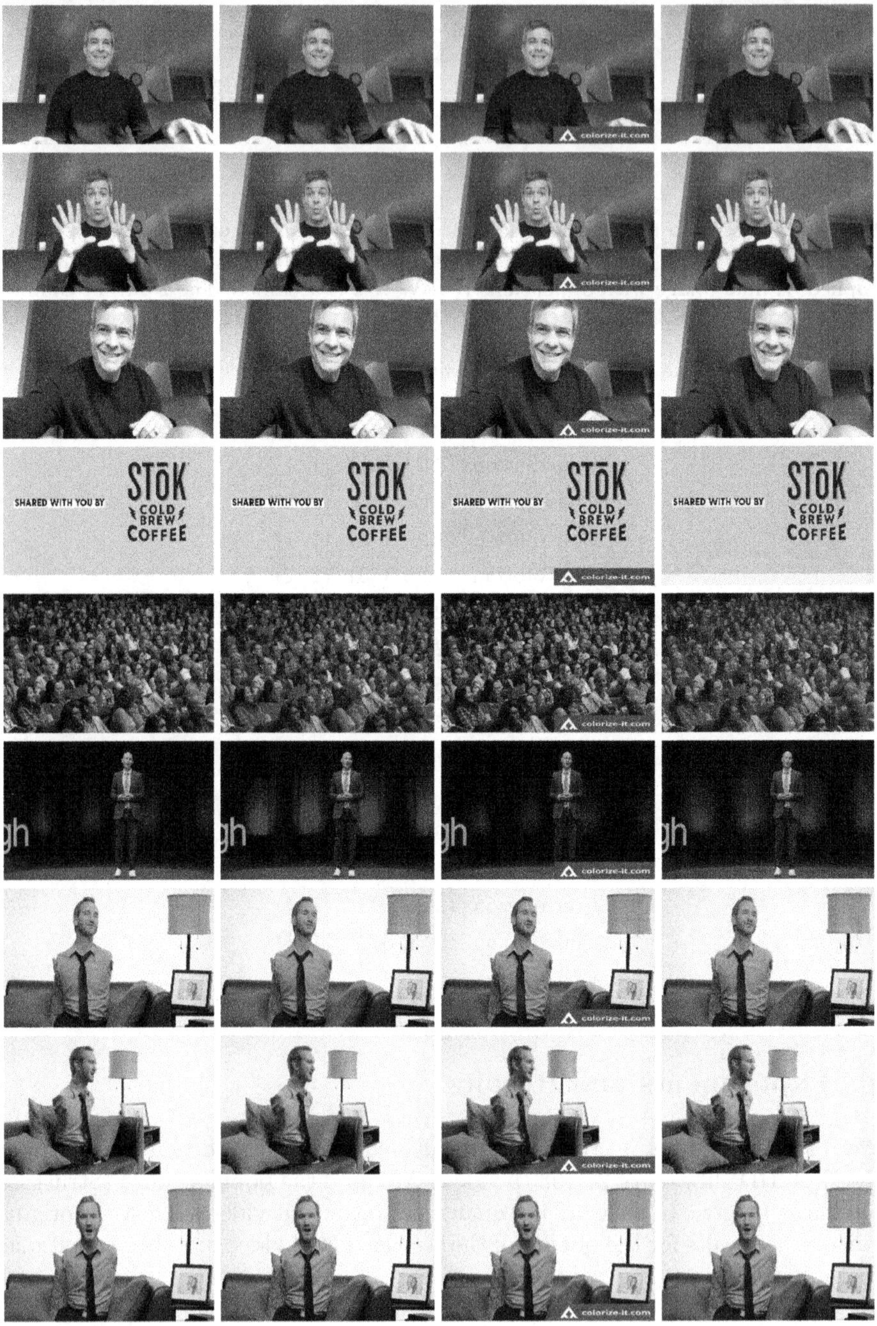

Table 3. Mean SSIM

Compression method	Mean SSIM (XiphHD dataset)
MPEG4	0.91
MPEG4-AVC	0.968
Ours	0.9695
HEVC HM	0.98

Table 4. Results of bandwidth saved in various videos through our approach

Input video frame size	Input video duration	Time taken to output colored video	Size of model	Bandwidth saved	Percentage bandwidth saved
256 × 256 (24 bit)	1 min	~4 min	30 MB	195 MB	57.78%
256 × 256 (24 bit)	15 min	~6 min	30 MB	3345 MB	66.07%
720 × 1280 (24 bit)	15 min	~6 h	45 MB	46.3 GB	66.6%

Since, our main aim was to reduce the model size so that we could save as much bandwidth as possible we kept our CNN model simple, without hampering the quality of the colored video that we output. The results turned out to be quite good for most of the videos. However, the videos in which there were drastic changes from one shot to another, our network was not able to produce that good results. We observed that although some results were quite good, some generated pictures tend to be low saturated, with the network producing a grayish color where the original was brighter.

For a 256 × 256 24 bit 15 min uncompressed colored video that has a size of around 5 GB, we could save a bandwidth of around 3.30 GB as our trained model had a size of only 30 MB. Also, it took us only around 6 min for the whole process starting from keyframes extraction to training a model and finally obtaining the colored output video for a 256 × 256 24 bit 15 min video on NVIDIA GeForce GTX 1080.

Table 2 shows some of the reconstructed images that are obtained through our model. The varying colored outputs of similar grayscale images has been clearly shown in the output of Zhang et al. model. The ground truth of the last and the second last images have the same colors. However, Zhang et al. outputs an image with a pinkish shade in one case and an image with a reddish tint in the other. We easily handle this anomaly since we already have a knowledge of the colors that have been used in the video, that we extract through the keyframes.

In Table 4 we have shown the bandwidth that is saved through our approach. The small size of the trained model helps in accomplishing our task to a great extent.

5 Conclusion

In this paper, we devised a new approach to save bandwidth to upto three times while transferring colored videos without losing data or hampering much the quality of the video. Usual compression algorithms are lossy, hence lose data while compressing videos. Lossy compressions are irreversible that use inaccurate estimations and discard some data to present the content. They are performed to decrease the size of data for storing, handling, and transmitting content. However the approach that we propose isn't irreversible in the sense that the quality of the output video is not hampered significantly. We also tackled the problem of varying colored outputs of a single grayscale frame of a video when tested with different colorization models, by using some colored keyframes of the video as reference images. Having a knowledge of the colors that have been used in a video help us in colorizing the rest of the frames of the video.

Our future work will focus on reducing the time taken to output the colored video through our model without trading off with the quality. We will also work on colorizing videos where drastic changes occur from one shot to another, in a better way.

References

1. Baldassarre, F., Morín, D.G., Rodés-Guirao, L.: Deep Koalarization: Image Colorization using CNNs and Inception-ResNet-v2. arXiv preprint arXiv:1712.03400 (2017)
2. Cheng, Y.: Mean shift, mode seeking, and clustering. IEEE Trans. Pattern Anal. Mach. Intell. **17**(8), 790–799 (1995)
3. Iizuka, S., Simo-Serra, E., Ishikawa, H.: Let there be color!: joint end-to-end learning of global and local image priors for automatic image colorization with simultaneous classification. ACM Trans. Graph. (TOG) **35**(4), 110 (2016)
4. Kingma, D.P., Ba, J.: Adam: a method for stochastic optimization. arXiv preprint arXiv:1412.6980 (2014)
5. Larsson, G., Maire, M., Shakhnarovich, G.: Learning representations for automatic colorization. In: Leibe, B., Matas, J., Sebe, N., Welling, M. (eds.) ECCV 2016. LNCS, vol. 9908, pp. 577–593. Springer, Cham (2016). https://doi.org/10.1007/978-3-319-46493-0_35
6. Springenberg, J.T., Dosovitskiy, A., Brox, T., Riedmiller, M.A.: Striving for simplicity: the all convolutional net. CoRR, abs/1412.6806 (2014)
7. Szegedy, C., Ioffe, S., Vanhoucke, V., Alemi, A.A.: Inception-v4, Inception-ResNet and the impact of residual connections on learning. In: AAAI, vol. 4, p. 12 (2017)
8. Welsh, T., Ashikhmin, M., Mueller, K.: Transferring color to greyscale images. In: ACM Transactions on Graphics (TOG), vol. 21, pp. 277–280. ACM (2002)
9. Zhang, R., Isola, P., Efros, A.A.: Colorful image colorization. In: Leibe, B., Matas, J., Sebe, N., Welling, M. (eds.) ECCV 2016. LNCS, vol. 9907, pp. 649–666. Springer, Cham (2016). https://doi.org/10.1007/978-3-319-46487-9_40
10. Zhuang, Y., Rui, Y., Huang, T.S., Mehrotra, S.: Adaptive key frame extraction using unsupervised clustering. In: Proceedings of 1998 International Conference on Image Processing, 1998. ICIP 1998, vol. 1, pp. 866–870. IEEE (1998)

Image Compression for Constrained Aerial Platforms: A Unified Framework of Laplacian and cGAN

A. G. J. Faheema$^{(\boxtimes)}$, A. Lakshmi, and Sreedevi Priyanka

Centre for AI and Robotics, DRDO Complex, Bangalore 560093, India
{faheema,lakshmi,sreedevip}@cair.drdo.in

Abstract. In this paper, we propose a new lossy image compression technique suitable for computationally challenged platforms. Extensive development in moving platforms create need for encoding images in real time with less computational resources. Conventional compression algorithms have potential to address this problem. However, the reconstruction accuracy of conventional encoders does not match that of deep learning based compression algorithms. In this paper, we have utilized best of both worlds by proposing a new compression method which combines conventional and deep learning based methods to sustain real time transmission and as well good reconstruction quality. We have validated our algorithm across a varied set of test images from EPFL mini drone dataset and Stanford drone dataset. The proposed algorithm exhibits better rate-distortion performance than conventional method. More importantly, our algorithm gives real time performance which has been substantiated by displaying a dramatic improvement in speed as against state-of-the-art deep learning compression method.

Keywords: Compression · Laplacian image · Conditional Generative Adversarial Network

1 Introduction

With continuing growth of multimedia technology, image compression has become a mandate technology of computer vision. Image compression effectively decreases the transmission bandwidth requirement, thus enabling streaming images in bandwidth constrained military scenarios. Conventional non-deep learning based methods are one-transform-fits-all: hard coded and cannot be customized based on the statistics of the images of particular scenario. In the recent past, Deep Learning (DL) has alleviated the above mentioned setback of conventional methods especially for lower bitrates. However DL methods come with the following bottlenecks, (i) run time inefficiency, (ii) high memory footprint and (iii) computation complexity, thus rendering them inefficient to be deployed in power and memory constrained situations.

© Springer Nature Singapore Pte Ltd. 2020
N. Nain et al. (Eds.): CVIP 2019, CCIS 1148, pp. 199–210, 2020.
https://doi.org/10.1007/978-981-15-4018-9_19

This paper introduces a novel framework which fuses the conventional and deep learning methods, thus enabling us to leverage the strengths of DL, while allowing real time realization of image streaming on resource-constrained platforms such as UAVs and drones. Deep learning has poised itself as state-of-the-art approach in different learning domains including image compression. DL based image compression algorithms consists of an encoder that transforms the input space x to a latent space representation $z = f(x)$ and a decoder that reconstructs the input space x' from the latent space representation z. Implementation of the image encoding techniques using deep learning becomes infeasible on aerial platforms due to the demanding computational overhead of DL methods. Hence, in this paper, we have leveraged conventional method to encode images from its higher dimensional space to lower dimensional space. This provides multitude of advantages in terms of runtime, memory and power. In the base station, where we can afford more computation, the images are reconstructed from its lower dimensional representation using conditional Generative Adversarial Network (cGAN). As per our knowledge, this is the first compression algorithm to combine the efficacy of conventional and DL methods for image compression. Our algorithms outperforms traditional methods in terms of reconstruction efficiency and surpasses DL methods [3,4] in terms of run time. We supplement the performance of our method with rate distortion curves (bits per pixel vs. commonly used metrics PSNR-HVS [1] and MS-SSIM [2]). We have also evaluated the performance of state-of-the-art object detection method on the images reconstructed from the proposed method.

The paper is organized as follows: Sect. 2 furnishes the state-of-the-art methods available for image compression in conventional domain and DL domain, whereas Sect. 3 gives a glimpse of GAN and cGAN. It is followed by Sect. 4, which gives the details of the proposed architecture, with in sights and justifications for each and every module. The algorithm is validated in Sect. 5 with experiments conducted to prove its supremacy over conventional and DL methods.

2   Related Work

This section gives an overview of the state-of-the-art technologies of image compression using conventional methods and deep learning based methods.

2.1   Traditional Compression Methods

Compression techniques discover the redundancy in the image and finds a more succinct representation. Traditional compression techniques are comprised of transformation, quantization and coding. Transformation is a pre-engineered compact representation with a fixed model across the images. These transformations are bijective. Hence, it restricts the ability to reduce redundancy before coding. The coding is engineered to suit the transformation process. In JPEG, run length encoding is employed to exploit the sparse nature of DCT coefficients, whereas JP2 employs an adaptive arithmetic coder to suit the multi-resolution wavelet transform. These traditional compression techniques has more room for improvement. This representation is not tied to the underlying visual information present in the images. The encode and decode cannot be optimized for a particular loss function.

2.2   DL Based Compression Methods

DL based image compression has emerged as an active area of research. Auto Encoders, Recurrent Neural Network (RNN), Generative Adversarial Networks (GAN) and conditional GAN are the most popularly used networks for the task of image compression. The deep networks compress the image into bit stream, which is then encoded by lossless huffman coding or arithmetic coding.

AutoEncoder. In [5], authors have proposed an encoder named Compressive AutoEncoder (CAE), which has three components, (i) encoder, $f : \mathbf{R}^N \rightarrow \mathbf{R}^M$, (ii) decoder, $g : \mathbf{R}^M \rightarrow \mathbf{R}^N$ and (iii) a probabilistic model, $Q : \mathbf{Z}^M \rightarrow [0,1]$, which aids in assigning bits to the representations based on their frequencies. The parameters of the three components were estimated by optimizing the trade off between small number of bits, $-\log_2 Q\left([f(x)]\right)$ and small distortion, $d\left(x, g\left([f(x)]\right)\right)$. Authors of [6] have proposed an autoencoder which in addition to compressing the input space to the feature space, z in the bottleneck layer of autoencoder, also maps z to a sequence of m symbols using a symbol encoder $E : z \in \mathbf{R}^d \rightarrow [L]^m$. These encoded symbols are reconstructed to $\hat{z} = D\left(E(z)\right)$ by a symbol decoder $D : [L]^m \rightarrow z \in \mathbf{R}^d$. The network is trained to optimize the trade off between expected loss and the entropy of $E(z)$. Authors of [7] has come up with a content weighted image compression framework. It is made up of four modules, (i) an encoder which converts the input image x to $E(x)$, (ii) a binarizer $B\left(E(x)\right)$ which assigns 1 to the encoder output which is greater than 0.5 and 0 otherwise, (iii) an importance map network that takes intermediate feature as input and provides a content weighted importance map $P(x)$, which is used to generate a mask $M\left(P(x)\right)$ to trim the binary code, (iv) a decoder to reconstruct the image.

RNN. In [3], RNN based compression network is proposed for image compression, which consists of an encoding network, binarizer and a decoding network. Encoder and decoder consists of RNN components. The input images are encoded, converted to binary codes, decoded via decoder. This process is repeated with residual images, which is the difference between input image and reconstructed image. The binary codes of the original image and the residual image are repetitively reconstructed to estimate the original image.

GAN. GAN has two frameworks [8,9], (i) generator framework that tries to produce images from a probability distribution and also tries to maximize the probability of making the discriminator to mistake its input as real, (ii) discriminator framework acting as judge that decides the quality of the images generated by generator in order to guide the generator to produce more realistic images. Authors of [4] have proposed two modes of compression, global generative compression and selective generative compression. Global generative compression tries to generate the whole image from the latent space variable whereas selective generative compression reconstructs only part of the image while preserving user defined regions with high degree of detail. The proposed network is the combination of conditional GAN and learned compression. Image x is encoded using an encoder and quantizer into a compressed representation, $\widehat{w} = q\left(E\left(x\right)\right)$. The latent vector z is formed by concatenating $\widehat{w}$ with noise v drawn from a fixed prior p_v. The decoder cum generator tries to generate (as a generator does) an image $\widehat{x}$, which is consistent with the image distribution p_x while recovering the specific encoded image (as a decoder). The discriminator part in similar to conventional GAN. In [4], authors proposed a autoencoder pyramidal analysis combined with Adversarial training. The autoencoder based feature extractor captures the structures at various scales. The extracted features are quantized and further compressed through bitplane decomposition and adaptive arithmetic coding. Reconstruction tries to minimize reconstruction loss and discriminator loss. Reconstrcution loss aids in reducing the distortions between reconstructed and original image, whereas discriminator helps in producing visually pleasing reconstruction by penalizing the difference in the distribution of target and reconstructed image.

3   Generative Adversarial Network

Generative Adversarial Network is the recent development in deep learning to train generative model. It has two Adversarial modules: a generative model G which captures the distribution hidden in the training data and a Discriminative model D which helps the generator to learn the data distribution by estimating the probability of the sample being from the data or from the generator. The generator tries to take a random input z from a probability distribution $p\left(z\right)$ and tries to generate data $G\left(z;\theta_g\right)$, where θ_g is the parameter of the generator. This data is fed into the discriminator, which tries to identify it as original or fake. Hence discriminator learns $D\left(G;\theta_d\right)$, where θ_d is the parameter of discriminator. $D\left(x;\theta_d\right)$ outputs a single scalar that represents the probability that x came from data rather than from the generator. The feedback from discriminator helps generator network to learn the mapping of z to data space x, which is defined as $G\left(z;\theta_g\right)$. The goal of generator and discriminator is achieved with a min max loss function as given in Eq. 1, where generator and discriminator competes with each other.

$$\max_{D} \min_{G} E_{x \sim p_{data}(x)}\left[logD\left(x\right)\right] + E_{z \sim p_z(z)}\left[log\left(1 - D\left(G\left(z\right)\right)\right)\right] \qquad (1)$$

The first term is the entropy that the real data passes through the discriminator. The second term is the entropy that the fake data generated by generator passes through the discriminator. The discriminator is trying to maximize this function so that it will be able to distinguish between real and fake data, whereas generator does the opposite of it, i.e. it tries to minimize the function so that the differentiation between real and fake data becomes less pronounced.

3.1   Conditional GAN

Conditional GAN [10] is extension of Generative Adversarial nets to a conditional model by conditioning generator and discriminator on some auxiliary information, such as class labels, text or images. The conditioning is generally performed by feeding the auxiliary information as additional input layer. The additional input will constrain the generator to generate output from a limited set. The loss function of cGAN is similar to that of GAN except that it is conditioned on the auxiliary information y

$$\max_{D} \min_{G} E_{x \sim p_{data}(x)} \left[log D \left(\frac{x}{y} \right) \right] \tag{2}$$

$$+ E_{z \sim p_z(z)} \left[log \left(1 - D \left(G \left(\frac{z}{y} \right) \right) \right) \right] \tag{3}$$

4   Proposed Method

This section describes the proposed image compression method. Our method compresses an image I by following the formulation, where we have used conventional method based encoder E, a deep learning based decoder D and a traditional quantizer q. The proposed conventional method based encoder E maps the input image I to latent feature map I_L. I_L is quantized to L levels using quantizer. The output of the quantizer is $\widehat{I_L} = q(E(x))$, which is then converted to bit stream $B\left(\widehat{I_L}\right)$ using lossless coding for transmission. The deep learning based decoder tries to reconstruct the image $\widehat{I} = D\left(\widehat{I_L}\right)$. The detailed description of each module of the proposed framework (Fig. 1) is furnished in the following sections.

4.1   Encoder Framework: Conventional Method

For the encoder part, we propose a method inspired from Burt-Laplacian pyramid [11]. We have generalized Burt-Laplacian pyramidal decomposition to encode our image. For the sake of completeness, the generation of Burt-Laplacian pyramid is briefly described as follows. Let G_0 be the original image and G_1 be the Gaussian image obtained by applying Gaussian filter to the original image and further sampling down by a factor of two. The Laplacian image L_0 is the error image obtained by upsampling G_1 and subtracting it from the original image G_0. The similar process is continued on Gaussian images to obtain the

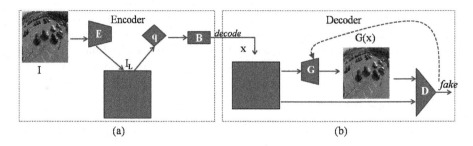

Fig. 1. Overall architecture of proposed framework. (a) Encoder, (b) Decoder. E is the encoder that generates Laplacian image. q is the quantizer and B is the bit stream generator. G is the generator that generates $G(x)$ from x sampled from the distribution of noise z. D is the decoder that sees either $G(x)$ and x or y (target image) and original image and identifies the probability of the image being from real image population

next levels of pyramid. Burt-Laplacian pyramid encoding is a well established image encoding technique as proposed in [11]. Burt-Laplacian encoding is a lossless encoding, which encodes Laplacian as well as Gaussian images. The novelty in this paper in the encoding framework is that our framework utilizes Laplacian image alone and discards the Gaussian image, which results in highly compressed image. The degree to which compression could be achieved when Laplacian image alone is used is far more than encoding Laplacian and Gaussian. This could be attributed to the following factors:

Entropy and Variance. It is a well known fact that in a statistically independent data, the bits per pixel required to encode the data depends on the entropy of the data. Much of the pixel to pixel correlation is removed in Laplacian image by subtracting predicted value of each pixel from its original value, whereas pixel to pixel correlation is increased in Gaussian image. This leads to crowding of the histogram of Laplacian image around zero as against that of Gaussian image (Fig. 2), thus leading to less variance and subsequently less entropy. The variance and entropy of four different images (from EPFL dataset) are given in the Table 1, which re-iterates the advantage of encoding Laplacian image alone.

Table 1. Left table depicts variance and right table depicts entropy of Guassian and Laplacian Images

Image	G_1	L_0
Image 1	6.223	1.8628
Image 2	6.2289	1.9001
Image 3	6.1423	1.7316
Image 4	6.0806	1.5213

Image	G_1	L_0
Image 1	7.0879	4.4120
Image 2	7.0011	3.9398
Image 3	7.1342	4.1430
Image 4	6.0783	3.3132

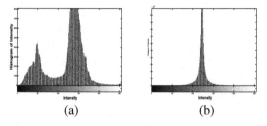

Fig. 2. Histogram of (a) Gaussian Image, (b) Laplacian Image

4.2   Quantization and Bit Stream Coding

The Laplacian images are quantized and encoded into bit stream similar to [11]. We have adapted quantization technique that involves k number of quantization levels. The laplacian image I_L is encoded into k quantization levels, $q(I_L)$. As histogram of Laplacian image is densely sampled around zero, higher number of bins could be afforded, hence keeping the lossy distortion introduced via quantization under limits. The quantized data is converted into bit stream using lossless compression binarizer, run length encoding. Run length encoder has been used in order to make use of the nature of Laplacian image, which has continuous zeros.

4.3   Reconstruction Framework: DL Based

Our reconstruction framework is inspired by pix2pix proposed in [12]. The decoder framework derived from pix2pix is as follows,

The architecture consists of a conditional GAN architecture. In contrast to a vanilla GAN which learns a mapping $G : z \rightarrow y$ from random noise vector z to output image y, the cGAN learns a mapping $G : \{x, z\} \rightarrow y$, which maps z to y, conditioned on the input image x. Hence, both the generator and discriminator sees the input image during the process of generation of output image. The objective of the conditional GAN is as follows

$$\max_{D} \min_{G} L_{cGAN}(G, D) + \lambda L_{L1}(G) \tag{4}$$

where, the first term of Eq. 4 gives the loss function of cGAN, through which the generator learns to fool the discriminator and discriminator learns to distinguish fake from real images.

$$\max_{D} \min_{G} E_{x \sim p_{data}(x)} [log D(x, y)] \tag{5}$$
$$+ E_{z \sim p_z(z)} [log(1 - D(G(z, x)))]$$

The first term of Eq. 5 is the loss incurred when the discriminator mistakes the target real image for fake image, when it sees the target real image and the input image. The second term of Eq. 5 is the loss that results when discriminator interprets fake image produced by generator as target real image, when it is presented with fake image and the input image. In addition to fooling the discriminator, generator has to try to mimic the original image which is achieved through the second term of Eq. 4, as L_1 loss between the generated image $G(x, z)$ and the target image y.

$$E_{x,y,z}\|y - G(x, z)\|_1 \tag{6}$$

We have used Adversarial framework because of their suitability for image reconstruction as they could be trained to reconstruct images that match the distribution of their original image counterparts, thus resulting in good reconstruction. Figures 3 and 4 shows the original image, Laplacian image of the input and the image reconstructed from the Laplacian image through the decoder under two different bpp scenarios for EPFL and Stanford dataset. It is evident that the reconstructed images are visually very similar to the original images.

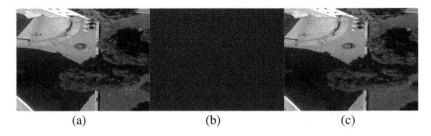

Fig. 3. Image reconstruction with Laplacian Image (0.6906 bpp) for EPFL dataset (a) Original Image, (b) Laplacian Image, (c) Reconstructed Image

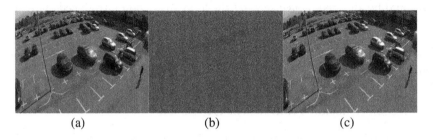

Fig. 4. Image reconstruction with Laplacian Image (0.02999 bpp) for Stanford dataset (a) Original Image, (b) Laplacian Image, (c) Reconstructed Image

5 Results

This section furnishes the results of the experiments carried out to prove the supremacy of the proposed method over conventional methods in terms of accuracy in reconstruction and over deep learning methods in terms of run time. To explore the efficiency of the proposed compression framework, we have tested the method on EPFL mini drone Video dataset [13] and Stanford drone dataset [16]. The proposed algorithm is best suited for aerial platforms, where the computation resource is very limited to transmit images in real time using DL based methods. Hence, the algorithm validation is carried out on Drone dataset. The algorithm is validated in terms of quantitative measures, run time complexity and performance of object detection algorithm.

5.1 Training Set-Up of DL Based Reconstruction Framework

In order to evaluate our method as described above, we have used large scale EPFL mini drone dataset [13] and Stanford drone dataset [16]. The EPFL mini drone dataset consist of 38 different full HD resolution videos with a duration of 16 to 24 s each. The videos are clustered into 3 different categories: normal, suspicious and illicit behaviors. A total of 15000 images comprises of images extracted from videos of EPFL mini drone dataset with heavy data augmentation. A subset of 12000, 2000 and 1000 were used for training, validation and testing. We have also made use of heavy data augmentation by ensuring sufficient invariance and robustness to avoid overfitting. We flipped, shifted and rotated EPFL dataset images to train our model. The Stanford Drone Dataset [16] consists of multiple aerial imagery comprising of 8 different locations around the Stanford campus and objects belonging to 6 different classes moving around. It has six different classes: pedestrian, bicycle, car, skateboard, cart and bus. A total of 20,000 images comprises of images extracted from the various categories of Stanford drone dataset with heavy data augmentation. A subset of 17000, 2000 and 1000 images were used for training, validation and test.

The cGAN network is trained from scratch with weights initialized using Guassian distribution with mean 0 and $\sigma = 0.02$. We trained the network for 200 epochs, batch size of 1 using minibatch SGD and Adam solver [14], with a learning rate of 0.0002 and with momentum parameters $\beta_1 = 0.5, \beta_2 = 0.999$. We have used convolution BatchNorm-RelU, binary cross entropy loss and L1 loss. The generator network remained same during inference and training time. We have trained our network on Nvidia DevBox. The network is trained with Laplacian images as input images and original images as target images. We have trained the network with different quantization of Laplacian images to generate different rate-distortion tradeoffs.

5.2 Comparison with Conventional Methods

This section furnishes the qualitative and quantitative comparison of the proposed compression method against conventional method JPEG. The quantitative

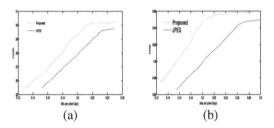

Fig. 5. (*a*) bpp vs. PSNR (*b*) bpp vs. SSIM

measures used are the popular metrics, Structural Similarity (SSIM) and Peak Signal to Noise Ratio (PSNR). A higher value represents closer match of reconstructed image with the original image.

The rate distortion curves (bpp vs. PSNR and bpp vs. SSIM) of comparison is shown in Fig. 5. The effect of using Laplacian image as encoded image can be seen as reduction in bits per pixel required for transmission. The advantage of using conditional GAN to reconstruct the image can be seen as higher PSNR and SSIM as compared to JPEG. As we claim better performance especially at lower bpp, the experiment result shows the comparison of the proposed method with JPEG at $bpp \leq 0.6$. We have also evaluated the proposed algorithm in terms of its usefulness for computer vision algorithms such as object detection. The state-of-the-art object detector Faster RCNN [15] is tested on the images compressed using the proposed method and JPEG. Figure 6 clearly indicates that the blocky nature of the JPEG compressed images deteriorates the performance of object detection efficacy at lower bpp, whereas the proposed method enables detection of objects with good accuracy. The blocky nature of JPEG compressed image has also generated few false positives such as truck in Fig. 6a and vase, bird in Fig. 6c.

Fig. 6. Object Detection on (a), (c) JPEG compressed image (0.15238, 0.09038 bpp) (b), (d) Proposed Compression (0.12421, 0.04461 bpp)

5.3   Comparison with DL Based Methods

The compression method that we have proposed is for computationally challenged platforms, where the encoder part of deep learning based methods could not be deployed because of their memory footprint and run time computational complexity. Hence, we have restricted the comparison of our method to the encoders of deep learning based compression methods in terms of run time, which proves low latency image transmission capability of our encoder. The encoders of the proposed method and deep learning based compression methods are tested on GTX 1080 Ti GPU. The encoder of our method comprises of Laplacian image generation, quantization and bit stream generation. Figure 7a shows the plot of run time of our encoder vs varied PSNR. As the operation involved in our encoder remains same across varied PSNR image generation, the run time ends up being 43 ms on an average. Figure 7b gives the plot of time taken vs SSIM of the encoder of the deep learning method proposed in [3]. It could be seen that the run time drastically shoots up for [3] as against the proposed method, especially at higher SSIM. In order to achieve better SSIM, the residual image generated as difference of original and encoded image was encoded repeatedly, thereby increasing the run time. It could be clearly inferred that the average run time of our encoder is far lesser than that of DL based encoders.

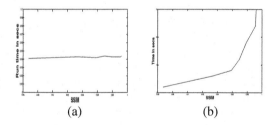

(a)   (b)

Fig. 7. SSIM vs. Run Time of Encoder of (a) Proposed method, (b) DL method

6   Conclusion

In this paper, we proposed a unified framework for image compression in computationally constrained platforms. Conventional compression techniques are known for its low latency, whereas deep learning compression techniques have demonstrated good PSNR and SSIM. Our framework elegantly captures the advantage of conventional and deep learning based compression techniques. Our framework can be easily used in any challenged scenarios which demands real time performance with good reconstruction accuracy. The efficacy of our algorithm over conventional and DL based methods is proved with exhaustive experiments conducted on public dataset, in terms of PSNR, SSIM and speed. We have also demonstrated better performance of object detection algorithm on the images reconstructed by the proposed method.

References

1. Gupta, P., Srivastava, P., Bhardwaj, S., Bhateja, V.: A modified PSNR metric based on HVS for quality assessment of color images. In: IEEEXplore 2011 (2011)
2. Wang, Z., Simoncelli, E.P., Bovik, A.C.: Multiscale structural similarity for image quality assessment. In: Conference on Signals, Systems and Computers, vol. 2, pp. 1398–1402 (2004)
3. Toderici, G., et al.: Full resolution image compression with recurrent neural networks. In: CVPR 2015 (2018)
4. Agustsson, E., Tschannen, M., Mentzer, F., Timofte, R., Gool, L.V.: Generative adversarial networks for extreme learned image compression, arXiv:1804.02958 (2018)
5. Theis, L., Shi, W., Cunningham, A., Huszar, F.: Lossy image compression with compressive autoencoders. In: International Conference on Learning Representations, ICLR-2017 (2017)
6. Agustsson, E., et al.: Soft-to-hard vector quantization for end-to-end learning compressible representations. In: ICLR-2017 (2017)
7. Li, M., Zuo, W., Gu, S., Zhao, D., Zhang, D.: Learning convolutional networks for content-weighted image compression, arXiv:1703.10553 (2017)
8. Radford, A., Metz, L., Chintala, S.: Unsupervised representation learning with deep convolutional generative adversarial networks, arXiv preprint 1511.06434 (2015)
9. Salimans, T., Goodfellow, I., Zaremba, W., Cheung, V., Radford, A., Chen, X.: Improved techniques for training GANS. In: Advances in NIPS 2016, pp. 2234–2242 (2016)
10. Mirza, M., Osindero, S.: Conditional generative adversarial Nets, arXiv:1411.1784 (2014)
11. Burt, P.J., Adelson, E.H.: The Laplacian pyramid as a compact image code. In: Readings in Computer Vision, pp. 671–679 (1987)
12. Isola, P., Zhu, J.Y., Zhou, T., Efros, A.A.: Image-to-image translation with conditional adversarial networks, arXiv prprint (2017)
13. Bonetto, M., Korshunov, P., Ramponi, G., Ebrahimi, T.: Privacy in mini-drone based video surveillance. In: Workshop on De-identification for Privacy Protection in Multimedia (2015)
14. Kingma, D.P., Adam, J.B.: A method for stochastic optimization, CoRR, abs/1412.6980 (2014)
15. Ren, S., He, K., Girshick, R., Sun, J.: Faster R-CNN: towards real-time object detection with region proposal networks. In: Advances in NIPS 2015 (2015)
16. Robicquet, A., Sadeghian, A., Alahi, A., Savarese, S.: Learning social etiquette: human "trajectory prediction in crowded scenes". In: ECCV 2016 (2016)

Multi-frame and Multi-scale Conditional Generative Adversarial Networks for Efficient Foreground Extraction

Himansu Didwania$^{(\boxtimes)}$, Subhankar Ghatak, and Suvendu Rup

Image and Video Processing Lab, Department of Computer Science and Engineering,
International Institute of Information Technology, Bhubaneswar,
Bhubaneswar 751003, India
hdidwania1997@gmail.com

Abstract. Alongside autonomous submissions, foreground extraction is considered to be the foundation for various video content analysis technologies, like moving object tracking, video surveillance and video summarization. This paper proposes an efficient foreground extraction methodology based on conditional Generative Adversarial Network. The proposed generator, which is made up of two networks working in series-Foreground Extractor and Segmentation Network, maps the video frames to corresponding foreground masks. The discriminator aids the learning of generator by learning to differentiate between seemingly real and fake foreground maps. The method used a multi-scale approach in order to capture robust features across multiple scales of input using the Feature Extractor Network, which are then used by the successive Segmentation Network to produce the final foreground map. In addition, a multi-frame approach is also used to facilitate capturing of appropriate temporal features. The performance of the proposed model is evaluated on CDnet 2014 Dataset and outperforms existing methods.

Keywords: Foreground extraction · Generative Adversarial Networks · Deep learning · Multimedia · Video surveillance

1 Introduction

Foreground extraction is a very important part in video content analysis pipeline. Many applications of video content analysis revolve around determining the foreground objects in the video and analyzing their behavior. These foreground objects are generally dynamic and move in space with respect to time, as opposed to the static and constant background. Overtime, many researchers have focused on this task to produce good results. Still, the problem remains a tough ground with room for improvements, as challenges are posed by factors like illumination changes, dynamic background objects, camera instability, etc.

The classical approaches modeled the background distribution in order to classify the regions. Gaussian Mixture Model (GMM) [14] was used to model

© Springer Nature Singapore Pte Ltd. 2020
N. Nain et al. (Eds.): CVIP 2019, CCIS 1148, pp. 211–222, 2020.
https://doi.org/10.1007/978-981-15-4018-9_20

variance in each pixel to classify it as foreground or background. Use of GMM for the task were further improved in methods [8,19]. Many non-parametric approaches have also been suggested for the task in works [2,12,13,15].

Recent advances in Deep Learning have introduced a new way to approach a variety of problem statements and resulted in a largely improved performance on many tasks. A lot of problems dealing with images and videos have been benefited from the use of Deep Learning. Foreground Extraction has also been touched by the advancements of Deep Learning in works [1,4,17].

Like [16], we approach the task as a binary segmentation problem. The segmentation is done using conditional Generative Adversarial Networks (GANs). GANs [5] have been very popular and successful deep learning technique for image generation and translation. DCGANs [10] and Conditional GANs [9] opened paths for powerful use of this technique in the domain of images for the tasks of generation, translation and manipulation. Image translation using GANs has been widely studied and presented in works [7,18]. Segmentation of images can be viewed as a specific case of image translation, where images are translated to a binary mask. Hence, we use the terms translation and segmentation interchangeably. Our work is heavily based on the working of pix2pix [7].

The contributions of this paper include the use of multi-frame and multi-scale approaches together with a conditional GAN for improved performance on the task of foreground extraction. Rest of the paper is presented as follows. Section 2 describes the method we propose, the architectures of models involved, the values of various hyperparameters used and the algorithms for training the model as well as generating the predictions using the trained model. Section 3 gives the experiment details and compares the results of our method against some popular baseline methods IUTIS-5 [3], DeebBS [1], PAWCS [13] and SuBSENSE [12]. Finally, concluding remarks are outlined in Sect. 4.

2   Proposed Methodology

We treat the problem of foreground extraction as binary segmentation in which every pixel of the input frame is to be classified as background or foreground. We assume that videos are continuous and the foreground objects have strong correlation in their spatial positions in consecutive frames. This is in the sense that between two consecutive frames, the foreground objects make very small movements. We wanted to take advantage of this property, hence our model tries to capture both spatial and temporal features. Using a single frame as input can be sub-optimal, since it fails to provide temporal features. To address this issue, unlike most current methods which make use of just a single frame, we incorporate a multi-frame input approach. The input to the model is a block of n consecutive frames concatenated together. The block allows the model to capture patterns from frames at different time steps at once, making the model aware of small movements taking place in the time frame. After a few experiments on some videos, we decided to use $n = 5$. Increasing n improved performance, but the improvements plateaued out after $n = 5$. Hence, for performing foreground extraction for the frame at time step t, $\mathbf{F}_t$, the model is given an input block $\mathbf{F}_{t-2:t+2}$ where $\mathbf{F}$ is the sequence of frames.

2.1   GAN Model

The GAN we use for the segmentation purpose is inspired from pix2pix [7]. The role of our generator is to learn a mapping from the input block of consecutive frames, to the corresponding foreground segmentation. The discriminator helps the learning of generator by determining whether the generated segmentation seems "fake" or "real". Fake or real for our task can be seen as how close the segmentation for a particular frame is to the actual ground truth. During training, the discriminator is fed both groundtruth segmentations and generated segmentations, which enables it to learn the differences between real and fake segmentations and forces the generator to generate segmentations which are more real. The loss function which guides the learning of our GAN is given below:

$$\min_{G} \max_{D} L(D,G) = \mathbb{E}_{x,y}[log(D(x,y)] + \mathbb{E}_{x,\hat{y}}[log(1 - D(x,\hat{y})] + \lambda L_{distance} \quad (1)$$

For some input block x with groundtruth y, the generator G predicts the foreground segmentation $\hat{y}$. Discriminator D takes generated segmentation $\hat{y}$ or groundtruth y as input, together with the block x to return the probability of $\hat{y}$ or y being a real foreground map for x. We add an additional term $L_{distance}$ to the loss, which corresponds to binary cross-entropy between predicted output and the groundtruth. Optimizing this distance term forces the generator to learn a mapping which is not only valid in terms of realness, but also close to groundtruth for a given frame. The term λ is used to set the relative weight of the distance term. The pipeline for the proposed method is given in Fig. 1.

The generator consists of two networks in series. To reflect their tasks, we name the first one as Feature Extractor Network and the second one as Segmentation Network. The Feature Extractor Network is a shared model, which extracts features from input provided in multiple scale. The same model is used for feature extraction from the multiple versions of input differing in scale. The motivation behind using this multi-scale approach is to attain a good balance between the coarse and fine features. Images of different scales when passed through the same network result in different kinds of feature extraction. The network captures the fine details better when working on a large input while missing the coarser details, which is addressed better by a small scale input at the cost of missing fine features. Hence multi-scale approach helps in capturing robust fine and coarse features from different scales. We considered three scales of input for the work. The Feature Extractor Network is implemented as a U-Net [11], which is an encoder-decoder network with skip connections between layers i and $l-i$, where l is the total number of layers. Although the outputs given by Feature Extractor Network are themselves plausible looking probability maps for segmentation, we pass these jointly through the Segmentation Network, which

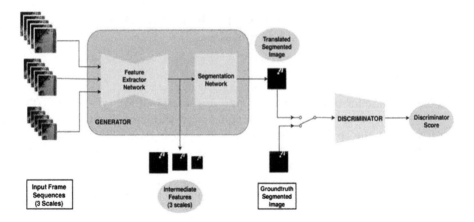

Fig. 1. Pipeline for proposed method - The elements represented in the ellipses are compared against groundtruth during training for learning purposes.

works on combining the features from the three scales to produce the final foreground mask. The architecture for both the networks in generator is shown in Fig. 2.

For the discriminator, we used PatchGAN as suggested in pix2pix [7]. PatchGAN is a convolution network with no dense layers in the end. The input to discriminator is the proabability map of segmentation (produced or groundtruth) concatenated to actual input frame. This concatenation is done in order to provide discriminator the information regarding the actual frame the segmentation must correspond to. This makes the discriminator able to differentiate between a real or a fake segmentation sample conditioned on the given frame. The output of the discriminator is a two-dimensional array of dimension $p \times p$, in which every element corresponds to the probability of the corresponding patch or region of the segmentation being real. The architecture of discriminator is given in Fig. 3.

LeakyRELU activation is used after every convolution layer. We also use Dropout tackle the problem of overfitting. We set the dropout parameter to 0.5 and LeakyRELU parameter to 0.2. Since the outputs of all three networks are probability maps, we use sigmoid activation after the final layers in each of the networks.

2.2   Training Phase

We used a fixed resolution of 240×240 pixels for our work. Hence all the video sequences are first rescaled to this dimension, followed by normalization of pixels values to range $[-1, 1]$. The last 20% of the frames for each considered video is separated to form a validation set, while the rest is used for training. Model is trained on every video sequence separately. During training, all the five frame blocks from the training set are fed in a random order for every epoch. Since the first part of the generator is a shared model, the same Feature Extractor

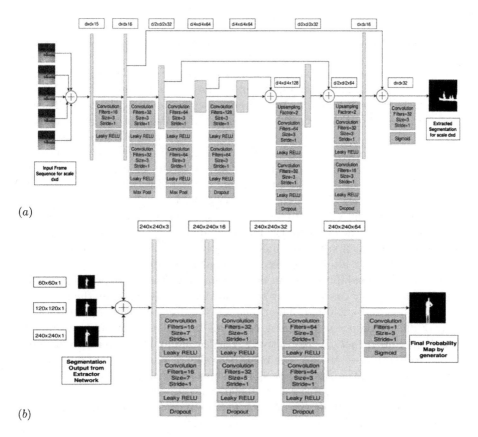

Fig. 2. Generator Architecture: (a) Feature Extractor Network (b) Segmentation Network

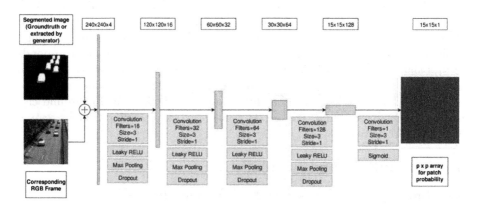

Fig. 3. Discriminator architecture

Network is used to extract features from input given in 3 different scales. Since 5 frame RGB input blocks are considered, the input dimensions to the Foreground Extractor are $240 \times 240 \times 15$, $120 \times 120 \times 15$ and $60 \times 60 \times 15$. For each of these three inputs, the Feature Extractor produces a single channel probability map having size same as that of the input scales. These three maps are then fed to the Segmentation Network jointly, after proper upscaling to dimension 240×240 and concatenation. The final output is the generated probability map of dimension 240×240 which is the predicted map for binary segmentation. There are a total of four outputs from the generator- three intermediate maps from the Feature Extractor Network and the fourth final output of the generator produced by Segmentation Network. All four of these are compared with the groundtruth to get the distance term of the loss. Binary crossentropy is used for the $L_{distance}$ term with the weight λ set to 10 for loss corresponding to final segmentation and 2 corresponding to the losses for the intermediate results. During each epoch, the Feature Extractor is first trained independently for all three scales. The discriminator is trained by alternately feeding actual groundtruth segmentations and the generated probability maps in every step. Since the input of discriminator is a block formed by concatenating segmentation and RGB frame, the input has the dimension $240 \times 240 \times 4$. The target for the discriminator training is a 2D array of dimension $p \times p$ containing either all 1s or all 0s. We call these targets $target_{valid}$ and $target_{fake}$ respectively. For the input resolution and discriminator architecture we employ, the resulting value of p is 15. Adam optimizer is used for training. The same hyper-parameters are used for training all of the networks. We set the learning rate to 0.0001 with β_1 as 0.5 and β_2 as 0.999. The model is trained for 30 epochs on each video sequence considered, keeping the batch size at 1. The algorithm for training is given in Algorithm 1.

2.3 Generating Segmentation

Once the model is trained, only the generator is needed for obtaining the foreground segmentation. The generator takes a 5 frame block in three different scales to produce the probability map for segmentation after passing through both the constituting networks. Value at every index in this map corresponds to probability of that pixel being a foreground. To convert this probability map to a binary mask, we select a threshold μ against which individual pixels are compared. Every pixel with probability above μ is classified as foreground else they are set as background. We use the value $\mu = 0.5$. The algorithm for testing is given in Algorithm 2.

3 Experiments and Results

3.1 Environment, Dataset, and Evaluation Metrics

We implemented the experiments on Python 3.7.3. For implementation of Deep Learning model, Keras framework with Tensorflow backend was used. The experiments were run on a laptop with GTX 1050 Ti GPU for increased processing speed.

Algorithm 1. Training the model

1: **function** TRAIN($GE, GS, D, train\_data$) ▷ GE - Feature Extractor of Generator, GS - Segmentation Network of Generator, G - Generator as a whole with GE and GS series, D - Discriminator, $train\_data$ - Dataset for training consisting of input frame blocks and corresponding groundtruth

2:      $patch\_dim \leftarrow$ G.output.shape

3:      $target_{valid} \leftarrow 1_{patch\_dim}$

4:      $target_{fake} \leftarrow 0_{patch\_dim}$

5:      **for** every epoch **do**

6:          Sample a batch of $ip1, ip2, ip3$ (input 5 frame blocks in 3 scales) and $gt1, gt2, gt3$ (corresponding groundtruth in 3 scales) from $train\_data$

7:          GE.train(ip1, target=gt1)

8:          GE.train(ip2, target=gt2)

9:          GE.train(ip3, target=gt3)

10:         $f1, f2, f3 \leftarrow GE$.predict(ip1), GE.predict(ip2), GE.predict(ip3)

11:         $f \leftarrow$ concatenate($f1, f2, f3$)

12:         $predicted\_fg \leftarrow GS$.predict(f)

13:         $Discriminator\_in\_real \leftarrow$ concatenate($gt1$, ip1.middle_frame)

14:         $Discriminator\_in\_fake \leftarrow$ concatenate($predicted\_fg$, ip1.middle_frame)

15:         D.train($Discriminator\_in\_real$, target = $target_{valid}$)

16:         D.train($Discriminator\_in\_fake$, target = $target_{fake}$)

17:         G.train([$ip1, ip2, ip3$], target = $target_{real}$)

18:     **end for**

19: **end function**

Algorithm 2. Testing for an image

1: **function** TEST(G, ip, μ)   ▷ Where G - Trained Generator, ip - 5 frame blocks in three scales, μ - Mask Threshold

2:      $f_g \leftarrow G$.predict(ip)

3:      **for** every pixel $\in f_g$ **do**

4:          **if** pixel.value $\geq \mu$ **then** pixel.class = 1

5:          **else if** pixel.value $\leq \mu$ **then** pixel.class = 0

6:          **end if**

7:      **end for**

8: **end function**

We trained and evaluated the performance of our model on a subset of CDNet 2014 dataset [6]. The dataset contains 11 video categories with 4 to 6 video sequences in each category. Due to limited computation ability, we restricted our experiments to the videos from baseline, dynamic background and camera jitter categories. The summary of video sequences used is given in Table 1.

For evaluating the performance of our model, we used the metrics precision, recall and F-measure. Precision gives a measure of how many pixels out of the predicted foreground pixels are actually correct. Recall gives the measure of correct foreground predictions our model makes as compared to total number of foreground. F-measure is the harmonic mean of precision and recall.

The formulae related the three metrics we use is given in the Table 2. Here TP = True Positives, TN = True Negatives, FP = False Positives and FN = False Negatives.

Table 1. Summary of video sequences used

Video	Category	Number of frames
pedestrians	baseline	1099
PETS2006	baseline	1200
highway	baseline	1700
office	baseline	2050
fountain01	dynamic background	1184
canoe	dynamic background	1189
overpass	dynamic background	2050
sidewalk	camera jitters	1200
traffic	camera jitters	1570
badminton	camera jitters	1150
boulevard	camera jitters	2500

Table 2. Metrics for result evaluation

Metric	Formula
Precision	$\frac{TP}{TP+FP}$
Recall	$\frac{TP}{TP+FN}$
F-measure	$\frac{2*Precision*Recall}{Precision+Recall}$

3.2   Baselines and Simulated Results

We compared our model to the performances of IUTIS-5 [3], DeepBS [1], PAWCS [13] and SuBSENSE [12]. These methods are standard baselines for foreground extraction and have been quite popular for this task. Out of these four, DeepBS uses Deep Learning approach. We present the performance of our model on the video sequences mentioned above for the three metrics discussed above, in Tables 3, 4, and 5. Representation of visual comparison the generated images is given in Fig. 4. The results for considered baselines methods have been taken from http://changedetection.net/. From the experimental results, it is evident that the proposed model outperforms the baselines in terms of foreground extraction accuracy and visual assessment.

Table 3. Comparison of precision

Video	IUTIS-5	DeepBS	PAWCS	SuBSENSE	Proposed method
pedestrians	0.954	**0.995**	0.931	0.971	0.945
PETS2006	0.906	0.932	0.918	0.912	**0.964**
highway	0.935	0.942	0.935	0.935	**0.974**
office	0.989	**0.992**	0.972	0.979	0.985
fountain01	0.783	0.818	0.803	0.659	**0.935**
canoe	0.991	0.979	0.928	**0.993**	0.974
overpass	**0.975**	0.968	0.956	0.943	0.956
sidewalk	0.904	**0.960**	0.905	0.827	0.912
traffic	0.782	0.812	0.794	0.747	**0.971**
badminton	0.906	**0.959**	0.874	0.843	0.923
boulevard	0.810	**0.993**	0.890	0.827	0.975

Table 4. Comparison of recall

Video	IUTIS-5	DeepBS	PAWCS	SuBSENSE	Proposed method
pedestrians	0.984	**0.989**	0.951	0.952	0.968
PETS2006	0.966	0.952	0.944	**0.971**	0.966
highway	0.972	**0.989**	0.951	0.952	0.983
office	0.949	0.963	0.905	0.946	**0.991**
fountain01	0.866	0.724	0.753	0.877	**0.940**
canoe	0.905	**0.979**	0.947	0.658	0.969
overpass	0.883	0.915	0.961	0.785	**0.982**
sidewalk	0.739	0.852	0.558	0.835	**0.980**
traffic	0.883	0.954	0.864	0.848	**0.984**
badminton	0.934	0.946	0.910	0.922	**0.957**
boulevard	0.729	0.761	0.802	0.690	**0.987**

Table 5. Comparison of F-measure

Video	IUTIS-5	DeepBS	PAWCS	SuBSENSE	Proposed method
pedestrians	**0.969**	0.945	0.946	0.954	0.957
PETS2006	0.935	0.942	0.931	0.940	**0.965**
highway	0.953	0.965	0.943	0.943	**0.978**
office	0.968	0.978	0.937	0.962	**0.988**
fountain01	0.822	0.768	0.777	0.753	**0.938**
canoe	0.946	**0.979**	0.937	0.792	0.971
overpass	0.927	0.941	0.959	0.857	**0.969**
sidewalk	0.813	0.903	0.690	0.831	**0.945**
traffic	0.830	0.877	0.827	0.795	**0.978**
badminton	0.920	**0.952**	0.892	0.881	0.940
boulevard	0.767	0.862	0.844	0.752	**0.981**

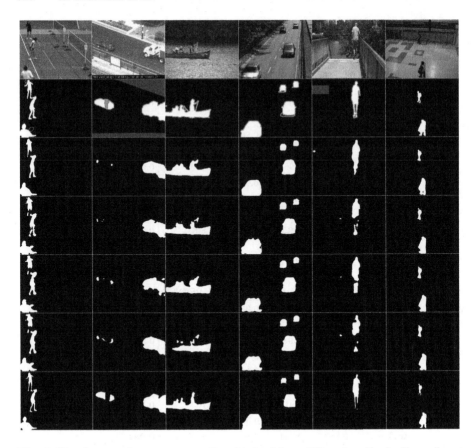

Fig. 4. Visual comparison among results, obtained from different methods- Rows: Input Frame, Groundtruth, DeepBS, IUTIS-5, PAWCS, SuBSENSE, Proposed Method. Columns- badminton, boulevard, canoe, highway, overpass, PETS2006. The grey area in ground truth are outside regions of interest. We decided to label them as background.

4   Conclusion

In this paper, we used a GAN with modified approach for foreground extraction. As per the simulated results, our method outperformed all the baselines and achieved quite high scores, especially in more difficult categories of videos. The use of multi-scale and multi-frame approach provided the model additional important features to learn from during the training. This work opens up scope for even more improved performance for the task of foreground extraction.

References

1. Babaee, M., Dinh, D.T., Rigoll, G.: A deep convolutional neural network for video sequence background subtraction. Pattern Recogn. **76**, 635–649 (2018)
2. Barnich, O., Van Droogenbroeck, M.: ViBe: a universal background subtraction algorithm for video sequences. IEEE Trans. Image Process. **20**(6), 1709–1724 (2010)
3. Bianco, S., Ciocca, G., Schettini, R.: Combination of video change detection algorithms by genetic programming. IEEE Trans. Evol. Comput. **21**(6), 914–928 (2017)
4. Braham, M., Van Droogenbroeck, M.: Deep background subtraction with scene-specific convolutional neural networks. In: 2016 International Conference on Systems, Signals and Image Processing (IWSSIP), pp. 1–4. IEEE (2016)
5. Goodfellow, I., et al.: Generative adversarial nets. In: Advances in Neural Information Processing Systems, pp. 2672–2680 (2014)
6. Goyette, N., Jodoin, P., Porikli, F., Konrad, J., Ishwar, P.: Changedetection.net: a new change detection benchmark dataset. In: 2012 IEEE Computer Society Conference on Computer Vision and Pattern Recognition Workshops, pp. 1–8, June 2012
7. Isola, P., Zhu, J., Zhou, T., Efros, A.A.: Image-to-image translation with conditional adversarial networks. In: 2017 IEEE Conference on Computer Vision and Pattern Recognition (CVPR), pp. 5967–5976, July 2017
8. KaewTraKulPong, P., Bowden, R.: An improved adaptive background mixture model for real-time tracking with shadow detection. In: Remagnino, P., Jones, G.A., Paragios, N., Regazzoni, C.S. (eds.) Video-based surveillance systems, pp. 135–144. Springer, Boston (2002). https://doi.org/10.1007/978-1-4615-0913-4_11
9. Mirza, M., Osindero, S.: Conditional generative adversarial nets. arXiv preprint arXiv:1411.1784 (2014)
10. Radford, A., Metz, L., Chintala, S.: Unsupervised representation learning with deep convolutional generative adversarial networks. arXiv preprint arXiv:1511.06434 (2015)
11. Ronneberger, O., Fischer, P., Brox, T.: U-Net: convolutional networks for biomedical image segmentation. In: Navab, N., Hornegger, J., Wells, W.M., Frangi, A.F. (eds.) MICCAI 2015. LNCS, vol. 9351, pp. 234–241. Springer, Cham (2015). https://doi.org/10.1007/978-3-319-24574-4_28
12. St-Charles, P.L., Bilodeau, G.A., Bergevin, R.: Subsense: a universal change detection method with local adaptive sensitivity. IEEE Trans. Image Process. **24**(1), 359–373 (2014)
13. St-Charles, P.L., Bilodeau, G.A., Bergevin, R.: A self-adjusting approach to change detection based on background word consensus. In: 2015 IEEE Winter Conference on Applications of Computer Vision, pp. 990–997. IEEE (2015)
14. Stauffer, C., Grimson, W.E.L.: Adaptive background mixture models for real-time tracking. In: Proceedings. 1999 IEEE Computer Society Conference on Computer Vision and Pattern Recognition (Cat. No PR00149). vol. 2, pp. 246–252. IEEE (1999)
15. Van Droogenbroeck, M., Paquot, O.: Background subtraction: experiments and improvements for vibe. In: 2012 IEEE Computer Society Conference on Computer Vision and Pattern Recognition Workshops, pp. 32–37. IEEE (2012)
16. Wang, Y., Luo, Z., Jodoin, P.M.: Interactive deep learning method for segmenting moving objects. Pattern Recogn. Lett. **96**, 66–75 (2017)

17. Xu, P., Ye, M., Li, X., Liu, Q., Yang, Y., Ding, J.: Dynamic background learning through deep auto-encoder networks. In: Proceedings of the 22nd ACM International Conference on Multimedia, pp. 107–116. ACM (2014)
18. Zhu, J.Y., Park, T., Isola, P., Efros, A.A.: Unpaired image-to-image translation using cycle-consistent adversarial networks. In: Proceedings of the IEEE International Conference on Computer Vision, pp. 2223–2232 (2017)
19. Zivkovic, Z., et al.: Improved adaptive Gaussian mixture model for background subtraction. In: ICPR, vol. 2, pp. 28–31. Citeseer (2004)

Ink Analysis Using CNN-Based Transfer Learning to Detect Alteration in Handwritten Words

Prabhat Dansena[1](✉)(iD), Rahul Pramanik[1](iD), Soumen Bag[1](iD),
and Rajarshi Pal[2](iD)

[1] Department of Computer Science and Engineering,
Indian Institute of Technology (ISM), Dhanbad, India
`p.dansena23@gmail.com`, `rahul.wbsu@gmail.com`, `bagsoumen@gmail.com`
[2] Institute for Development and Research in Banking Technology,
Hyderabad, India
`iamrajarshi@yahoo.co.in`

Abstract. Alteration of words in handwritten financial documents such as cheques, medical claims, and insurance claims may lead to monetary loss to the customers and financial institutions. Hence, automatic identification of such alteration in documents is a crucial task. Therefore, an ink color based analysis using Convolutional Neural Network (CNN) automation method has been introduced for alteration detection. Pre-trained AlexNet and VGG-16 architectures have been used to study the effect of transfer learning on the problem at hand. Further, two different shallow CNNs have been employed for recognition. A data set has been created using ten blue and ten black pens to simulate the word alteration problem. The dataset captures the word alteration by addition of the characters (or even pen strokes) in the existing word. Experiments have revealed that the transfer learning based deep CNN architectures have outperformed the shallow CNN architectures on both blue and black pens.

Keywords: Convolutional Neural Network · Document alteration · Handwritten forensics · Transfer learning

1 Introduction

Document forgery is a serious concern for financial institutions around the globe. Many of these document forgery cases are related to bank cheques, insurance claims, etc. Document forgery can be classified into two categories, namely counterfeiting and alteration of handwritten entries in documents. Counterfeiting of documents are easily and effectively handled by security feature identification and verification [3]. But detection of altered handwritten entries in documents is comparatively a much difficult task. This is due to the fact that alteration in documents mostly requires addition of few pen strokes (or even complete characters) with perceptually similar color. Such small addition in terms of pen stroke is

© Springer Nature Singapore Pte Ltd. 2020
N. Nain et al. (Eds.): CVIP 2019, CCIS 1148, pp. 223–232, 2020.
https://doi.org/10.1007/978-981-15-4018-9_21

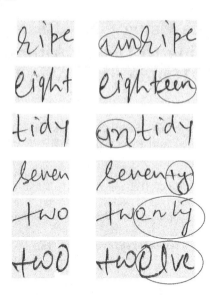

Fig. 1. Few samples of altered word images; **Left Column:** Genuine word images; **Right Column:** Corresponding altered images marked with red circles indicating the altered portions. (Color figure online)

mostly invisible to the naked eye. Sample set of words and corresponding altered versions are presented in Fig. 1. Advancements in technology and communication system have helped most ledger-based systems to reshape the processing techniques. Physical paper-based clearing systems like claims of medical or insurance bills, and bank cheques are not untouched from these technological development. Digitally scanned copy of these documents are currently used for processing. This makes the identification of alteration in these documents more difficult. If this kind of alteration is not detected effectively, then it may lead to huge financial loss to the consumer as well as financial institutions. Therefore, there is a need to develop an alteration detection method for handwritten document forensic analysis.

Alteration of handwritten entries in documents using perceptually similar ink can be broadly classified into two categories: **(i)** addition of new words in the document; and **(ii)** alteration of existing words. Most of the existing works based on destructive techniques [8] and special hardware equipment [17,21,24] require physical copy of the document for aforementioned problems. Since, majority of the documents are digitally scanned and processed, we do not have the physical copy of the document. Hence, these methods are infeasible for the said task. Only a handful of research works have focussed towards the first category, i.e., addition of new words for alteration of handwritten entries in documents. Dasari and Bhagvati [5] and Gorai *et al.* [7] have dealt with the forgery issues related to adding new words in the handwritten documents and have used handcrafted feature sets with threshold value to tackle the same. Moreover, Dansena *et al.* [4]

have extracted handcrafted features and presented classification based approach for identification of alteration in the form of addition of new words in the document. Contrary, for the second category of the problem, i.e., alteration in existing words, Kumar *et al.* [11,12] have presented a classification based approach to detect alteration of existing handwritten word using handcrafted features. But, these methods require manual selection of two pen-stroke regions for detecting the possible alteration of a word in documents. In such scenario, correct identification of word alteration depends on right selection of the pen strokes which is not always possible due to the several possibilities of word alterations as shown in Fig. 1. To this end, this paper skips the selection of pen stroke region and focuses on automatic detection of possible word alterations. To the best of our knowledge, no research work has been carried out on automatic detection of alterations in digital copy of handwritten documents till now.

In the past few years, Convolutional Neural Network (CNN) has captured the attention of researchers by providing exceptional results in comparison with contemporary classification. Image classification [16,19] is one such domain where CNN has been very effective. But, CNN has two major limitations: (i) it requires a large amount of data for training and (ii) it also requires a large amount of resource and time for training. CNN fails to perform well when very less amount of data is available for training. Transfer learning has emanated as a solution to this problem. Transfer learning has been recently applied in varieties of domains like medical imaging [14,15], character recognition [2,18,23], plant disease identification [22], etc., where the availability of training data is quite low and outperformed majority of the contemporary classification solutions.

Hence, in this paper, transfer learning has been applied for the task of detecting alterations in handwritten words using ink analysis. The main contribution of this paper is as follows:

a. This is the first attempt to employ transfer learning (AlexNet [10] and VGG-16 [20]) for the automatic classification of genuine and altered word in which the alteration is done by modifing any alphabet or by adding few alphabets within existing words as shown in Fig. 1.
b. Further, two shallow CNN architectures have been designed in order to provide a basic comparison with transfer learning based approaches. It also demonstrates that the transfer learning based CNN performs better than a shallow CNN trained from scratch, when applied on a small dataset.
c. A new dataset has been created for word alteration consisting of 720 words written using 10 blue and 10 black pens. This dataset is made freely available to the entire research community.

The rest of the paper is organized as follows. Proposed methodology of alteration detection in handwritten document is delineated in Sect. 2. Section 3 discusses the creation of the data set. Experimental results are discussed in Sect. 4. Finally, Sect. 5 concludes the paper.

2   Proposed Methodology

In this method, whether a word is written by single pen or not in a handwritten document is formulated as a two class classification problem. If two different pens are used to write a word then it is labelled as altered; otherwise labelled as genuine. Four different CNN architectures have been used to differentiate altered and genuine words. First, pre-trained Alex-Net [10] and VGG-16 [20] architectures have been used to analyse the effect of transfer learning on word alteration detection. Both AlexNet and VGG-16 architectures are trained on ImageNet dataset [6]. Second, two different shallow CNN architectures have been designed and trained them from scratch for identification of word alteration in document. Our proposed method consists of two major steps: (i) background suppression and (ii) word alteration detection using CNN as discussed next.

2.1   Background Suppression

At first, the handwritten words are cropped manually from the dataset. After the extraction, k-means clustering $(k = 2)$ is performed to separate foreground ink pixels from background. To reduce the effect of background in classification process, each background pixel is assigned a value 255 for all three color channels. Thus, background suppression minimizes the noise in the form of background information. Sample handwritten segmented words and corresponding background suppressed version as obtained using k-means clustering are shown in Fig. 2.

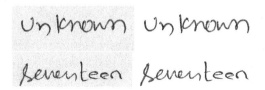

Fig. 2. Background suppression using k-means clustering; **Left Column:** Sample images from our own dataset; **Right Column:** Corresponding background suppressed version.

2.2   Word Alteration Detection Using CNN

Effect of CNN to identify word alteration has been analyzed using four different architectures: two pre-trained deep CNN architectures and two shallow CNNs.

Pre-trained CNN: In order to investigate transfer learning, AlexNet [10] and VGG-16 [20] architectures have been used for word alteration detection. AlexNet architecture consists of five convolution layers, three max-pooling layers, three fully connected layers, followed by a softmax and a classification layer as shown in Fig. 3a. VGG-16 architecture comprises of thirteen convolution layers, four max-pooling layers, three fully connected layers, followed by a softmax, and a classification layer as shown in Fig. 3b. Last three layers of these architectures have been replaced with a new fully connected layer, a softmax, and a classification layer with two nodes to fine-tune the architectures. These changes are performed to adapt the architectures for the current two-class classification problem. These newly added layers in the architecture are initialized with random weights with Gaussian distribution.

Shallow CNN: The two shallow CNNs, viz. $CNN_{32 \times 32}$ and $CNN_{64 \times 64}$ are trained from scratch. $CNN_{32 \times 32}$ comprises of three convolution layers, three max-pooling layers, followed by two fully connected layers, a softmax layer, and a classification layer. Each convolution layer comprises of convolution operations of kernel size 5×5. Max-pooling layers use a 3×3 kernel to reduce the feature map size. An illustration of shallow architecture for $CNN_{32 \times 32}$ is depicted in Fig. 3c. For $CNN_{64 \times 64}$, similar architecture is considered with input image size of $64 \times 64 \times 3$. Shallow CNN architectures have been initialized with random weights with Gaussian distribution.

Due to the fact that CNN works on predetermined input receptor size, the input images are resized using bi-cubic interpolation to make them compatible to the input image receptor size of the CNN. In order to avoid overfitting, data augmentation as well as dropouts are performed while training the CNN architectures. Taking word alteration scenario into account, horizontal and vertical flips of input images have been performed with 0.5 probability in each iteration of the training process. Moreover, dropout (0.5 probability) of random neurons makes the proposed CNN architecture more robust against overfitting. All four architectures are trained using stochastic gradient descent with momentum solver [13]. Due to memory constraint, training of these CNNs are performed with mini-batch size of 20. All these architectures are trained with 30 epochs with learning rate 0.001 and L2 regularization of 0.004. Moreover, during training of pre-trained architectures, learning rate of last fully connected layer has been kept 20 times higher than the rest of the layers.

3   Creation of the Data Set

A dataset has been created to carry out the experiments in this paper. This word alteration dataset consists of 720 handwritten word samples. The dataset contains two types of word samples, namely genuine and altered. Word alteration samples have been created based on the assumption that alteration of any word in a particular document is made by a different person using a different pen with similar color. Therefore, two similar colored pens are used by different volunteers for the creation of altered word samples. This data set is created with the help of 10 volunteers representing various age groups, educational backgrounds, and genders. Similarly, 10 blue and 10 black pens are used to write these words. Each volunteer is associated with two pens, one from blue and one from black pen set to create the data set. Each volunteer writes 72 words in which half of

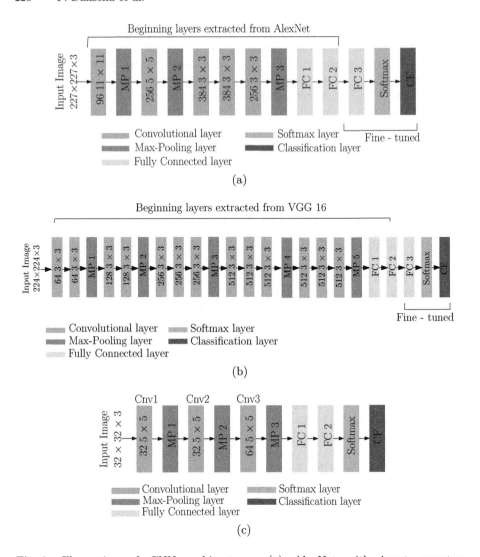

Fig. 3. Illustration of CNN architectures: (a) AlexNet with input receptor $227 \times 227 \times 3$, (b) VGG-16 with input receptor $224 \times 224 \times 3$, (c) $CNN_{32 \times 32}$ with input receptor $32 \times 32 \times 3$.

the words (i.e., 36 words) are written by blue pen and remaining 36 words are written by black pen. Hence, at the beginning a total of 720 ($= (36 + 36) \times 10$) words are written by all volunteers. Then, half of those words ($36/2 = 18$) written by each volunteer using each pen are considered in the *genuine* set. Remaining half of the words (i.e., 18 words) are altered using similar color pen by other volunteers, i.e., words written using a blue pen are altered using another blue pen. Similarly, words written using a black pen are altered using another black pen by remaining 9 volunteers. There are 10 pens of a particular ink color type (either blue or black). Among them, one pen is originally used by a particular volunteer to write a word. So, remaining 9 pens, which are associated with remaining 9 volunteers, are used to alter these 18 words (2 words are altered by each volunteer with associated similar colored pen, i.e., $2 \times 9 = 18$) of a particular color type. This process is also repeated for each pen being used for initial writing. Thus, total $18 \times 10 = 180$ altered word samples are created using blue pens. Similarly, another 180 altered word samples are created for black pens. The number of words (ink-wise) in *genuine* and *altered* sets are summarized in Table 1.

Table 1. Numbers of words (ink-wise) in *genuine* and *altered* sets.

Word type	Pen color		
	Blue	Black	Total
Genuine	180	180	360
Altered	180	180	360
Total	360	360	720

It is to be noted that JK CMAX A4 sheets (72 GSM) are used as a paper on which these words are written. All these word are scanned using a normal scanning device (Cannon E560) with a resolution of 600 dpi. One of the primary contribution of the present work is that the database used in this work is made freely available to the entire research community. The data set can be accessed using the URL as provided in [1].

4   Experimental Results and Discussion

This section reports the performance of the proposed CNN architectures in the context of alteration detection in handwritten words. For performance evaluation of proposed CNN architectures, the dataset is partitioned into two sets (i.e., training and testing datasets) using 20-fold cross-validation scheme. In each fold i, if a word is either originally written or altered using pen P_i, then it is included in testing data set. Otherwise, the word is included into training data set. As 20 pens are used to create the dataset, a 20-fold cross validation scheme has been adopted.

The average classification accuracies by $CNN_{32 \times 32}$, $CNN_{64 \times 64}$, AlexNet, and VGG-16 for blue pen are 66.85%, 70.19%, 78.88% and 76.67%, respectively. Similarly, in case of black pens, the average accuracies are 69.30%, 74.33%, 85.69%, and 79.13% for $CNN_{32 \times 32}$, $CNN_{64 \times 64}$, AlexNet, and VGG-16, respectively. Combining all pens, the classification accuracies are 68.07%, 72.26%, 82.28%, and 77.90% for $CNN_{32 \times 32}$, $CNN_{64 \times 64}$, AlexNet, and VGG-16, respectively. Comparative performance analysis of different CNN architectures based on average accuracy is presented in Table 2. Experimental results reveal that pre-trained AlexNet outperforms the shallow CNN architectures with minimum

Table 2. Accuracy achieved on word alteration dataset using different CNN architectures.

Pen set	Average accuracy (%)			
	$CNN_{32 \times 32}$	$CNN_{64 \times 64}$	Pre-trained AlexNet	Pre-trained VGG-16
Blue pens	66.85	70.19	**78.88**	76.67
Black pens	69.30	74.33	**85.69**	79.13
Overall average	68.07	72.26	**82.28**	77.90

Table 3. Statistical significance test of pre-trained AlexNet with VGG-16 and shallow CNN architectures.

Hypothesis (h_0)	at $\alpha =$	Result
Pre-train AlexNet and $CNN_{32 \times 32}$ accuracies belong to the same median	0.01	$h = 1$
Pre-train AlexNet and $CNN_{64 \times 64}$ accuracies belong to the same median	0.01	$h = 1$
Pre-train AlexNet and VGG-16 accuracies belong to the same median	0.01	$h = 1$

10.02% ($82.28 - 72.26 = 10.02$) on average classification accuracy. Moreover, pre-trained AlexNet performers better than the pre-treined VGG-16 with 4.38% ($82.28 - 77.90 = 4.38$) more on average accuracy. Statistical significance tests on classification accuracies are performed by means of two-tailed Wilcoxon signed rank test [9] with significance level 1% ($\alpha = 0.01$), and presented in Table 3. In this table, null hypothesis is represented by h_0, whereas acceptance and rejection of the hypothesis are identified by {0,1}. Statistical significance test results show the superiority of AlexNet based transfer learning over the pre-trained VGG-16 and shallow CNN architectures. Based on the experimental results, it has been observed that transfer learning based CNN performs better than the shallow CNNs due to the following reasons:

- AlexNet and VGG-16 being denser than shallow CNN architectures are able to extract more features.
- Already optimized network weights trained from millions of images demonstrate better performance in comparison with shallow CNN architectures trained on a very small dataset.

It has also been observed that gradual increase in image input size enhances the classification performance as it helps the CNN to grasp much finer features.

At the best of our knowledge, this is the first of its kind of completely automated approach to detect alteration of handwritten words by adding few pen strokes (or even characters). Hence, the obtained results cannot be compared with any other technique.

5   Conclusion

With the advent of technology, image-based clearing systems are currently used for automatic processing of handwritten financial documents. This calls for developing automatic forgery detection system to safeguard the handwritten financial documents from being altered or manipulated. In the present work, a CNN-based automatic alteration detection technique has been proposed. Two shallow CNN architectures trained from scratch and pre-trained AlexNet and VGG-16 architectures have been evaluated for this task. On experimentation, it has been observed that the pre-trained AlexNet architecture outperforms the pre-trained VGG-16 and shallow CNN architectures trained from scratch.

References

1. DIAL word alteration dataset. https://sites.google.com/site/diafcse/resources/dial-word-alteration-dataset
2. Boufenar, C., Kerboua, A., Batouche, M.: Investigation on deep learning for off-line handwritten Arabic character recognition. Cognit. Syst. Res. **50**, 180–195 (2018)
3. Chhabra, S., Gupta, G., Gupta, M., Gupta, G.: Detecting fraudulent bank checks. In: Advances in Digital Forensics XIII, pp. 245–266 (2017)
4. Dansena, P., Bag, S., Pal, R.: Differentiating pen inks in handwritten bank cheques using multi-layer perceptron. In: International Conference on Pattern Recognition and Machine Intelligence, pp. 655–663 (2017)
5. Dasari, H., Bhagvati, C.: Identification of non-black inks using HSV colour space. In: International Conference on Document Analysis and Recognition, pp. 486–490 (2007)
6. Deng, J., Dong, W., Socher, R., Li, L.J., Li, K., Fei-Fei, L.: ImageNet: a large-scale hierarchical image database. In: International Conference on Computer Vision and Pattern Recognition, pp. 248–255 (2009)
7. Gorai, A., Pal, R., Gupta, P.: Document fraud detection by ink analysis using texture features and histogram matching. In: International Joint Conference on Neural Networks, pp. 4512–4517 (2016)
8. Harris, J.: Developments in the analysis of writing inks on questioned documents. J. Forensic Sci. **37**(2), 612–619 (1992)
9. Hollander, M., Wolfe, D.A., Chicken, E.: Nonparametric statistical methods, vol. 751. Wiley, Hoboken (2013)
10. Krizhevsky, A., Sutskever, I., Hinton, G.E.: Imagenet classification with deep convolutional neural networks. In: Advances in Neural Information Processing Systems, pp. 1097–1105 (2012)
11. Kumar, R., Pal, N.R., Chanda, B., Sharma, J.: Forensic detection of fraudulent alteration in ball-point pen strokes. IEEE Trans. Inf. Forensics Secur. **7**(2), 809–820 (2012)
12. Kumar, R., Pal, N.R., Sharma, J.D., Chanda, B.: A novel approach for detection of alteration in ball pen writings. In: International Conference on Pattern Recognition and Machine Intelligence, pp. 400–405 (2009)
13. LeCun, Y.A., Bottou, L., Orr, G.B., Müller, K.R.: Efficient backprop. In: Neural Networks: Tricks of the Trade, pp. 9–48 (2012)

14. Lei, L., Zhu, H., Gong, Y., Cheng, Q.: A deep residual networks classification algorithm of fetal heart CT images. In: International Conference on Imaging Systems and Techniques, pp. 1–4 (2018)
15. Liu, X., Wang, C., Hu, Y., Zeng, Z., Bai, J., Liao, G.: Transfer learning with convolutional neural network for early gastric cancer classification on magnifiying narrow-band imaging images. In: International Conference on Image Processing, pp. 1388–1392 (2018)
16. Meyer, B.J., Harwood, B., Drummond, T.: Deep metric learning and image classification with nearest neighbour gaussian kernels. In: International Conference on Image Processing, pp. 151–155 (2018)
17. Pereira, J.F.Q., et al.: Projection pursuit and PCA associated with near and middle infrared hyperspectral images to investigate forensic cases of fraudulent documents. Microchem. J. **130**, 412–419 (2017)
18. Pramanik, R., Dansena, P., Bag, S.: A study on the effect of CNN-based transfer learning on handwritten Indic and mixed numeral recognition. In: Workshop on Document Analysis and Recognition, pp. 41–51 (2018)
19. Roy, S., Sangineto, E., Sebe, N., Demir, B.: Semantic-fusion gans for semi-supervised satellite image classification. In: International Conference on Image Processing, pp. 684–688 (2018)
20. Simonyan, K., Zisserman, A.: Very deep convolutional networks for large-scale image recognition (2014). arXiv preprint arXiv:1409.1556
21. Suzuki, M., Akiba, N., Kurosawa, K., Akao, Y., Higashikawa, Y.: Differentiation of black writing ink on paper using luminescence lifetime by time-resolved luminescence spectroscopy. Forensic Sci. Int. **279**, 281–287 (2017)
22. Too, E.C., Yujian, L., Njuki, S., Yingchun, L.: A comparative study of fine-tuning deep learning models for plant disease identification. Comput. Electron. Agric. **161**, 272–279 (2018)
23. Tushar, A.K., Ashiquzzaman, A., Afrin, A., Islam, M.R.: A novel transfer learning approach upon Hindi, Arabic, and Bangla numerals using convolutional neural networks. In: Computational Vision and Bio Inspired Computing, pp. 972–981 (2018)
24. Wang, X.F., Yu, J., Xie, M.X., Yao, Y.T., Han, J.: Identification and dating of the fountain pen ink entries on documents by ion-pairing high-performance liquid chromatography. Forensic Sci. Int. **180**(1), 43–49 (2008)

Ensemble Methods on Weak Classifiers for Improved Driver Distraction Detection

A. Swetha$^{(\boxtimes)}$![ORCID], Megha Sharma ![ORCID], Sai Venkatesh Sunkara,
Varsha J. Kattampally, V. M. Muralikrishna, and Praveen Sankaran

Department of Electronics and Communication Engineering,
NIT Calicut, Calicut, India
swethaaprabhu@gmail.com, meghasharma07@gmail.com,
venkateshsunkaras@gmail.com, varshajk1612@ymail.com,
vm.muralikrishna9@gmail.com, psankaran@nitc.ac.in

Abstract. Distracted driving is the act of driving when engaged in other activities which takes the driver's attention away from the road. It is one among the main causes of road accidents. Majority of these accidents occur because of momentary negligence, thus a driver monitoring system which executes and analyzes in real time is necessary. The required dataset will be prepared and this data will be utilised to train the neural network. The major areas of focus are extreme head poses, yawning detection, facial expressions, head nodding and hand movements that are indicators of distraction. Emphasis will also be given in exploring different ensemble methods for combining various existing models.

Keywords: Distracted driving · Dataset · Neural networks · Ensemble

1 Introduction

The number of traffic accidents due to distracted and fatigued state of the driver has been on an upward trend in the recent years. Road crashes and accidents have turned out to be a major health crisis all over the world by being one among the major causes of death. The World Health Organisation categorizes distracted driving as one of the prime causes of road accidents [1]. Along with the impacts of alcohol and speeding, nowadays distractions while driving has also added itself to the list of leading factors in fatal and serious crashes.

In general, distracted driving can be considered as any act that will divert the attention of the driver from their primary task of driving. This distraction usually happens when the driver temporarily performs some secondary task which could lead to a diversion in attention such as usage of mobile phones for calling and messaging, focusing on events in the surrounding environment, eating, being drowsy and sleeping and so on [1]. There are mainly four categories of distractions which are mutually inclusive: auditory, visual, physical/manual and cognitive [1].

© Springer Nature Singapore Pte Ltd. 2020
N. Nain et al. (Eds.): CVIP 2019, CCIS 1148, pp. 233–242, 2020.
https://doi.org/10.1007/978-981-15-4018-9_22

Visual distractions are those that tend to take away the attention of the driver from the road whereas the manual distractions include those that make the driver let go off the steering wheel and engage in other activities [1]. The masking of the sounds that are crucial to be heard by the driver while driving forms the auditory distractions whereas cognitive distraction induces the driver's mind to wander away and give thought and attention to things other than driving. As different brands of car manufacturers incorporate more and more intelligent vehicle systems to satisfy the customer's demand for comfort, navigation or communication, more are the number of sources of distraction and the level of cognitive stress on the driver. Hence developing a system that could monitor the driver's level of vigilance using visual behaviour and artificial intelligence and alerting them when necessary is of prime importance.

In the past few years, many researchers have been working in the area of developing techniques to detect drowsiness as well as distraction in drivers. More accurate methods involve physiological measures [7] such as brain waves, heart rate, respiration, pulse rate etc. However these processes being intrusive can cause annoyance to the drivers, hence other non-intrusive methods were explored. This includes observing vehicle behaviour like lateral position, steering wheel movements, inter-vehicle distance measurement etc. [6]. However this is not that universal because of the diversity in the vehicle types, driver expertise, condition of the roads and also the longer time required to analyze user behaviour. People under distraction or experiencing fatigue explicitly show certain changes in their facial features such as head pose, mouth in the form of yawns, eyes etc. [4]. Hence computer vision can be a non-intrusive technique to extract characteristic features and estimate the degree of alertness of the driver from images taken by a camera.

Many studies on developing distraction and drowsiness detection systems have been reported in literature.

A web-cam based system is proposed to detect driver's fatigue from the face image using only image processing and machine learning techniques. In this, the face is detected using Histogram of Oriented Gradients (HOG) technique and linear Support Vector Machine (SVM) for detection of facial landmarks like eyes and nose [4]. Here a linear SVM is used for the object classification task along with negative hard mining. Classification of the driver state using the features is done using Bayesian classifier, Fisher's linear discriminant analysis and Support Vector Machine and the observations for the different methods are compared.

Detecting the driver's visual and cognitive states by fusing stereo vision and lane tracking data [9], both visual and cognitive distractions could be detected to some extent. The visual distraction detection was implemented using an attention mapping algorithm utilising the head and gaze directions. Cognitive distraction detection could be done using SVM techniques which are suited for momentary changes. The features used included gaze angles, head rotations and lane positions.

A tuned convolutional neural network is developed in this paper to classify face pose and this information is used for estimating the extent of driver distraction. Based on the position and angle of the varying head pose, classifications are made using CNN architecture. Different performance measures like Classification Accuracy, Recall, Precision, F-Measure, Error rate etc. are evaluated [8].

2 Proposed Methodology

The proposed solution throws light on the real time implementation of a system using a webcam for the image acquisition and extracting frames. The position of the driver's face is localised using Viola and Jones [10] algorithm and coordinates returned are used for cropping the face. This image can now be given to three existing CNN architecture models:Alexnet [3], LeNet [12] and VGG16 [11] and these outputs are combined using ensemble methods such as stacking, majority voting and weighted voting to obtain the classification labels (Fig. 1).

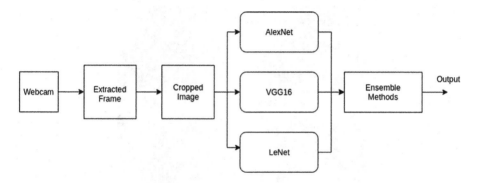

Fig. 1. Proposed implementation for detecting driver distraction

2.1 Dataset Creation

As far as the area of research related to driver distraction detection is concerned, the dataset which is widely used is the State Farm dataset which consists of 10 classes. One main observation is that this data consists of drivers driving car with left hand drive which is not suited in countries such as India having right hand drive.

This led to the motivation to create a custom dataset that will capture the frontal face of the driver rather than the side view. The state of the driver was captured through a camera placed at the dashboard of the car and the driving data of 11 individuals(6 female and 5 male) was taken under different lighting conditions. Care was taken to include people with beard and glasses so that diversity is introduced and hence the model generalises well. The major distractions were divided into the following classes-left, right, yawn, down and distraction due to phone. The frames extracted from the driving data is manually annotated into the various classes to create the custom dataset. Better accuracy was obtained when the face was localised from the entire image as the network would be otherwise learning from a lot of unnecessary details. Hence a mask tailor made to each person was utilised for extracting the facial region and the training data was prepared. Some sample images of the custom dataset are given below (Fig. 2).

Fig. 2. Sample images from custom dataset

Approximately 10 h of driving data was captured using a Noise Play Action camera. From the frames extracted from the videos, it was observed that the number of images corresponding to different classes were not the same. Despite the disparity in the amount of data present, for training the neural network models the images corresponding to each of the classes was taken in equal proportion.

These classes however have a degree of overlap, that is for instance, a person holding a mobile phone can belong to more than one classes of distraction- left, right, phone etc. This is therefore a challenge to overcome.

The proportion of the total data belonging to different classes can be approximately summarized as follows (Fig. 3),

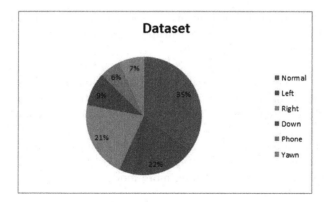

Fig. 3. Distribution of images belonging to the several classes in the driving data

2.2   Ensemble Methods

The accuracy could decrease due to the degree of overlap between the different classes of distraction and this led to the need to combine the decisions taken by several of these weak classifiers that are performing above average.

The reasoning behind using an ensemble is that by stacking different models representing different hypotheses about the data, we can find a better hypothesis that is not in the hypothesis space of the models from which the ensemble is built. By using a very basic ensemble, a lower error rate can be achieved than when a single model was used in most cases [5]. This proves the effectiveness of ensembling.

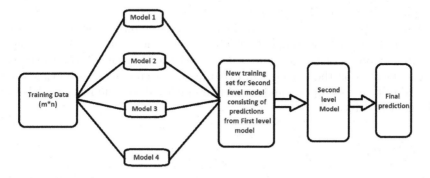

Fig. 4. Stacking approach

Stacking. Stacking is an ensemble learning technique that combines multiple classification or regression models with the help of a meta-classifier or a meta-regressor. The base level models are trained on the whole training set and then the meta-model is trained on the outputs of the base level model as features. The base level often consists of different learning algorithms and therefore stacking ensembles are often heterogeneous [5].

As such there is no limit as to how many models can be stacked. But a plateauing is observed after a number of models. That is, initially it will have some significant uplift in whatever metric is being tested every time we run the model [2]. But after some point, the increments will be fairly small. There's no way to know apriori as to exactly what is the number of models with which the plateauing will start but it is seen to be affected by how many features are there in the data and how much diversity is brought into the models. That means in general, there is a point where adding more models actually does not add that much value (Fig. 4).

As the meta model will only use predictions of other models, it can be assumed that the other models would have done a deep work to analyse and intrepret the data [2]. Therefore the meta model doesn't need to be so deep and complicated. That is, the ensemble method basically deals with training a model with the predictions of a number of other models which have been trained deeply and this model needs to just relate between the correct predictions made by each model. Therefore, the meta model is generally simpler.

Implementation: The three models were separately trained on a dataset containing images corresponding to the six classes for a fixed number of epochs and the accuracy vs epochs was observed. In order to understand the effect of the overlap of the classes such as yawn and phone, the models were also trained with lesser number of classes i.e. left, right and normal.In both the scenarios, the three models were evaluated on a test set and parameters such as the accuracy and precision were computed for comparison. Confusion matrices were also plotted to understand the disparity in the classes predicted and which classes were wrongly predicted as which.

Since the accuracy of the individual models were only above average, there was a need for implementing ensemble methods which could combine the accurate predictions of each model and hence reduce the error rate. Initially averaging is implemented on the models. On exploring other ensemble methods, stacking seemed to be a good option and stacking is employed to build meta models from the models trained.

3 Results

The overlap in classes and its effect on training a proper model could be evaluated by training models based on different number of classes. It is seen that many predictions in down, phone and yawn gets distributed in other classes such as normal, left and right. The accuracy parameter is also found to improve with decrease in the number of overlapping classes (refer Table 1). This is an effect of the overlap in classes. The confusion matrices corresponding to an ensembled average of VGG, LeNet and AlexNet is shown with the models trained using 6,4 and 3 classes respectively (Fig. 5).

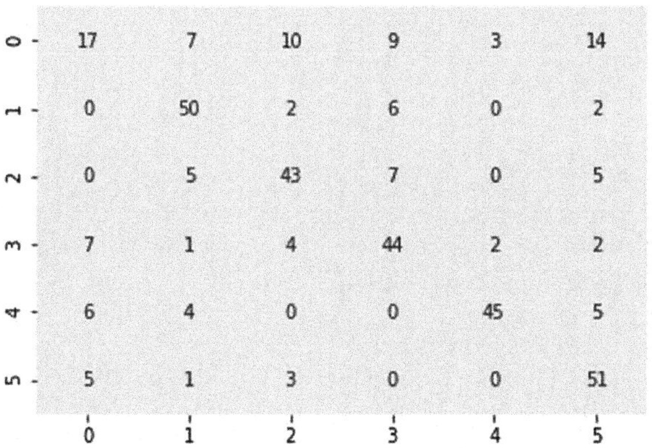

Fig. 5. Model with 6 classes (0: Down, 1: Left, 2: Normal, 3: Phone, 4: Right, 5: Yawn)

It is observed that the accuracy and precision of the models increases with decrease in number of overlapping classes (Fig. 6).

Table 1. Averaging ensemble based on different number of classes

Models	Accuracy	Precision
6 classes	0.68056	0.71885
4 classes	0.7875	0.84036
3 classes	0.85	0.86017

The individual parameters of the base models are summarised (Table 2) based on a test set. As a crude approach to improve the accuracy of the model, averaging the output predictions is implemented.

Among ensemble methods, stacking was found to give good results. The three models discussed above were combined and meta model was created. Meta models were trained on the predictions of these base models. The metamodels

Table 2. Evaluation parameters for the models trained with 6 classes

Models	Accuracy	Precision
VGG16	0.64723	0.69495
Lenet	0.67223	0.66892
AlexNet	0.60833	0.67406
Averaged model	0.68055	0.71885
Weighted average	0.69445	0.68779

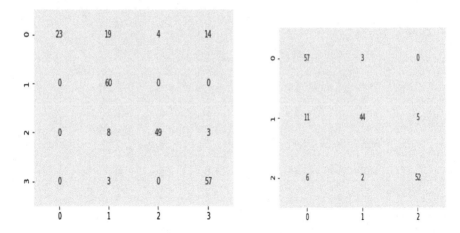

Fig. 6. Model trained with 4 and 3 classes respectively (0: Down, 1: Left, 2: Normal, 3: Right) and (0: Left, 1: Normal, 2: Right)

Table 3. Different metamodels from the base models

MetaModels	Accuracy	Precision
Logistic regression (LR)	0.835	0.85123
KNN classifier (KNN)	0.78667	0.82386
Support vector classifier (SVC)	0.765	0.82032

were developed using Logistic Regression, KNN Classifier as well as Support Vector Classifier as the algorithms and the parameters are as follows (Table 3).

In order to explore the possibility of multi level stacking improving accuracy to greater levels, the idea of stacking a meta model of the meta models were implemented in various combinations. All these meta-metamodels were implemented using Logistic Regression and the results are tabulated as follows (Fig. 7 and Table 4).

It is observed that the accuracy does not greatly improve, rather the models were found to saturate at an accuracy of around 80%. Hence bi-level stacking

Table 4. Stacking the metamodels

SubModels	Accuracy	Precision
LR, KNN and SVC	0.795	0.83813
LR and SVC	0.78	0.82525
LR and KNN	0.79833	0.83161
KNN and SVC	0.785	0.83276

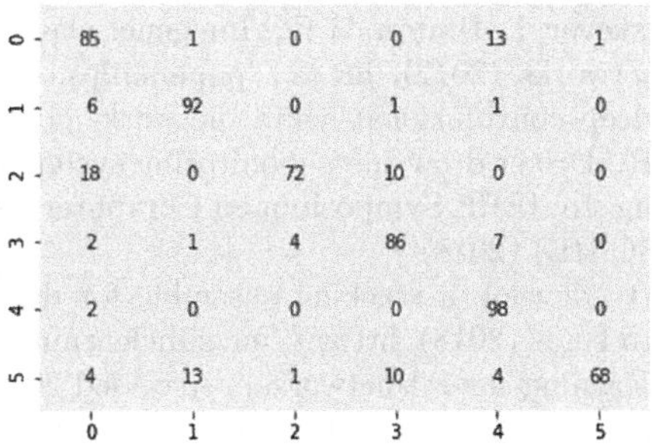

Fig. 7. Logistic regression metamodel

does not prove to be a good option here. The reason could be due to lesser number of features and over-fitting.

4   Conclusion

The custom dataset developed capturing the frontal face of the driver is suitable for universal scenarios irrespective of the left/right hand drive and the localisation of the face helps to take only the necessary information. A real time model can be implemented using the frames extracted from the real time video captured using webcam as input to the final model. Stacking proves to be a very effective method for improving the overall performance of the model by combining the capabilities of the lower performing base classifiers. The method of stacking based on training a metamodel over the predictions of the base model could greatly improve the accuracy. Since the metamodel was being trained on lesser number of features, simpler methods like Logistic regression was adopted whereas more complex Convolutional Neural Network models were implemented for perfecting the base models. However stacking the stacked metamodels could not improve the accuracy further rather it was saturated. Further, the challenge to train models to overcome the effect of overlapping classes with high accuracy as compared to the non-overlapping classes could be explored. Other methods for multi level stacking that can increase the overall performance of the model can also be explored.

References

1. Distracted driving in India: A study on mobile phone usage pattern and behaviour (2017). http://savelifefoundation.org/wp-content/uploads/2017/04/Distracted-Driving-in-India_A-Study-on-Mobile-Phone-Usage-Pattern-and-Behaviour.pdf. Accessed 20 Apr 2019
2. Stacking-ensembling— coursera (2018). https://www.coursera.org/lecture/competitive-data-science/stacking-Qdtt6. Accessed 20 Apr 2019

3. Krizhevsky, A., Sutskever, I., Hinton, G.E.,: Imagenet classification with deep convolutional neural networks (2012). https://papers.nips.cc/paper/4824-imagenet-classification-with-deep-convolutional-neural-networks.pdf

4. Kumar, A., Patra, R.: Driver drowsiness monitoring system using visual behaviour and machine learning. In: IEEE Symposium on Computer Applications and Industrial Electronics (ISCAIE) (2018)

5. Brownlee, J.: How to develop a stacking ensemble for deep learning neural networks in python with keras (2018). https://machinelearningmastery.com/stacking-ensemble-for-deep-learning-neural-networks/. Accessed 20 Apr 2019

6. Siegmund, G.P., King, D.J., Mumford, D.K.: Correlation of heavy-truck driver fatigue with vehicle-based control measures. J. Commer. Veh. **104**, 441–468 (1995). www.jstor.org/stable/44612140

7. Kawanaka, H., Miyaji, M., Bhuiyan, M., Oguri, K.: Identification of cognitive distraction using physiological features for adaptive driving safety supporting system. Int. J. Veh. Technol. (2013). http://dx.doi.org/10.1155/2013/817179

8. Kumari, M., Hari, C.V., Sankaran, P.: Driver distraction analysis using convolutional neural networks. In: International Conference on Data Science and Engineering (ICDSE) (2018)

9. Kutila, M., Jokela, M., Markkula, G., Rué, M.R.: Driver distraction detection with a camera vision system. In: ICIP 2017 (2017)

10. Viola, P., Jones, M.: Robust real-time object detection. Int. J. Comput. Vision **4**, 4 (2001). citeseerx.ist.psu.edu/viewdoc/summary?doi=10.1.1.110.4868

11. Simonyan, K., Zisserman, A.: Very deep convolutional networks for large scale image recognition. In: International Conference on Learning Representations (2015). https://arxiv.org/pdf/1409.1556.pdf

12. LeCun, Y., Bottou, L., Bengio, Y., Haffner, P.: Gradient based learning applied to document recognition. In: Proceedings of IEEE (1998). http://vision.stanford.edu/cs598_spring07/papers/Lecun98.pdf

DeepRNNetSeg: Deep Residual Neural Network for Nuclei Segmentation on Breast Cancer Histopathological Images

Mahesh Gour$^{(\boxtimes)}$ ⓘ, Sweta Jain ⓘ, and Raghav Agrawal ⓘ

Maulana Azad National Institute of Technology, Bhopal 462003, India
maheshgour0704@gmail.com, shweta1008@gmail.com, raghavagrawal999@gmail.com

Abstract. Nuclear segmentation in the histopathological images is a very important and prerequisite step in computer aided breast cancer grading and diagnosis systems. Nuclei segmentation is very challenging in the complex histopathological images due to uneven color distribution, cell overlapping, variability in size, shape and texture. In this paper, we have developed a deep residual neural network (DeepRNNetSeg) model for automatic nuclei segmentation on the breast cancer histopathological images. DeepRNNetSeg learns high-level of discriminative features of the nuclei from the pixel intensities and produces probability maps. Annotated image mask is applied to the image in order to obtain the image patches, which are then fed to DeepRNNetSeg, which classify each image patches as nuclei or non-nuclei. We evaluate our proposed model on publicly available 143 H&E stain images of estrogen receptor positive (ER+) breast cancer. DeepRNNetSeg model has achieved an improved mean F1-score of 0.8513 and a mean accuracy of 86.87%.

Keywords: Deep learning · Residual neural network · Histopathological images · Breast cancer · Nuclei segmentation

1 Introduction

Breast cancer (BC) is a deadly disease among women worldwide and it is the second main cause of cancer related death after lung cancer. Pathology plays an important role in the detection and diagnosis of BC. In pathological analysis pathologist visually examine tissues under the microscope to see the structure and topological features (such as phenotype distributions) of nuclei of a cancerous tumor. This process is very time consuming and error prone and it also depends on the expertise level of the pathologist. With the advent of digital pathology, tissue slides are stored in digital image form [1], it is now possible to develop a computer aided diagnosis (CAD) system for automatic, faster and accurate diagnosis of breast cancer from the pathological images.

To develop a high throughput CAD system usually needs an accurate nuclei segmentation as the prerequisite step. However, automatic nuclei segmentation

© Springer Nature Singapore Pte Ltd. 2020
N. Nain et al. (Eds.): CVIP 2019, CCIS 1148, pp. 243–253, 2020.
https://doi.org/10.1007/978-981-15-4018-9_23

is a very challenging and complicated task due to the inherent complexity of histopathological images such as cell overlapping, uneven colour distribution, the complex appearance of the tissues and staining differences. These problems makes very difficult to design an efficient algorithm for nuclei segmentation that works satisfactory for all these cases. Recently, deep learning have shown significant improvements and achieved state-of-the-art performance in the field of pattern recognition, voice recognition, object detection, etc. [2–4]. In this paper, we have developed a Deep Residual Neural Network for nuclei Segmentation (DeepRNNetSeg) on histopathological images of breast cancer. The main contribution of this work is as follows:

– DeepRNNetSeg is capable to learn the high level of structure information (discriminative features) of nuclei at the multiple level of abstractions from the unlabelled image patches. Unlike existing hand-crafted feature based methods where they rely on low-level features such as color, texture and shape.
– DeepRNNetSeg model employs the hierarchical architecture on the fed image patches and produces probability maps of nuclei or non nuclei structured corresponding to pixel intensities of image patches.

Rest of the paper is organized as follows: Sect. 2 presents the previous related works of nuclei segmentation. Proposed method and details of developed Deep-RNNetSeg model is discussed in the Sect. 3. Section 4 presents the implementation details and experimental results. Conclusive remarks of this work have discussed in the Sect. 5.

2   Related Work

Nuclei segmentation methods can be divided into two categories as: one that are based on image processing methods such morphological operations, color thresholding, region growing, clustering, graph-cut, etc. [5–11]. Second that are based on deep learning approach, in which deep convolutional neural networks (CNNs) are used to detect and classify nuclei in the histopathological images [12–16]. The summary of literature review of nuclei segmentation is represented in the Table 1 and detailed descriptions of literature has given as follows.

Chang et al. [6] have proposed a method for nuclear segmentation using a multi-reference graph cut (MRGC). This technique overcomes the technical variations related to sample preparation by integrating previous information from manually annotated source images and native image options. Nielsen et al. [7] have proposed a method for the automatic segmentation of the nucleus from the Feulgen-stained histological images of prostate cancer. They have used local adaptive thresholding which was combined with an active contour model to enhance the convergence of the nuclei segmentation. Vink et al. [8] have proposed machine learning based approach to detect a nucleus. A modified version of AdaBoost was applied in order to make two detectors, which focus on various features of nucleus. An optimal active contour algorithm was used to combine the results of two detectors. Fatakdawala et al. [9] have presented a scheme to automatically detect and segment the lymphocyte nucleus using the expectation-maximization algorithm. This algorithm was driven by geodesic active contour with overlap resolution. To resolve the overlapping structure of nuclei, heuristic splitting of contours was used.

Table 1. Summary of literature review

Author(s)	Nuclei segmentation approach(s)	F1-Score	Accuracy
Chang et al. [6]	Multi-reference graph cut	0.80	–
Nielsen et al. [7]	Local adaptive thresholding	–	73%
Vink et al. [8]	Modified AdaBoost	–	95%
Fatakdawala et al. [9]	EMaGACOR	–	90%
Al-Kofahi et al. [10]	Graph-cut-based algorithm	–	86%
Qi et al. [11]	Mean-shift clustering	0.84	–
Xu et al. [12]	Stacked sparse autoencoder	0.84	–
Sirinukunwattana et al. [13]	Spatially constrained CNN	0.69	–
Kumar et al. [14]	Deep learning	0.83	–
Janowczyk et al. [15]	AlexNet	0.83	–
Janowczyk et al. [16]	RADHicaL, AlexNet	0.82	–

Al-Kofahi et al. [10] have presented a concept to automatically detect and segment the cell nuclei based on graph-cut-based algorithm. Nuclear seed points were marked using adaptive scale selection and multi-scale Laplacian Gaussian filter and the segmented nucleus was refined using graph colouring and alpha-expansion. Qi et al. [11] have proposed an approach that can separate overlapping cell in histopathology images. The algorithm were applied in two steps. First, mean shift clustering was applied and then the contour of each nucleus was acquired using level-set algorithm. Xu et al. [12] have presented stacked sparse autoencoder (SSAE) which is a deep learning method to detect the nuclei. SSAE identify features of nuclei from pixel intensity. Sirinukunwattana et al. [13] have presented a Spatially Constrained Convolutional Neural Network (SC-CNN) model for nuclei detection. The possibility of a pixel being the centre of a nucleus was confirmed via SC-CNN. The high probability values was domestically forced in the proximity of centres of the nucleus. This method was applied on colorectal adenocarcinoma images, made up of 20000 annotated nuclei. Kumar et al. [14] have introduced a large dataset based on H&E stained images consisting of more than 21,000 annotated nuclei. The author has applied a deep learning technology strategy to detect the boundaries of the nucleus.

Janowczyk et al. [15] have presented a comprehensive study on deep learning for pathology image analysis, in which they have used 13-layered AlexNet network for nuclei segmentation and reported F1-score of 0.83 with 12000 manually annotated nuclei. Janowczyk et al. [16] proposed resolution adaptive deep hierarchical learning (RADHicaL) approach, in which authors multi-resolution images are obtained from the input image and applied pre-trained AlexNet separately on the each resolutions. Each test images have fed to the lowest resolution network to higher resolution networks to get the probability of nuclei or non-nuclei image patches.

3   Proposed Method

In this section we present our proposed method for nuclei segmentation on pathology images and its block diagram is depicted in the Fig. 1. As shown in the recent literature [12–17], Convolutional Neural Network (CNN) in the field of medical image analysis achieved significant improvement in the performance. CNNs are also used for region of interest (ROI) segmentation in the medical images. We have proposed a deep residual neural network based approach for nuclei segmentation on histopathological images of breast cancer. The proposed approach can be divided into three folds: in the first fold, image patches are generated from the input image and obtained the training and validation set. In order to generate image patches, an annotated image mask of a positive class

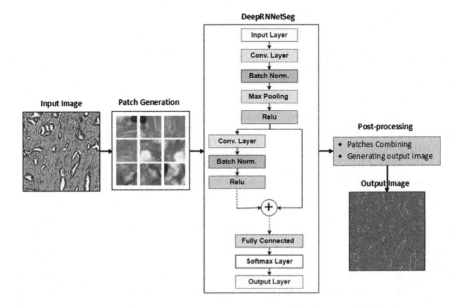

Fig. 1. Block diagram of proposed nuclei segmentation approach.

has been applied for positive class (nuclei) patches and similarly negation of the mask has been applied for negative class patches (non-nuclei) [15]. In the second fold, the DeepRNNetSeg model has been trained in a training set of image patches to learn the discriminating features or structured information of the nuclei. DeepRNNetSeg produces probability maps with respect to pixel intensities of image patches and finally, thresholding has been applied on the probability maps to classify the image patches into nuclei or non-nuclei. In the last fold, image patches are combined to generate the nuclei segmented output image. The algorithm of the proposed approach is as follows:

Algorithm : Nuclei Segmentation on breast cancer hispathological images.
Input: Histopathological image
Output: Nuclei segmented image

Step 1: Take histopathological image I_k
Step 2: Generate patches $I_{k,p}$ of size 32×32 by applying annotated mask and assigning label (0 or 1) to each image patch.
Step 3: Train DeepRNNetSeg on the training set $T(I_t, L_t)$ of labelled image patches.
Step 4: a) Take test set images and generate patches using step 2.
      b) Feed the patches into the trained DeepRNNetSeg model, to obtain probability maps.
      c) Apply thresholding on the probability maps and classify them as nuclei or non-nuclei.
Step 5: Combine the labelled patches and obtain the nuclei segmented image.

---

where I_k : H&E stained breast cancer image of size 2000×2000, $1 \leq k \leq 143$
   $I_{k,p}$: Generated patches for the kth image of size 32×32
$T(I_t, L_t)$: I_t represents training instance and L_t represents corresponding labels.

3.1 DeepRNNetSeg Model

DeepRNNetSeg model is a 30-layer residual neural network which consists of 9 learnable layers (8 Convolutional layers and 1 Fully Connected layer), where the weights and biases are learned during the training of the network. The layered architecture of the developed DeepRNNetSeg model is represented in Table 2. While keeping an increasing number of layers in the network the accuracy of the network starts to saturate or even degrade at some point due to the vanishing gradient problem. This problem can be resolved by using residual connections over the layers [4]. A residual block is represented in Fig. 2. We have experimentally determined the values of hyper parameter such as number of hidden layers, filter size in Convolution layer, number of neurons that are best suitable to our problem (see Table 2). We have employed a Convolution layer which convolutes input images with kernel to produce activation for next layer. Followed by, we have included the Batch Normalization layer which normalized activation for

Table 2. DeepRNNetSeg architecture for segmenting nuclei

Layer no.	Layer name	Type of layer	Filter size/no. of filters	Activations
1	inputimage	Image input	–	$32 \times 32 \times 3$
2	conv_1	Convolution	$3 \times 3/8$	$32 \times 32 \times 8$
3	bn_1	Batch normalization	–	$32 \times 32 \times 8$
4	maxpool_1	Max pooling	2×2	$16 \times 16 \times 8$
5	relu_1	ReLu	–	$16 \times 16 \times 8$
6	conv_2	Convolution	$1 \times 1/16$	$16 \times 16 \times 16$
7	bn_2	Batch normalization	–	$16 \times 16 \times 16$
8	conv_4	Convolution	$3 \times 3/16$	$16 \times 16 \times 16$
9	bn_4	Batch normalization	–	$16 \times 16 \times 16$
10	conv_5	Convolution	$3 \times 3/16$	$16 \times 16 \times 16$
11	bn_5	Batch normalization	–	$16 \times 16 \times 16$
12	relu_2	ReLu	–	$16 \times 16 \times 16$
13	addition_2	Addition	–	$16 \times 16 \times 16$
14	relu_5	ReLu	–	$16 \times 16 \times 16$
15	conv_7	Convolution	$3 \times 3/16$	$16 \times 16 \times 16$
16	conv_6	Convolution	$1 \times 1/16$	$16 \times 16 \times 16$
17	bn_6	Batch normalization	–	$16 \times 16 \times 16$
18	conv_8	Convolution	$3 \times 3/16$	$16 \times 16 \times 16$
19	bn_8	Batch normalization	–	$16 \times 16 \times 16$
20	relu_4	ReLu	–	$16 \times 16 \times 16$
21	bn_7	Batch normalization	–	$16 \times 16 \times 16$
22	addition_1	Addition	–	$16 \times 16 \times 16$
23	maxpool_2	Max pooling	2×2	$8 \times 8 \times 16$
24	conv_3	Convolution	$3 \times 3/16$	$8 \times 8 \times 16$
25	bn_3	Batch normalization	–	$8 \times 8 \times 16$
26	maxpool_3	Max pooling	2×2	$4 \times 4 \times 16$
27	relu_3	ReLu	–	$4 \times 4 \times 16$
28	fc	Fully connected	–	$1 \times 1 \times 2$
29	softmax	Softmax	–	$1 \times 1 \times 2$
30	classoutput	Classification output	–	–

minimizing the internal covariate shift and speed up the learning process. Rectified Linear Unit (ReLu) layer have been used as an activation function, which introduces non-linearity in the network. We have used the Max pooling layer to reduce the activation dimension for minimizing the computation. At the last, we have used the Softmax layer which is responsible for the classification of image patches in nuclei or non-nuclei. At the Output layer Cross-entropy loss function have been used and network weights are updated using ADAM optimizer. To prevent the network from the overfitting, we have employed L2-regularization.

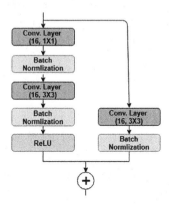

Fig. 2. Residual block.

4 Implementation and Results

4.1 Data Set and Experimental Setup

The performance of proposed method is evaluated at the publicly available dataset of estrogen receptor positive (ER+) breast cancer (BCa) images [15]. The dataset has 143 H&E stained images, with 12,000 annotated nuclei, which are scanned at 40X. Size of each image is 2000 × 2000. Dataset is divided in a ratio of 80:20 to obtain training and test set. There is no overlapping image in the training set and test set. Some patients are associated with more than 1 image, so we have to be sure to divide them into training sets and test sets at the patient level, not image level. We have followed the same protocol to obtain 5-fold of the dataset and results present in this work is an average of 5-fold. Patches generation mechanism has been separately applied on the training set and as well as on the test set to get the training and set of image patches.

To train the DeepRNNetSeg model training set of image patches are introduced to the network. The initial weight of kernel has initialized by the Gaussian distribution with standard deviation 0.01 and hyperparameters values have represented in the Table 3. We experimentally find out that these hyperparameter values are well suitable for our problem. All experiments are performed on Quadro K5200 GPU with CUDA 10.0 using cuDNN.

Table 3. Hyper-parameter values

Hyper-parameter	Values
Learning rate	0.001
Regularization parameter	0.0005
Gradient decay	0.9
Mini-batch size	1024

4.2   Results

To evaluate the performance of the proposed approach we have used specificity, precision, sensitivity/recall, F1-score and accuracy as evaluation metrics.

$$Specificity = \frac{TN}{TN + FP} \times 100 \quad (in \ \%) \tag{1}$$

$$Precision/PPV = \frac{TP}{(TP + FP)} \times 100 \quad (in \ \%) \tag{2}$$

$$Sensitivity/Recall/TPR = \frac{TP}{(TP + FN)} \times 100 \quad (in \ \%) \tag{3}$$

$$F_1 - Score = \frac{2.Precision.Recall}{(Precision + Recall)} \times 100 \quad (in \ \%) \tag{4}$$

$$Accuracy = \frac{TP + TN}{(TP + FP + FN + TN)} \times 100 \quad (in \ \%) \tag{5}$$

where FN, TN, FP and TP represent the false negative, true negative, false positive, and true positive respectively and TPR represents true positive rate and PPV represents positive predictive value.

Table 4. Results of the proposed nuclei detection method.

Fold(s)	Specificity (in %)	Sensitivity/Recall (in %)	Precision (in %)	F1-Score (in %)	Accuracy (in %)
Fold1	88.72	86.79	85.11	85.94	87.89
Fold2	87.71	85.73	83.74	84.72	86.87
Fold3	88.15	87.50	84.15	85.79	87.88
Fold4	88.07	81.61	84.99	83.27	85.15
Fold5	86.96	85.96	82.53	84.21	86.54
Mean	**87.92**	**85.68**	**84.62**	**85.13**	**86.87**

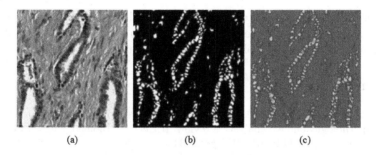

(a)   (b)   (c)

Fig. 3. Nuclei segmentation by our proposed approach; (a) input image; (b) ground truth image and (c) nuclei segmented output image

Table 5. Performance comparison with existing methods

Approach	F1-Score	Recall/TPR	Precision/PPV
Janowczyk et al. [15]	0.83	0.85	0.86
Janowczyk et al. [16]	0.8218	0.8061	**0.8822**
Proposed approach	**0.8513**	**0.8568**	0.8462

The quantitative results of the proposed method for nuclei detection is shown in Table 4. Here, we have presented a performance on the individual folds. Our proposed approach have achieved mean accuracy of 86.87% with 1.13 standard deviation and mean F1-score of 85.13%. We can visually examine the nuclei segmentation results of the proposed method in Fig. 3. The test image as in Fig. 3a are divided in the patches and image patches are fed to the trained DeepRNNetSeg model which produce probability maps corresponding to pixel intensities. The threshold value of 0.5 is applied on the probability maps to obtain the binary results as nuclei or non-nuclei and finally, patches are combined to obtain the nuclei segmented image as in Fig. 3c.

The performance comparison of proposed DeepRNNetSeg model with state-of-the-art methods has shown in the Table 5. It can be observed from the table the proposed model shown is better performance compared to the method present in [15,16]. Proposed method achieved higher true positive rate and F1-score with compared to other state-of-the-art method.

5   Conclusion

In this paper, we have developed a deep residual neural network for nuclei segmentation on histopathological images of breast cancer. DeepRNNetSeg model learned the discriminative feature from pixel intensities of image patches and generate probability maps. We have applied post-processing steps on the probability maps and combined the image patches for obtaining the nuclei segmented output image. Our model has shown improved performance compared to existing state-of-the-art nuclei segmentation algorithm.

In future, the performance of proposed model need to be tested on other publicly available dataset. The proposed model is a generalized method of nuclei segmentation that can be applied to other applications.

References

1. Gurcan, M.N., Boucheron, L., Can, A., Madabhushi, A., Rajpoot, N., Yener, B.: Histopathological image analysis: a review. IEEE Rev. Biomed. Eng. **2**, 147 (2010). https://doi.org/10.1109/RBME.2009.2034865

2. Krizhevsky, A., Sutskever, I., Hinton, G.E.: Imagenet classification with deep convolutional neural networks. In: Advances in neural information processing systems, pp. 1097–1105 (2012). https://doi.org/10.1145/3065386

3. He, K., Zhang, X., Ren, S., Sun, J.,: Deep residual learning for image recognition. In: Proceedings of the IEEE conference on computer vision and pattern recognition, pp. 770–778 (2016). https://doi.org/10.1109/CVPR.2016.90

4. Lu, H., Li, Y., Chen, M., Kim, H., Serikawa, S.: Brain intelligence: go beyond artificial intelligence. Mobile Netw. Appl. **23**(2), 368–375 (2017). https://doi.org/10.1007/s11036-017-0932-8

5. Irshad, H., Veillard, A., Roux, L., Racoceanu, D.: Methods for nuclei detection, segmentation, and classification in digital histopathology: a review-current status and future potential. IEEE Rev. Biomed. Eng. **7**, 97–114 (2013). https://doi.org/10.1109/RBME.2013.2295804

6. Chang, H., et al.: Invariant delineation of nuclear architecture in glioblastoma multiforme for clinical and molecular association. IEEE Trans. Med. Imaging **32**(4), 670–682 (2013). https://doi.org/10.1109/TMI.2012.2231420

7. Nielsen, B., Albregtsen, F., Danielsen, H.E.: Automatic segmentation of cell nuclei in Feulgen-stained histological sections of prostate cancer and quantitative evaluation of segmentation results. Cytometry Part A **81**(7), 588–601 (2012). https://doi.org/10.1002/cyto.a.22068

8. Vink, J.P., Van Leeuwen, M.B., Van Deurzen, C.H.M., De Haan, G.: Efficient nucleus detector in histopathology images. J. Microsc. **249**(2), 124–135 (2012). https://doi.org/10.1111/jmi.12001

9. Fatakdawala, H., et al.: Expectation-maximization-driven geodesic active contour with overlap resolution (emagacor): application to lymphocyte segmentation on breast cancer histopathology. IEEE Trans. Biomed. Eng. **57**(7), 1676–1689 (2010). https://doi.org/10.1109/TBME.2010.2041232

10. Al-Kofahi, Y., Lassoued, W., Lee, W., Roysam, B.: Improved automatic detection and segmentation of cell nuclei in histopathology images. IEEE Trans. Biomed. Eng. **57**(4), 841–852 (2009). https://doi.org/10.1109/TBME.2009.2035102

11. Qi, X., Xing, F., Foran, D.J., Yang, L.: Robust segmentation of overlapping cells in histopathology specimens using parallel seed detection and repulsive level set. IEEE Trans. Biomed. Eng. **59**(3), 754–765 (2012). https://doi.org/10.1109/TBME.2011.2179298

12. Xu, J., et al.: Stacked sparse autoencoder (SSAE) for nuclei detection on breast cancer histopathology images. IEEE Trans. Med. Imaging **35**(1), 119–130 (2016). https://doi.org/10.1109/TMI.2015.2458702

13. Sirinukunwattana, K., Raza, S.E.A., Tsang, Y.W., Snead, D.R., Cree, I.A., Rajpoot, N.M.: Locality sensitive deep learning for detection and classification of nuclei in routine colon cancer histology images. In: IEEE Trans. Med. Imaging, vol. 35(5), pp. 1196–1206, (2016). https://doi.org/10.1109/TMI.2016.2525803

14. Kumar, N., Verma, R., Sharma, S., Bhargava, S., Vahadane, A., Sethi, A.: A dataset and a technique for generalized nuclear segmentation for computational pathology. IEEE Trans. Med. Imaging **36**(7), 1550–1560 (2017). https://doi.org/10.1109/TMI.2017.2677499

15. Janowczyk, A., Madabhushi, A.: Deep learning for digital pathology image analysis: A comprehensive tutorial with selected use cases. J. Pathol. Inform. 7 (2016). https://doi.org/10.4103/2153-3539.186902

16. Janowczyk, A., Doyle, S., Gilmore, H., Madabhushi, A.: A resolution adaptive deep hierarchical (RADHicaL) learning scheme applied to nuclear segmentation of digital pathology images. Comput. Methods Biomech. Biomed. Eng.: Imaging Visual. **6**(3), 270–276 (2018). https://doi.org/10.1080/21681163.2016.1141063

17. Lam, C., Yu, C., Huang, L., Rubin, D.: Retinal lesion detection with deep learning using image patches. Invest. Ophthalmol. Vis. Sci. **59**(1), 590–596 (2018). https://doi.org/10.1167/iovs.17-22721

Classification of Breast Tissue Density

Kanchan Lata Kashyap[1($\boxtimes$)], Manish Kumar Bajpai[2],
and Pritee Khanna[2]

[1] VIT University, Bhopal, India
kanchan.k@vitbhopal.ac.in
[2] Indian Institute of Information Technology,
Design and Manufacturing, Jabalpur, India

Abstract. Breast density classification plays an important role in breast cancer screening. Radiologists visually evaluate mammograms to classify it according to breast tissue density. In this work, automatic breast tissue density classification is presented which consists of preprocessing of mammograms, breast tissue segmentation, feature extraction from the segmented breast tissue and its classification based on the density. Mammogram preprocessing includes breast region extraction and enhancement of mammograms. Partial differential equation based variational level set method is applied to extract the breast region. Enhancement of mammograms is done by anisotropic diffusion. Further, breast tissues are segmented by applying clustering based technique. Texture based Local Ternary Pattern (LTP) and Dominant Rotated Local Binary Pattern (DRLBP) features are extracted in the subsequent step. Breast tissues are classified into 4- classes by applying support vector machine. The proposed algorithm has been tested on the publicly available 500 sample mammograms of Digital Database of Screening Mammography (DDSM) dataset.

Keywords: Anisotropic diffusion · Variational level set · Dominated rotated local binary pattern component · Local Ternary Pattern

1 Introduction

Mammography is widely used screening image modality for early detection of breast cancer. Various types of breast lesions such as micro-calcification, mass, architectural distortion, and bilateral asymmetry can be analyzed by using mammograms. The risk of breast cancer increases as the density of breast tissue increases. The main issue with mammography is that it is difficult to detect the abnormal and normal tissue in a dense breast. Many times, radiologists face difficulty in analyzing small lesions hidden in the dense breast tissue. Therefore, breast tissue density classification plays an important role in analysis of lesions in dense breast tissue. Computer aided breast tissue density classification is of significant importance due to various reasons such as (a) to analyze the risk of breast cancer, (b) to enhance the sensitivity and specificity of computer aided breast cancer detection system, and (c) to decrease time interval for screening. The American College of Radiology proposed Breast Imaging Reporting and Data System (BIRADS) which follows Wolfe mammographic breast tissue density classification to classify breast tissue density as fatty (BIRADS I), fibro-glandular dense tissue (BIRADS II),

© Springer Nature Singapore Pte Ltd. 2020
N. Nain et al. (Eds.): CVIP 2019, CCIS 1148, pp. 254–265, 2020.
https://doi.org/10.1007/978-981-15-4018-9_24

heterogeneously dense tissue (BIRADS III), and extremely dense tissue (BIRDS IV) [1]. Qualitative and quantitative evaluation of breast tissue density has been done by various authors by applying various approaches [2, 3]. Motivated with these observations, an automatic breast tissue density classification approach for 4-class breast tissue density is proposed in this work.

The structure of the paper is articulated as follows. A discussion on the related work and contribution of the present work are discussed in Sect. 2. Methodology used in the present work is explained in Sect. 3. The experimental results and discussions are presented in Sect. 4. Finally, the work is concluded in Sect. 5.

2 Related Work

Petroudi et al. classified breast tissue density by using statistical distribution of texons [4]. Classification accuracy of 76% is obtained on 132 mammograms from Oxford database. Breast tissue density classification of 4 and 2-class is done by Bovis et al. by using Gray-Level Co-occurrence Matrix (GLCM), histogram, intensity, and wavelet based features. Recognition rate of 96.7% is obtained with bagging method on 377 mammograms of Digital Database of Screening Mammography (DDSM) dataset [5]. Scale-Invariant Feature Transform (SIFT) and texon based texture feature are used by Bosch et al. for 4-class breast tissue classification [6]. Classification accuracy of 95.39% and 84.75% are achieved with SVM classifier on 322 and 500 images of MIAS and DDSM dataset, respectively. Oliver et al. used Fuzzy c-means (FCM) clustering algorithm for breast tissue segmentation [7]. GLCM and morphological features are extracted from segmented breast tissue. Highest 77% and 86% classification accuracy are achieved with the combination of Bayesian and k-nn classifier on 831 and 322 images of DDSM and Mammographic Image Analysis Society (MIAS) dataset, respectively.

Liu et al. presented 4-class breast tissue density classification based on multi-scale analysis and statistical features [8]. Subashini et al. applied thresholding technique for breast tissue segmentation [9]. Classification between fatty, Glandular, and dense breast tissue is done by applying statistical features with SVM classifier and obtained 95.44% and 86.4% classification accuracy on 43 and 88 mammograms of MIAS and Full-Field Digital Mammography (FFDM) dataset, respectively. Weighted voting tree classifier for 4-class density classification using histogram, GLCM, Local Binary Pattern (LBP), Chebyshev moments, and Gabor features are proposed by Vállez et al. [10]. Classification accuracy of 91.75% and 91.58% are obtained on 322 and 1137 images from MIAS and FFDM dataset, respectively. He et al. proposed binary model and Bayes classifier for tissue segmentation and mammographic risk assessment [11]. Classification accuracy of 78% and 88% are obtained for 4 and 2-class classification, respectively. Li et al. used texon based and Gabor features for 2-class breast tissue density classification [12] and observed 76% accuracy on 160 mammograms. Sharma et al. proposed correlation-based feature selection for classification between fatty and dense mammogram by applying GLCM, intensity, Law's, and fractal features [13]. Highest 96.46% classification accuracy is achieved on 322 images of MIAS dataset. Chen et al. used various texture-based features and achieved classification accuracy of 78% and 90% for 4 and 2-class classification, respectively. Tzikopoulos et al. used statistical feature for classification [14]. Mohamed et al. employed deep learning approach for two class classification of breast tissue density [15]. Deep convolution neural network is applied by Wu et al. to classify the four class breast tissue density [16]. Classification of breast tissue density into fatty, glandular, and dense by using Dominant Rotated Local Binary Pattern (DRLBP) and LBP features is presented by Kashyap et al. and highest 94.21% classification accuracy is achieved with DRLBP features on MIAS dataset [17]. Here mammogram enhancement has been done using fractional order based technique. Morphological operations are applied to extract breast region and FCM is used to segment breast tissues.

In the present work, four class breast tissue classifications as compared to three class classification of breast tissue in [17] is targeted. Fractional order based enhancement used in [17] depends on the fractional order parameter which needs to be tuned for a particular application. Therefore, this work uses PDE based anisotropic diffusion equation for mammogram enhancement. Variational level set method is applied for better extraction of breast region and fast fuzzy c-means (FFCM) clustering is used to segment breast tissues. Also, evolution of variational level set method is done using RBF based mesh-free technique which is faster than mesh-based technique [19]. Performance obtained with DRLBP features was found better as compared to LBP in [17]. To further establish the usefulness of DRLBP features, the performance is obtained and compared with another texture based Local Ternary Pattern (LTP) features as it can handle the limitation of LBP discussed later.

3 Proposed Technique

Proposed methodology consists of preprocessing, breast tissue segmentation, feature extraction, and classification steps. Detailed description of each step is given below.

3.1 Preprocessing

Preprocessing includes breast area segmentation and mammogram enhancement. Breast area segmentation is done by partial differential equation (PDE) based variational level set method. This step eliminates unnecessary background information and preserves the breast region containing useful information. The evolution of level set method is done by globally supported mesh-free based radial basis function [18]. Low contrast mammogram images are enhanced by PDE based anisotropic diffusion equation which is represented as:

$$I_{EnBreast} = \frac{\partial I_{breast}(x, y, t)}{\partial t} = \nabla.(D(x, y, t)\nabla I_{breast}(x, y, t)) \tag{1}$$

here $I_{breast}(x, y, t)$ denotes extracted breast area, $D(x, y, t)$ denotes diffusion conductance, ∇ represents gradient operator, $\nabla.$ is a divergent operator, and $I_{EnBreast}$ represents enhanced breast area. Mathematically diffusion conductance $D(x, y, t)$ is represented as:

$$D(x, y, t) = \frac{1}{1 + \frac{|\nabla MI_{br}|^2}{K^2}} \tag{2}$$

$$D(x, y, t) = \exp(-\frac{|\nabla MI_{br}|^2}{2K^2}) \qquad (3)$$

here K is conductance parameter. It serves as gradient magnitude threshold which controls the diffusion rate.

3.2   Breast Tissue Segmentation

Breast tissue segmentation is an important step for an automatic CAD system for breast tissue density classification. Breast tissue segmentation is performed by FFCM clustering algorithm in which histogram of the pixel values is generated at first and then it is updated in subsequent iterations of FCM clustering [19]. The steps are given in tissue segmentation algorithm.

3.3   Feature Extraction

Spatial relationship and variations among pixel values of the breast tissue can be obtained by textural features. Texture based DRLBP and LTP features extracted from the segmented breast tissue are discussed here [22, 23].

Dominant Rotated Local Binary Pattern (DRLBP)
LBP code is calculated in the local circular region by considering the difference between the central pixel with its neighboring pixels. LBP code changes upon object rotation due to permanent ordering of weights. The rotation effect of LBP code is shown in Fig. 1 [20]. Pixel values are rotated based on the rotation of the object. The resultant LBP code on the rotated image is different due to fixed ordering of weight as depicted in Fig. 1(d) with yellow color. The arrangement of weights can be changed based on the reference direction. Change in the reference direction also changes according to the image rotation. Only the sign of the difference between center pixel and its neighboring pixels is considered for computation of LBP code instead of the magnitude of the differences. The magnitude of the difference which provides complimentary information is applied in rotated local binary pattern (RLBP) to locate the dominant direction which enhances the discriminative power of LBP code. Maximum difference denoting the index in the circular neighborhood is defined as the dominant direction. Rotation invariant LBP is generated by circularly shifting the weight based on the dominant direction given as:

$$DD = \max_{a \in (0,1,2...P-1)} |x_a - x_c| \qquad (4)$$

here x_c and x_a represent gray value of central pixel and neighborhood pixels, respectively. In a circular symmetric neighborhood, local neighborhood of pixel is determined by the radius, R, and the number of points P. Weights are given with respect to DD which is used as the reference in the circular neighborhood. LBP code computed by this technique is called RLBP which is expressed as:

$$RLBP_{R,P} = \sum_{a=0}^{P-1} f(x_a - x_c).2^{\mathrm{mod}(a-DD,P)} \tag{5}$$

here mod represents modulus operation and weight terms $2^{\mathrm{mod}(a-DD,P)}$ depend on the dominant direction DD.

Algorithm Tissue_Segmentation: FFCM (I_{sub_Breast}, I_{tisuue})

Input: I_{sub_Breast}

Output: I_{tisuue}

I_{sub_Breast} : *subtracted breast image*

I_{sub_Breast} : $x_1, x_2, x_3.... \subseteq R^m$: *set of pixel values in vector space of m dimension*

n : *Number of pixel values*

h : *Intensity histogram of enhanced image* I_{En_Breast}

n_c : *Number of cluster center*

u_{rj} : *Degree of fuzzy membership of pixel x_j in the rth cluster*

k : *Weighting exponent to control fuzziness of the member function*

v_r : *Prototype of cluster center r*

$d^2(x_j, v_r)$: *Distance between pixel x_j and cluster center v_r*

I_{tisuue} : *Segmented breast tissue*

begin

1. Compute intensity histogram $h = hist(I_{sub\_Breast})$

2. Initialize $U = [u_{rj}]$ matrix, U^0

3. At step k : Calculate center vectors $v^k = [v_j]$ with U^k

$$v_j = \frac{\sum_{j=1}^{n} u_{rj}^k . x_j}{\sum_{j=1}^{n} u_{rj}^k}, 1 \le r \le n_c$$

4. Update U^k and U^{k+1} and

$$u_{rj} = \left(\sum_{s=1}^{n_c} \left(\frac{d^2(x_j, v_r)}{d^2(x_j, v_s)} \right)^{\frac{2}{k-1}} \right)^{-1}, 1 \le j \le n,\ 1 \le r \le n_c$$

5. if $u_{rj}^{k+1} - u_{rj}^k < \varepsilon$ then STOP ; otherwise return to step 2

end

Weights are circularly shifted based on the dominant direction and depending on the neighborhood. The gray and red values in Fig. 1 represent the pixel above the threshold and the pixel value which is related to dominant direction, respectively. The bit value of index DD is fixed and it is always 1; and other remaining weights are rotated corresponding to it. The same weight is assigned in both original and rotated segmented tissue. So computed DRLBP codes are same for segmented and rotated ROI as shown in Fig. 1(d) and (h).

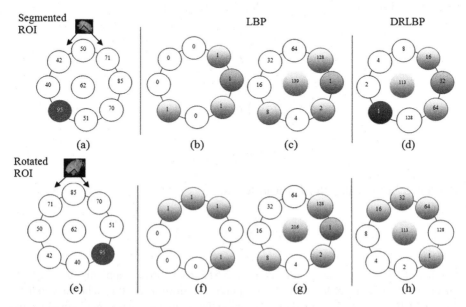

Fig. 1. Computation of LBP and DRLBP codes: (a) neighborhood pixel values in the segmented tissue, (b) thresholded neighboring pixel value which are above threshold value presented in gray color, (c) weights given to the thresholded neighbors and computed LBP code presented in yellow color, (d) weights given to the thresholded pixel values for RLBP with reference presented in red color, (e) pixel values of neighborhood for rotated segmented tissue given in (a), (f) thresholded neighboring pixel values above threshold presented in gray color, (g) weights given to thresholded neighbors and computed LBP code presented in yellow color and (h) weights given to thresholded neighbors and the same RLBP codes computed for original and rotated segmented tissue are depicted in yellow color [20]. (Color figure online)

Local Ternary Pattern (LTP)

LBP is robust for variational illumination and contrast but sensitive to noise and small pixel value variations. LTP can remove this limitation of LBP. Mathematical expression for computing LTP for pixel position (x, y) can be given as:

$$LTP_{x,y} = \sum_{a=0}^{B-1} s'(x_a - x_c)3^a \tag{6}$$

$$s'(z) = \begin{cases} 1, z \geq TH \\ 0, -TH < z < TH, \\ -1, z \leq -TH \end{cases} \tag{7}$$

where TH is a user defined threshold. LTP has three states resulting in 3^B bin block histogram. Histogram of 6651 bins is generated for $(B = 8)$. This high dimensional data is converted into low dimensional by splitting LTP code into "Lower" and "upper" LBP codes. The "upper" LBP code LBP_U is computed as:

$$LBP_U = \sum_{a=0}^{B-1} g(x_a - x_c)2^a \tag{8}$$

$$g'(z) = \begin{cases} 1, z \leq -T \\ 0, otherwise \end{cases} \tag{9}$$

$$g(z) = \begin{cases} 1, z \geq T, \\ 0, otherwise \end{cases} \tag{10}$$

The "lower" LBP code LBP_L is calculated as:

$$LBP_L = \sum_{a=0}^{B-1} g'(x_a - x_c)2^a \tag{11}$$

Number of bins in the histogram is reduced to 512 from 6561 by applying this method. This number is further reduced to 118 by applying uniform LBP. It is to be noticed here that DRLBP codes are computed by converting the segmented tissue in 2^B bin block by using single threshold value, whereas LTP codes are computed in 3^B bin block by using two threshold values.

3.4 Classification

Multi-class SVM classifier, which is a well-known supervised machine learning tool, has been used to classify breast tissue density [24]. SVM with four kernel functions, i.e., linear, polynomial, RBF, and MLP are used to classify 4-class breast tissue.

4 Experimental Results

The proposed method is validated on publicly available DDSM dataset [21]. The information about patient age, screening exam date, date on which mammograms have been digitized, and breast tissue density are given in the dataset. Breast tissue density of DDSM dataset is divided into 4-classes, i.e., BIRADS-I, BIRADS-II, BIRADS-III, and BIRADS-IV. Total 500 randomly chosen mammograms include 90 BIRADS-I, 130

BIRADS-II, 140 BIRADS-III, and 140 BIRADS-IV types of breast tissue. Examples of mammograms from DDSM dataset with different breast tissue density are shown in Fig. 2.

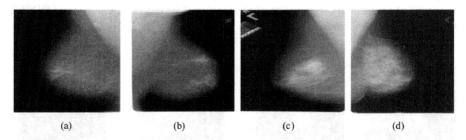

(a)   (b)   (c)   (d)

Fig. 2. Sample mammograms with different breast tissue densities from DDSM dataset: (a) BIRADS-I (D_4010), (b) BIRADS-II (A_1169), (c) BIRADS-III (A_1316), and (d) BIRADS-IV (B_3072).

4.1 Results of Preprocessing

Pre-processing step starts with breast region extraction. Intermediate results of breast region extraction of randomly chosen mammograms with dense breast tissue are shown in Fig. 3. Original mammogram, evolution of level set method to extract breast region, and extracted breast region are shown in Fig. 3(a), (b), and (c) respectively. It can be seen that patient labels are eliminated properly. Further, PDE based linear diffusion method has been applied on extracted breast region for enhancement. Original extracted breast region is inverted in the subsequent step and the same PDE based linear diffusion method is applied on the inverted image. Original enhanced image is subtracted with the enhanced inverted image to get the subtracted image.

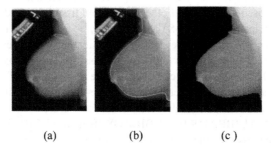

(a)   (b)   (c)

Fig. 3. Results obtained from different operations to segment the breast area of a mammogram from DDSM dataset: (a) Sample mammogram, (b) Evolution of level set function using RBF method, and (c) Extracted breast region.

4.2   Results of Tissue Segmentation

In further step, FFCM algorithm is utilized on subtracted image to segment the breast tissue. Figure 4 depicts the outcome of segmented breast tissue on randomly chosen sample images of DDSM datasets. Figure 4(a) and (b) present the subtracted mammogram and segmented breast tissue of cluster map 3 on the sample images of DDSM dataset with different tissue densities.

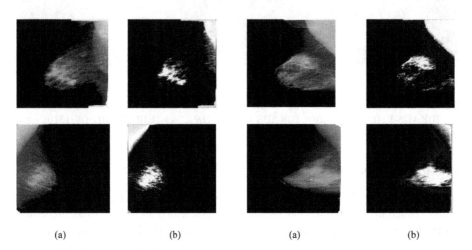

(a)   (b)   (a)   (b)

Fig. 4. Segmentation results for sample mammograms from DDSM dataset with different tissue density: (a) Subtracted mammogram and (b) segmented breast tissue.

4.3   Results of Density Classification

Connected component labeling is executed to extract the segmented breast tissue. Largest extracted component is excluded as pectoral muscle and not used for feature extraction in the subsequent step. All other segmented tissue are used for extraction of DRLBP and LTP features. Multi-SVM classifier, a supervised machine learning method, is utilized for 4-class tissue density classification with linear, RBF, polynomial, and MLP kernel functions. In multi-SVM classifier one-versus-all classification technique is used. Training of SVM classifier is done by two-third of sample images of mammograms whereas testing is performed on one-third samples images. Total 500 sample mammograms of DDSM dataset are used for validation of algorithm. Out of this, 30 BIRADS-I, 44 BIRADS-II, 44 BIRADS-III, and 47 BIRADS-IV type mammograms are used for testing. The classification performance is evaluated by employing 10-fold cross-validation 10 times, and calculating average of these runs. In the case of 10-fold cross validation technique, whole database is divided into 10-folds. Training of SVM is performed using 9-folds and remaining 1-fold is used for testing. Classification performance of SVM classifier with RBF, linear, polynomial, and MLP kernel function on DDSM datasets are presented in Table 1. Irrespective of the classifier, performance obtained with DRLBP features is better as compared to that obtained with LTP features. SVM classifier with RBF kernel function gives better performance on DRLBP features. Performance of the classifier is measured in terms of sensitivity, specificity and accuracy which are defined as:

Sensitivity, also called True Positive Rate (TPR), is defined as the ratio of the actual positive which is correctly classified as abnormal. It is also known as the true positive fraction (TPF).

$$Sensitivity = \frac{TP}{TP + FN} \tag{12}$$

Specificity, also called False Positive Rate (FPR), is defined as the ratio of the actual negative which is correctly classified as normal. It is also known as the false positive fraction (FPF).

$$Specificity = \frac{TN}{TN + FP} \tag{13}$$

Accuracy is defined as the ratio of the correct classification to the total number of test cases.

$$Accuracy = \frac{TP + TN}{TP + TN + FP + FN} \tag{14}$$

Table 1. Classification performance (in %) using SVM classifier with different kernel function.

	Kernel function	Sensitivity	Specificity	Accuracy
DRLBP	RBF	**90.4**	**88.6**	**89.5**
	Linear	84.7	82.3	83.5
	Polynomial	86.8	84.4	85.6
	MLP	83.4	85.5	84.4
LTP	RBF	84.2	80.7	82.45
	Linear	81.23	79.12	80.17
	Polynomial	83.2	82.7	82.95
	MLP	84.9	80.1	82.5

4.4   Comparison with the Existing Techniques

To compare the proposed approach with the existing approaches, results on the average sensitivity, specificity and accuracy are summarized in Table 2. Correct classification accuracy is computed by averaging the highest classification accuracy for each class using each features. For fair comparison, the approach in [9] is validated on the same 500 mammograms of DDSM dataset. Breast tissues are segmented using thresholding technique and six first order statistical features are extracted from segmented ROI. Average sensitivity, specificity and accuracy of 82.8%, 79.6%, and 81.2%, respectively, is obtained for 4-class breast tissue density on DDSM dataset with SVM classifier and RBF kernel function. The methodology used by [14] is also applied on the same 500 mammograms of DDSM dataset. Minimum cross entropy thresholding is used for breast tissue segmentation and the first order statistical features are extracted from segmented ROI. Average sensitivity, specificity and accuracy of 83.6%, 80.8%, and 82.20% have been obtained on DDSM dataset, respectively. It can be observed that the proposed approach gives better correct classification rate as compared to the existing automatic density classification techniques.

Table 2. Comparison of the present work with the existing approaches on DDSM dataset.

Reference	Features	Tissue type	Sensitivity	Specificity	Accuracy
Subashini et al. [9]	Statistical	4	82.8%	79.6%	81.2%
Tzikopoulos et al. [14]	Statistical	4	83.6%	80.8%	82.20%
Proposed work	DRLBP	4	**90.4%**	**88.6%**	**89.5%**

5 Conclusions

Fully automatic classification of breast tissue density is presented in this work. Breast tissue region is segmented by applying FFCM clustering technique. Texture based DRLBP and LTP features are extracted from the segmented breast tissue for its classification. Classification has performed by SVM classifier with various kernel functions and it is analyzed from the results that the classification accuracy obtained for DRLBP feature is better as compared to LTP feature with RBF kernel function. It can be observed that there is a lot scope of improvement in the observed performance here. It is aimed to improve classification accuracy by applying deep learning based approaches in future.

References

1. Wolfe, J.N.: Breast patterns as an index of risk for developing breast cancer. Am. J. Roentgenol. **126**(6), 1130–1137 (1976)
2. Ho, W.T., Lam, P.W.T.: Clinical performance of computer-assisted detection (CAD) system in detecting carcinoma in breasts of different densities. Clin. Radiol. **58**(2), 133–136 (2003)
3. Assi, V., Warwick, J., Cuzick, J., Duffy, S.W.: Clinical and epidemiological issues in mammographic density. Nat. Rev. Clin. Oncol. **9**(1), 33–40 (2012)
4. Petroudi, S., Kadir, T., Brady, M.: Automatic classification of mammographic parenchymal patterns: a statistical approach. In: Proceedings of the 25th Annual International Conference of the IEEE Engineering in Medicine and Biology Society, pp. 798–801 (2003)
5. Bovis, K., Singh, S.: Classification of mammographic breast density using a combined classifier paradigm. In: 4th International Workshop on Digital Mammography, 177–180 (2002)
6. Bosch, A., Munoz, X., Oliver, A., Marti, J.: Modeling and classifying breast tissue density in mammograms. In: IEEE Computer Society Conference on Computer Vision and Pattern Recognition, pp. 1552–1558 (2006)

7. Oliver, A., et al.: A novel breast tissue density classification methodology. IEEE Trans. Inf Technol. Biomed. **12**(1), 55–65 (2008)

8. Liu, Q., Liu, L., Tan, Y., Wang, J., Ma, X., Ni, H.: Mammogram density estimation using sub-region classification. In: 4th International Conference on Biomedical Engineering and Informatics (BMEI), pp. 356–359 (2011)

9. Subashini, T.S., Ramalingam, V., Palanivel, S.: Automated assessment of breast tissue density in digital mammograms. Comput. Vis. Image Underst. **114**, 33–43 (2010)

10. Vállez, N., et al.: Breast density classification to reduce false positives in CADe systems. Comput. Methods Programs Biomed. **113**(2), 569–584 (2014)

11. He, W., Denton, E.R.E., Zwiggelaar, R.: Mammographic segmentation and risk classification using a novel binary model based bayes classifier. In: Maidment, A.D.A., Bakic, P.R., Gavenonis, S. (eds.) IWDM 2012. LNCS, vol. 7361, pp. 40–47. Springer, Heidelberg (2012). https://doi.org/10.1007/978-3-642-31271-7_6

12. Li, T., et al.: The association of measured breast tissue characteristics with mammographic density and other risk factors for breast cancer. Cancer Epidemiol. Prev. Biomark. **14**(2), 343–349 (2005)

13. Sharma, V., Singh, S.: CFS–SMO based classification of breast density using multiple texture models. Med. Biol. Eng. Comput. **52**(6), 521–529 (2014)

14. Tzikopoulos, S.D., Mavroforakis, M.E., Georgiou, H.V., Dimitropoulos, N., Theodoridis, S.: A fully automated scheme for mammographic segmentation and classification based on breast density and asymmetry. Comput. Methods Programs Biomed. **102**(1), 47–63 (2011)

15. Wu, N., et al.: Breast density classification with deep convolutional neural networks. In: proceeding of 2018 IEEE International Conference on Acoustics, Speech and Signal Processing (ICASSP), Calgary, Canada (2018)

16. Mohamed, A.A., Berg, W.A., Peng, H., Luo, Y., Jankowitz, R.C., Shandong, W.: A deep learning method for classifying mammographic breast density categories. Med. Phys. **45**(1), 314–321 (2018)

17. Kashyap, K.L., Bajpai, M.K., Khanna, P.: Breast Tissue Density Classification in mammograms based on supervised machine learning technique. In: Proceedings of 10th Annual ACM COMPUTE Conference(ACM COMPUTE), India, pp. 131–135 (2017)

18. Kashyap, K.L., Bajpai, M.K., Khanna, P.: Globally supported radial basis function based collocation method for evolution of level set in mass segmentation using mammograms. Comput. Biol. Med. **87**, 22–23 (2017)

19. Kashyap, K.L., Bajpai, M.K., Khanna, P.: An efficient algorithm for mass detection and shape analysis of different masses present in digital mammograms. Multimed. Tools Appl. **77**(8), 9249–9269 (2017)

20. Kashyap, K.L., Bajpai, M.K., Khanna, P., Giakos, G.: Mesh free based variational level set evolution for breast region segmentation and abnormality detection using mammograms. Int. J. Numer. Methods Biomed. Eng. **34**(1), 1–20 (2018)

21. Rose, C., et al.: Web services for the DDSM and digital mammography research. In: Astley, S.M., Brady, M., Rose, C., Zwiggelaar, R. (eds.) IWDM 2006. LNCS, vol. 4046, pp. 376–383. Springer, Heidelberg (2006). https://doi.org/10.1007/11783237_51

22. Mehta, R., Egiazarian, K.: Dominant rotated local binary patterns (DRLBP) for texture classification. Pattern Recogn. Lett. **71**, 16–22 (2016)

23. Satpathy, A., Jiang, X., Eng, H.L.: LBP-based edge-texture features for object recognition. IEEE Trans. Image Process. **23**(5), 1953–1964 (2014)

24. Cortes, C., Vapnik, V.: Support-vector networks. Mach. Learn. **20**, 273–297 (1995)

Extreme Weather Prediction Using 2-Phase Deep Learning Pipeline

Vidhey Oza[1]([✉]), Yash Thesia[1], Dhananjay Rasalia[1], Priyank Thakkar[1], Nitant Dube[2], and Sanjay Garg[1]

[1] Computer Science and Engineering Department, Institute of Technology, Nirma University, Ahmedabad, Gujarat, India
`15bce130@nirmauni.ac.in`
[2] Space Application Centre, Indian Space Research Organisation, Ahmedabad, Gujarat, India

Abstract. Weather nowcasting is a problem pursued by scientists for a long time. Accurate short-term forecasting is helpful for detecting weather patterns leading to extreme weather events. Adding the dimension of nowcasting to extreme weather prediction increases the ability of models to look for preliminary patterns ahead in time. In this paper, we propose a two-stage deep learning pipeline that fuses the usability of nowcasting to the high value of extreme events prediction. Our experiments are performed on INSAT-3D satellite data from MOSDAC, SAC-ISRO. We show that our pipeline is modular, and many events can be predicted in the second phase based on the availability of the relevant data from the first phase. Testing for extreme events like the Chennai floods of 2015 and Mumbai floods of 2017 validates the efficacy of our approach.

Keywords: Weather prediction · Deep learning · Convolutional LSTM · U-Net · Two stage learning

1 Introduction

Early detection of extreme weather events like extreme rain or heat events is highly valuable not only to the general public but also to meteorological departments as well as disaster relief organizations for preliminary actions. Existing methods for prediction and detection of extreme events primarily rely upon the adeptness of human researchers to accurately define parameters and thresholds for identifying such events.

Recent works in deep learning have shown great promises in various prediction tasks, including computer vision and pattern recognition tasks. Looking at weather prediction from this perspective, the ultimate goal of a predictive system in this scenario is to accurately define hotspots of extreme weather events based on past activities like movement of convective systems etc.

If extreme events are extracted from features predicted in the very short term, the benefit of analysing patterns will be of benefit to the general public and

© Springer Nature Singapore Pte Ltd. 2020
N. Nain et al. (Eds.): CVIP 2019, CCIS 1148, pp. 266–282, 2020.
https://doi.org/10.1007/978-981-15-4018-9_25

scientists alike. Hence, an important scientific goal for operational meteorology is weather nowcasting, which can be used effectively to predict extreme weather events. This problem of weather nowcasting can be formulated as the prediction of the next data point or frame in sequence, given a fixed window of sequence that precedes the said data point. Using these short-term predictions, the system should be able to identify areas where the cloud systems or humidity levels can lead to extreme conditions in the near future.

We propose a 2 phase deep learning pipeline. In the first phase, we nowcast frames captured by the INSAT-3D satellite sensors. In the second phase, we use these predictions to identify hotspots of meteorological variables that signal towards genesis of extreme weather events.

The rest of the paper is structured as follows. Related work and literature survey is discussed in Sect. 2, and required preliminaries are discussed in Sect. 3. Then the methodology and implementation details are elaborated in Sect. 4, followed by discussion of results in Sect. 5. Section 6 ends with concluding remarks and potential future exploration.

2   Literature Survey

Weather prediction is a problem researchers have been trying to solve using machine learning and deep learning. Authors of [1] surveyed different methods of deep learning techniques that can be used for weather forecasting by contrasting different aspects of the algorithm. They analyze the learning speed, amount of data required, etc. to gauge the capability of recurrent networks with time-delay networks and conventional ANNs. Forecasting using statistical models have been previously explored by researchers [2,3].

[4–8] presented different techniques and perspectives that can be used to predict various meteorological variables. Authors of [4] proposed an efficient learning procedure that made use of genetic programming and DBNs to boost the predictions of the decision tree based models. [5] used the widely adopted LSTM models for sequence-based weather forecasting. [6] compared two commonly used deep learning approaches to forecast the weather of Nevada: sparse denoising autoencoders and feedforward neural networks. Unique applications of forecasting were explored in [7]. They forecasted the ground visibility of airports, and tested their algorithms with positive results at the Hang Nadim airport. Such applications inspire researchers to explore new ways in which forecasting a particular aspect of weather can help the society in a more direct manner. Research in [8] compared 5 different machine learning algorithms to perform supervised lithology classification, based on geophysical data that is spatially constrained and remotely sensed. They find that Random forests are a good choice for first-pass predictions for practical geomapping applications.

Since weather data is generally extracted through a series of satellite images or an array of ground sensors, weather prediction can be thought of as a computer vision problem. [9–14] encourage research in this domain through the lenses of convolutional neural networks. Weather nowcasting can be viewed as a computer vision problem, where the goal is to predict the next image frame based on a previous sequence. Authors of [10] trained a deep CNN model to classify for common extreme weather events like tropical cyclones, weather fronts and atmospheric rivers. For each event, a separate model was trained which was capable of achieving accuracy of more than 90% in most cases. [9] predicted the onset of extreme precipitation cluster using meteorological variables like zonal and meridional wind. Their proposition was that prediction of a long-lead cluster (5–15 days ahead) can be approached by designing a spatio-temporal convolutional network. Similar concept was used in [11], where hourly windspeed near the Rhode Island area was predicted by looking at sequence data in the form of sparse images gathered from 57 METAR stations. Authors of [13] proposed a novel Convolutional LSTM network that fuses the architectural benefits of the LSTM modules with the spatial information preservation of convolutional networks. They used this architecture to nowcast radar echo images using the past image sequence, making their contribution not only elegant, but also generalized in potential applications. A unique architecture was proposed in [14] for short-range weather prediction. They developed a Dynamic Convolution layer, which change their weights even during test time based on the value of the inputs themselves, and thus increases the learnability of the algorithm.

With rising trend in solving weather forecasting and extreme events prediction, researchers in [12] had proposed a repository of extreme climate datasets. It includes data in 16 different channels with a temporal resolution of 3 h and a spatial resolution of 25 km. This repository has the potential to support numerous upcoming as well as ongoing research in many different fields adhering to this domain of weather forecasting.

In this paper, we explore a novel way to predict extreme weather events. Through our literature survey, we found that there are different deep learning models that predict whether a particular image frame is premonition to a specific weather event. However, research is usually carried out for a specific kind of event or set of events only. We propose a two-stage pipeline that is designed to be modular, and can hence predict any weather event given the availability of data and the appropriate dependencies with the first-stage predictions. We discuss this in detail in Sect. 4.

3   Preliminaries

In this section, we discuss the preliminary knowledge that we build upon for our implementation. Mainly, we discuss The Convolutional LSTM network (and its own prerequisites) that was proposed for precipitation nowcasting, and U-Net, which was developed for image segmentation but is being tweaked and modified to perform different kinds of image-to-image predictions.

3.1   Convolutional LSTM (ConvLSTM)

The Long Short-Term Memory (LSTM) network was introduced for the temporal information prediction. The idea was to represent the sequence data in 1-D space and predicting the future element based on the stacked LSTM layers. They are designed in form to carry the required information which can hold the sequential information.

This approach was modified by [13], with novel approach of integration of convolution operation and LSTM for extending the functionality of temporal with spatial specific details. So, the dimension of input is transformed from 1-D to 3-D, as it has spatial information as a two-dimensional vector and third dimension specifies the time-series related information for prediction of next 3-D tensor. The 2D image will be applied convolution resulting it into the 3D tensor whose last two dimensions signifies rows and columns of the image. The first dimension is the result of convolution and keeping the spatial characteristic of the information, since weather holds significance in location as much as time-series. So, there will be layers consisting of the 3D tensors each with having both temporal and spatial features The purpose of integration of convolutional layers is for integrating location specific details. This convolutional kernel is selected in the manner for having small-scale and large-scale information preservation simultaneously. The important gate equations related to ConvLSTM are listed in Eq. 1.

$$
\begin{aligned}
i_t &= \sigma(W_{xi} * \chi_t + W_{hi} * H_{t-1} + W_{ci} * C_{t-1} + b_i) \\
f_t &= \sigma(W_{xf} * \chi_t + W_{hf} * H_{t-1} + W_{cf} * C_{t-1} + b_f) \\
C_t &= f_t \circ C_{t-1} + i_t \circ tanh(W_{xc} * \chi_t + W_{hc} * H_{t-1} + b_c) \\
o_t &= \sigma(W_{xo} * \chi_t + W_{ho} * H_{t-1} + W_{co} * C_t + b_o) \\
H_t &= o_t \circ tanh(c_t)
\end{aligned}
\tag{1}
$$

Here, $*$ and $\circ$ are denoting the Hadamard product and convolution operator on the parameters respectively. We can also consider the fully connected LSTM as the particular case of ConvLSTM, which will be having only feature representation in 1-D space.

3.2   U-Net

U-Net is fully convolutional network proposed by [15]. It consists of convolutional layers, max pooling layers and up-convolutional layers. U-Net has the capability to preserve the structure of an image. It was introduced for medical image segmentation. U-Net has 9 convolutional group in which 4 of them are for encoding purpose, 1 is lower dense layer and remaining 4 are used for decoding of the image. Encoding path of U-Net is termed as contraction path and decoding path of it was termed as expansion path.

Convolutional groups in the encoder consists of two 3×3 convolutional layers with ReLU activation followed by a 2×2 max-pooling layer. Input of the convolutional group of expansion path is concatenation of output of previous layer that is up-convolutional and output of mirrored layer of contraction path. Concatenation is done for image localization purpose. The bottom most convolutional group has two 3×3 convolutional layers with ReLU activation function followed by 2×2 up-convolutional layer. The last layer of U-Net is 1×1 convolutional layer with sigmoid activation function. They used elastic deformation techniques for data augmentation. Number of features channels in up-sampling part are more to preserve image context.

4   Our Proposed Approach

In this section, we present our 2-phase pipeline, which is used to nowcast extreme events. In the first phase, we nowcast INSAT-3D 6 channel frames using separate ConvLSTM models, and in the second phase, we use these predictions to forecast extreme events.

4.1   Framework

We now discuss the framework of our approach in detail, including which kind of models are used in what phase, and what the data flow is from one phase to another. Basic framework is given in Fig. 2, and details are as follows.

Data Preprocessing. The complete data that we have worked on was fetched using the INSAT-3D imaging sensors. The standard L1B products in the database correspond to the first level of processing over the raw files from the satellite sensors.

For training purposes, we convert the full frames into patches of $p \times q$ with a stride of s. During reconstruction, we compute the pixel by overlapping the patches and averaging the values based on its occurrences on different patches.

With such a mechanism of strided patching, we solve three problems. First, that any frame prediction model can be trained with very limited resources. With increasing availability of resources, the model can be adjusted to accommodate a larger patch. Second, the model can be trained with fairly low data. This is explained in detail in Sect. 4.2. Finally, the same model is trained on various patches of the full frame, and thus the model undergoes generalized training as well. This means, instead of the network learning on a single patch from the full frame, multiple patches are fed with the same importance and hence instead of learning specifically about the design of a single patch, it learns the shifting of brightness temperature or radiation of the target channel.

Phase 1. In the 1st Phase, we use $(t-4)^{th}$ to t^{th} frame as input, and extrapolate $(t+1)^{th}$ frame. This prediction is done for all 6 channels. This prediction helps in generating derived products for the predicted future. To make this possible, we use a ConvLSTM model that takes 5 temporal images as input and forecasts the 6th image. As discussed in Sect. 3, ConvLSTM specializes in introducing a temporal dimension to convolutional networks, and hence learn sequence information in a more robust way. Hence, we train 6 different ConvLSTM models for the 6 channels as given in Table 1, and generate $(t+1)^{th}$ frames for each channel. Our selection for this is done through GridSearch as discussed in Sect. 4.3.

Phase 2. Using these predicted frames, and use the appropriate ones as input to a separate model and forecast the derived product frame in Phase 2. As the parameter, Hydro-Estimator Precipitation which is used for the detection of extreme rainfall events in near future can be derived from TIR1 and TIR2 channels. For extreme heat events, we make use of Land Surface Temperature which are generally derived from VIS and SWIR channels.

The extreme weather events prediction can be interpreted as an image segmentation problem. Our requirement is pixel-wise classification, which will denote the existence of extreme changes in different weather conditions. With this idea in place, U-Net can be interpreted as a type of autoencoder which downsample our image to generate a latent representation, and then reconstruct it using upsampling and concatenation. The main functionality of U-Net that we exploit is image-to-image prediction for more accurate changes in the edges, which needs to be addressed because the complex derived product of the extreme events are very much threshold-based events. This is inspired from different papers using U-Net in their own way for image-to-image prediction [16,17] (Fig. 1).

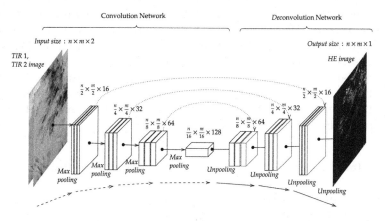

Fig. 1. Phase 2 - extreme event prediction using U-Net. The example used here is for hydro-estimator prediction using TIR1 and TIR2 frames.

The following subsections describe the data preparation and the model description for the complete approach, and then the specific details of how the models are structured and aligned to build the 2-phase pipeline.

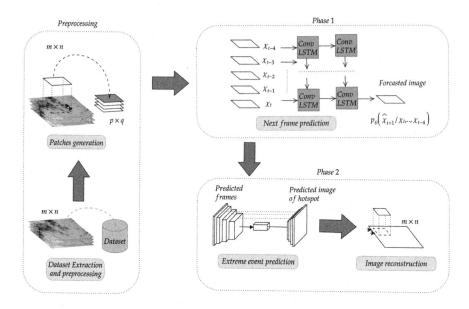

Fig. 2. Two-phase deep learning pipeline framework.

4.2 Dataset Details

In the first phase, we used the 6 channel standard disk products as given in Table 1. This data consists of the L1B standard products that are used to derive geophysical products like hydro-estimator precipitation or land surface temperature. These 6 channels refer to 6 different spectra of light from the INSAT-3D Imager sensors for the same time-stamp and position.

The spatial resolution of the complete image in coordinates is 60 °N to 60 °S and 30 °E to 130 °E, with varying pixel based resolution as given in Table 1. The temporal resolution of the data is 30 min. In the table, B.T. refers to brightness temperature values of the given channel.

Table 1. L1B standard products (INSAT-3D). Used in Phase 1.

Sr. no.	Channel name	Spatial resolution	Description
1	VIS	1 km	Radiance for visible channel
2	SWIR	1 km	Radiance for shortwave infrared channel
3	TIR1	4 km	B.T. for thermal infrared channel 1
4	TIR2	4 km	B.T. for thermal infrared channel 2
5	MIR	4 km	B.T. for middlewave channel
6	WV	8 km	B.T. for water vapor channel

These standard products are then used to derive different derived products like hydro-estimator precipitation, fog, snow, temperature, etc. These are computed using different complex algorithms designed by the scientists and researchers at ISRO. We worked on different derived products to predict 2 different sets of extreme weather events, viz. extreme rainfall events and extreme heat events. To create the training set, we make use of the derived products of the relevant meteorological variables. The complete description of this data is given in Table 2.

Table 2. Derived products. Used in Phase 2.

Sr. no.	Product name	Resolution	Application	Dataset period
1	Hydro-estimator precipitation	4 km	Used for extreme rainfall events	July to August 2018
2	Land surface temperature	4 km	Used for extreme heat events	April to May 2018

The hydro-estimator method is one of the ways of precipitation forecasting, and is comprehensively compared in [18]. They show that products derived from this hydro-estimator rainfall are able to capture heavy rainfall episodes that were not as predictable with previous techniques. Heat events prediction is done by data from land surface temperature.

Since neural networks work best when the data is properly normalized, we use different normalization techniques in the two phases. In Phase 1, we make use of Min-Max scaling, given by Eq. 2a. This choice was made mainly due to the fact that the range of data is well defined in the data retrieval system, and hence it is best to make use of this range for making sure the data stays in the range of 0–1.

In Phase 2, the interpretation of data changes from a quantitative to a qualitative viewpoint. In other words, extreme weather events are extracted from the derived products using a certain threshold defined by the Meteorology Department for each product. This means that after a certain value, any event is an extreme weather event. To implement this effectively while normalizing, we use tanh normalization, as given in Eq. 2b. After a certain value, tanh returns 1 for any further increase in the input. This efficiently implements the act of thresholding while still being continuous and differentiable. While for Land Surface Temperature, we follow Min-Max Scaling as usual for a complete temperature analysis.

$$X_{norm} = \frac{X - X_{min}}{X_{max} - X_{min}} \tag{2a}$$

$$X_{norm} = tanh(X) \tag{2b}$$

4.3   Implementation Details

Figure 2 shows our complete implementation pipeline structure. As discussed in the previous sections, our approach is divided into 2 stages.

In the next-frame prediction stage, we train a standard ConvLSTM model on five 1800×1800 images as input, and one 1800×1800 image as expected output. With this mechanism, we essentially feed the model with data from the past 2.5 h, and hence the model learns the movement of pixels in a particular direction and uses it to identify the next position of the given pixel, and augment this information in a complete frame of next time step (+30 min). As discussed in Sect. 4.1, the frames are converted into overlapping patches and then used to train the frame prediction model (Table 3).

Table 3. GridSearch values for hyper-parameters.

Feature tuning	Parameter values
Frame size for Phase-1	3, 4, ... 8

In the extreme events prediction stage, we use the predicted frames as input to the model, and train a modified version of U-Net to predict the values of various parameters that correspond to different extreme events (as mentioned in Table 2). Using a given threshold value, we classify each pixel as having the extreme weather event or not. For different weather events, we input different 1800×1800 frames, and train the model for the expected 1800×1800 image of the corresponding meteorological variable.

We chose a patch size of 40×40 with stride of 20. This choice of shape and stride was done for divisibility with the dimensions of the full frame. We chose an overlapping stride to make up for the square grid noise inherent with the prediction mechanism of the convolutional network. This strided patching was inspired by [19], where they used strided patching to solve problems like inadequate dataset volume and discussed auxiliary benefits like generalized training.

Since each data point of 1800×1800 frame is converted to such patches, each such data point is converted to 7921 patch-based data points. Hence, data of approximately one month is enough to train a model with adequate robustness.

5   Results and Discussion

For a comprehensive evaluation of our implementation, we use PSNR and SSIM as metrics. These are benchmark accuracy measures used in full image prediction, in frame interpolation as well as frame extrapolation.

5.1   Performance Parameters

PSNR refers to Peak Signal-to-Noise Ratio, which is the ratio of the maximum possible signal strength with the noise affecting the original signal representation. Mathematically, it is described in Eq. 3.

$$PSNR = 10 \cdot \log_{10}\left(\frac{MAX_I^2}{MSE}\right) \tag{3}$$

Structural Similarity Index (SSIM) refers to a more robust image similarity metric that combines the concept of luminance, contrast and structure into a single, more holistic, index. It is fully given by Eq. 4. With its component metrics representing unique relative characteristics of an image, the two images being compared must be similar in all the ways to have a high SSIM. Conversely, an above average SSIM can also lead to a deceptively low PSNR, leading to wrong conclusions if proper analyses are not carried out.

$$\text{SSIM}(x,y) = \frac{(2\mu_x\mu_y + c_1)(2\sigma_{xy} + c_2)}{(\mu_x^2 + \mu_y^2 + c_1)(\sigma_x^2 + \sigma_y^2 + c_2)} \tag{4}$$

Hence, we use both these metrics to evaluate and represent our method. With this in mind, results are given in Tables 4 and 5. For sample reference, image sequences as input and the corresponding output are given in Fig. 4 (Phase 1) and Fig. 5 (Phase 2).

5.2   Results

We achieve on average PSNR of 28.804 which suggests an MSE of around, and a corresponding mean SSIM of 0.7666. With these values for $t + 1^{th}$ frame (+00:30), we are confident in using these predicted frames for forecasting extreme events using different meteorological variables. The high values of SSIM in Table 4b suggests that the images are structurally more similar to each other. This SSIM accuracy directly affect Phase 2 of the pipeline.

When we successively predict the next frame using from predicted frame it is noticed that every time when we predict the next frame with some percentage of error. So the error of the frame derived from the recursive procedures cumulatively adds the error into the predicted image. Figures 3a and b suggest that the predicted frame error increases exponentially through time.

Phase 2 contains the extreme weather event identification. The hydro-estimator rain, predicted using the TIR1 and TIR2 images, accomplish SSIM of 0.923 and PSNR of 24.36. And the land surface temperature, processed from SWIR and VIS frames, achieve 0.803 SSIM and 22.96 PSNR. These values suggest structurally very accurate predicted frame. This results got reflected in the images very properly and show the significance of this high SSIM and PSNR.

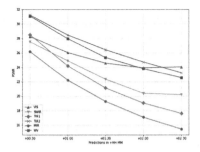

(a) PNSR values for successive predictions

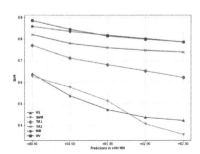

(b) SSIM values for successive predictions

Fig. 3. Graphs for metrics of successive predictions. Note how with each successive prediction, the values deteriorate consistently.

Table 4. Phase 1 results.

Channels	+00:30	+01:00	+01:30	+02:00	+02:30
VIS	28.264	26.053	24.598	23.929	24.066
SWIR	27.538	24.921	22.358	20.404	20.200
TIR1	28.504	24.202	21.146	19.046	17.604
TIR2	31.239	28.481	26.447	24.747	23.256
MIR	26.178	22.210	19.275	17.084	15.443
WV	31.100	27.970	25.384	23.822	22.565
Avg.	28.804	25.639	23.201	21.505	20.522

(a) PSNR values.

Channels	+00:30	+01:00	+01:30	+02:00	+02:30
VIS	0.6389	0.5367	0.4727	0.4370	0.4225
SWIR	0.6273	0.5767	0.5133	0.4070	0.3577
TIR1	0.7707	0.7114	0.6809	0.6521	0.6218
TIR2	0.8203	0.7794	0.7595	0.7473	0.7393
MIR	0.8579	0.8347	0.8170	0.8016	0.7854
WV	0.8847	0.8442	0.8134	0.7990	0.7863
Avg.	0.7666	0.7139	0.6761	0.6406	0.6188

(b) SSIM values.

Table 5. Phase 2 results (+00:30).

Extreme event parameters	SSIM	PSNR
Hydro-estimator	0.923	24.36
Land surface temperature	0.803	22.96

5.3   Validation

For the visual validation we have taken the case study of 2015 Chennai floods and the 2017 Mumbai floods.

2015 Chennai Floods. In the period from 8th November to 15th December, more than 500 people were killed and over 1.8 million people were displaced. With estimates of damages and losses ranging from nearly Rs. 200 billion (US$ 3 billion) to over Rs. 1 trillion (US$ 14 billion), the floods were the costliest to have occurred in 2015, and were among the costliest natural disasters of the year [20]. So the prediction of this type of event into early stage can save lot's of life as well as the assets.

The Extreme flood event can be extracted from the hydro-estimator parameter. We took the data of 8th November 2015 for validating our approach. This was the day the floods struck Chennai for the first time. We predicted TIR1

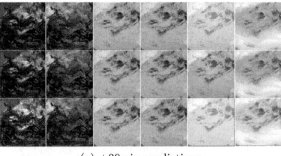

(a) +30min predictions

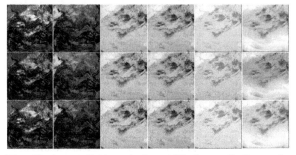

(b) +60min predictions (using +30min prediction as input).

Fig. 4. Sample images of predictions from Phase 1. Top to bottom: last image in 5-image sequence input, ground truth image, predicted image. Left to right: VIS, SWIR, TIR1, TIR2, MIR, WV. It is important to note how last sequence input in Fig. 4b is the output of Fig. 4a.

and TIR2 frames for that time-stamp in Phase 1, and further used to predict extreme hydro-estimator rain for the same.

5.4   2017 Mumbai Floods

The 2017 Mumbai flood refers to the flooding that occurred on 29 August 2017 following heavy rain on 29 August 2017 in Mumbai. Transport systems were unavailable through parts of the city as trains and roadways were shut. Power was shut off from various parts of the city to prevent electrocution. The International Federation of Red Cross and Red Crescent Societies (IFRC) called the South Asian floods one of the worst regional humanitarian crises in years. This event can be compared with the 2005 floods in Mumbai, which recorded 944 mm (37.17 in.) of rainfall within 24 h on 26 July [21].

Figures 6 and 8 clearly present the importance of visual significance of output. The scale of the data given in the figure is normalized in such a way that 0.6 onwards the predictions are more than 25 mm/h, which is considered as a stan-

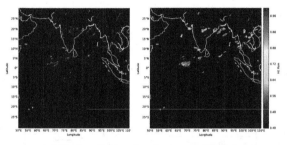

(a) Hydro-Estimator frame predictions

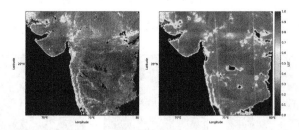

(b) Land Surface Temperature patch predictions

Fig. 5. Sample images of predictions from Phase 2. Left to right: actual, predicted.

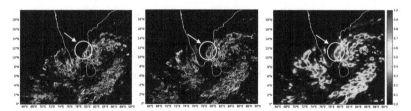

Fig. 6. Left to right: $(t-4)^{th}$ input image, actual image and predicted image of hydro-estimator rain at Chennai.

dard extreme rain threshold. The circled region is near Chennai, where extreme rain led to the infamous deluge, with each frame described in the caption.

The robustness of our predictions is apparent with Figs. 7 and 9, which represent the mean normalized precipitation rate for every time-step of the given day for the graph (represented by the x-axes). Over the course of the entire day, our mean-squared error for the 40×40 pixel frame over the Chennai and Mumbai area is respectively 0.025 and 0.037. This means for the day of the flood event in the cities and the surrounding area, our model could in fact predict the extremely high precipitation rate.

As evident from the figures, the first image (time-stamp 8 November 00:00 for Chennai, 29 August 10:30 for Mumbai) shows the dense rain formation approaching the region. The second image is the actual hydro-estimator precipitation on

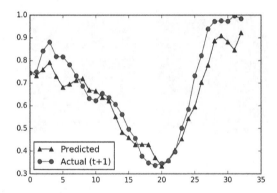

Fig. 7. Average precipitation over Chennai area for the day of 8th November 2015.

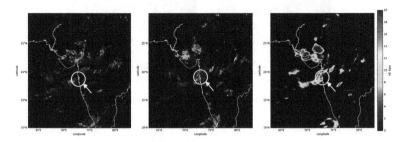

Fig. 8. Left to right: $(t-4)^{th}$ input image, actual image and predicted image of hydro-estimator rain at Mumbai.

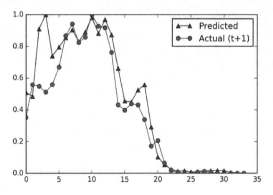

Fig. 9. Average precipitation over Mumbai area for the day of 29th August 2017.

time-stamp 02:30, while the third is the predicted precipitation on that time-stamp using frames from 00:00 to 02:00 for Chennai, 10:30 to 12:30 for Mumbai (input - $(t-4)$ to t, output - $(t+1)$). Even with a subjective visual perception, our approach is able to predict the next frame of hydro-estimator using 5 frames of TIR1 and TIR2 images.

This should be noted, since it shows that using the standard channel frames only, we can predict next frames for any derived product without giving any information about how the said derived product is formulated.

6   Conclusion and Future Work

The results of these predicted weather events seem very promising in the field of nowcasting. The extreme rainfall events and heat events have been calculated using the derived product hydro estimator precipitation and land surface temperature respectively. The full human interaction with identification of features has been avoided using our method, as it will directly try to learn the pattern based on ConvLSTM and U-Net phases. So, the goal of the extreme weather events have been achieved using the nowcasting of the 6 standard channel predication i.e. Radiance for Visible, Radiance for shortwave infrared, Brightness Temperature for Thermal Infrared 1–2 and brightness temperature for middle-wave. The subsequent 1 h will be predicted to identify the dependent event. These derived products associated with extreme events are then compared for the deviation with actual values and bring some essential find-outs in meteorology field as more focus is given on the prediction patterns.

As future work, we can focus on the efficiency of the ConvLSTM as a three-dimensional network with time-series mechanisms. So, we can replace the ConvLSTM model with a U-Net model, which is highly efficient in training and testing speed but may sacrifice accuracy.

Acknowledgements. This work was supported by the Satellite Meteorology and OceAography Research and Training (SMART) program at Space Application Centre, ISRO. We are thankful to SAC-ISRO for providing delightful opportunity and also giving us the relevant data and facilitating environment.

Conflict of Interest. The authors declare that they have no conflict of interest.

References

1. Darji, M.P., Dabhi, V.K., Prajapati, H.B.: Rainfall forecasting using neural network: a survey. In: 2015 International Conference on Advances in Computer Engineering and Applications, pp. 706–713 (2015)
2. Agrawal, K., Garg, S., Sharma, S., Patel, P., Bhatnagar, A.: Fusion of statistical and machine learning approaches for time series prediction using earth observation data. Int. J. Comput. Sci. Eng. **14**(3), 255–266 (2017)
3. Jönsson, P., Eklundh, L.: TIMESAT - a program for analyzing time-series of satellite sensor data. Comput. Geosci. **30**, 833–845 (2004)
4. Grover, A., Kapoor, A., Horvitz, E.: A deep hybrid model for weather forecasting. In: KDD (2015)
5. Zaytar, M.A., Amrani, C.E.: Sequence to sequence weather forecasting with long short-term memory recurrent neural networks. Int. J. Comput. Appl. **143**(11), 7–11 (2016)

6. Hossain, M., Rekabdar, B., Louis, S.J., Dascalu, S.M.: Forecasting the weather of Nevada: a deep learning approach. In: 2015 International Joint Conference on Neural Networks (IJCNN), pp. 1–6 (2015)
7. Salman, A.G., Heryadi, Y., Abdurahman, E., Suparta, W.: Weather forecasting using merged long short-term memory model (LSTM) and autoregressive integrated moving average (ARIMA) model. JCS **14**, 930–938 (2018)
8. Cracknell, M.J., Reading, A.M.: Geological mapping using remote sensing data: a comparison of five machine learning algorithms, their response to variations in the spatial distribution of training data and the use of explicit spatial information. Comput. Geosci. **63**, 22–33 (2014)
9. Zhuang, W., Ding, W.: Long-lead prediction of extreme precipitation cluster via a spatiotemporal convolutional neural network. In: Proceedings of the 6th International Workshop on Climate Informatics: CI (2016)
10. Liu, Y., et al.: Application of deep convolutional neural networks for detecting extreme weather in climate datasets. CoRR. arXiv:1605.01156 (2016)
11. Ghaderi, A., Sanandaji, B.M., Ghaderi, F.: Deep forecast: deep learning-based spatio-temporal forecasting. CoRR. arXiv:1707.08110 (2017)
12. Racah, E., Beckham, C., Maharaj, T., Kahou, S.E., Prabhat, M., Pal, C.: ExtremeWeather: a large-scale climate dataset for semi-supervised detection, localization, and understanding of extreme weather events. In: Guyon, I., et al. (eds.) Advances in Neural Information Processing Systems 30, pp. 3402–3413. Curran Associates, Inc. (2017). http://papers.nips.cc/paper/6932-extremeweather-a-large-scale-climate-dataset-for-semi-supervised-detection-localization-and-understanding-of-extreme-weather-events.pdf
13. Shi, X., Chen, Z., Wang, H., Yeung, D.Y., Wong, W.K., Woo, W.C.: Convolutional LSTM network: a machine learning approach for precipitation nowcasting. In: NIPS (2015)
14. Klein, B.E., Wolf, L., Afek, Y.: A dynamic convolutional layer for short range weather prediction. In: 2015 IEEE Conference on Computer Vision and Pattern Recognition (CVPR), pp. 4840–4848 (2015)
15. Ronneberger, O., Fischer, P., Brox, T.: U-Net: convolutional networks for biomedical image segmentation. In: Navab, N., Hornegger, J., Wells, W.M., Frangi, A.F. (eds.) MICCAI 2015. LNCS, vol. 9351, pp. 234–241. Springer, Cham (2015). https://doi.org/10.1007/978-3-319-24574-4_28
16. Batson, J., Royer, L.: Noise2self: blind denoising by self-supervision. arXiv preprint arXiv:1901.11365 (2019)
17. Mao, X., Shen, C., Yang, Y.B.: Image restoration using very deep convolutional encoder-decoder networks with symmetric skip connections. In: Advances in Neural Information Processing Systems, pp. 2802–2810 (2016)
18. Kumar, P., Varma, A.K.: Assimilation of INSAT-3D hydro-estimator method retrieved rainfall for short-range weather prediction. Q. J. R. Meteorol. Soc. **143**(702), 384–394 (2017). https://doi.org/10.1002/qj.2929
19. Agostinelli, F., Anderson, M.R., Lee, H.: Adaptive multi-column deep neural networks with application to robust image denoising. In: NIPS 2013 (2013)
20. Wikipedia Contributors: 2015 south Indian floods – Wikipedia, the free encyclopedia (2019). https://en.wikipedia.org/w/index.php?oldid=896549980. Accessed 11 May 2019
21. Wikipedia Contributors: 2017 Mumbai flood – Wikipedia, the free encyclopedia (2018). https://en.wikipedia.org/w/index.php?oldid=864880459. Accessed 11 May 2019

Deep Hybrid Neural Networks for Facial Expression Classification

Aakash Babasaheb Jadhav$^{(\boxtimes)}$, Sairaj Laxman Burewar,
Ajay Ashokrao Waghumbare, and Anil Balaji Gonde

Shri Guru Gobind Singhji Institute of Engineering and Technology,
Nanded, India
2017mec023@sggs.ac.in

Abstract. Facial Expression evaluation has become necessary for human machine interaction, behavior analysis and also forensic and clinical evaluation. Deep convolutional neural networks (CNN) have been largely used for facial expression recognition but due to locality of convolution, CNNs results in lower accuracy when trained with facial expression data of varying ethnicity and emotion intensity. Recurrent neural networks (RNN) are used to work with sequential data and used to predict the sequences. We propose CNN-RNN network approach, a hybrid network, wherein the outputs from CNN and RNN have been concatenated to predict the final emotion, similarly a CNN model followed by a RNN layers has been designed that gives promising results. The proposed hybrid models are evaluated on two publically available datasets CK+ and JAFFE which provide us with variation in ethnicity and emotion intensity. Promising results have been obtained with this hybrid approach when compared to various machine learning and deep learning methods.

Keywords: Image blending · CNN · RNN · Hybrid network · Expression classification

1 Introduction

Facial expressions have universal acceptance, hence could be used for signaling and interaction. Facial expressions can be used for reliable communications for human machine interactions. Facial expression analytics can be of great assistance in automated driver assistance systems (ADAS), clinical and behavioral studies, security surveillance and interactive gaming. Ekman et al. [3] defined six basic and most common facial expressions, anger, disgust, fear, happiness, sadness, and surprise. Friesen [4] and Ekman introduced facial action units. Specific expressions activate specific sections of the face known as facial action units. Earlier works were based on detection of activated facial action units, as each expression simulated certain sections of the face. Detection of action units used hand engineered features like Local Binary Pattern (LBP), SIFT features, Haar Features and also methods involving combinations of hand crafted features. Deep neural networks, CNNs [9] and increased processing abilities of GPUs have immensely benefitted common computer vision tasks of object and face recognition, object localization and object detection. Convolution operation in

© Springer Nature Singapore Pte Ltd. 2020
N. Nain et al. (Eds.): CVIP 2019, CCIS 1148, pp. 283–293, 2020.
https://doi.org/10.1007/978-981-15-4018-9_26

CNNs is localized and handles mostly spatial data. Hence various data fusion methods have been developed. Deep learning based methods have achieved highly accurate and state of the art on several benchmarks. Deep learning models use a visual feature based aggregation strategy for learning.

This work proposes and evaluates deep learning techniques that involve integration of convolutional neural networks (CNNs) and recurrent neural networks (RNNs). This work proposes a facial expression recognition approach wherein convolution-pooling operations and recurrent operations are performed independently and sum the results for the expression prediction. Also a network involving recurrent operation after convolution operation has been designed.

2   Related Work

Facial expression recognition has been carried out on static images and dynamic sequence of images. Static images comprise of single emotion at peak intensity, includes images captured randomly, snapshots from videos or laboratory controlled and images captured under supervision. Dynamic sequences on the other hand represent a variation in intensity of emotion from low to peak and subsequently back to normal. Before the advent to deep learning methods, most methods used hand crafted features like Histogram of Oriented Gradients (HOG) [5], Gabor Filter [8], Local Binary Pattern (LBP) [6] and Facial Historic Points [7]. These methods were specialized for certain datasets and performed well where those handcrafted features were present.

Deep Learning methods mostly employ a standard pipeline of Input → Preprocess data → Training Neural Networks → Classification [5, 21]. Preprocessing includes standard filtering operations, face alignment, data augmentation, face and pose normalization and changes in illumination. CNN based methods involve transfer learning and fine tuning of VGG-16 [20], VGG-19 [20], AlexNet [9], Google Net [10] networks and the derivatives of these networks have promising results. This has been the mainstream approach for classification problems. Handcrafted and CNN features were combined for Facial Expression Recognition (FER), Connie et al. [21] employ scale invariant feature transform (SIFT). Kaya et al. [22] uses SIFT, HOG and local gabor binary patterns (LGBP). RNNs recently have gained success when dealing with sequential data, audio [17], and video [22]. RNNs scan the images into sequences in certain directions; RNNs recollect information from past inputs and learns relative dependencies between sequential data [19].

RNNs are generally combined to overcome the shortcomings of CNNs in learning overall dependencies [10–14]. CNN-RNN can be combined in two ways: the unified combination and the cascaded combination. The unified combination attempts to introduce a recurrent property into the traditional CNN structure, image input is processed separately by CNN and RNN and final fully connected layers unified for predictions. The cascaded combination, the input image undergoes convolution in the CNN and then the feature maps and feature activations are sequentially correlated for final prediction. CNN-RNN used for Image recognition [16] and Image segmentation [29]. Khorrami et al. used CNN-RNN design for action recognition in videos [15]. Recurrent neural networks are mostly used for sequential and time series data prediction. However, vanishing gradient and inability to learn long term patterns makes RNNs difficult to train. Long Short Term Memory (LSTM) and Gated Recurrent Units (GRU) were developed to tackle the vanishing gradient problem in RNNs. We use RNN with ReLU activation function [24]. Jain et al. [30] have used CNN-RNN model for facial expression recognition. Choi et al. [35] have used generative adversarial networks for generation of facial expressions. Similarly, [30, 33, 34, 36] used unpaired data for image to image translation. The concept of [31, 32] could further assist for emotion recognition in dynamic image sequences.

3 Datasets

3.1 CK+

The Extended Cohn Kanade (CK+) database is laboratory-controlled facial expression dataset. CK+ contains 593 video sequences from 123 subjects. The sequences vary range in between 10 to 60 frames per subject for each expression, and show a gradual change from a neutral facial expression to the peak expression. Based on facial action coding system (FACS) 327 sequences from 118 subjects are categorized into six basic expression labels as per Ekman et al. [3] and neutral as an additional expression.

3.2 JAFFE

The Japanese Female Facial Expression (JAFFE) database is a laboratory-posed image database that contains 213 samples of expressions from 10 Japanese females. Each subject was asked to pose for 7 expressions (6 Basic expressions and Neutral). The database is challenging because it contains less samples and provides variation in ethnicity and also expression intensity.

4 Proposed Work

The Objective of our work is to avoid the use of image standardization like adjusting the mean shape, subject focus, standard deviation and mean picture. Additionally, the intention is to better the performance with lesser data and inconsistent data distribution across classes and improve overall classification accuracy across entire dataset and not just the validation split (Fig. 1).

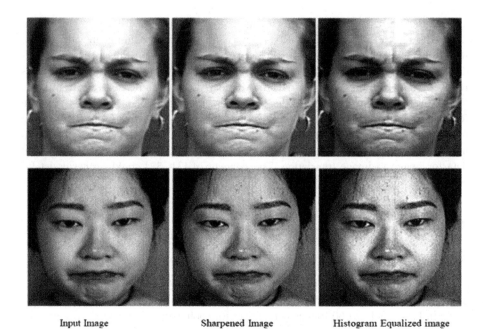

Input Image   Sharpened Image   Histogram Equalized image

Fig. 1. Pre-processed images from dataset

4.1 Preprocessing

Initially, we combine the images of same expression together under an expression label. Since CK+ is a sequence of images we consider the last few frames with peak expression. Further processing in done accordingly:

1. Face detection using Haar cascade
2. Convert the images to grayscale
3. Sharpen the Grayscale image
4. Blend the images obtained in 2 and 3
5. Histogram equalization of image obtained in 4
6. Resize the images to 64 × 64 (Table 1 and Fig. 2).

Table 1. Images from datasets for training.

Expression	CK+	JAFFEE	Combined
Angry	132	30	162
Disgust	180	29	209
Fear	75	32	107
Happiness	204	31	235
Sadness	87	31	118
Surprise	249	30	279
Neutral	115	30	145
	1042	213	1255

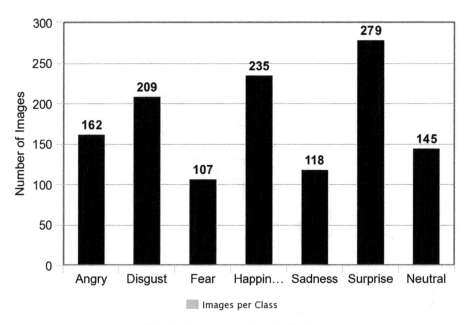

Fig. 2. Inconsistent data distribution

4.2 Convolutional Neural Networks

CNNs comprises of convolutional blocks. A typical convolutional block consists of convolutional layer, activations layer, max pool/average pool layer. The convolutional layer comprises of learnable spatial filters. These filters convolve through the image and highlight distinguishable features. These features maps are further refined using the activation functions. Down sampling (Max Pooling) reduces the spatial size of feature maps and eventually reduces computational complexity. The fully connected layer carries over the activations from all nodes in previous layer to the nodes of the next layer. The weights on these connectionist layers are fine tuned in the learning process. ReLUs activation function has been used extensively to get rid of the non-linearity in feature maps. CNNs are limited by the locality of convolution operation and don't consider temporal dependencies (Fig. 3).

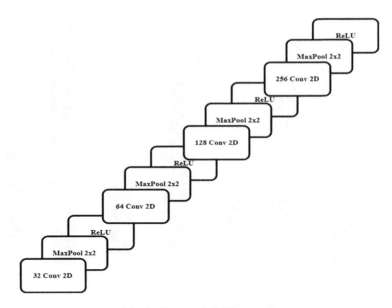

Fig. 3. Proposed CNN network

4.3   Recurrent Neural Networks

A recurrent neural network (RNN) is a feed forward deep neural network model that correlates sequences of temporal information of arbitrary lengths. RNNs are suited mostly for sequential data. RNNs include recurrent edges that span adjacent time steps and share the same parameters across all steps. Back propagation through time (BPTT) algorithm is used to train the RNN. RNNs suffer from vanishing gradient and exploding problems. Hence we use ReLU activation function for RNN.

4.4   Combined CNN-RNN Model

Cascaded Combination Model: The Images are fed to the CNN and undergoes the usual convolutional block operations. The feature maps from the final convolutional block are fed to the RNN with hidden units having ReLU activation function. The final layer used softmax activation for prediction which converts the node activation in to class probability (Fig. 4).

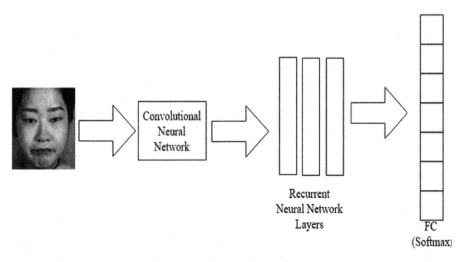

Fig. 4. Cascaded CNN-RNN network

Unified CNN-RNN Model: The images are fed simultaneously to the CNN and RNN, a fully connected layer connected to both CNN and RNN having 7 nodes, same as the number of classes. These final two layers, output of CNN and output of RNN are concatenated to form a single strand and further connected to another fully connected layer having 7 output nodes which are converted into probabilities by a softmax function (Fig. 5).

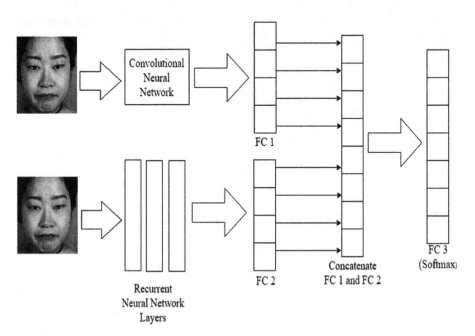

Fig. 5. Unified CNN-RNN network

RNN can take in 1xN data where data is fed in the form of chunks. RNN learns inter dependencies between the feature maps, since the number of feature maps are 256 in the last layer the RNN is used after the last convolutional block. A vanilla RNN classifier gives poor results for the JAFFE and CK+ datasets. For images having a certain common pattern amongst them the vanilla RNN gives much better results, like MNIST (above 95% accuracy on validation set). Here cascade network could prove to be useful. A simple CNN classifier and a vanilla RNN classifier give lesser classification accuracy.

5   Results and Evaluation

The models have been trained with 80% of the dataset used for training and 20% for validation against the training. Further the trained model has been used to predict the class of each sample image in the combined dataset. The weights for the network are randomly initialized with 80%–20% data split into training data and validation data respectively.

We experimented using the cascaded style and unified type architecture by varying two parameters, that is number of hidden units and number of recurrent layers. The best results were obtained for cascaded style network with 75 hidden units and 3 recurrent layers. Almost equivalent results were obtained for unified network with 2 recurrent layers and 125 hidden units. Comparatively, the unified network is very slow to train and equally slow to respond also the numbers of trainable and non-trainable parameters are high for unified networks. The cascade type of network takes very less time to train and to respond. The training losses for all cases were less than 3.5%. Networks were trained sequentially one expression at a time using the preprocessed image data and tested with the original data and not the pre-processed image data (Tables 2, 3 and 4).

Table 2. Result of cascaded CNN-RNN models

RNN layers	Hidden units	Total accuracy (%)	CK+ accuracy (%)	JAFFE accuracy (%)
1	75	97.57	97.61	97.18
	100	97.82	98.01	97.21
	125	96.72	97.21	94.37
2	75	97.41	98.33	94.41
	100	97.47	97.90	96.29
	125	97.45	97.24	98.13
3	75	**98.60**	**98.82**	**98.15**
	100	97.98	97.82	97.58

Table 3. Result of unified CNN-RNN models

RNN layers	Hidden units	Total accuracy	CK+ accuracy	JAFFE accuracy
1	75	93.31	93.98	90.60
	100	97.08	97.81	94.32
	125	96.81	96.01	95.70
2	75	97.67	97.14	88.99
	100	95.65	97.08	90.86
	125	**98.60**	**98.84**	**97.66**
3	75	97.61	98.06	96.24
	100	97.68	97.94	94.85

Table 4. Comparative results

Method	Accuracy of JAFFE	Accuracy of CK+
Minaee et al. [27]	92.8	98
Shan et al. [28]	76.74	80.303
Kim et al. [29]	91.27	96.46
Proposed cascade network	**98.15**	**98.82**
Proposed unified network	**97.66**	**98.84**

6 Conclusion

We proposed a unified CNN-RNN network and a cascade CNN-RNN network for facial expression classification and compare the performances of the networks. Convolution is a filtering process leaving behind important features of image information. The cascaded model used the RNN to calculate the dependency and continuity features from the feature maps. The unified model combined features from the CNN model with the dependencies and continuity from the input images. Hence the improvement in classification accuracy has been observed over CNN. The networks vary in terms of training and response time but they manage to yield similar accuracies. The unified network takes relatively more time to train and respond to test samples. The cascade method would fare better for real time applications due to, much better response time and when trained with even larger and varied dataset.

References

1. Lucey, P., Cohn, J.F., Kanade, T., Saragih, J., Ambadar, Z., Matthews, I.: The extended Cohn-Kanade dataset (CK+): a complete dataset for action unit and emotion-specified expression. In: 2010 IEEE Computer Society Conference on Computer Vision and Pattern Recognition Workshops (CVPRW), pp. 94–101. IEEE (2010)
2. Lyons, M.J., Akamatsu, S., Kamachi, M., Gyoba, J., Budynek, J.: The Japanese female facial expression (JAFFE) database (1998)
3. Ekman, P.: Pictures of Facial Affect. Consulting Psychologists Press, USA (1976)

4. Ekman, P.: Facial expression and emotion. Am. Psychol. **48**(4), 384 (1993)
5. Dalal, N., Triggs, B.: Histograms of oriented gradients for human detection. In: International Conference on Computer Vision and Pattern Recognition (CVPR 2005), San Diego, United States, pp. 886–893 (June 2005)
6. Shan, C., Gong, S., McOwan, P.W.: Facial expression recognition based on local binary patterns: a comprehensive study. Image Vis. Comput. **27**(6), 803–816 (2009)
7. Cootes, T.F., Edwards, G.J., Taylor, C.J.: Active appearance models. In: Burkhardt, H., Neumann, B. (eds.) ECCV 1998. LNCS, vol. 1407, pp. 484–498. Springer, Heidelberg (1998). https://doi.org/10.1007/BFb0054760
8. Ou, J., Bai, X.B., Pei, Y., Ma, L., Liu, W.: Automatic facial expression recognition using Gabor filter and expression analysis. In: IEEE International Conference on Computer Modeling and Simulation, pp. 215–218 (2010)
9. Krizhevsky, A., Sutskever, I., Hinton, G.E.: ImageNet classification with deep convolutional neural networks. In: Advances in Neural Information Processing Systems, pp. 1097–1105 (2012)
10. Szegedy, C., et al.: Going deeper with convolutions. In: CVPR (2015)
11. Donahue, J., et al.: Long-term recurrent convolutional networks for visual recognition and description. In: Proceedings of the IEEE Conference on Computer Vision and Pattern Recognition, pp. 2625–2634 (2015)
12. Liang, M., Hu, X.: Recurrent convolutional neural network for object recognition. In: Proceedings of the IEEE Conference on Computer Vision and Pattern Recognition, pp. 3367–3375 (2015)
13. Wang, J., Yi, Y., Mao, J., Huang, Z., Huang, C., Xu, W.: CNN-RNN: a unified framework for multi-label image classification. In: Proceedings of the IEEE Conference on Computer Vision and Pattern Recognition, pp. 2285–2294 (2016)
14. Zuo, Z., et al.: Convolutional recurrent neural networks: learning spatial dependencies for image representation. In: Proceedings of the IEEE Conference on Computer Vision and Pattern Recognition Workshops, pp. 18–26 (2015)
15. Khorrami, P., Paine, T.L., Brady, K., Dagli, C., Huang, T.S.: How deep neural networks can improve emotion recognition on video data. In: IEEE Conference on Image Processing (ICIP) (2016)
16. Visin, F., Kastner, K., Cho, K., Matteucci, M., et al.: ReNet: a recurrent neural network based alternative to convolutional networks. arXiv:1505.00393 (2015)
17. Graves, A., Mohamed, A.R., Hinton, G.: Speech recognition with deep recurrent neural networks. In: Proceedings of the IEEE International Conference on Acoustics, Speech and Signal Processing, pp. 6645–6649 (2013)
18. Sanin, A., Sanderson, C., Harandi, M.T., Lovell, B.C.: Spatiotemporal covariance descriptors for action and gesture recognition. In: IEEE Workshop on Applications of Computer Vision (2013)
19. Jain, S., Hu, C., Aggarwal, J.K.: Facial expression recognition with temporal modeling of shapes. In: Proceedings of the IEEE International Conference on Computer Vision Workshops, pp. 1642–1649 (2011)
20. Simonyan, K., Zisserman, A.: Very deep convolutional networks for large-scale image recognition. arXiv preprint arXiv:1409.1556 (2014)
21. Connie, T., Al-Shabi, M., Cheah, W.P., Goh, M.: Facial expression recognition using a hybrid CNN–SIFT aggregator. In: Phon-Amnuaisuk, S., Ang, S.-P., Lee, S.-Y. (eds.) MIWAI 2017. LNCS (LNAI), vol. 10607, pp. 139–149. Springer, Cham (2017). https://doi.org/10.1007/978-3-319-69456-6_12
22. Kaya, H., Gürpınar, F., Salah, A.A.: Video-based emotion recognition in the wild using deep transfer learning and score fusion. Image Vis. Comput. **65**, 66–75 (2017)

23. Lowe, D.G.: Distinctive image features from scale-invariant key points. Int. J. Comput. Vis. **60**(2), 91–110 (2004)
24. Le, Q.V., Jaitly, N., Hinton, G.E.: A simple way to initialize recurrent networks of rectified linear units. arXiv:1504.00941 (2015)
25. Minaee, S., Abdolrashidi, A.: Deep-emotion: facial expression recognition using attentional convolutional network. arXiv preprint arXiv:1902.01019 (2019)
26. Shan, K., Guo, J., You, W., Lu, D., Bie, R.: Automatic facial expression recognition based on a deep convolutional-neural-network structure. In: 2017 IEEE 15th International Conference on Software Engineering Research Management and Applications (SERA), pp. 123–128 (2017)
27. Kim, J.-H., Kim, B.-G., Roy, P.P., Jeong, D.-M.: Efficient facial expression recognition algorithm based on hierarchical deep neural network structure. IEEE Access **7**, 41273–41285 (2019)
28. Jain, N., et al.: Hybrid deep neural networks for face emotion recognition. Pattern Recogn. Lett. **115**, 101–106 (2018)
29. Visin, F., Kastner, K., Courville, A., Bengio, Y., et al.: ReSeg: a recurrent neural network for object segmentation arXiv:1511.07053 (2015)
30. Patil, P., Murala, S.: FgGAN: a cascaded unpaired learning for background estimation and foreground segmentation. In: 2019 IEEE Winter Conference on Applications of Computer Vision (WACV), pp. 1770–1778. IEEE (2019)
31. Patil, P., Murala, S., Dhall, A., Chaudhary, S.: MsEDNet: multi-scale deep saliency learning for moving object detection. In: 2018 IEEE International Conference on Systems, Man, and Cybernetics (SMC), pp. 1670–1675. IEEE (2018)
32. Patil, P.W., Murala, S.: MSFgNet: a novel compact end-to-end deep network for moving object detection. IEEE Trans. Intell. Transp. Syst. **20**(11), 4066–4077 (2018)
33. Dudhane, A., Murala, S.: CDNet: single image de-hazing using unpaired adversarial training. In: 2019 IEEE Winter Conference on Applications of Computer Vision (WACV), pp. 1147–1155. IEEE (2019)
34. Dudhane, A., Murala, S.: C^2MSNet: a novel approach for single image haze removal. In: 2018 IEEE Winter Conference on Applications of Computer Vision (WACV), pp. 1397–1404. IEEE (2018)
35. Choi, Y., Choi, M., Kim, M., Ha, J.-W., Kim, S., Choo, J.: StarGAN: unified generative adversarial networks for multi-domain image-to-image translation. In: The IEEE Conference on Computer Vision and Pattern Recognition (CVPR), pp. 8789–8797 (2018)
36. Zhu, J.-Y., Park, T., Isola, P., Efros, A.A.: Unpaired image-to-image translation using cycle-consistent adversarial networks. In: The IEEE International Conference on Computer Vision (ICCV), pp. 2223–2232 (2017)

SCDAE: Ethnicity and Gender Alteration on CLF and UTKFace Dataset

Praveen Kumar Chandaliya$^{(\boxtimes)}$, Vardhman Kumar, Mayank Harjani,
and Neeta Nain

Malaviya National Institute of Technology Jaipur, Jaipur, India
{2016rcp9511,2015ucp1429,2015ucp1482,nnain.cse}@mnit.ac.in

Abstract. Global face attributes like Gender, Ethnicity, and Age are attracting attention due to their specific explanation of human faces. Mostly prior face attribute alteration works are on large-scale CelebA and LFW dataset. We address more challenging problem called global face attribute alteration on data sets like CLF and UTKFace. Our approach is based on sampling with global condition attribute. It consists of five components Encoder (E_Z), Encoder (E_Y), Sampling (S), Latent Space (ZL), and Decoder (D). The E_Z with S component is responsible to generate structured latent vector Z and E_Y produces condition vector L which we modify according to desired condition, latent vector Z and modified condition vector L are concatenated to make Latent Space ZL to help global face attribute alteration and Decoder D is used to generate modified images. We trained our SCDAE (Sampling and Condition based Deep AutoEncoder) model for gender and ethnicity alteration on CLF and UTKFace dataset. Both qualitative and quantitative experiments show that our approach can alter untouched global attributes and generates more realistic faces in term of person identity and age uniformity which is comparable to human observation.

Keywords: Global face attribute alteration · Deep Auto Encoder · Sampling · Latent Space

1 Introduction

Face image attribute editing is a daunting task. For example, altering the attributes of a face like smile, hair color, facial hair, remove or put eyeglasses, even change the gender and ethnicity. Earlier this task required a person well acquainted with image editing software and was time consuming, but in recent times deep learning has significantly improved face image attribute editing by employing generative models to produce plausible and realistic images. Variants of Deep Auto Encoder [1], Variational auto encoder [2] and Generative adversarial networks [3] have led to favorable results in image generation [4], image editing [5], super-resolution [6], image inpainting [7,8], and text to image synthesis [9], in recent years. In addition to this VAE and GAN can explicitly control

© Springer Nature Singapore Pte Ltd. 2020
N. Nain et al. (Eds.): CVIP 2019, CCIS 1148, pp. 294–306, 2020.
https://doi.org/10.1007/978-981-15-4018-9_27

the features or latent vector with categorical, binary and text descriptions, and landmarks type condition extensions using Conditional Generative Adversarial Network (cGAN) [10]. Despite that, the Auto Encoder framework is an insufficient inference technique, in other words finding the latent representation of an input image, which is an essential step for being able to rejuvenate and alter real images. To achieve this in our proposed model we Incorporate Sampling and Condition base Encoders (E_Z) and (E_Y). Generally, face attribute are categorized in two ways based on global and local attributes. In Table 1 we summarize the overview of facial attribute data-sets based on global and local face attributes, further this attributes are categorized into categorical and binary attributes.

Table 1. Global and local face attributes of various face attribute data sets.

Dataset:	Global attribute	Categorical	**Age:** Baby, Child, Youth, Middle Aged, Senior **Ethnicity:** Asian, White, Black, Indian, other
CelebA [11] LFW [12] PubBig [13]		Binary	**Gender:** Male, Female
FaceTrace [14] CLF [15] UTKFace [16]	Local attribute	Categorical	**Hair Type:** Black Hair, Blond Hair, Gray Hair, Bald Bangs, Curly Hair, Wavy Hair, Straight Hair, Bald, Receding Hairline **Hair Color:** Straight Hair, Wavy Hair, Brown Hair, Indian Gray Hair **Face Shape:** Oval Face, Chubby, Double Chin, Rosy Cheeks, Square Face, Round Face, Round Jaw, Double Chin, High Cheekbones **Eyebrows Type:** Arched Eyebrows, Bushy Eyebrows **Illumination:** Blurry, Harsh Lighting, Flash, Soft Lighting, Outdoor, Flushed Face **Forehead Type:** Fully Visible, Partially Visible, Obstructed Forehead
		Binary	5_o_Clock_Shadow, Attractive, Bags Under Eyes, Goatee, Big Nose, Blurry, Eyeglasses, Heavy Makeup, Mouth Slightly Open, Mustache, No Beard, Big Lips, Receding Hairline, Sideburns, Mouth Wide Open, Wearing Earrings, Wearing Hat, Wearing Lipstick, Smiling, Wearing Necklace, Wearing Necktie, Eyeglasses, Sunglasses, Narrow Eyes, Eyes Open, Mouth Closed, Posed Photo, Teeth Not Visible, Wearing Hat, Color Photo, Pointy Nose Attractive Man, Attractive Woman, Brown Eyes, Sideburns, Shiny Skin, Pale Skin, No Eyewear

To add up, our contribution is four fold:

1. A novel Deep Auto Encoder based model is developed for global facial attribute transfer by combining sampling base latent vector and repetition base conditional vector to generate regulated latent space.

2. Sampling S component to make compact latent vector, resulting in better image quality generated image with identity preservation. Compact latent vector also stabilizes the training process.
3. To the best of our knowledge first time ethnicity alteration on face is addressed on CLF [17] and UTKFace [18].
4. We are addressing children age and gender alteration in the age group of 2 - 18 years, where the physiological changes are most prominent.

The remainder of the paper is organized as follows. Section 2 gives a brief explanation of Deep Face Attribute Manipulation methods. Section 3 gives detail explanation of our proposed model SCDAE. Section 4 describes the loss functions used in our model. Section 5 details the algorithm. Section 6 reports the qualitative and quantitative experimental results. Finally, Sect. 7 ends this work with conclusions and future work.

2   Related Work

The objective of facial attribute alteration or generation is to manipulate certain attributes, either binary or categorical, of a given face image, while keeping other attributes unchanged. It has multifarious applications such as in entertainment, face recognition, and is an emerging area in deep learning, with applications to computer vision and graphics [19]. A large number of existing model and framework is based on local binary attributes, for example, "eyeglasses" "no eyeglasses", "mustache" or "no mustache" and many more. Deep Face Attribute Manipulation (DFAM) method generally categories in Model Based and Extra condition-based Methods. Model based approach follows Deep GAN and Extra condition based model follows Deep cGAN and VAE.

2.1   Model-Based Approach

Model-based approach maps a face image in the source group to target group, and apply adversarial loss is used to distinguish between generated image and reference image. This approach is task specific because no extra condition is added and results in visual fidelity and plausible images. InfoGAN [20], DIAT [21], UNIT [22], Residualimage [23], Wang *et al.* [24], SaGAN [25] address these as follows: DIAT: Conveys reference face image to each reference face attribute label while preserving identity for Identity Aware Transfer of facial attribute by applying adversarial loss. InfoGAN: Information theory concept is combined with GAN to learn disentangling of latent vector to maximize the interactive information between the subspace of the latent vector and the observation. UNIT: Unsupervised image-to-image translation problem is addressed by introducing sharing of latent vector between GAN with VAE architecture. Residual Base Image translation model does not represent entire face image as input to avoid computation in learning residual image. Wang *et al.* combined VGG based perceptual loss with adversarial loss to generate more realistic images with identity preservation. SaGAN is based on mixing sparsely grouped data set with training data and very few labelled data to transfer pattern from one group to other group using the GAN network.

It is observed from the above that, when multiple face attributes are added with Model based approach then the training becomes very unstable and time-consuming.

2.2 Extra Condition-Based Models

On the other hand, when continuous and discrete conditions are combined in the form of one-hot vectors to manipulate images in latent space then they are called extra condition-based methods. In this approach the architecture of the model is based on AE, VAE, and GAN. VAE/GAN [26], CVAE [27], IcGAN [28], Fader [29], CAAE [18], CPAVAE [15,30], cCycelGAN [31], StarGAN [32]. VAE/GAN: In this paper VAE is combined with GAN to learn feature representation with feature wise loss to capture data distribution in better way. CVAE: Learning approach is based on Condition Variational Auto Encoder with energy minimization algorithm to make better latent variables to generate novel image. IcGAN: Two encoders are used one for latent vector and other for condition vector generation, they are concatenated and passed to cGAN. cCycleGAN: Expands the cycleGAN conditioned on facial attributes with the adversarial loss and cycle consistency loss. StarGAN: does multiple domain image to image translations using single model. In addition to GAN, as well as VAE and their variants, Deep auto encoder is also capable to generate more realist results with face manipulation. CAAE: Zhang *et al.* stated a conditional adversarial autoencoder for face aging. CAAE first transfers reference image to a latent code through deep encoder. After this transfer, latent code tiling with an age and gender label one-hot vector is fed into decoder or generator for manifold learning. Condition one-hot vector is to control the alteration of age and latent code conserved personalized face features. Fader Networks architecture is based on auto encoder that generates image by extricating the salient information of the attributes directly in the latent space, alteration of the attribute value is done by sliding knobs. One-hot vectors indicate the presence of corresponding facial attributes. During the training, the conditional vectors are concatenated with the to-be-manipulated image in latent spaces.

All aforementioned models of DFAM methods use two common data sets CelebA [11] and LFW [12] but none of them addressed ethnicity and gender alteration in child face. In our proposed model we have used child specific data set CLF [15,30] and Ethnicity based dataset UTKFace [16] for experimental analysis. Our proposed model SCDAE relies on a deep encoder-decoder architecture.

3 Model Architecture

Given the reference input image X and the reference attribute C, our framework aims to generate a facial image that owns the reference attribute as well as keeps the same or similar identity to the input image. In general, our framework consists of Condition based deep Auto-Encoder With Sampling concept to alter global facial attributes which work in synergy to generate facial image. In our model condition vector is concatenation of age, gender and ethnicity by repetition of conditions. The reason of repetition of conditions is to give significant information to decoder for generated realistic face.

The Deep Encoder-Decoder model is enhanced using the VAE based Sampling concept which helps in generating a compressed latent vector Z and repetition of condition face attribute L enhances the quality of images with identity preservation. We further modify this L according to our target output L'. This is concatenated to latent vector Z to generate latent space ZL' for Decoder. The proposed model, which is given in Fig. 1 is generic as it can work on both binary and categorical attributes. Our model has an advantage that we can selectively change face attributes while keeping all other attributes intact.

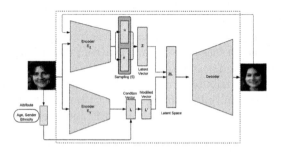

Fig. 1. SCDAE: the proposed model

3.1 Encoder E_Z

Table 2 gives the list of operations performed on the input images X of size 128×128 to find the latent vector z. It is generated from the input X using stack of 6 Convolution-ReLU layers and the kernels used have width and height of 5×5 and stride used is 2. Also, the output size is down-sampled by a factor of 2 in each layer, and number of feature map is increased by a factor of 2 in each convolution layer. The last $Convolution_6$ layer is then followed by a Flatten layer generating output of 4096 nodes. Two parallel dense layers are used to generate the vectors μ and σ.

In our network E_Z is performing as $Q(Z|X)$ and D is performing as $P(X|Z)$. In practice for majority of X, $P(X|Z)$ will be close to 0, hence this leads to very less contribution of Z in estimating $P(X)$. By sampling we attempt to sample those values, which will have high impact on $P(X)$. Sampling is then applied by generating two parallel layers of Mean (μ) and Standard Deviation (σ) which are of size 60 each. Finally, latent vector Z is sampled from these two layers which is of size 60. Sampling (S) also helps to converge faster and reduce complexity.

Table 2. Encoder E_z network architecture

Operations	Kernel	Stride	Filters	BN	Activation	Output Shape
Conv1	5×5	2×2	32	No	LReLU	$64 \times 64 \times 32$
Conv2	5×5	2×2	64	No	LReLU	$32 \times 32 \times 64$
Conv3	5×5	2×2	128	No	LReLU	$16 \times 16 \times 128$
Conv4	5×5	2×2	256	No	LReLU	$8 \times 8 \times 256$
Conv5	5×5	2×2	512	No	LReLU	$4 \times 4 \times 512$
Conv6	5×5	2×2	1024	No	LReLU	$2 \times 2 \times 1024$
Flatten	-	-	-	No	-	4096
Dense_1	-	-	-	Yes	LReLU	60
Dense_2	-	-	-	Yes	LReLU	60
Sampling	-	-	-	Yes	LReLU	60

Table 3. Encoder E_y network architecture

Operations	Kernel	Stride	Filters	BN	Activation	Output Shape
Conv1	5×5	2×2	32	No	LReLU	$64 \times 64 \times 32$
Conv2	5×5	2×2	64	No	LReLU	$32 \times 32 \times 64$
Conv3	5×5	2×2	128	No	LReLU	$16 \times 16 \times 128$
Conv4	5×5	2×2	256	No	LReLU	$8 \times 8 \times 256$
Conv5	5×5	2×2	512	No	LReLU	$4 \times 4 \times 512$
Conv6	5×5	2×2	1024	No	LReLU	$2 \times 2 \times 1024$
Flatten	-	-	-	No	-	4096
Dense	-	-	-	Yes	LReLU	40

3.2 Encoder E_Y

The architecture of the Encoder E_Y is represented using Table 3. A latent vector L is generated from the input X using 6 Convolution-ReLU layers. The kernels used have width and height of 5×5 and stride used is 2. Also, the output size is up-sampled by a factor of 2 in each layer and number of feature map is increased by a factor of 2 in each layer. The last Convolution$_6$ layer is then followed by a Flatten layer generating output of 4096 nodes. Finally, we get latent vector Y of size 40.

3.3 Latent Space (ZL)

Here global modified condition attribute vector L' is append to Z which is responsible for personality transition with respect to global target face attribute. Thus, we could tweak gender or ethnicity while preserving age. ZL' so obtained is of size 100. As the size of Z is 60, rest 40 is made up by tiling Ethnicity (one hot vector of length 5) 4 times, gender (one hot vector of length 2) 5 times and age (one hot vector of length 10) without repetition resulting in a vector of length 100.

3.4 Decoder D

The architecture of the Decoder is represented using Table 4. A Latent Space ZL is passed into the Decoder and through a fully connected layer Dense$_1$ converted to 32768 node. This output is then reshaped to $4 \times 4 \times 2048$. It is then passed through 5 Convolution-ReLU layers to generate an image $\bar{X}$. The kernels used have width and height of 5×5 and stride used is 2. Also, the output size is up-sampled by a factor of 2 in each layer and number of feature maps is decreased by a factor of 2 in each layer. The final image $\bar{X}$ is reconstructed which is called the generated image.

Table 4. Decoder network architecture

Operations	Kernel	Stride	Filters	BN	Activation	Output shape
Dense	–	–	–	No	–	32768
Reshape	–	–	–	No	–	$4 \times 4 \times 2048$
Conv1	5×5	2×2	1024	Yes	LReLU	$8 \times 8 \times 1024$
Conv2	5×5	2×2	512	Yes	LReLU	$16 \times 16 \times 512$
Conv3	5×5	2×2	256	Yes	LReLU	$32 \times 32 \times 256$
Conv4	5×5	2×2	128	Yes	LReLU	$64 \times 64 \times 128$
Conv5	5×5	2×2	64	Yes	LReLU	$128 \times 128 \times 64$
Conv6	5×5	2×2	3	No	Tanh	$128 \times 128 \times 3$

4   Loss Functions

In our proposed model we have used three loss functions: identity loss L_{Ez} and two mean square losses for condition vector L and on generated image $\bar{X}$. MSE and identity both losses are important to capture how face generally look and local details like eyes, face or hair respectively, making interesting application of deep auto encoder based models described as follows.

4.1   Latent Vector Z Optimization

Our Encoder E_Z finds latent vector Z from the input image X but the problem is we do not know what Z should be generated by the E_Z so we take help of the generated image $\bar{X}$ to find Z from this recreated image which is $S(E_Z(D(Z, L')))$ and then we minimize this loss as given in the Eq. 1 to train E_Z. Sampling S component of E_Z make latent vector Z very compact, and hence the generated image quality is better with better identity preservation. This also stabilized the training process.

$$L_{Ez} = E_{Z \sim P_Z, L' \sim p_y} ||Z - S(E_Z(D(Z, L')))||_2^2 \tag{1}$$

4.2   Condition Vector Optimization

Global facial attribute alteration with better results can be obtained by using optimization of encoder E_Y based on mean square loss, as this function captures large changes on face. We have used the actual condition label L and label vector generated by $E_Y(X)$, and the loss is calculated by the Eq. 2.

$$L_{Ey} = E_{X, L \sim P_{data}} ||L - E_y(X)||_2^2 \tag{2}$$

4.3   Pixel-Wise Reconstruction Loss

Deep auto encoder training is done by finding mean square loss between the generated image $\bar{X}$ and the original input image X, as given by the Eq. 3.

$$L_D = ||X - G(Z, L)||_2^2 \tag{3}$$

5   Algorithm

Required Batch size b, Real data P_d, initial parameters of E_z, E_y, and D are θ_z, θ_y and θ_d respectively.

#Training Part
1:     **while** θ_D as not converged **do**
2:          Sample $\{X, C\} \sim P_d$, a batch from real data
3:          $L \leftarrow E_y(X, C)$
4:          $L_{Ey} \leftarrow ||L - L'||^2$, where L' is modified condition vector
5:          $\theta_y \xleftarrow{+} -\nabla_{\theta y}(L_{Ey})$
6:          $Z \leftarrow S(E_Z(X, C))$
7:          $\bar{X} \leftarrow D(Z, L)$
8:          $Z' \leftarrow S(E_z(\bar{X}, C))$
9:          $L_z \leftarrow ||Z - Z'||^2$
10:        $\theta_z \xleftarrow{+} -\nabla_{\theta z}(L_z)$
11:        $L_D \leftarrow ||X - \bar{X}||^2$
12:        $\theta_G \& \theta_{Ez} \& \theta_{Ey} \xleftarrow{+} -\nabla_{\theta_D \& \theta_{Ez} \& \theta_{Ey}}(L_D)$
13:    **end while**
#Testing Part
14:        Sample $\{X, C\} \sim P_d$, a batch from real data
15:        $L \leftarrow E_Y(X, C)$
16:        $Z \leftarrow S(E_Z(X, C))$
17:        $\bar{X} \leftarrow D(Z, L')$

6   Experiments

In this section, we will present two datasets on which all of our global face attribute alteration experiments are carried out and describe the prepossessing steps and our model implementation details during training SCDAE. At last we show our model qualitative and quantitative analysis results.

6.1   Experiment on the CLF and UTKFace Data Set

The two datasets used during the training and testing of the model are UTK-Face [18] and private child specific dataset CLF [17]. We used these datasets as they provide global face attributes like age, gender, and ethnicity of each age group. We have applied MTCNN [33] on CLF and UTKFace dataset. MTCNN detects the face and applies proper alignment on the images. We cropped the images to the resolution of 128×128 pixels, according to the distance between eyes and nose to keep head with hair. To make the learning easier, we preprocessed the data by normalizing the pixel values to the range $[-1, 1]$. All components are trained with batch size 64 using ADAM optimizer with hyperparameter $\alpha = 0.0001$ and $\beta = (0.5, 0.999)$. The output of Decoder D is restricted to [−1,1] using tanh activation function. After 50 epochs we were able to achieve competent results.

6.2 Qualitative Results

We studied three global attribute alterations, i.e., gender, ethnicity, and age. For gender alteration, we evaluate our model for male-to-female and vice versa. For ethnicity conversion, we test the our model by White, Black, Asian, and Indian ethnicity conversion. For age alteration, evaluated the model only on CLF data set by age regression on child face.

Gender Alteration Results. In Fig. 2 we see that females are modified to males. The change in the facial features is prominently visible as there is addition of moustache for male, and the jawline becomes a bit broader in male as compared to female.

Fig. 2. Gender alteration Female to Male results from our SCDAE. The first and second row is the input facial image and the alteration results, respectively.

Fig. 3. Gender alteration Female to Male results from our SCDAE model. The first and second row is the input images and the output images, respectively.

Ethnicity Alteration Results. Ethnicity is very import with respect to increasing the dataset size or to reduce augmentation. Now we show our results obtained from ethnicity alterations. In Fig. 4 while conversion to White ethnicity we observe that color of face has turned light, and cheek bone structure is also similar to people of white ethnicity. In Fig. 5 for conversion to Black ethnicity the modified faces have dark color. In Fig. 6 for conversion to Asian ethnicity, we can see clearly that eyes have become narrower. In Fig. 7 for conversion to Indian ethnicity, we see that eyes have become wide open and color of face has turned into typical Indian color.

Age Alteration. Here, in Fig. 8 we see two relatively young children belonging to age group of [5–7] are converted to boys of age group [15–17]. The change in the facial features is visible as there is a bit of moustache which starts growing as

Fig. 4. The first row is the input image and second row for each column is generated image whose ethnicity is White.

Fig. 5. The first row is the input image and second row for each column is generated image whose ethnicity is Black.

boys attain the age of puberty. Also, the face is broader and taller as compared to small children. In Fig. 9 two relatively old children belonging to age group of [15 − 17] are converted to children of age group [5 − 7]. The change in the facial features is visible as the moustache in the old boy is removed when it is reconstructed as a younger one. The size of the face also reduces in both the images. The eyebrows become light in the younger ones as compared to the old ones.

Fig. 6. The first row is the input image and second row for each column is generated image whose ethnicity is Asian.

Fig. 7. The first row is the input image and second row for each column is generated image whose ethnicity is Indian.

6.3 Quantitative Analysis

We used faceplusplus [34] application to analyze the results obtained from our model. This application recognizes the gender and ethnicity of faces. For finding our accuracy for gender modification we experimented on 50 modified faces and observed whether the alteration is successful. Thus, our conversion ratio is given by $\frac{Gender\,Altered\,Successfully}{Total\,Images\,Tested}$ which was $\frac{35}{70}$ i.e,. 70%. For finding accuracy for ethnicity modification we used a similar approach and ethnicity conversion success was only 58%.

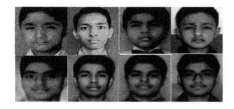 

Fig. 8. Results showing aging on child faces.

Fig. 9. Results showing de-ageing on child faces.

7   Conclusions and Future Work

In this paper we proposed a novel model SCDAE to perform global facial attributes alteration namely gender, ethnicity and age using two deep encoder, sampling and decoder components. Experiments demonstrate that our SCDAE approach can alter untouched global attributes along with more realistic faces in term of person identity and age uniformity while aligning better with human observation. We have presented the first significant results on ethnicity and child face image alterations. In the future, we will do experimental analysis on the complete dataset and explore further to improve the generated image quality results.

Acknowledgments. We gratefully acknowledge the support of NVIDIA Corporation with the donation of the TITAN V GPU used for this research.

References

1. Hinton, G., Salakhutdinov, R.: Reducing the dimensionality of data with neural networks. Science **313**(5786), 504–507 (2006)
2. Kingma, D.P., Welling, M.: Auto-encoding variational bayes. In: ICLR (2014)
3. Goodfellow, I., et al.: Generative adversarial nets. In: Advances in Neural Information Processing Systems, vol. 27, pp. 2672–2680. Curran Associates Inc. (2014)
4. Radford, A., Metz, L., Chintala, S.: Unsupervised representation learning with deep convolutional generative adversarial networks. In: ICLR (2016)
5. Zhu, J., Krähenbühl, P., Shechtman, E., Efros, A.A.: Generative visual manipulation on the natural image manifold. CoRR, vol. abs/1609.03552 (2016)
6. Xu, Z., Yang, X., Li, X., Sun, X.: The effectiveness of instance normalization: a strong baseline for single image dehazing. CoRR, vol. abs/1805.03305 (2018)
7. Gatys, L.A., Ecker, A.S., Bethge, M.: A neural algorithm of artistic style. CoRR, vol. abs/1508.06576 (2015)
8. Shi, Y., Debayan, D., Jain, A.K.: WarpGAN: automatic caricature generation (2018)
9. Reed, S.E., Akata, Z., Yan, X., Logeswaran, L., Schiele, B., Lee, H.: Generative adversarial text to image synthesis. CoRR, vol. abs/1605.05396 (2016)
10. Mirza, M., Osindero, S.: Conditional generative adversarial nets. CoRR, vol. abs/1411.1784 (2014)

11. Liu, Z., Luo, P., Wang, X., Tang, X.: Deep learning face attributes in the wild. In: Proceedings of International Conference on Computer Vision (ICCV) (2015)

12. Huang, G.B., Mattar, M., Lee, H., Learned-Miller, E.: Learning to align from scratch. In: NIPS (2012)

13. Kumar, N., Berg, A.C., Belhumeur, P.N., Nayar, S.K.: Attribute and simile classifiers for face verification. In: IEEE International Conference on Computer Vision ICCV (2009)

14. Kumar, N., Belhumeur, P., Nayar, S.: FaceTracer: a search engine for large collections of images with faces. In: Forsyth, D., Torr, P., Zisserman, A. (eds.) ECCV 2008. LNCS, vol. 5305, pp. 340–353. Springer, Berlin (2008). https://doi.org/10.1007/978-3-540-88693-8_25

15. Chandaliya, P.K., Garg, P., Nain, N.: Retrieval of facial images re-rendered with natural aging effect using child facial image and age. In: The 14th International Conference on Signal Image Technology and Internet Based System, Spain, 26–29 November 2018, pp. 457–464 (2018)

16. Song, Y., Zhang, Z., Qi, H.: Age progression/regression by conditional adversarial autoencoder. In: IEEE Conference on Computer Vision and Pattern Recognition (CVPR). IEEE (2017)

17. Deb, D., Nain, N., Jain, A.K.: Longitudinal study of child face recognition. In: 2018 International Conference on Biometrics, ICB 2018, Gold Coast, Australia, 20–23 February 2018, pp. 225–232 (2018)

18. Zhang, Z., Song, Y., Qi, H.: Age progression/regression by conditional adversarial autoencoder. In: CVPR, pp. 4352–4360. IEEE Computer Society (2017)

19. Zheng, X., Guo, Y., Huang, H., Li, Y., He, R.: A survey to deep facial attribute analysis. CoRR, vol. abs/1812.10265 (2018)

20. Chen, X., Duan, Y., Houthooft, R., Schulman, J., Sutskever, I., Abbeel, P.: InfoGAN: interpretable representation learning by information maximizing generative adversarial nets. CoRR, vol. abs/1606.03657 (2016)

21. Li, M., Zuo, W., Zhang, D.: Deep identity-aware transfer of facial attributes. CoRR, vol. abs/1610.05586 (2016)

22. Liu, M., Breuel, T., Kautz, J.: Unsupervised image-to-image translation networks. CoRR, vol. abs/1703.00848 (2017)

23. Shen, W., Liu, R.: Learning residual images for face attribute manipulation. CoRR, vol. abs/1612.05363 (2016)

24. Wang, Y., Wang, S., Qi, G., Tang, J., Li, B.: Weakly supervised facial attribute manipulation via deep adversarial network. In: WACV, pp. 112–121. IEEE Computer Society (2018)

25. Zhang, G., Kan, M., Shan, S., Chen, X.: Generative adversarial network with spatial attention for face attribute editing. In: Ferrari, V., Hebert, M., Sminchisescu, C., Weiss, Y. (eds.) ECCV 2018. LNCS, vol. 11210, pp. 422–437. Springer, Cham (2018). https://doi.org/10.1007/978-3-030-01231-1_26

26. Larsen, A.B.L., Sønderby, S.K., Winther, O.: Autoencoding beyond pixels using a learned similarity metric. CoRR, vol. abs/1512.09300 (2015)

27. Yan, X., Yang, J., Sohn, K., Lee, H.: Attribute2Image: conditional image generation from visual attributes. In: Leibe, B., Matas, J., Sebe, N., Welling, M. (eds.) ECCV 2016. LNCS, vol. 9908, pp. 776–791. Springer, Cham (2016). https://doi.org/10.1007/978-3-319-46493-0_47

28. Perarnau, G., van de Weijer, J., Raducanu, B., Álvarez, J.M.: Invertible conditional GANs for image editing. CoRR, vol. abs/1611.06355 (2016)

29. Lample, G., Zeghidour, N., Usunier, N., Bordes, A., Denoyer, L., Ranzato, M.: Fader networks: manipulating images by sliding attributes, pp. 5969–5978 (2017)

30. Chandaliya, P.K., Nain, N.: Conditional perceptual adversarial variational autoencoder for age progression and regression on children face. In: The 12th IAPR International Conference On Biometrics, Crete Greece, 4–7 June 2019, pp. 200–208 (2019)
31. Lu, Y., Tai, Y.-W., Tang, C.-K.: Attribute-guided face generation using conditional cycleGAN. In: Ferrari, V., Hebert, M., Sminchisescu, C., Weiss, Y. (eds.) ECCV 2018. LNCS, vol. 11216, pp. 293–308. Springer, Cham (2018). https://doi.org/10.1007/978-3-030-01258-8_18
32. Choi, Y., Choi, M., Kim, M., Ha, J., Kim, S., Choo, J.: StarGAN: unified generative adversarial networks for multi-domain image-to-image translation. In: 2018 IEEE Conference on Computer Vision and Pattern Recognition, CVPR 2018, Salt Lake City, UT, USA, 18–22 June 2018, pp. 8789–8797 (2018)
33. Zhang, K., Zhang, Z., Li, Z., Qiao, Y.: Joint face detection and alignment using multitask cascaded convolutional networks. IEEE Signal Process. Lett. **23**(10), 1499–1503 (2016)
34. Faceplus. https://www.faceplusplus.com

Manipuri Handwritten Character Recognition by Convolutional Neural Network

Sanasam Inunganbi[1]([⊠]), Prakash Choudhary[2], and Khumanthem Manglem[1]

[1] National Institue of Technology, Manipur, Imphal, India
inung.sam@gmail.com, manglem@gmail.com
[2] National Institute of Technology, Hamirpur, Himachal Pradesh, India
choudharyprakash87@gmail.com

Abstract. Handwritten character recognition is an essential field in pattern recognition. Its popularity is increasing with the potential to thrive in various applications such as banking, postal automation, form filling, etc. However, developing such a system is a challenging task with the diverse writing style of the same character, and present of visually similar characteristics. In this paper, a recognition system is proposed using a deep neural network. The performance of the network is investigated on a self-collected handwritten dataset of Manipuri script contributed by 90 different people of varying age and education. A total of 4900 sample images is considered for the experiment and recorded a recognition rate of 98.86%.

Keywords: Handwritten dataset · Character recognition · Manipuri script · Convolutional neural network

1 Introduction

Recently, CNN has gained great success in large scale image and video processing. One of the core problem in computer vision problem is image classification where unknown images are given a labeled from a fixed set of categories based on its visual contents. Handwritten character recognition has emerged as one of the critical areas with a varied range of application. The applications stretch from zip code identification to writer recognition, from recognizing numerals and alphabets in the number plate of traffics to bank check processing, etc. However, designing an architecture for handwritten character recognition impose numerous challenges to the researchers due to the problems that prevail in the nature of unconstrained handwritten characters or words. The shape of the same character that may differ depending upon the writers, some may write with large structure while others may complete in small-scale version. Hence, the overlap area of ink trace of the character may be very less. Further, depending on the acquisition device, pen width, and ink color may impose variation on writing style. Moreover, handwritten Manipuri characters are complicated due to their

N. Nain et al. (Eds.): CVIP 2019, CCIS 1148, pp. 307–318, 2020.
https://doi.org/10.1007/978-981-15-4018-9_28

structure and shape. They include a significant character set with more curves, loops, and other details in the characters. There are many character pairs which are quite similar in shape. All these issues demand attention and solution with the help of an efficient recognition system.

Generally, a recognition algorithm trains on a dataset of known characters with own label to determine the characters included in test set accurately. In this paper, a deep learning approach for handwritten Meitei Mayek (Manipuri) character recognition is proposed. Manipuri is the official language of Manipur, and Meitei Mayek is its script. Being a regional language, research on this script is still at an early stage. Therefore, a convolutional neural network-based app-roach with six convolution layer with two max-pooling layers (applied on every third convolution layers) to downsize the image volume is investigated. Every convolution layer has filter size of 5×5 while max-pooling has filter size of 2×2 and stride 2. In this work, the normalized handwritten Meitei Mayek charac-ter images of size $32 \times 32 \times 1$ are provided into the network to classify them accurately. From the experimental results, it can be shown that CNN method has outperformed other conventional methods in classifying handwritten Meitei Mayek characters.

Manipuri script consist of rich set of characters which involve 10 numerals called *Cheising Eeyek*, 27 basic alphabets (which is further comprise of 18 original letters called *Eeyek Eepee* and 9 additional called *Lom Eeyek*), 8 derived letters called *Lonsum Eeyek* and 8 associating symbol called *cheitap Eeyek*. Their archi-tecture is given in Fig. 1. In this paper, we have considered the 27 basic alphabet and 8 derived letters only for the experiment.

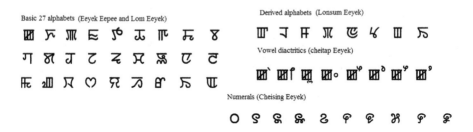

Fig. 1. Meitei Mayek alphabets for recognition system

1.1   Related Work

Numerous character recognition systems have flourished for diverse languages in the world. One of the primary methods is pixel-based method [3], which straightly uses the pixel intensities of the character image as a feature. The feature size is usually dependent on the image size and is generally high. The authors have also articulated the black and white scale down pixel-based method to reduce the feature where the image is decomposed into non-overlapping blocks, and foreground pixel pattern is considered from each block for computing the feature element. In contemporary years, deep learning architecture [4,5] has gotten more attention for pattern recognition and other machine learning problems.

The advent of deep learning acts as the onset of the convolutional neural network (CNN) in computer vision. A multilayer artificial neural network has been proposed in [6] for character recognition and other computer vision problems. One of the classic systems for document recognition is proposed by Lecun et al. with gradient-based learning methods of convolutional neural network (CNN) popularly known as LeNet [7]. Another network similar to Lenet but quite bigger and powerful has been proposed in [8]. The network used ReLu, and multiple GPUs with an exclusive layer called local response normalization (LRN). This deep CNN has been used to classify ImageNet dataset. A remarkable procedure is proposed in [9] where instead of having as many hyperparameters, the focus has been made on the evaluation of simpler networks where convolution layer of (3×3) filters is fixated with increasing depth.

In [10], a system for recognizing handwritten word expressed by broken letters using statistical features of broken letters of Persian alphabets is proposed using a fuzzy neural network. The proposed system acquired high accuracy and feasible to extend dataset. A piecewise feature extraction technique for handwritten devnagiri character recognition is proposed [11]. The image partitioning technique is used for piecewise histogram of oriented gradient (HOG) features extraction, and the feature vector has been trained by a neural network, achieving a maximum accuracy of 99.27%. A deep belief neural network is investigated for Arabic handwritten character and word recognition in [12]. The authors have taken a raw image as input and proceed with a greedy layer-wise unsupervised learning algorithm. The method is implemented on two different, namely, HABCD for handwritten characters and ADAB of 946 different town names. Character level recognition has resulted in 2.1%, but word-level recognition has room for improvement with an error rate of 41%. In [13], a convolutional deep neural network model has been presented to recognize Bengali handwritten character. A compelling set of features has been extracted using kernels and local receptive fields and then forwarded to densely connected layers for the classification task. The experiment is conducted on BanglaLekha Isolated dataset with an overall accuracy of 89.93%. Another recognition system on isolated handwritten Bangla character has also proposed using CNN in [18] where a modified model of ResNet-18 architecture has been stated. A framework of recurrent neural network (RNN) serves as a discriminant model for Chinese character recognition has been proposed and further extends the model for generating the characters in [14]. Here, the RNN based model acts an end-to-end system which directly handles with the sequential structure without any specific domain knowledge. Mathematical models based on the semi-Markov conditional random field were also explored in literature for character or text recognition [15]. Combining the artificial neural network (ANN) with a hidden Markov model (HMM) has been exploited extensively in the field of character recognition for different languages [16,17]. A simple technique for Lampung handwritten character recognition has been proposed in

[19] where the training algorithm is a back-propagation neural network. In this approach, the hierarchical network system has been performed to optimize the training and recognition algorithm. Persian handwritten character recognition has been investigated in [20] using CNN. Along with the conventional methods, two different types of CNN has been stated: one with simple CNN (SCNN) which has been implemented based on LeNet-5 and another with extension into ensemble CNN (ECNN). Accuracies of 97.1% and 96.3% have been obtained by the SCNN and ECNN methods, respectively. Very few research works have been performed on Meitei Mayek script and have been summarized in Table 1.

Table 1. Existing research works on Meitei Mayek script.

Papers	Methods	Recognition rate
[21]	Probabilistic and fuzzy features with ANN	90.03%
[22]	Chain code, directional feature, aspect ratio and the longest vertical run with SVM RBF kernel classifier	96%
[23]	NN with back propagation	80%
[24]	NN using pixel density of binary pattern as feature vector	85%
[25]	Features from Gabor filter using SVM classifier	89.58%

The rest of the paper is organized as follows: Sect. 2 describes in detail about the database with dimension and acquisition process, and Sect. 3 explains the CNN in simplified terms. Further, Sect. 4 presents out CNN model for character recognition with experimental results and analysis; lastly Sect. 5 concludes this paper with the critical and findings.

2   Description About Dataset

Data acquisition plays a significant role in the research area. It accounts for gathering and estimating relevant information to develop a target system, here handwritten character recognition system. There is no publicly available dataset for handwritten Meitei Mayek characters. Therefore, we have manually collected isolated characters from various people who can read and write Meitei Mayek for development and evaluation of efficient character recognition. Previously, an isolated handwritten Meitei Mayek dataset has been proposed in [1], but it consists of the only 27 classes of *Eeyek eepee*. In this paper, we have included the 8 letters called *Lonsum Mayek* which are derived from distinct *Eeyek Eepee*. So, in total, there are 35 classes of Meitei Mayek characters consider for recognition in this paper. The derived characters are very similar to their respective original, which further add to the challenge in recognizing them.

The 35 characters have been collected in a set of 4 in 140 pages of the A4 sheet. The isolated characters are raised in a tabular format where a cell is occupied by one handwritten character sample. Figure 2 illustrates an instance of a filled form of Meitei Mayek dataset. Each page is comprised of a printed character and 35 empty slots for various individuals to inscribe the written character in their writing style. Since every character has been sampled 35 instances in a set, a sub-total of 1225 (35 × 35) isolated characters are collected for a set. Therefore, considering all the four sets, there are a total of 4900 Meitei Mayek character available for experimentation in this work. To complete the data acquisition process, 90 people have contributed to their writing habit. These people have a different educational background and have a mixed age group between 6 to 40 years. The writers also record their demographic information in the dataset form such as name, address, occupation, qualification signature, etc. so that other application like signature verification can utilize the data. The preprocessing methods performed in this paper is similar to the approach described in our previous work in [1, 2].

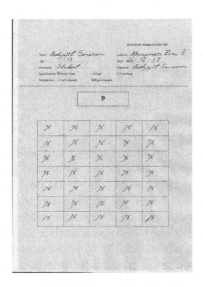

Fig. 2. A sample filled form of Meitei Mayek dataset

3   Convolutional Neural Network

Convolutional neural networks (CNN) is a genre of deep, feed-forward artificial neural networks which are successfully adapted to various application of investigating visual imagery. Generally, images are vectors of high dimension and would take a large parameter to describe the network. Therefore, to address the problem of an extensive parameter, bionic convolutional neural networks are designed to curtail the number of parameters and prepared the network design and architecture categorically to diverse vision tasks. The CNNs are commonly sequenced by a set of layers that can be aggregated by their functionalities, as illustrated in Fig. 3 and interpreted as follows:

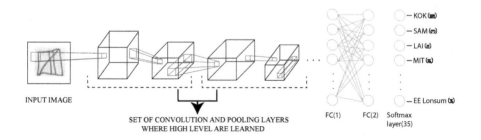

Fig. 3. A sample architecture of convolutional neural network

3.1   Convolution Layer

It performs 2D convolution to the input, which is modestly the dot product with the weight of filter and integrating them across the various channels. The weights of the filter are distributed across the corresponding fields. The filter has an equal number of layers as the input image channels while the output volume has the same depth as the number of filters.

The non-linearity mapping between the input and the output variables is introduced by a critical mechanism called 'activation function.' The operation is imposed to make the network more robust and boost strength to it to determine something complicated and useful from the imagery data. The idea is transforming an input signal of a node to output which will subsequently serve as input in the next layer on the stack. Another significant characteristic of an activation function is differentiability to perform backpropagation escalation procedure. The process computes gradients of error(loss) concerning weights and then accordingly optimize weights using gradient descent. Therefore, the activation layer increased the non-linearity of the network without affecting respective fields of convolution layer. The Sigmoid and Rectify linear Unit (ReLu) are the popular types of activation functions given by Eqs. 1 and 2.

$$\sigma(z) = \frac{1}{(1 + e^{-z})} \tag{1}$$

$$R(z) = max(0, z) \tag{2}$$

Machine learning and computer science notice that most consistent and straightforward techniques and methods are only preferred and are the best. Hence, it avoids and rectifies the vanishing gradient problem. Almost all deep learning Models use ReLu nowadays. However, it should only be used within the hidden layers of a Neural Network Model.

Hence for the output layer, a special kind of activation function called soft-max layer is used at the end of fully connected layer output to compute the probabilities for the classes (for classification). For a given sample vector input x and weight vectors w_i, the predicted probability of $y = j$ is given by Eq. 3.

$$P(y = i|x) = \frac{e^{x^T w_j}}{\sum_{k=1}^{K} e^{x^T w_k}} \tag{3}$$

3.2 Pooling Layer

It can be inferred from the previous section that the convolution layers provide activation mapping between input and output variables, pooling layers employ non-linear downsampling on activation maps. This layer is aggressive with the discard of information; the trend is to use smaller filter size and abandon pooling. It uses two hyperparameters, filter size F and stride S. For an image of size, $M \times N \times D$, pooling results in $[(M - F)/S + 1 \times (N - F)/S + 1 \times D]$ size output.

The intuition behind pooling activity is that a max operation does a lot of features detected anywhere in any of the quadrants; it then remains preserved in the output of max-pooling.

3.3 Fullyconnected Layer

It is usually a regular neural network which can be perceived as the ultimate learning stage with all the acquired visual features to relate to the appropriate output labels. Fully-connected (FC) layers are usually adaptive to classification or encoding task with the standard output of a vector which, when assigned to the softmax layer, display the confidence level for classification.

4   Proposed Method

The model of CNN architecture in this approach consists of six convolution layers with two pooling layers, each one for downsizing after the three convolution layer, as illustrated in Fig. 4. Finally, there are three fully connected layers, with the last being the Softmax layer (Eq. 4). As described in Eq. 3, this layer postulates a probability distribution over a fixed number of categories and select the category that has the maximum probability designated by the network.

$$S(y_i) = \frac{e^{y_i}}{\sum_j e^{y_j}} \tag{4}$$

The first convolution layer accepts a character image of size 32 × 32 for the start of distinctive feature extraction for recognition. Every convolution layer has the same filter size of 5 × 5, but the number of filters varies. In the first

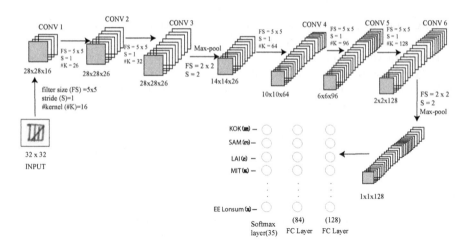

Fig. 4. The proposed convolutional network model

layer, we have employed 16 filters followed by 26 and 32 filters and hence the number of units per layer increase thereby boosting representational power of the network. Except for the first convolution layer, the other two have maintained the output size as the same as input size and hence the resulting size after three convolution layer is 28×28 with 32 filters volume (as there are 32 filters on the third convolution layer). Each convolution layer is passed through ReLu activation layer, and final ReLu layer output is downsized to half (14×14) by a max-pooling layer of size 2×2 with stride 2.

Further, the resulting mapping from the max-pooled layer is subsequently passed through three convolution layer where we have used 64, 96 and 128 kernels respectively looking for 5×5 size active filter. This time the padding is set to 1 and hence the size of each the three convolution layer is respectively, $10 \times 10 \times 64$, $6 \times 6 \times 96$ and $2 \times 2 \times 128$. Here also, every convolution layer is applied with ReLu activation function for increasing non-linearity of the network. The final results ($2 \times 2 \times 128$) is passed to second max-pooling of 2×2 of stride two to get a deep network of size $1 \times 1 \times 128$. The second max-pooling layer instigates the features that would help the final softmax layer in the classification procedure. This mark the closure of the feature extraction process and consequently recognize the actual class from these features.

Finally, it is connected to three fully connected layers of size 128, followed by 84 and 35. The last one is the softmax layer, which generates a probability distribution over the classes for a given input. In this architecture, we have applied the regularization technique, batch normalization on every layer to facilitate network training and reduce the sensitivity to network initialization.

4.1   Experimental Results and Analysis

The CNN model is tested on self-collected handwritten Meitei Mayek isolated character dataset of size 4900 sample images. All images are normalized to an equivalent size of 32 × 32 before the operation. The number of training images to train the network is 105 from each class label totaling to 3675 of 35 classes. This CNN model works well even with not so large number of sample images. An accuracy of 98.86% of correct recognition is obtained on 10th epoch. It took about 207 s to reach until 10th epoch and met the validation criteria. A graph of training and validation accuracy is provided against the number of the epoch is given in Fig. 5.

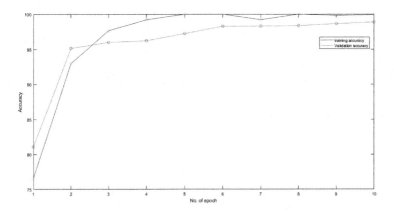

Fig. 5. The growth of training and validation accuracy as the number of epoch increases

Further, as the network goes more in-depth with the higher number of convolution layers and filters, complex and detail information are gained. It can be illustrated visually in Fig. 6; more meaningful information is perceived as we go deeper into the network. The CONV layer one can be seen as pure black and white blocks stack together. However, as the network advances toward CONV Layer 4, CONV Layer 5 and CONV Layer 6, more purposeful image can be seen. These images are focused on the features cultivated by the network.

The proposed character recognition work has been compared with the previous work in literature, and the results have been summarized in Table 2. It can be observed from the table that a deep neural network has provided with higher recognition rate as compare to the other neural network methods and techniques existing in the literature. Further, the proposed method is also compared with a classic network called Lenet-5 on the developed dataset.

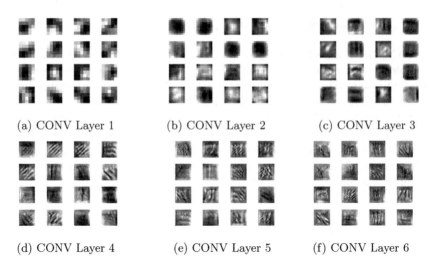

(a) CONV Layer 1   (b) CONV Layer 2   (c) CONV Layer 3

(d) CONV Layer 4   (e) CONV Layer 5   (f) CONV Layer 6

Fig. 6. Images that activate the channels within the network through the layers (for simplicity 16 images from each CONV layer are shown)

Table 2. Comparison of accuracy of the proposed CNN model with the existing methods in literature.

Papers	Recognition methods	Accuracy
[21]	Probabilistic and fuzzy feature with ANN	90.3%
[23]	Binary pattern as vector and NN with back propagation	80%
[24]	NN using pixel density of binary pattern as feature	85%
[25]	SVM using Gabor filter	89.58%
[7]	Lenet-5	96.02%
Proposed method	CNN model	**98.86%**

5   Conclusion and Future Work

In this investigation, we have proposed a convolutional neural network for recognition of handwritten Meitei Mayek character set. It can be found from the experimental analysis that the method is highly effective for the recognizing problem and performs superior as compare to the existing methods in the literature. Further, it has been noted that a higher level of distinctive features can be derived by growing the capacity of the networks with larger filter or kernel size. It is found that not very large epoch is required to train this CNN model, around 10 epochs are successfully enough.

In the future, the experiment can be performed with more data samples and across different languages. Further, recognition of word or sentence level can be taken up with efficient network model.

References

1. Inunganbi, S., Choudhary, P.: Recognition of handwritten meitei mayek script based on texture feature. Int. J. Nat. Lang. Comput. (IJNLC) **7**(5), 99–108 (2018)
2. Inunganbi, S.C., Choudhary, P.: Recognition of handwritten Meitei Mayek and English Alphabets using combination of spatial features. In: Abraham, A., Cherukuri, A.K., Melin, P., Gandhi, N. (eds.) ISDA 2018 2018. AISC, vol. 940, pp. 1133–1142. Springer, Cham (2020). https://doi.org/10.1007/978-3-030-16657-1_106
3. Surinta, O., Schomaker, L., Wiering, M.: A comparison of feature and pixel-based methods for recognizing handwritten bangla digits. In: 2013 12th International Conference on Document Analysis and Recognition. IEEE (2013)
4. Schmidhuber, J.: Deep learning in neural networks: an overview. Neural Netw. **61**, 85–117 (2015)
5. LeCun, Y., Bengio, Y., Hinton, G.: Deep learning. Nature **521**(7553), 436 (2015)
6. Fukushima, K.: Neocognitron: a self-organizing neural network model for a mechanism of pattern recognition unaffected by shift in position. Biol. Cybern. **36**(4), 193–202 (1980)
7. LeCun, Y., et al.: Gradient-based learning applied to document recognition. In: Proceedings of the IEEE 86.11, pp. 2278–2324 (1998)
8. Krizhevsky, A., Sutskever, I., Hinton, G.E.: ImageNet classification with deep convolutional neural networks. In: Advances in Neural Information Processing Systems (2012)
9. Simonyan, K., Zisserman, A.: Very deep convolutional networks for large-scale image recognition. arXiv preprint arXiv:1409.1556 (2014)
10. Kia, M.M.M., et al.: A novel method for recognition of Persian alphabet by using fuzzy neural network. IEEE Access **6**, 77265–77271 (2018)
11. Singh, N.: An efficient approach for handwritten devanagari character recognition based on artificial neural network. In: 2018 5th International Conference on Signal Processing and Integrated Networks (SPIN). IEEE (2018)
12. Elleuch, M., Tagougui, N., Kherallah, M.: Arabic handwritten characters recognition using deep belief neural networks. In: 2015 IEEE 12th International Multi-Conference on Systems, Signals & Devices (SSD15). IEEE (2015)
13. Purkaystha, B., Tapos D., Islam, M.S.: Bengali handwritten character recognition using deep convolutional neural network. In: 2017 20th International Conference of Computer and Information Technology (ICCIT). IEEE (2017)
14. Zhang, X.-Y., et al.: Drawing and recognizing Chinese characters with recurrent neural network. IEEE Trans. Pattern Anal. Mach. Intell. **40**(4), 849–862 (2017)
15. Zhou, X.-D., et al.: Handwritten Chinese/Japanese text recognition using semi-Markov conditional random fields. IEEE Trans. Pattern Anal. Mach. Intell. **35**(10), 2413–2426 (2013)
16. Espana-Boquera, S., et al.: Improving offline handwritten text recognition with hybrid HMM/ANN models. IEEE Trans. Pattern Anal. Mach. Intell. **33**(4), 767–779 (2010)

17. Kishna, N.P.T., Francis, S.: Intelligent tool for Malayalam cursive handwritten character recognition using artificial neural network and Hidden Markov Model. In: 2017 International Conference on Inventive Computing and Informatics (ICICI). IEEE (2017)
18. Alif, M.A.R., Ahmed, S., Hasan, M.A.: Isolated Bangla handwritten character recognition with convolutional neural network. In: 2017 20th International Conference of Computer and Information Technology (ICCIT). IEEE (2017)
19. Fitriawan, H., Setiawan, H.: Neural networks for lampung characters handwritten recognition. In: 2016 International Conference on Computer and Communication Engineering (ICCCE). IEEE (2016)
20. Alizadehashraf, B., Roohi, S.: Persian handwritten character recognition using convolutional neural network. In: 2017 10th Iranian Conference on Machine Vision and Image Processing (MVIP). IEEE (2017)
21. Thokchom, T., et al.: Recognition of handwritten character of manipuri script. JCP 5(10), 1570–1574 (2010)
22. Ghosh, S., et al.: An OCR system for the Meetei Mayek script. In: 2013 Fourth National Conference on Computer Vision, Pattern Recognition, Image Processing and Graphics (NCVPRIPG). IEEE (2013)
23. Laishram, R., et al.: A neural network based handwritten Meitei Mayek alphabet optical character recognition system. In: 2014 IEEE International Conference on Computational Intelligence and Computing Research (ICCIC). IEEE (2014)
24. Laishram, R., et al.: Simulation and modeling of handwritten Meitei Mayek digits using neural network approach. In: Proceedings of the International Conference on Advances in Electronics, Electrical and Computer Science Engineering-EEC (2012)
25. Maring, K.A., Dhir, R.: Recognition of Cheising Iyek/Eeyek-Manipuri digits using Support Vector Machines. IJCSIT 1(2) (2014)

Design and Implementation of Human Safeguard Measure Using Separable Convolutional Neural Network Approach

R. Vaitheeshwari, V. Sathiesh Kumar, and S. Anubha Pearline[✉]

Department of Electronics Engineering, Madras Institute of Technology,
Anna University, Chennai 600044, India
vaitheeshwarir@gmail.com, sathieshkumar@annauniv.edu,
anubhapearl@mitindia.edu

Abstract. Smart surveillance system is designed and developed to mitigate the occurrence of crime scenarios. Traditional image processing methods and deep learning approaches are used to identify the knife from camera feed. On identification of knife, the identity of person holding the knife is obtained using SSD ResNet CNN model. Also, an awareness alarm is generated by the system to caution the people in the surroundings. Experimental investigation clearly shows that the method of fine-tuned Xception deep learning model based on Separable Convolutional Neural Network (SCNN) with Logistic Regression (LR) classifier resulted in highest accuracy of 97.91% and precision rate of 0.98. Face detection is employed using a conditional face detection model based on SSD ResNet. The result obtained using deep learning approach is high compared to that of traditional image processing method. Real time implementation result shows that the model effectively detects the knife and identifies the person holding knife.

Keywords: Knife detection · Face detection · Deep learning · Finetuning · Smart surveillance

1 Introduction

Safety measure is an important constraint of human being for living a peaceful life. It is better to prevent the crime action rather than analyzing the footages after the crime incident. As per the statistics report by National Crime Records Bureau (NCRB-2016 and 2017), India, out of all violent crimes, murder occupies 7.1% and kidnapping about 20.5% of total population in which women are being highly targeted compared to men [1].

Mostly these violent crime involve knifes and firearms (guns) to threaten the person. Also, several criminal attack happen in public places and crowded areas. These actions are recorded in surveillance cameras. Police investigation often assists the help of surveillance camera footages to identify the offender as well as the defender. There are several steps taken by the Government, Researchers and Innovators to provide safety solutions for humans. Devices such as, Foot Wear Chip and SHE (Society Harnessing Equipment) has been implemented and safety measure application such as,

© Springer Nature Singapore Pte Ltd. 2020
N. Nain et al. (Eds.): CVIP 2019, CCIS 1148, pp. 319–330, 2020.
https://doi.org/10.1007/978-981-15-4018-9_29

Raksha- women safety alert, VithU:V Gumrah Initiative and Shake2Safety are incorporated. Recently, the Government of Tamil Nadu, India, has launched a new application called "Kavalan". This application tracks the location of the victim in real time. All these precautious measure directly or indirectly involves the person's attention who is having the device or application to trigger the system. On the other hand, object detection and identification techniques are rapidly increasing using deep learning approach for several applications.

Thus, this paper aims to create a warning system based on deep learning concept is used to minimize the occurrence of crime incident by identifying the knife from the video feed and generates the alert sound to mitigate the crime action.

2 Related Work

Numerous work are reported by the researchers to detect the object in camera for safety purpose. Grega et al. [2] proposed an algorithm that is able to alert the human operator when a firearm or knife is visible in the image [2]. The authors implemented MPEG-7 feature extractor with Support Vector Machine (SVM) classifier and Canny edge detection, with MPEG-7 classifier for knife and firearm detection respectively.

Buckchash et al. [3] proposed a robust object detection algorithm. This proposed approach has three stages, foreground segmentation, Features from Accelerated Segment Test (FAST) based prominent feature detection for image localization and Multi-Resolution Analysis (MRA) [3]. The authors utilized Support Vector Machine (SVM) classifier for image classification and target confirmation. This method achieved about 96% accuracy in detecting the object.

Kibria et al. [4] proposed a comparative analysis of various methods for object detection, it involves HOG-SVM (Histogram of Oriented Gradients- Support Vector Machine), CNN (Convolutional Neural Network), pre-trained AlexNet CNN and the deep learning CNN methods are analyzed to detect object in the images. Authors reported that among all those methods CNN achieved the highest accuracy in detecting objects.

Yuenyong et al. [5] trained a deep neural network is on natural image (GoogleNet dataset) and fine-tuned to classify the IR images as person, or person carrying hidden knife [5]. By fine-tuning the GoogleNet trained on ImageNet dataset achieved 97% accuracy in predicting its classes.

Mahajan et al. [6] proposed a rescue solution for the safety of women as a wearable device-using microcontroller. The wearable device involves switch to trigger the shock circuit. An on-body camera and audio recorder is used to store the data in a SD card attached to the device. A GPS module is attached with the microcontroller device to track the location.

Harikiran et al. [7] proposed a security solution for women. The authors used a microcontroller based smart band and it is connected to a smart phone. The smart band proposed in the work consists of the several sensors to monitor the status of human and sends intimation to registered phone number in case of emergency.

From the literature, it is observed that researchers concentrated on detecting knives as a precautionary measure for ensuring people safety. So far, identification of the person holding the knife has not been carried out. Also, the researchers have not used conventional image processing techniques such as segmentation and windowing methods to identify objects in CCTV cameras. The observed sensitivity rate in the existing methods is less. Safety measures reported in the literature review resulted as wearable device. Therefore, any damages to the device disqualify the reliability of that device.

Hence, in the proposed a system is designed using deep learning neural network approach. The system automatically detects the dangerous objects such as knife in CCTV images and alerts about the hazardous situation with improved accuracy and precision rate. The proposed framework detects the knife using fine-tuned Xception deep learning model and identifies the person involved in the crime through face detection algorithm (SSD-Resnet CNN) from CCTV footages.

3 Methodology

Knife detection is employed by performing comparative analysis of various traditional image processing method and Xception deep learning model. After analyzing the various algorithms for the sensitivity and accuracy, the model with highest sensitivity value is selected for the real time implementation. The face identification system is implemented by using the conditional face detection algorithm based on SSD ResNet model.

The overall system workflow is shown in Fig. 1. The work involves using two different approaches, namely, traditional image processing and deep learning. In traditional image processing method, feature extraction is carried out using Local Binary Pattern (LBP), Haralick feature and Histogram of Oriented Gradients (HOG). The extracted features are classified using Machine Learning (ML) classifiers such as Support Vector Machines (SVM), Logistic Regression (LR) and Random Forest (RF). In deep learning method, Xception CNN model is utilized as feature extractor and classifier.

3.1 Traditional Image Processing Approach

Traditional image processing method involves extraction of features from the input images and classifying the images using ML classifier. At first, the input image is fed into the preprocessing unit. Then, the preprocessed image is fed to feature extraction block. After extraction of features from the images, it is flattened into a 1D array. The combined 1D feature vector Haralick-LBP feature and HOG features of the images in dataset are fed into three machine learning classifiers individually.

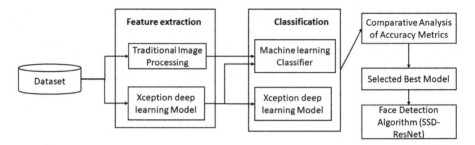

Fig. 1. Workflow for human safeguard measure.

Feature Extraction. To detect the knife in images, it is essential to extract its features such as texture, corner, edges etc. In this approach, the description of images in terms of its features is performed using three different methods.

Local Binary Pattern (LBP). It is a texture descriptor of an image introduced by Ojala [8]. For detecting knife, it is highly essential to find out its texture. Hence, the LBP pattern analysis is used. The standard LBP involves thresholding the center pixel (g_c) of 3×3 gray level matrix with its neighbor gray level intensities (g_i). LBP thresholding is given in Eq. (1). If g_i is lesser than g_c, the binary result of the pixel is set to 0 otherwise it is set to 1 [9].

$$LBP = \sum\nolimits_{p=0}^{p-1} s(g_i - g_c)2^i \tag{1}$$

Where, 'p' is the number of neighbourhood points and 'i' is the neighbour pixel position.

Haralick Feature. Haralick feature is a texture feature extractor of an image introduced by Haralick [10]. It is obtained from Gray Level Co-occurrence Matrix (GLCM) of the image. GLCM computes the relationship between the intensities of neighboring pixel values [11]. GLCM is used to find the region of interest of an image by computing its gray level pixel intensities.

Histogram of Oriented Gradients (HOG). HOG descriptor is a feature extraction method that detects corners and edges of the object in the image. It is achieved using extraction of HOG feature [12]. The steps involved in computation of HOG are as follows.

Algorithm 1: HOG Feature Extraction

Input : Image, I
Output : Feature vector (f_1, f_2,... f_N)

Step 1: Normalization of Image – square root of color channels
Step 2: Calculation of Gradients –Computing contour and silhouette
Step 3: Cell formation and histogram computation
Step 4: Normalizing local cell blocks (HOG feature values)
Step 5: Converts all the hog descriptor for all blocks and combine as a HOG feature vector.

Machine Learning Classifiers

Logistic Regression. Logistic regression (LR) is a supervised binary classifier. Its performance is similar to SVM with linear kernel. It uses the logistic function (sigmoid function) to determine the probability of the predicted class. It predicts the probability by using Eq. (2) [13],

$$y = e^{b0 + b1*x} / \left(1 + e^{b0 + b1*x}\right) \tag{2}$$

where, y is the predicted output, b0 is the bias or intercept term, b1 is the coefficient for the single input value (x) [13] and e is the Euler's value.

Support Vector Machine. Support Vector Machine (SVM) [14, 15] is a supervised learning algorithm, for classifying binary classes. SVM classifier is accomplished using Radial Basis Function (RBF) kernel. Linear kernel considers the hyperplane as a line. While the RBF kernel is based on Eq. (3) for creating the hyperplane to separate the classes [16]. In Eq. (3), $x^{(e)}$ and $x^{(k)}$ represent the feature value of class empty and class knife, respectively. γ is the boundary decision region.

$$K(x^{(e)}, x^{(k)}) = \exp\left(-\gamma \left||x^{(e)} - x^{(k)}\right|\right)^2, \gamma > 0 \tag{3}$$

Random Forest. Random forest (RF) is based on decision tree algorithm. It is an ensemble algorithm utilizing two or more methods for predicting the class [17]. The number of trees used in the work is 500. Random forest generates random subsets of tree, and aggregates the votes from the nodes for best selected feature values. It then averages the votes and the highest voted feature value class is considered as destination class [17].

3.2 Deep Learning Approach

Xception Deep Learning Model as Feature Extractor and Classifier. This is the second approach used in the studies. The performance of the traditional model with ML classifier resulted in lowest accuracy. In order to improve its accuracy a powerful deep learning model called 'Xception' is fine-tuned for feature extraction and classification. In this approach, the analysis is carried out in two different ways. Previous layer trainable parameter is set as a false or true. Setting Layer-trainable as 'false' considers the pre-trained weights from the Xception model and trains only the last three fine-tuned layers. On the other hand, setting layer-trainable as 'true' involves training the model from scratch.

Xception Deep Learning Model with ML Classifier. In this approach, the Xception pre-trained model is used for feature extraction. The extracted features are flattened to 1D vector and classified using ML classifiers such as random forest and logistic regression.

Fine-Tuned Xception Deep Learning Model. Xception model is the Extreme version of the Inception model. There are about 36 convolutional layers in Xception model followed by one fully connected layer, Global Average Pooling (GAP) and one output layer predicts the classes (Knife or Empty) using sigmoid activation function. The sigmoid function maps the feature values in the range between 0 and 1 [18].

It contains several depthwise separable convolution. This depthwise separable convolution is channel-wise nxn spatial convolution [18] and is followed by pointwise convolution. The mathematical representations of the convolution and depthwise separable convolutions are represented in Eqs. (4), (5) and (6) [19].

$$Conv(W,y)_{(i,j)} = \sum_{k,l,m}^{K,L,M} W_{(k,l,m)} y_{(i+k,j+l,m)} \tag{4}$$

$$PointwiseConv(W,y)_{(i,j)} = \sum_{k,l,m}^{K,L,M} W_{(m)} y_{(i,j,m)} \tag{5}$$

$$DepthwiseConv(W,y)_{(i,j)} = \sum_{k,l}^{K,L} W_{(k,l)} y_{(i+k,j+l)} \tag{6}$$

where, W is the weight matrix, y(i, j) is the image pixel coefficient, k, l, m is width, height and channel of the image, respectively.

3.3   Face Detection Algorithm

Once the knife in image is detected, the person holding the knife has to be identified. In the proposed work, a conditional face detection algorithm is used to identify the person holding knife. Hence, to incorporate the face detection method, SSD (Single Shot Detector) model is implemented. SSD is used for object detection. It is the fastest known model since it eliminates the need for region proposal of the object [20, 21]. SSD performs two operations. One is extracting the feature values and another one is to detect the object based on the convolution filter [21]. The architecture of face detection algorithm is shown in Fig. 2.

The steps involved in Conditional face detection algorithm are described in Algorithm 1. By implementing this algorithm, the model is effective in predicting the face of the attacker. This is due to conditional approach of the probability rate produced by the classifier. By doing so, only the faces in image where, the knife is detected in attacking position is highlighted with a bounding box.

Algorithm 1: Conditional Face detection

Input: Knife detected image(k), probability of the image p(k)
Output: Assaulter identified image

Step 1: Extract feature from the input image using Xception
      model
Step 2: Inspect the image for knife

Step 3: Once knife in image is detected, check its probability rate.
      If (P (k) > 0.85)
   i.   Detect the faces in image.
   ii. Implement the bounding box, to highlight the face of the assaulter. If the assaulter face is
       not clear, detect all the faces that are clearly identified as face by the model.
      Else if (P (k)<0.85)
   i.   The face detection is not implemented
   ii. Print only the probability occurrence of the knife and display the label as knife.
Step 4: Repeat step 1 to 3 for all the consecutive frames.

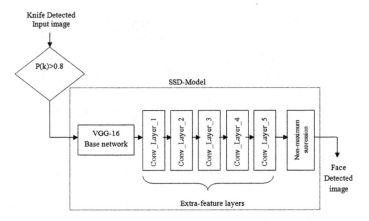

Fig. 2. Block diagram for conditional face detection algorithm using SSD-model

3.4 Dataset Description

The knife dataset is collected from two sources. The first source is Katedra Teleko-munikacji [22], a university in Poland and another one is the datasets created by the students of Department of Computer Science, IIT Roorkee [23]. In addition to that, custom created real- time dataset is appended. One class consists of images with knife and another class consists of images without knife i.e., empty hand images. Thus, the three different dataset sources are collectively named as Dataset-weapon and used in the studies. The sample images from Dataset-weapon are shown in Fig. 3. The number of images in the datasets before and after augmentation is listed in Table 1. The analysis is carried for augmented dataset with train-test split ratio of 7:3 is considered.

Table 1. Dataset-weapon description

Classes	Number of images	
	Without augmentation	With augmentation
Positive (with knife)	3753	10891
Negative(without knife)	9750	10891
Total	13503	21782

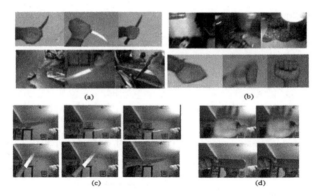

Fig. 3. Sample dataset (a) Positive images from database. (b) Negative images from database. (c) Positive images from real time dataset. (d) Negative image from real time dataset.

4 Results and Discussions

The reliability of the model depends on the performance metrics such as sensitivity, specificity and accuracy. The performance metrics considered in this paper are precision, recall, f1-score and accuracy.

4.1 Results of Traditional Image Processing Approach

The resulted performance metrics for the traditional feature extractors, Haralick and LBP with various ML classifiers are tabulated in Table 2. It is observed from Table 2, that the random forest classifier resulted in the highest accuracy of 86.7% compared to other classifiers.

Table 2. Accuracy metrics of Haralick-LBP features with ML classifiers

Classifier	Precision	Recall	F1 score	Accuracy (%)	Loss
SVM	0.74	0.74	0.73	74.02	0.45
LR	0.72	0.75	0.73	72.13	0.35
RF	**0.87**	**0.86**	**0.86**	**86.70**	**0.27**

Table 3. Accuracy metric of HOG features with ML classifiers

Classifier	Precision	Recall	F1-score	Accuracy (%)	Loss
SVM	0.65	0.67	0.65	68.08	0.49
LR	0.66	0.66	0.66	64.97	0.44
RF	**0.82**	**0.81**	**0.81**	**80.12**	**0.33**

Similarly, for HOG features, the performance metrics are tabulated in Table 3. From the Table 3, it observed that RF classifier resulted in highest accuracy with increased precision rate when compared to other classifiers. A good model should not only have highest accuracy but it must have highest sensitivity (precision) rate. Thus, precision value reveals how accurate the model is, while detecting the knife in images. Though considerable precision is achieved in Tables 2 and 3, the precision value is not sufficient to develop a precise prototype for real time implementation. Hence, the fine-tuned deep learning model is considered for both feature extraction and classification.

4.2   Results of Fine-Tuned Xception Deep Learning Model

In this approach, Fine-tuned Xception deep learning model is used as both feature extractor and classifier. The analysis is carried out using binary cross-entropy as loss function and ReLU as an activation function for 50 epochs. Adam is used as an optimizer with the learning rate of 0.001. The result of this approach is tabulated in Table 4.

Table 4. Analysis of Xception deep learning model

Method	Accuracy	Loss	Computation time	Number of parameters
Layer trainable false	86%	0.3	215 s/Epoch	54,528
Layer trainable true	**99.9%**	**0.02**	510 s/Epoch	2,29,07,128

It is observed that making layer trainable as 'true' attained global minima with increased accuracy of 99.9% and reduced loss of 0.02. The accuracy and loss plot of the model with respect to epochs is shown in Fig. 4(a) and (b).

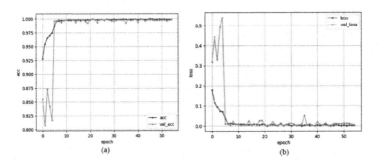

(a)   (b)

Fig. 4. (a) Accuracy plot (b) Loss plot of the Fine tuned Xception model

It is noticed from Fig. 4(a), the validating accuracy suddenly increasing from 4th epoch. This is due to the fact of optimization landscape. Hence, the activation function and optimizer attains global maxima in fourth epoch. Similarly, in Fig. 4(b), loss plot measures the inconsistency between the predicted outcomes with the true class. It is observed from the plot that the model generates very low loss value of about 0.01. Precision and recall values for the fine-tuned model are 0.56. This is due to the output sigmoid layer is sensitive to the texture feature and falsely detects knife in the image.

4.3   Results of Fine-Tuned Xception Model with ML Classifier

In this approach, the last output layer with sigmoid activation function is replaced using random forest and logistic regression classifier. From Table 5, it is observed from the table that for LR classifier, the precision rate is 0.98 that indicates the highest reliability of the model. This is because of logistic regression that is meant for speeded confluence due to being zero-centered. The accuracy gained by the model is 97.84%.

Table 5. Performance metrics of Xception model with LR classifier and RF classifier

Class	LR classifier			RF classifier		
	Precision	Recall	F1 score	Precision	Recall	F1 score
Empty	**0.97**	0.98	0.98	0.92	0.96	0.94
Knife	**0.98**	0.97	0.98	0.96	0.92	0.94

4.4   Real Time Implementation

From the analysis, it is observed that fine-tuned Xception model with LR classifier resulted in the highest accuracy and precision rate. Thus, it is selected for real time testing. The block diagram involved in real time prediction is shown in Fig. 5. It is carried out using Raspberry pi-3 board that consists of 4xARM cortex A53 processor with RAM about 1 GB.

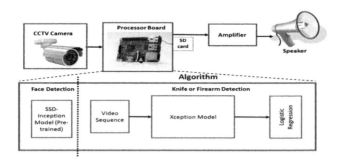

Fig. 5. Real time prediction of the CCTV frame.

Fig. 6. Experimental setup of real time knife detection using Raspberry pi-3 board.

The real time implementation result is shown in Fig. 6. The camera lively monitors the environment. Once the knife is detected, the faces in the images are identified and the pre-recorded police siren sound start to alert the environment.

5 Conclusion

Knife detection in public places is an effective safety measure for human. The proposed method detects both knife and the person holding the knife (probably the offender). The work achieves highest accuracy of 97.84% utilizing the fine-tuned Xception model with LR classifier. The sensitivity rate achieved is 0.98. The prediction time of the model for real time data is less than a second. As a future work, the model will be implemented in real-time with increased number of images for different classes like gun, sword and other sharp objects.

Acknowledgement. The authors would like to thank NVIDIA for providing NVIDIA TITAN X GPU under University Research Programme.

References

1. Crimes in India-2016 Statistics - National crime records Bureau-Ministry of Home Affairs. https://timesofindia.indiatimes.com/realtime/Crime_in_India_2016_Complete_PDF.PDF
2. Grega, M., Matiolański, A., Guzik, P., Leszczuk, M.: Automated detection of firearms and knives in a CCTV image. Sensors **16**(1), 47 (2016)
3. Buckchash, H., Balasubramanian, R.: A robust object detector: application to detection of visual knives. In: IEEE International Conference on Multimedia & Expo Workshops (ICMEW), pp. 633–638. Hong Kong, China (2017)
4. Kibria, S.B., Hasan, M.S.: An analysis of feature extraction and classification algorithms for dangerous object detection. In: 2nd International Conference on Electrical & Electronic Engineering (ICEEE), December, pp. 1–4 (2017)
5. Yuenyong, S., Hnoohom, N., Wongpatikaseree, K.: Automatic detection of knives in infrared images. In: 2018 International ECTI Northern Section Conference on Electrical, Electronics, Computer and Telecommunications Engineering (ECTI-NCON), pp. 65–68 (2018)
6. Mahajan, M., Reddy, K.T.V., Rajput, M.: Design and implementation of a rescue system for safety of women. In: 2016 International Conference on Wireless Communications, Signal Processing and Networking (WiSPNET), pp. 1955–1959 (2016)

7. Harikiran, G.C., Menasinkai, K., Shirol, S.: Smart security solution for women based on Internet Of Things (IOT). In: 2016 International Conference on Electrical, Electronics, and Optimization Techniques (ICEEOT), pp. 3551–3554 (2016)
8. Ojala, T., Pietikäinen, M., Mäenpää, T.: Multiresolution gray-scale and rotation invariant texture classification with local binary patterns. IEEE Trans. Pattern Anal. Mach. Intell. 24(7), 971–987 (2002)
9. Meena, K., Suruliandi, A. Local binary patterns and its variants for face recognition. In: 2011 International Conference on Recent Trends in Information Technology (ICRTIT), pp. 782–786 (2011)
10. Haralick, R.M., Shanmugam, K., Dinstein, I.: Textural features for image classification. IEEE Trans. Syst. Man Cybern. 6, 610–621 (1973)
11. Salhi, K., Jaara, E.M., Alaoui, M.T., Alaoui, Y.T.: GPU implementation of Haralick texture features extraction algorithm for a neuro-morphological texture image segmentation approach. In: 2018 International Conference on Electronics, Control, Optimization and Computer Science (ICECOCS), pp. 1–4 (2018)
12. Zhang, S., Wang, X.: Human detection and object tracking based on Histograms of Oriented Gradients. In: 2013 Ninth International Conference on Natural Computation (ICNC), pp. 1349–1353 (2013)
13. Dreiseitl, S., Ohno-Machado, L.: Logistic regression and artificial neural network classification models: a methodology review. J. Biomed. Inform. 35(5–6), 352–359 (2002)
14. Xiong, S.W., Liu, H.B., Niu, X.X.: Fuzzy support vector machines based on FCM clustering. In: 2005 International Conference on Machine Learning and Cybernetics, vol. 5, pp. 2608–2613 (2005)
15. Marsland, S.: Machine Learning: An Algorithmic Perspective. Chapman and Hall/CRC, Boca Raton (2011)
16. Non-linear SVM classification with kernels (2011). https://www.google.com/url?q=http://openclassroom.stanford.edu/MainFolder/DocumentPage.php?course%3DMachineLearning%26doc%3Dexercises/ex8/ex8.html
17. Liaw, A., Wiener, M.: Classification and regression by random forest. R News 2(3), 18–22 (2002)
18. Chollet, F.: Xception: deep learning with depthwise separable convolutions. In: Proceedings of the IEEE Conference on Computer Vision And Pattern Recognition, pp. 1251–1258 (2017)
19. Kaiser, L., Gomez, A.N., Chollet, F.: Depthwise separable convolutions for neural machine translation (2017). arXiv preprint. arXiv:1706.03059
20. Karpathy, A.: CS231n: Convolutional Neural Networks for Visual Recognition. http://cs231n.github.io/convolutional-networks
21. Liu, W., et al.: SSD: single shot multibox detector. In: Leibe, B., Matas, J., Sebe, N., Welling, M. (eds.) ECCV 2016. LNCS, vol. 9905, pp. 21–37. Springer, Cham (2016). https://doi.org/10.1007/978-3-319-46448-0_2
22. Knives images database. http://kt.agh.edu.pl/~matiolanski/KnivesImagesDatabase/
23. Knives Dataset. https://www.sites.google.com/site/kdsdataset/

Tackling Multiple Visual Artifacts: Blind Image Restoration Using Conditional Adversarial Networks

M. Anand[2], A. Ashwin Natraj[1], V. Jeya Maria Jose[2],
K. Subramanian[2], Priyanka Bhardwaj[1], R. Pandeeswari[1],
and S. Deivalakshmi[1(✉)]

[1] Department of Electronics and Communication Engineering,
National Institute of Technology, Tiruchirappalli, Tiruchirappalli, India
deiva@nitt.edu
[2] Department of Instrumentation and Control Engineering,
National Institute of Technology, Tiruchirappalli, Tiruchirappalli, India

Abstract. Restoring images that are degraded by visual artifacts like noise, blurness and other environmental visual artifacts like shadow, snow, rain, and haze is a challenging task. From literature, it can be seen that there are many model-based as well as blind restoration methods that have been proposed to restore an image degraded by a single artifact. In most practical cases, the image is degraded by more than one artifact. Complexity arises while trying to estimate degradation function using conventional techniques where images are degraded by multiple visual artifacts. To the best of our knowledge, there has not been any generalized method proposed to tackle this problem. In this paper, we propose a methodology using conditional adversarial networks for blind image restoration of images that are degraded by multiple artifacts. To analyze the performance, ISTD dataset (meant originally for shadow removal) is used by augmenting it with different types of noises and blurness. The network has been trained on this data and has been analyzed how it behaves during the addition of each artifact. Various image quality metrics like Peak signal-to-noise ratio (PSNR), Mean squared error (MSE), Structural similarity index (SSIM), Blind/reference-less image spatial quality evaluator (BRISQUE) and Naturalness image quality evaluator (NIQE) have been evaluated to validate the performance of the proposed method.

Keywords: Blind image restoration · Conditional adversarial networks · Shadow removal · Deblurring · Denoising · Multiple visual artifacts

1 Introduction

Image degradation is a very prevalent issue while dealing with the processing of image data. It is caused due to the error in image capturing systems, processing procedures or camera misfocus due to improper handling. Typical degradation examples during image acquisition are blurring and noise. The quality reduction is also sometimes caused by uncontrollable environmental factors like haze, snow, rain or shadow. Visual artifacts

© Springer Nature Singapore Pte Ltd. 2020
N. Nain et al. (Eds.): CVIP 2019, CCIS 1148, pp. 331–342, 2020.
https://doi.org/10.1007/978-981-15-4018-9_30

are these anomalies that cause image degradation during visual representation of imagery. It is a general term that deals with many anomalies ranging from noise, rainbow effect, color banding and degradation due to blurring. Degradation in an image due to artifacts not only affects the visual appearance but also causes a loss in information. This causes computer vision algorithms like detection, classification, and tracking to fail miserably on the degraded image. Thus, any further processing of the image gets affected as well. Image restoration [1] is the method of undoing the effects on an image brought by degradation. The main goal of restoration is to bring back the quality of the degraded image to that of the original image. Complete restoration of the image is not possible when the intensity of degradation is severe. During such scenarios, restoration of the image to some extent is itself plausible. Degraded image is represented mathematically as

$$\delta(x,y) = \alpha(x,y) * \sigma(x,y) + \dot{\epsilon}(x,y) \tag{1}$$

where the $\alpha(x,y)$ is input image function, $\sigma(x,y)$ is image degradation function, $\delta(x,y)$ is degraded image and $\dot{\epsilon}(x,y)$ is the channel transmission noise. General image restoration techniques first estimate the degradation function. The image is then restored from the degraded image by applying the inverse of the degradation function on the degraded image.

Complexity arises when there are multiple artifacts present in the same image. Estimating the degradation function by conventional techniques for restoration is not very effective when the degradation happens due to a mixture of various artifacts. When degradation function estimation is not possible, methods which do not require the degradation function are used to restore the image. Such type of methods are called as Blind restoration methods [2, 3] as they do not need the degradation model to remove artifacts from the image. Blind restoration algorithms come with their own disadvantages as they are generally time-consuming and unreliable. Since the advent of Deep Learning, it has achieved amazing breakthroughs by solving several problems in the fields of imagery, robotics [4], health care [5, 6], audiovisual enhancements and other computational sciences. Generative Adversarial Networks (GANs) [7] was introduced by Goodfellow et al. to synthesize images by effectively training the network to learn the feature distribution of training images. It also dealt with the concept of adversarial training for the first time where two separate deep convolutional networks were pitted against each other to improve the performance of each other. The generation of realistic images by GANs motivated many to use it for various other applications. Adversarial Networks began to be widely popular as they had several applications ranging from style transfer [8], inpainting [9], super-resolution [10], deblurring, denoising [11], image translation [12] and medical image segmentation [13]. Conditional Adversarial Network [14] was introduced by Isola et al. which used GANs in a conditional setting for the image to image translation tasks. Conditional GAN (cGAN) learns the mapping between the input and the target image to perform image to image translation.

The main contribution of the proposed work is to use cGANs to restore images that are corrupted by multiple artifacts since it is highly complex to restore images that are corrupted by various artifacts using conventional restoration techniques. ISTD dataset which is affected only by shadow is used as the base dataset. Those images are further corrupted with different noises like Salt and Pepper noise, Gaussian noise and Speckle noise. The images are also further degraded by adding motion blurness. So, the network is trained with these images which are afflicted by shadow, different types of noises and blurring; both separately and with different combinations of the artifacts. To the best of our knowledge, we are the first to propose a method that can perform shadow removal, denoising and deblurring all at once. cGANs proposed by Isola et al. have been used for this propose. The performance of the network is validated by evaluating quantitative metrics using the restored image from multiple artifacts and comparing it with the ground truth. Qualitative results have also been shown by illustrations. The rest of the paper discusses the following: Sect. 2 deals with the related work that are available in literature separately for Shadow Removal, Denoising, and Deblurring. Section 3 explains how the dataset was augmented to include the artifacts. It also explains the architecture of cGANs that is used in the paper along with how the model is trained and tested. Section 4 describes about the various experiments that were carried out and the visualization results, performance metrics that were evaluated for validation. Section 5 concludes the work and discusses its future scope.

2   Related Work

2.1   Denoising and Deblurring

During the early stages of research on denoising, images were transformed into other appropriate domains to filter out the effect of noise on the signal. Some of the most popular transforms include Discrete Cosine Transform and Discrete Fourier Transform. Discrete Cosine Transform works in such a way that the decorrelated RGB channel's DCT filter coefficients are thresholded, averaged and aggregated over a sliding window patch to remove noise. Variants of the DFT like Fibonacci Fourier transforms are used over double sliding window filters to remove noise. Donoho et al. [15] introduced the idea of using Discrete Wavelet Transform to adaptively threshold the coefficients in his Sure Shrink Algorithm. Although effective at removing noise, these frequency domains local filters tend to introduce new image artifacts such as ringing effect.

This work deals with image restoration by the elimination of multiple artifacts especially the mixture of Gaussian noise, Salt and Pepper Noise (SP), Speckle noise and Motion blurring. Salt and pepper noise occur due to sudden impulses scattered non uniformly throughout the image. Standard methods of removing the same make use of different variants of Median Filters (MF) such as Adaptive MF with regularization and Fuzzy switching MF to name a few [16]. The variety of methods discussed above have all been successful to an extent in gaussian denoising. The environmental conditions of the surroundings of the imaging device at the time of operation can introduce granular noise called speckle noise which is most prevalent in medical images such as ultra-sound and synthetic aperture radar images. Extensive research has been carried out on Non-local means despeckling [17]. DeblurGAN produced the state of the art results for deblurring using a cGAN architecture.

2.2   Shadow Removal

The majority of the already existing shadow removal techniques are mainly composed of 2 phases: detection and removal. There are also techniques directly taking the shadow detected images as input for implementing the shadow removal algorithm. There are computationally intensive methods like reintegration methods which work by identifying the image gradient along the shadow boundaries, nullifying them and reintegrating them back for shadow removal. In addition to the computational complexity, it also suffers from disadvantages like the need for the presence of strong shadow edges. Methods based on relighting works by finding a suitable scale factor to lessen the difference in illumination between shadow and non-shadow regions [18] while it is tougher to find a suitable scale factor along shadow boundary which makes it a bit unpopular.

Some recent shadow removal techniques make use of Otsu's thresholding technique to detect shadow regions and re-illuminate the shadow regions based on information regarding the amount of direct light obstructed for each pixel. There are also techniques which work given the location of shadow. They work by detecting the difference between the edges present in the original image and invariant image, which results in the production of shadow edges of the image. Once shadow edges are obtained, reintegration methods may be used for removal of the shadow region [19]. The recent method based on illumination recovering optimization works by decomposing the input image into patches and based on texture similarity. Equivalence between shadow and lit patches are established, using which optimization illumination operator is found out thereby eliminating the shadow patches and gets back the texture details of the same. Then coherent optimization among neighboring patches results in a shadow-free image [20, 21].

3   Methodology

3.1   Dataset

The ISTD dataset which contains 1870 sets of images degraded by shadow along with their respective ground truth has been used for both training and testing. This is split into 1330 training images and 540 test images. For investigating the performance of the network on multiple artifacts, the training dataset should contain other artifacts as well. This is achieved by augmenting the training dataset with different types of noises and blurness. Data augmentation is done using MATLAB for this purpose. The different types of noises added are salt and pepper noise, speckle noise and gaussian noise. Further degradation is introduced by motion blurring. To understand the variation in performance of the network during the addition of each artifact, various combinations of the different artifacts as well as a final dataset that contains all the artifacts have been created. To improve the robustness of the network, every single training image is added with noise whose noise factor is obtained from a uniform discrete probability distribution function with the lower limit as 10% and the upper limit is 50%. This ensures that the algorithm learns different degrees of noise and all the noise types. This thereby improves our training dataset size to 11970 images (9 different combinations of arti-facts for every 1330 images). All the training images are resized to 256×256 and normalized with mean 0 and standard deviation 1 to reduce the training time.

3.2 Architecture

The conditional Generative Adversarial Networks (cGANs) are a special type of GANs that takes input as an image and gives an output image rather than having input as a random vector. The cGANs have two different convolutional neural networks called Generator and Discriminator which compete against each other to get the best possible output.

The generator used for this cGAN is based on the U-Net architecture [22] which was initially used for biomedical applications. The discriminator used for this cGAN is based on PatchGAN Architecture mostly used for image processing applications. Figure 1 shows the basic block diagram of a cGAN.

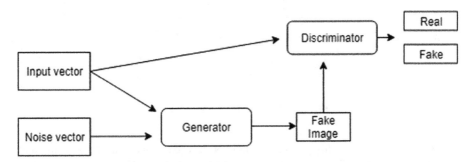

Fig. 1. Block diagram of cGAN [23]

(1) Generator: The Generator takes the image degraded by multiple artifacts as the input and tries to generate a restored image as an output using U-Net. U-Net is very helpful in these kinds of applications because they perform very well in Image to Image translation applications. The U-Net architecture heavily resembles the autoencoder architecture. The only additions are the skip connections. U-Net has many encoder blocks that have convolutional layers, batch normalization layers and activation layers (ReLU activation). Following the encoder blocks, there are many decoder blocks. The decoder blocks consist of a deconvolutional layer, batch normalization layer and activation layer (ReLU activation). The encoder blocks are used to decrease the dimensions of the input image and the decoder blocks are used to increase the dimensions of the input image. Shrinking and enlarging the input images causes loss of information. To rectify this, skip connections have been used. Here the activation of the current layer is computed using the activations of more than one previous layer. This eliminates vanishing gradients thereby helping in retaining prior information about the image as we go down the layers. Figure 2 illustrates the architecture of the generator and Fig. 3 shows the blocks inside the generator.

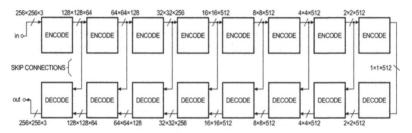

Fig. 2. Architecture of generator [23]

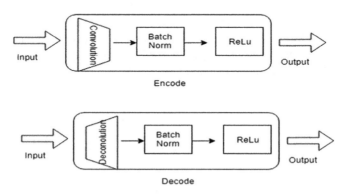

Fig. 3. Blocks inside a generator [23]

(2) Discriminator: In this work, PatchGAN has been used as a discriminator. PatchGAN doesn't compare the entire images but checks N × N patches of them (the patch resolution here being 30 × 30), computes L1 losses thereby trying to reduce them. Patch wise computation aids in faster runtime and lesser parameters since it uses convolutional layers. This method considers each patch as a Random Markov Field such that each patch is independent of each other. Figure 4 shows the architecture of discriminator.

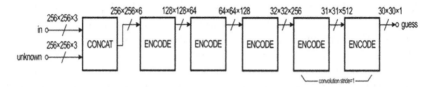

Fig. 4. Architecture of discriminator [23]

3.3 Objective Function

cGANs are trained to learn the mapping from observed image x and random noise vector z to the ground truth y, i.e. G: {x,z} → y. The Generator G tries to generate indistinguishable images that closely resemble the "real" images that cannot be differentiated by the trained discriminator D which is good at detecting fakes.

The objective function of the cGAN is

$$L_{cGAN} = E_{x,y}[logD(x,y)] + E_{x,z}[log(1 - D(x,G(x,z)))] \qquad (2)$$

where G tries to minimize the objective function against discriminator D that tries to maximize it, i.e.,

$$G^* = arg\ min_G\ max_D\ L_{cGAN} \qquad (3)$$

Since the generator has to be trained to fake the discriminator and also train it to match the ground truth, L1 loss has been chosen rather than L2 loss. It also helps to avoid blurred generated outputs.

$$L_{L1}(G) = E_{x,y,z}[||y - G(x,z)||] \qquad (4)$$

Thus, the final objective function is

$$G^* = arg\ min_G\ max_D\ L_{cGAN}\ (G,\ D) + \lambda L_{L1}(G) \qquad (5)$$

3.4   Training and Testing

While training, the generator, and discriminator are trained simultaneously using the loss calculated at both the networks. The discriminator D learns to differentiate between fake (output from the generator) and real image (a tuple made of the input and ground truth). The Generator G learns to fool the discriminator D by learning to generate better images that are similar to the ground truth. The network alternates between one gradient descent step on D, then one step on G. We used a mini-batch SGD and Adam optimizer with the same hyper parameter values as used in [14].

A Nvidia 1080 Ti GPU was used for the training process. It took less than 3 h to complete training. However, with a better computation power, training can be done in a lot less duration for even a huge dataset. The coding part is done using the PyTorch framework in Python. During testing, it took less than 2 s to generate an output for a single image using the same GPU.

4   Experiments and Results

The network is trained separately for data degraded by different combinations of artifacts as explained in Sect. 3. The experiments are conducted on different combinations of artifacts to analyze how the performance of the network changes with the addition of each artifact. The order in which the combinations of artifacts are added in the dataset are as follows: Shadow, Shadow + Salt and Pepper Noise, Shadow + Speckle Noise, Shadow + Gaussian Noise, Shadow + Blur, Shadow + Salt and Pepper

Noise + Blur, Shadow + Speckle Noise + Blur, Shadow + Gaussian Noise + Blur, Shadow + All Noises + Blur.

The image quality metrics that are evaluated to assess the performance of the network are as follows.

4.1 Peak Signal-to-Noise Ratio

Peak Signal-to-Noise ratio (PSNR) is the ratio between the maximum possible value of the signal to the noise that affects the quality of the signal. PSNR is usually expressed in the logarithmic decibel scale. The mathematical expression of PSNR is as follows:

$$PSNR = 20\log_{10}(MAX_f/\sqrt{MSE}) \tag{6}$$

where MAX_f is the maximum value of the signal and MSE is the Mean Squared Error. A higher value of PSNR indicates lesser noise and a higher quality of an image.

4.2 Mean Squared Error

Mean Squared Error (MSE) is expressed as

$$MSE = 1/mn \sum \sum ||f(i, j) - g(i, j)||^2 \tag{7}$$

where f represents the intensity values of the original image, g represents the intensity values of the predicted image, m, n represents the number of rows and columns of the images and i, j represents the index of that row and column respectively. The lesser the value of MSE, the higher the quality of the image.

4.3 Structural Similarity Index

Structural Similarity Index (SSIM) is a perceptual metric that quantifies visual image degradation. Unlike PSNR, SSIM relies on the visible structures of the image. It is computed by three terms namely the luminance, the contrast, and the structural term. The overall index is a multiplicative combination of the three terms.

4.4 Blind/Reference-Less Image Spatial Quality Evaluator

Blind/Reference-less Image Spatial Quality Evaluator (BRISQUE) is a no reference image quality performance metric. It extracts the pointwise statistics of local normalized luminance signals and measures image naturalness based on measured deviations from a natural image model. A smaller BRISQUE score indicates better perceptual quality.

4.5 Naturalness Image Quality Evaluator

Naturalness Image Quality Evaluator (NIQE) extracts 5 different natural scene or statistical (NSS) characteristics from the original natural picture collection, to learn the multivariate Gaussian model (MVG) of the original image. For a given test image, the quality of the patches is evaluated. Then the patch quality scores are averaged and a general quality score is obtained. A smaller NIQE score indicates better perceptual quality.

Table 1 shows the quantitative results of the reconstructed images (of the test dataset) that are degraded by different combinations of artifacts. The tabulated quantitative metrics are the average of the individual metrics computed on every single test image with the respective artifact considered for evaluation. The results, however, could not be compared with any previous related work since there were no works that considered all the above artifacts. Figure 5 shows the results of the network trained only for Shadow removal. Figures 6, 7 and 8 shows the results of the trained network which is used for removal of shadows from images affected by noises like salt and pepper, speckle and gaussian respectively. Figures 9, 10, 11 and 12 depict the results of the trained network which is used for removal of shadow from motion-blurred images affected by noises like salt and pepper, speckle and gaussian respectively.

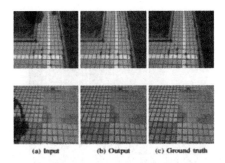

(a) Input   (b) Output   (c) Ground truth

Fig. 5. Results of network trained only for Shadow removal

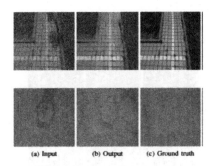

(a) Input   (b) Output   (c) Ground truth

Fig. 6. Results of network trained for shadow and salt and pepper noise

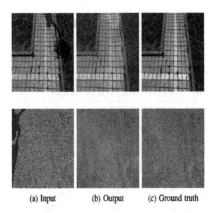

(a) Input   (b) Output   (c) Ground truth

Fig. 7. Results of network trained for removal of shadow and speckle noise

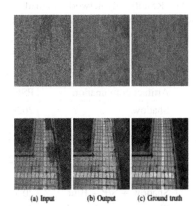

(a) Input   (b) Output   (c) Ground truth

Fig. 8. Results of network trained for removal of shadow and Gaussian noise

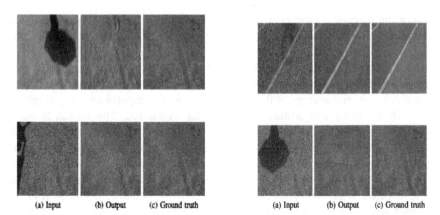

(a) Input   (b) Output   (c) Ground truth   (a) Input   (b) Output   (c) Ground truth

Fig. 9. Results of network trained for removal of shadow and motion blur

Fig. 10. Results of network trained for shadow, motion blur and salt and pepper

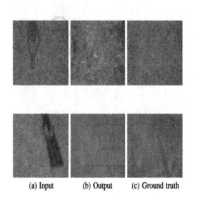

(a) Input   (b) Output   (c) Ground truth

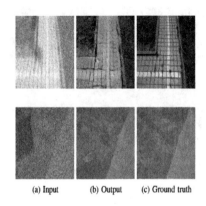

(a) Input   (b) Output   (c) Ground truth

Fig. 11. Results of network trained for removal of shadow, motion blur and speckle noise

Fig. 12. Results of network trained for removal of shadow, motion blur and Gaussian noise

Table 1. Performance metrics

Artifact combination	PSNR(dB)	MSE	SSIM	BRISQUE	NIQE
Shadow	26.9002	143.57	0.9223	34.22	10.38
Shadow + S/P noise	22.5532	423.87	0.6304	26.52	7.20
Shadow + speckle	23.1735	354.96	0.6764	26.89	6.27
Shadow + Gaussian	20.1207	832.86	0.4923	26.93	6.13
Shadow + blur	20.4108	770.89	0.4862	32.87	8.299
Shadow + S/P + Blur	20.7897	774.31	0.5017	22.81	5.632
Shadow + speckle + blur	21.4953	624.81	0.5628	23.45	5.869
Shadow + Gaussian + blur	19.4628	942.65	0.4184	26.99	5.66
Shadow + All noise + blur	19.7769	865.61	0.4559	23.94	5.868

5   Conclusion and Future Work

The usage of conditional adversarial networks for restoring images degraded by multiple visual artifacts has been successfully demonstrated in this paper. Visualizing and evaluating certain image quality metrics validate the closeness of the appearance of the generated output with that of the ground truth. This work can be further extended by training the network with all the possible known visual artifacts for creating a universal image restoration network. This universal image restoration network can also be deployed in various medical imaging applications to support physicians for better diagnosis of diseases. Although the network discussed in this work is found to give decent quantitative and qualitative results for multiple visual artifact removal, there is scope for improvement. The architecture of the generator and discriminator can be improved instead of using the UNet and PatchGAN. Various recent architectures like PSPNet, ICNet, LEDNet, etc. can be plugged in as generators and can be further modified for this problem statement by analyzing the gradient flow. Also, various other datasets can be augmented and added to the ISTD dataset while training to better generalize the work.

References

1. Banham, M.R., Katsaggelos, A.K.: Digital image restoration. IEEE Sig. Process. Mag. **14**(2), 24–41 (1997)
2. Kaur, A., Verma, D.: Blind and non-blind image restoration techniques. Int. J. Adv. Res. Comput. Sci. **4**(8), 315–317 (2013)
3. LeCun, Y., Bengio, Y., Hinton, G.: Deep learning. Nature **521**(7553), 436 (2015)
4. Sünderhauf, N., et al.: The limits and potentials of deep learning for robotics. Int. J. Robot. Res. **37**(4-5), 405–420 (2018)
5. Bakas, S., et al.: Identifying the best machine learning algorithms for brain tumor segmentation, progression assessment, and overall survival prediction in the brats challenge. arXiv preprint arXiv:1811.02629 (2018)
6. Islam, M., Jose, V.J.M., Ren, H.: Glioma prognosis: segmentation of the tumor and survival prediction using shape, geometric and clinical information. In: Crimi, A., Bakas, S., Kuijf, H., Keyvan, F., Reyes, M., van Walsum, T. (eds.) BrainLes 2018. LNCS, vol. 11384, pp. 142–153. Springer, Cham (2019). https://doi.org/10.1007/978-3-030-11726-9_13
7. Goodfellow, I., et al.: Generative adversarial nets. In: Advances in Neural Information Processing Systems, pp. 2672–2680 (2014)
8. Junginger, A., Hanselmann, M., Strauss, T., Boblest, S., Buchner, J., Ulmer, H.: Unpaired high-resolution and scalable style transfer using generative adversarial networks. arXiv preprint arXiv:1810.05724 (2018)
9. Demir, U., Unal, G.: Patch-based image inpainting with generative adversarial networks. arXiv preprint arXiv:1803.07422 (2018)
10. Ledig, C., et al.: Photo-realistic single image super-resolution using a generative adversarial network. In: Proceedings of the IEEE Conference on Computer Vision and Pattern Recognition, pp. 4681–4690 (2017)
11. Kupyn, O., Budzan, V., Mykhailych, M., Mishkin, D., Matas, J.: DeblurGAN: blind motion deblurring using conditional adversarial networks. In: Proceedings of the IEEE Conference on Computer Vision and Pattern Recognition, pp. 8183–8192 (2018)

12. Islam, M., Vaidyanathan, N.R., Jose, V.J.M., Ren, H.: Ischemic stroke lesion segmentation using adversarial learning. In: Crimi, A., Bakas, S., Kuijf, H., Keyvan, F., Reyes, M., van Walsum, T. (eds.) BrainLes 2018. LNCS, vol. 11383, pp. 292–300. Springer, Cham (2019). https://doi.org/10.1007/978-3-030-11723-8_29

13. Tripathi, S., Lipton, Z.C., Nguyen, T.Q.: Correction by projection: denoising images with generative adversarial networks. arXiv preprint arXiv:1803.04477 (2018)

14. Isola, P., Zhu, J.-Y., Zhou, T., Efros, A.A.: Image-to-image translation with conditional adversarial networks. In: Proceedings of the IEEE Conference on Computer Vision and Pattern Recognition, pp. 1125–1134 (2017)

15. Donoho, D.L., Johnstone, I.M.: Adapting to unknown smoothness via wavelet shrinkage. J. Am. Stat. Assoc. **90**(432), 1200–1224 (1995)

16. Chan, R.H., Ho, C.-W., Nikolova, M.: Salt-and-pepper noise removal by median-type noise detectors and detail-preserving regularization. IEEE Trans. Image Process. **14**(10), 1479–1485 (2005)

17. Zhong, H., Li, Y., Jiao, L.: SAR image despeckling using bayesian nonlocal means filter with sigma preselection. IEEE Geosci. Remote Sens. Lett. **8**(4), 809–813 (2011)

18. Finlayson, G.D., Hordley, S.D., Drew, M.S.: Removing shadows from images. In: Heyden, A., Sparr, G., Nielsen, M., Johansen, P. (eds.) ECCV 2002. LNCS, vol. 2353, pp. 823–836. Springer, Heidelberg (2002). https://doi.org/10.1007/3-540-47979-1_55

19. Fredembach, C., Finlayson, G.D.: Fast re-integration of shadow free images. In: Twelfth Color Imaging Conference: Color Science and Engineering Systems, Technologies, and Applications, pp. 117–122 (2004)

20. Zhang, L., Zhang, Q., Xiao, C.: Shadow remover: image shadow removal based on illumination recovering optimization. IEEE Trans. Image Process. **24**(11), 4623–4636 (2015)

21. Wang, J., Li, X., Yang, J.: Stacked conditional generative adversarial networks for jointly learning shadow detection and shadow removal. In: Proceedings of the IEEE Conference on Computer Vision and Pattern Recognition, pp. 1788–1797 (2018)

22. Ronneberger, O., Fischer, P., Brox, T.: U-Net: convolutional networks for biomedical image segmentation. In: Navab, N., Hornegger, J., Wells, W.M., Frangi, A.F. (eds.) MICCAI 2015. LNCS, vol. 9351, pp. 234–241. Springer, Cham (2015). https://doi.org/10.1007/978-3-319-24574-4_28

23. https://affinelayer.com/pix2pix/

Two-Stream CNN Architecture for Anomalous Event Detection in Real World Scenarios

Snehashis Majhi$^{(\boxtimes)}$, Ratnakar Dash, and Pankaj Kumar Sa

National Institute of Technology Rourkela, Rourkela, Odisha, India
majhisnehashis@gmail.com

Abstract. Anomalous event detection in any surveillance system has become an important area of research to make the surveillance effective and real time. In recent years, deep learning schemes are predominant to improve the detection accuracy. However, due to high computational complexity associated in deep learning architectures, it becomes a challenge to implement them in real-time scenarios. In this paper we propose a scheme to detect anomalous event in real time surveillance video. A database pre-processing algorithm has been proposed to capture the spatial and temporal frames in every second, which is subsequently utilized in two-stream 2D-CNN architecture for feature extraction and classification. A standard dataset, UCF-crime has been used to validate the proposed method. Finally, a comparative analysis has been made and it is observed that the classification accuracy and area under curve (AUC) of the suggested scheme is superior as compared to the recently proposed competent schemes.

Keywords: Anomalous event detection · Intelligent surveillance system · Two-stream CNN

1   Introduction

Increased crime rate across the globe have been responsible to the loss of life and property during past few decades. To combat the scenario, intelligent surveillance is one of the most favored solutions. As a result, massive installations of surveillance cameras have taken place in public places like airport, railway stations, shopping malls, parks etc., to ensure public safety. In order to monitor these surveillance cameras on a 24×7 basis, there is a requirement of huge and continuous manpower and again the accuracy of such systems depend on the alertness of the supervising entities. To improve the purpose of surveillance system an real time anomalous event detection is the utmost need of the hour.

Detecting anomalous events from surveillance videos is a challenging area in the domain of computer vision due to huge dimensions of data, complexity of the scene, availability of unwanted entities, variation in illumination condition, presence of noise and occlusion [2]. The complexity of the problem increases with

© Springer Nature Singapore Pte Ltd. 2020
N. Nain et al. (Eds.): CVIP 2019, CCIS 1148, pp. 343–353, 2020.
https://doi.org/10.1007/978-981-15-4018-9_31

dense entities under the field of view of camera and becomes more complex when it deals with crowd [15]. As it is an important area several research has been accelerated in the development of computer vision algorithms for automated anomalous event detection in last two decades.

In general, an automated surveillance system consists of four major steps: (1) database pre-processing, (2) feature extraction, (3) modeling of scene and entity behavior, and (4) designing the decision module. The dataset creation step includes collection of video data containing both anomalous and normal behavior of entities to train the system. Feature extraction deal with representing behavior in real value attributes. Feature can be local or global, the local feature are extracted from the predefined region in the frame which is represented by interest points where as global features are computed to describe motion in the entire frame. Subsequently, a model is trained using the extracted features where training can be supervised, unsupervised or semi-supervised. Now, the trained model is responsible for detecting anomaly in the query video under consideration.

In this paper, we propose a anomalous event detection scheme in the surveillance videos using two-stream CNN architecture. The suggested scheme works in two phases, namely,

- A database pre-processing phase that captures the spatial and temporal information for every second in a video, so that a whole video can be represented by very few number of frames.
- A selection of two-stream CNN architecture to detect anomalous event in surveillance video.

The remainder of the paper is organized as follows. In Sect. 2, an overview of the related work is given. Section 3 portrays the working procedure of the proposed method. In Sect. 4, experimental results and discussion are presented. Finally, Sect. 5 deals with the concluding remarks.

2   Related Work

Anomalous event detection is one of the most challenging and long-standing problems in the area of computer vision. In the last decade, several methodologies have been proposed by a diverse group of researchers for anomalous event detection in videos. Mu et al. [13] used motion vector as feature and classified with support vector machines (SVM) for recognition of suspicious behavior in HD videos. Gnanavel et al. [4] computed the feature vector from Multi scale Histogram of Optical Flow and used K-NN as classifier for abnormal event detection in crowded scene. Zhang et al. [21] used a Gaussian model of optical flow (GMOF) and orientation histogram of optical flow as a feature representation and classified with SVM for group violence detection in surveillance scene.

Kratz *et al.* [9] computed local spatio-temporal motion patterns and used it for feature representation followed by a Hidden Markov Model (HMM) for anomaly detection in extremely crowded scene. Jeong *et al.* [7] employed trajectories for feature extraction and applied LDA+GMM technique for traffic anomaly detection. Piciarelli *et al.* [16] adopted a trajectory sub-sampling with Gaussian kernel based approach for feature extraction and used one class SVM clustering for anomalous event detection. Medioni *et al.* [12] computed grids of HOF for feature representation followed by one class SVM for visual abnormal events detection in crowded scene.

Recently, deep learning based approaches have achieved success in detection of anomalous event. Xu *et al.* [20] used stacked denoising auto encoders (SDAE) to learn the appearance and motion feature based on a early fusion mechanism to detect anomalous event detection. Similarly, Chong*et al.* [1] used convolutional spatio-temporal auto encoders in order to learn deep feature from videos to detect abnormal events from the normal ones. Hasan *et al.* [5] combined handcrafted feature (HOG and HOF) with convolutional auto encoder (CAE) to extract deep feature for learning temporal regularity in video sequence. Vu *et al.* [19] used Restricted Boltzmann Machine (RBM) for anomaly detection in videos. Tran *et al.* [18] used C3D architecture for learning spatio-temporal features for action recognition in UCF101 dataset. Sultani *et al.* [17] used deep multiple instance based learning framework using C3D architecture followed by SVM classifier for real world anomaly detection in surveillance videos. Medel *et al.* [11] used convolutional long short term memory for anomaly detection in videos. Moreover, Hinami *et al.* [6] designed Multi test fast R-CNN for detection and recounting of abnormal events in videos.

The literature review reveals that various complex deep learning approaches like auto encoder (AE), RBM, 3D-CNN, R-CNN and multi-modal architectures have been utilized to detect anomalous events. Most of the reported schemes have high computational complexity and not suitable for real time implementation. Thus, there is a scope for further improvement in deep learning techniques with lower computational overhead. The present work deals with a two-stream 2D-CNN architecture for the purpose.

3   Proposed Work

The proposed framework comprises of three key stages: database pre-processing, feature extraction and fusion with the two-stream CNN and classification using fully connected (FC) neural network. Figure 1 depicts the overall flow diagram of the proposed method. Each stage of the scheme is elaborated below.

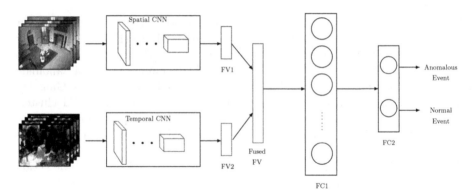

Fig. 1. Overview of the proposed two-stream CNN approach for anomalous event detection.

Algorithm 1. Data-set Pre-processing Algorithm

Input: Video Dataset $V = \{(v_1, y_1), (v_2, y_2), \ldots, (v_N, y_N)\}$ where
$v_i = i^{th}$ video, $y_i = class\ label\ of\ i^{th}\ video \in [0,1]$, $N = no.\ of$
$videos$

Output: Pre-processed Dataset
$DV = \{(DF_1, OF_1, Y_1), (DF_2, OF_2, Y_2), \ldots (DF_K, OF_K, Y_K)\}$ where
$DF_i = i^{th}\ frame$, $OF_i = i^{th}\ optical\ flow$, $y_i = i^{th}\ frame\ class\ label$

begin
    Set k=1;
    for $i = 1\ to\ N$ **do**
        Read (v_i, y_i);
        Set v_i to 30fps where v_i is a set of frames $\{F_1, F_2, \ldots F_M\}$;
        for $j = 1\ to\ M$ **do**
            if $j\ modulo\ 30==1$ **then**
                $T = \sum_{p=1}^{30} Dense\_optical\_flow(F_{j+p}, F_{j+p+1})$;
                $DF_k = F_j$;
                $OF_k = T$;
                $Y_k = y_i$;
                $k \leftarrow k + 1$;
            end
        end
    end
    $DV = \{(DF_i, OF_i, Y_i)\}$ where $i = 1, 2, \ldots K$;
end

3.1 Database Processing

The objective of the database processing is to extract spatial and temporal information among the frames in overall video sequence $(v_1, v_2, \ldots, v_n)$ collected at different instances from a single surveillance camera. It is assumed that the videos are in different frames per second (fps). Initially, all the video sequences are set to 30 fps and the frames for video sequence v_i be denoted as v_{i1}, v_{i2}, ...; i=1, 2, ..., n. The temporal information in terms of dense optical flow [3] are collected in one second as T_{vi1}, T_{vi2}, ..., T_{vi29} and aggregated in OF_{vi1}. Similarly, the frame DF_{vi1} represents the spatial information for first 30 frames of first video sequence. In the same way, we generate OF_{vi1}, OF_{vi31}, ... and DF_{vi1}, DF_{vi31}, ...; i=1, 2, ..., n. The generalized algorithms for spatial and temporal frame generation is given in Algorithm 1 which reduces the M-frame video to K-frame sequence in each category, where $K \approx M \div 30$. Thus, Algorithm 1 achieves a phenomenal dimensionality reduction of the problem in hand.

3.2 Feature Extraction Using Two-Stream CNN Architecture

It may be observed from the Fig. 1 that the spatial and temporal frames generated from the set of video sequences are passed through two separate 2D-CNNs to generate FV1 and FV2 corresponding to DF_{vi1} and OF_{vi1}. These two vectors FV1 and FV2 are concatenated to generate Fused Feature Vector (FFV). Similarly, subsequent FFVs are generated from corresponding DF_{vi31} and OF_{vi31} and so on. We have considered three pre-trained 2D-CNN architecture namely VGG16, ResNet50 and InceptionV3 trained in ImageNet challenge dataset, in order to have nine distinct combinations of two-stream CNN frameworks. The intuition behind choosing these three pre-trained architecture is due to the fact that, they achieve a phenomenal accuracy on ImageNet dataset. The architectural specification of each network is given below.

- **VGG16:** It comprises of 16 layers, out of which 13 are convolutional layers and 3 are dense FC layers. The 13 convolutional layers are partitioned into 6 groups and each layer uses $3 \times$ convolutional layers are partitioned into3 kernels followed by ReLU [14] activation function.
- **ResNet-50:** It comprises of 50 layers, out of which 49 are convolutional layers and only one FC layer. It uses residual units to allow learning in deeper models by using shortcut connections in order to avoid vanishing gradient problem during training.
- **InceptionV3:** It is a 48 layers of architecture including the convolutional and fully connected layers. It comprises of several inception module which makes the network more wider rather than deeper and avoids over fitting.

In order to extract the features from CNNs, we have removed the FC layers from the pre-trained models and kept the weights unchanged up to the flatten layers. The fused feature vector (FFV) is applied to a 2-layer fully connected (FC) neural network. The first FC layer contains 2048 units followed by 2 units in the second FC layer for classification of anomalous video from normal ones. ReLU [14] and Softmax activations are applied in the first and second FC layers respectively.

4   Experiments

To validate the performance of the proposed architecture, experiments are conducted on a workstation PC with 2.80 GHz Xeon processor and 32 GB of RAM, running under ubuntu 16.04 LTS Operating system. The GPU card used for this experiment is NVIDIA GeForce GTX1080 Ti which has 3584 CUDA cores. Python with Keras API has been used with Tensor-flow backend. We employ Adam [8] optimizer with hyper parameters as given in Table 1. The details of the dataset and experiments are discussed below.

Table 1. Hyper parameters for Adam optimizer.

Name	Value
Learning rate	0.00001
First momentum decay	0.9
Second momentum decay	0.999

4.1   Dataset

One benchmark dataset, "UCF-crime dataset" [17] has been used for the validation of proposed scheme. The dataset consists of 1900 surveillance videos that covers 13 real world anomalous events such as abuse, arrest, arson, burglary, explosion, fighting, road accidents, robbery, shooting, stealing, vandalism along with the normal events. It also provides a list of videos to be considered during training and testing phase. Thus, we have used the training videos to train our proposed model and validated on test videos. The sample frames containing anomalous events from UCF-crime dataset are shown in Fig. 2. The distribution of videos from each class is listed in Table 2. Among the total video counts of 1900, the 950 anomalous events videos from 13 class are grouped into one category as "anomalous" whereas rest videos containing normal events are categorized as "normal".

Table 2. Distribution of videos in the UCF-crime dataset [17].

Event	Video count	Event	Video count	Event	Video count
Abuse	50	Explosion	50	Shoplifting	50
Arrest	50	Fighting	50	Stealing	100
Arson	50	Road accidents	150	Vandalism	50
Assault	50	Robbery	150	Normal event	950
Burglary	100	Shooting	50		

Subsequently, Algorithm 1 is applied to create pre-processed database followed by feature extraction using two-stream CNN architecture. These feature sets are utilized for further classification in FC neural network.

Fig. 2. Sample frames of different anomalous and Normal events: (a) Abuse, (b) Arrest, (c) Arson, (d) Assault, (e) Burglary, (f) Explosion, (g) Fighting, (h) Road accident, (i) Robbery, (j) Shooting, (k) Shoplifting, (l) Stealing, (m) Vandalism, (n)–(r) Normal

4.2   Results and Discussion

The two-stream CNN architecture considers three pre-trained models (i.e., VGG16, ResNet50, InceptionV3) trained in ImageNet dataset. So, there are nine possible combinations between these three CNN models shown in Table 3. It may be observed the nine combinations of two-stream CNN models can be divided into 3 groups based on their dimensions of feature vector (i.e.,4096, 2560, 1024). Group-1 generates 4096D FFVs, Group-2 generates 2560D FFVs, where as Group-3 generates 1024D FFVs through two-stream CNN frameworks. It is evident from Table 3, that ResNet50-ResNet50 two-stream model from group-1, InceptionV3-VGG16 two-stream CNN model from group-2 and VGG16-VGG16 two-stream CNN model from group-3 achieve the highest accuracy among the group members. Between these three two-stream CNN models we treat InceptionV3-VGG16 two-stream model as our proposed scheme since it achieves maximum accuracy of 88.74% with a feature vector of dimension 2560. Moreover, it is observed that ResNet50-ResNet50 two-stream CNN model with a feature vector of dimension 4096 achieves only 88.71% accuracy which is less than the proposed model. It is also observed that VGG16-VGG16 two-stream model achieves 84.64% accuracy with a feature vector of dimension 1024. However, the accuracy achieved VGG16-VGG16 two-stream model is lesser than the proposed InceptionV3-VGG16 two-stream model.

The convergence of accuracy and loss plots and the confusion matrix of the proposed InceptionV3-VGG16 two-stream model are shown in Figs. 4 and 5 respectively. It is observed that the training accuracy increases with a high thrust between first 100 epochs and it attains a 98.12% training accuracy after

Table 3. Accuracy and AUC obtained by different two-stream CNN networks.

Number of groups	Spatial stream	Temporal stream	Dimension of FV	Test accuracy (%)	AUC (%)
Group-1	ResNet50	ResNet50	4096	88.71	95.18
	InceptionV3	InceptionV3	4096	86.90	94.38
	ResNet50	InceptionV3	4096	85.86	95.64
	InceptionV3	ResNet50	4096	87.92	94.42
Group-2	ResNet50	VGG16	2560	86.81	95.96
	VGG16	ResNet50	2560	87.34	94.34
	VGG16	InceptionV3	2560	84.92	93.53
	InceptionV3	**VGG16**	**2560**	**88.74**	**94.47**
Group-3	VGG16	VGG16	1024	84.64	95.03

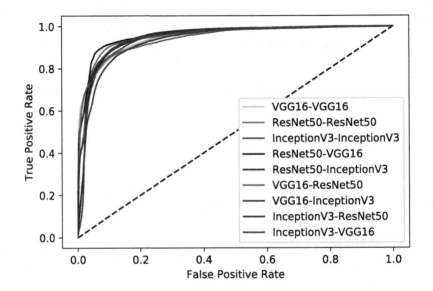

Fig. 3. ROC comparison of different two-stream models

Table 4. Comparison with state-of-the-art methods in terms of AUC

Existing methods	AUC (%)
Hasan *et al.* [5]	50.6
Lu *et al.* [10]	65.51
Sultani *et al.* [17]	75.41
InceptionV3-VGG16 Two-stream 2D-CNN (Proposed)	**94.47**

500 epochs. Similarly, the loss decays drastically between first 100 epochs and attains a 0.0419 after 500 epochs.

The comparative analysis of the proposed method using UCF-crime dataset with state-of-the art methods in terms of area under curve (AUC) is listed in Table 4. The methods mentioned in Table 4 are validated on the UCF-crime

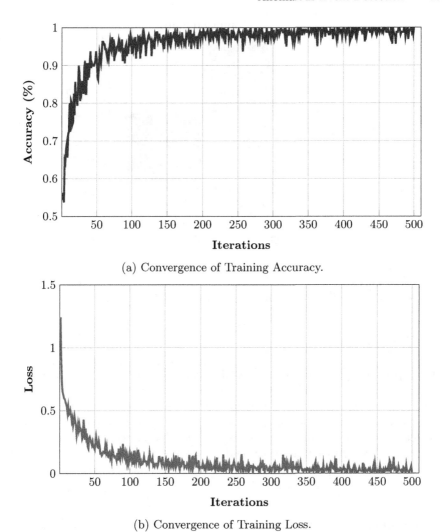

(a) Convergence of Training Accuracy.

(b) Convergence of Training Loss.

Fig. 4. Convergence plots of InceptionV3-VGG16 two-stream 2D-CNN.

dataset. It is evident that the proposed method yields superior performance compared to other methods. Given a test video, the execution time per frame to take a decision is 346 ms. However, the time needed for training the two-stream CNN model is not considered during time analysis.

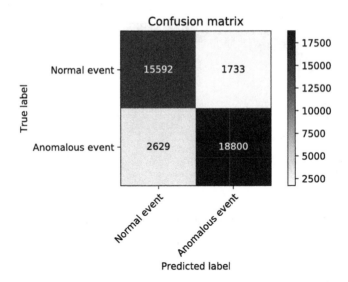

Fig. 5. Confusion matrix of proposed model

5   Conclusion

This paper proposes an efficient method to detect anomalous events in a surveillance video in real time. The proposed scheme comprises of three steps namely, video pre-processing, feature extraction and classification. Pre-processing strategy extracts temporal and spatial frames from a sequence of videos. Feature extraction is achieved through two-stream 2D-CNN and generates fused feature vector. The classification accuracy and area under curve (AUC) are considered for performance comparison with the existing competent schemes. In general, the proposed scheme is computationally efficient and achieves superior performance.

References

1. Chong, Y.S., Tay, Y.H.: Abnormal event detection in videos using spatiotemporal autoencoder. In: Cong, F., Leung, A., Wei, Q. (eds.) ISNN 2017. LNCS, vol. 10262, pp. 189–196. Springer, Cham (2017). https://doi.org/10.1007/978-3-319-59081-3_23
2. Dick, A.R., Brooks, M.J.: Issues in automated visual surveillance. In: International Conference on Digital Image Computing: Techniques and Applications (2003)
3. Farnebäck, G.: Two-frame motion estimation based on polynomial expansion. In: Bigun, J., Gustavsson, T. (eds.) SCIA 2003. LNCS, vol. 2749, pp. 363–370. Springer, Heidelberg (2003). https://doi.org/10.1007/3-540-45103-X_50
4. Gnanavel, V.K., Srinivasan, A.: Abnormal event detection in crowded video scenes. In: Satapathy, S.C., Biswal, B.N., Udgata, S.K., Mandal, J.K. (eds.) Proceedings of the 3rd International Conference on Frontiers of Intelligent Computing: Theory and Applications (FICTA) 2014. AISC, vol. 328, pp. 441–448. Springer, Cham (2015). https://doi.org/10.1007/978-3-319-12012-6_48

5. Hasan, M., Choi, J., Neumann, J., Roy-Chowdhury, A.K., Davis, L.S.: Learning temporal regularity in video sequences. In: Proceedings of the IEEE Conference on Computer Vision and Pattern Recognition, pp. 733–742 (2016)
6. Hinami, R., Mei, T., Satoh, S.: Joint detection and recounting of abnormal events by learning deep generic knowledge. In: Proceedings of the IEEE International Conference on Computer Vision, pp. 3619–3627 (2017)
7. Jeong, H., Yoo, Y., Yi, K.M., Choi, J.Y.: Two-stage online inference model for traffic pattern analysis and anomaly detection. Mach. Vis. Appl. **25**(6), 1501–1517 (2014). https://doi.org/10.1007/s00138-014-0629-y
8. Kingma, D.P., Ba, J.: Adam: a method for stochastic optimization. arXiv preprint arXiv:1412.6980 (2014)
9. Kratz, L., Nishino, K.: Anomaly detection in extremely crowded scenes using spatio-temporal motion pattern models. In: 2009 IEEE Conference on Computer Vision and Pattern Recognition, pp. 1446–1453. IEEE (2009)
10. Lu, C., Shi, J., Jia, J.: Abnormal event detection at 150 FPS in MATLAB. In: Proceedings of the IEEE International Conference on Computer Vision, pp. 2720–2727 (2013)
11. Medel, J.R., Savakis, A.: Anomaly detection in video using predictive convolutional long short-term memory networks. arXiv preprint arXiv:1612.00390 (2016)
12. Medioni, G., Cohen, I., Brémond, F., Hongeng, S., Nevatia, R.: Event detection and analysis from video streams. IEEE Trans. Pattern Anal. Mach. Intell. **23**(8), 873–889 (2001)
13. Mu, C., Xie, J., Yan, W., Liu, T., Li, P.: A fast recognition algorithm for suspicious behavior in high definition videos. Multimed. Syst. **22**(3), 275–285 (2015). https://doi.org/10.1007/s00530-015-0456-7
14. Nair, V., Hinton, G.E.: Rectified linear units improve restricted Boltzmann machines. In: Proceedings of the 27th International Conference on Machine Learning, ICML 2010, pp. 807–814 (2010)
15. Pawar, K., Attar, V.: Deep learning approaches for video-based anomalous activity detection. World Wide Web **22**(2), 571–601 (2018). https://doi.org/10.1007/s11280-018-0582-1
16. Piciarelli, C., Micheloni, C., Foresti, G.L.: Trajectory-based anomalous event detection. IEEE Trans. Circ. Syst. Video Technol. **18**(11), 1544–1554 (2008)
17. Sultani, W., Chen, C., Shah, M.: Real-world anomaly detection in surveillance videos. In: Proceedings of the IEEE Conference on Computer Vision and Pattern Recognition, pp. 6479–6488 (2018)
18. Tran, D., Bourdev, L., Fergus, R., Torresani, L., Paluri, M.: Learning spatiotemporal features with 3D convolutional networks. In: Proceedings of the IEEE International Conference on Computer Vision, pp. 4489–4497 (2015)
19. Vu, H., Nguyen, T.D., Travers, A., Venkatesh, S., Phung, D.: Energy-based localized anomaly detection in video surveillance. In: Kim, J., Shim, K., Cao, L., Lee, J.-G., Lin, X., Moon, Y.-S. (eds.) PAKDD 2017. LNCS (LNAI), vol. 10234, pp. 641–653. Springer, Cham (2017). https://doi.org/10.1007/978-3-319-57454-7_50
20. Xu, D., Ricci, E., Yan, Y., Song, J., Sebe, N.: Learning deep representations of appearance and motion for anomalous event detection. arXiv preprint arXiv:1510.01553 (2015)
21. Zhang, T., Yang, Z., Jia, W., Yang, B., Yang, J., He, X.: A new method for violence detection in surveillance scenes. Multimed. Tools Appl. **75**(12), 7327–7349 (2015). https://doi.org/10.1007/s11042-015-2648-8

3D CNN with Localized Residual Connections for Hyperspectral Image Classification

Shivangi Dwivedi[1], Murari Mandal[2], Shekhar Yadav[1],
and Santosh Kumar Vipparthi[2(✉)]

[1] Madan Mohan Malaviya University of Technology, Gorakhpur, India
`shivangid48@gmail.com`, `syee@mmmut.ac.in`
[2] Vision Intelligence Lab, Malaviya National Institute of Technology,
Jaipur, India
`murarimandal.cv@gmail.com`, `skvipparthi@mnit.ac.in`

Abstract. In this paper we propose a novel 3D CNN network with localized residual connections for hyperspectral image classification. Our work chalks a comparative study with the existing methods employed for abstracting deeper features and propose a model which incorporates residual features from multiple stages in the network. The proposed architecture processes individual spatio-spectral feature rich cubes from hyperspectral images through 3D convolutional layers. The residual connections result in improved performance due to assimilation of both low-level and high-level features. We conduct experiments over Pavia University and Pavia Center dataset for performance analysis. We compare our method with two recent state-of-the-art methods for hyperspectral image classification method. The proposed network outperforms the existing approaches by a good margin.

Keywords: Hyperspectral · Residual · 3D CNN · Remote sensing

1 Introduction

Advancements in sensing technologies have led to improved spectral resolution of hyperspectral images. Digitization and automatic analysis of hyperspectral images have enabled exploration of remote areas in an efficient way. It has enabled the possibility of forecasting geographical events before their occurrence and to map the effects of any geographical phenomena. Moreover. hyperspectral images also find application in various fields like biological threat detection, fire-tracking problem, landcover usage applications like agriculture, construction, etc. [1, 2]. However, automatic analysis of these images such as pixel-wise classification is still a challenging task. Existing methods designed for classification of spatio-spectral images lack the desired efficacy. Thus, special modifications are required in existing methods to improve the quality of information abstraction from these images.

Satellite images are captured by different type of sensors (Aviris, Enmap, Hyperion, etc.) in different spectral formats. The spectral formats can be subdivided into 3 types based on the spectral information they carry, namely, panchromatic, multi-spectral and hyperspectral. All three can have same spatial information but vary in their respective

© Springer Nature Singapore Pte Ltd. 2020
N. Nain et al. (Eds.): CVIP 2019, CCIS 1148, pp. 354–363, 2020.
https://doi.org/10.1007/978-981-15-4018-9_32

spectral information. Panchromatic images are gray scale images whereas multispectral and hyperspectral images contain tens or hundreds of spectral bands respectively. These spectral bands range from continuous and narrow (for hyperspectral images) to discrete and wider bands for a pixel (for multi-spectral images).

Analyzing classification accuracy of hyperspectral images is the main concern of this paper. Hyperspectral images captured by sensors are rich in spatio-spectral information but suffer from curse of dimensionality as well. Therefore, the conventional methods have used pre-processing [3–7] before classifying these images. It involves removal of unwanted, irrelevant bands from each pixel of the image. The existing approaches for hyperspectral image classification (HSIC) can be grouped into traditional [3–7, 15, 16] and deep learning-based methods [17–24]. Deep learning models have been much more effective in HSIC as compared to the traditional approaches. We investigated some recent CNN models for HSIC and identified some of the network design enhancements to improve classification accuracy.

In this paper we proposed a new 3D-CNN network with localized residual connections for HIC. The proposed network learns abstract representation from raw spectral signals at each pixel location through gradual reduction in number of 3D convolutional kernels. Moreover, at multiple convolution stages, we add residual features from previous layers to reinforce the low-level features. The spatio-spectral features learned from both low-level and high-level abstractions led to improved performance over both PaviaU and PaviaC datasets.

2 Related Work

Hyperspectral (HS) images comprise of a contiguous, narrow spectrum of informational bands. Hence feature extraction and classification of regions in a scene is a very challenging task. In [4], each pixel is smoothened by a weighted mean filter. Then spectral regularized scatter matrix is integrated with spatial neighborhood scatter matrix to find local similarity pattern for classification. The authors in [5] used mahalanobis distance metric to calculate intra-class and inter-class distance between pixels. Koonsanit et al. [6] use combination of PCA and integrated gain to select relevant band. Similarly, Li et al. [7] used Fisher's discriminant analysis to get rid of redundant bands for hyperspectral image analysis. Spatial filtering technique [15] is one of the conventional methods in which image is divided into its constituent spatial frequency. Then selected altering of certain frequency is done to suppress some frequencies and emphasize other features. Moreover, PCA features with SVM classifiers are used in [16].

Recently, many deep learning models have also been designed for hyperspectral image classification. We give a brief description of different model architectures proposed by various researchers. Li et al. [8] proposed a simple 3D CNN whereas, Chen et al. [9] added L2 regularization and dropouts to address the problem of overfitting. Mei et al. [10] enhanced the HS image by designing a network for spatial super resolution. Liu et al. [11] utilized bidirectional recurrent connections to abstract minute details form the image. Zhu et al. [12] presented a generative adversarial network comprised of two parts 1D and 2D convolutions. He et al. [13] used five layered multi-scale 3D CNN and

Chen et al. [14] used logistic regression as final classifier for combined data of LIDAR and HS image.

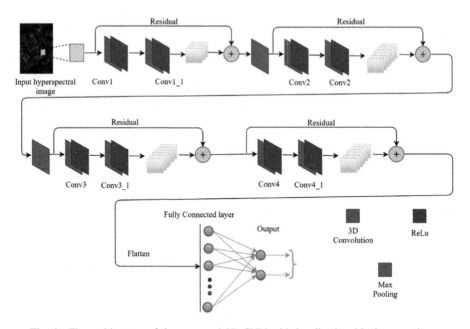

Fig. 1. The architecture of the proposed 3D CNN with localized residual connections.

Lee et al. [17] proposed a contextual deep CNN by concurrently extracting multiple 3-dimensional local convolutional features with different sizes jointly exploiting spatial and spectral features of HS image. Hamida et al. [18] proposed a 3D CNN prototype to facilitate joint study of spatial and spectral features of HS image. Similarly, many other CNN models have been designed in the literature for HSIC and aerial images analysis [19–24].

3 Proposed Method

This paper introduces a novel 3D CNN architecture using residual blocks for HSIC. Residual blocks are used to retain the features from earlier convolutional layers. We use four residual blocks to assimilate both primitive and abstract information. Similarly, the spatio-spectral are features learned from both low-level and high-level abstractions leading to robust performance. We depict the proposed 3D CNN architecture in Fig. 1. Each pixel in a hyperspectral (HS) image can be seen as a cube containing spatial and spectral data. Each 3-dimensional cube is configured into $1 \times 1 \times S$ structure. Of this 1×1 is spatial data dimension and S is taken as continuous, narrow spectrum bands. These spectrum bands are digitally represented with multiple channels in HS images.

Table 1. Detailed network architecture of the proposed network

Layer name	Kernel	Stride	Inputs	Outputs
Conv1	$3 \times 3 \times 3$	1, 1, 1	102	20
Conv1_1	$1 \times 1 \times 1$	1, 1, 1	20	20
Pool1	$1 \times 1 \times 3$	1, 1, 2	20	20
Conv2	$3 \times 3 \times 3$	1, 1, 1	20	35
Conv2_1	$1 \times 1 \times 1$	1, 1, 1	35	35
Pool2	$1 \times 1 \times 3$	1, 1, 2	35	35
Conv3	$1 \times 1 \times 3$	1, 1, 1	35	35
Cov3_1	$1 \times 1 \times 1$	1, 1, 1	35	35
Conv4	$1 \times 1 \times 2$	1, 1, 2	35	35
Conv4_1	$1 \times 1 \times 1$	1, 1, 1	35	35

Let $HSI_S(a, b)$ be an input HS image of size $P \times Q \times S$, $a \in [1, P]$, $b \in [1, Q]$ having S channels. From S spectral channels, the 3D convolutional features (3DCF) are computed using Eq. (1)–(5).

$$3DCF(HSI_S) = \psi_4(\psi_3(\psi_2(\psi_1(HSI_S)))) \tag{1}$$

$$\psi_1(z) = ap_{1,1,3}(\kappa_{20,1,1,1}(\Re(\kappa_{20,3,3,3} \otimes z))) \tag{2}$$

$$\psi_2(z) = ap_{1,1,3}(\kappa_{30,1,1,1}(\Re(\kappa_{35,3,3,3} \otimes z))) \tag{3}$$

$$\psi_3(z) = \kappa_{35,1,1,1}(\Re(\kappa_{35,1,1,3} \otimes z)) \tag{4}$$

$$\psi_4(z) = \kappa_{35,1,1,1}(\Re(\kappa_{35,1,1,2} \otimes z)) \tag{5}$$

where $\kappa_{n,h,w,d}$ denotes 3D convolution with parameters n, h, w and d representing the number of kernels, height, width and depth of the kernels respectively. We use strides (0, 0, 0), (0, 0, 0) and (0, 0, 1) in three layers of ψ_1 and ψ_2 blocks. Similarly, we use strides (0, 0, 1) and (0, 0, 0) in two layers of ψ_3 and ψ_4 blocks. The $ap_{h,w,d}$ represents 3D average pooling and $\Re(\cdot)$ denotes the rectified linear unit (ReLu) activation function.

The proposed model convolves input HS image with filter *Conv1* of shape of $3 \times 3 \times 3$ taking input channel as 102 or 103 and output channel as 20. Here, spatial window of 7×7 is selected for feature extraction at each pixel. Then further convolving it with filter *Conv1_1* ($1 \times 1 \times 1$), taking input and output channels to be 20. The identical value of input is then added to output of the second convolving layer after applying ReLu non-linearity. In short, residue between input to first convolving layer and output of second convolving layer is calculated. The pooling layer then comes in function reducing spectral depth of the image with filter size of $1 \times 1 \times 3$ at stride of 2 along depth, 1 along height and width; adding zero padding only across depth.

Similar convolution layers with residual block is replicated as shown in Fig. 1 and Eqs. (1)–(5). The final stack of convolution blocks (*Conv4, Conv4_1*) applies the localized residuals as follows. *Conv4* has filter of size $1 \times 1 \times 2$ at stride of 1 along height, width and of 2 along depth of the image. Zero padding is also added along depth of the image. The *Conv4_1* convolution involves use of filter of size of $1 \times 1 \times 1$ at stride of 1 along all the directions. This is further connected to a fully connected (FC) layer after flattening the output of convolution stack. The detailed network architecture of the proposed methods is tabulated in Table 1.

4 Experiments and Discussions

In this section, we conduct multiple experiments to demonstrate the effectiveness of the proposed residual based 3D CNN design. We first discuss about the HS image datasets used in our experiments. Moreover, the hyperparameters and other training configurations are also explained. Finally, we discuss experimental results of the proposed and existing state-of-the-art approaches in detail.

Table 2. Data division for model training in PaviaU dataset

PaviaU classes	Train samples	Test samples
Asphalt	200	6431
Meadows	200	18449
Gravel	200	1899
Trees	200	2864
Sheets	200	1154
Bare soil	200	4829
Bitumen	200	1130
Bricks	200	2482
Shadows	200	747

Table 3. Data division for model training in PaviaC dataset

PaviaC classes	Train samples	Test samples
Water	200	624
Trees	200	620
Asphalt	200	616
Self-blocking bricks	200	608
Bitumen	200	608
Tiles	200	1060
Shadows	200	276
Meadows	200	624
Bare soils	200	620

Table 4. Number of trainable parameters at each layer of the proposed model

Network layers	# Parameters
Conv1	560
Conv1_1	420
Conv2	18,935
Conv2_1	1,260
Conv3	3,710
Conv3_1	1,260
Conv4	2,485
Conv4_1	1,260
Total	29,890

4.1 Datasets

The proposed network has been trained and evaluated on two publicly available datasets Pavia University (PaviaU) and Pavia Center (PaviaC). PaviaU and PaviaC HS images were acquired by ROSIS sensor in a campaign over northern Italy. Each of these scenes comprise 1.3 m spatial resolution. PaviaU is 610×340 resolution image and contains 103 band spectrums. Its ground truth differentiates the entire scene in 9 classes. PaviaC is 1096×1096 resolution image and consists of 102 band spectrums. Its ground truth groups the entire scene in 9 classes. Composed of rich spatial and spectral information, these datasets offer excellent platform for testing and evaluating performance of proposed network prototype. Training and testing sample size for each class in both the datasets is stated in Tables 2 and 3.

4.2 Training Configuration

The proposed model is trained using stochastic gradient descent with momentum of 0.9 at learning rate of 0.02. Weight decay rate is set to 0.0005. We also experimented with Adam optimizer which did not produce good result. Zero padding is added at some layers to maintain input and output shape of the image while abstracting features of the input spatial-spectral rich image. Table 4 represents number of trainable parameters generated at each layer of the proposed model.

4.3 Results and Discussions

The proposed method has been trained and evaluated over PaviaC and PaviaU datasets. The performance is compared with recent state-of-the-art approaches [17, 18] in terms of classification accuracy. We trained our model with different percentage of training data (4.4%, 5%, 9%, 15%). The accuracy for different training sizes is depicted in Fig. 2. From Fig. 2, it is evident that the proposed method performs well even with very small training size. More specifically, it achieves 94.19%, 94.57%, 94.87% and 95.26% accuracy over PaviaU with 4.4%, 5%, 9% and 15% training data respectively. Similarly, the proposed method also achieves 98.46%, 98.52%, 98.63% and 98.7%

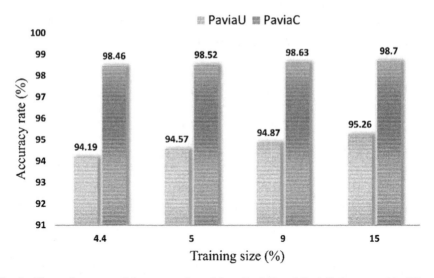

Fig. 2. The performance of the proposed model on PaviaU and PaviaC datasets with different percentages of training data.

Predicted Label

	Asphalt	Meadows	Gravel	Trees	Sheets	Bare Soil	Bitumen	Bricks	Shadows
Asphalt	6041	7	15	0	1	0	21	70	0
Meadows	0	16457	0	6	0	20	0	0	0
Gravel	9	2	1660	0	0	0	0	260	0
Trees	0	70	0	2800	0	16	0	0	0
Sheets	0	0	0	0	1286	0	0	0	0
Bare Soil	6	32	0	0	0	4770	0	0	0
Bitumen	64	0	0	0	0	0	1202	2	3
Bricks	26	19	59	0	0	3	0	3413	0
Shadows	8	5	0	0	0	1	0	0	891

Fig. 3. Confusion matrix of the proposed method over PaviaU dataset

Predicted Label

	Asphalt	Meadows	Gravel	Trees	Sheets	Bare Soil	Bitumen	Bricks	Shadows
Asphalt	62319	0	0	0	0	0	0	0	0
Meadows	0	6837	338	0	0	0	0	0	0
Gravel	0	52	2697	0	5	0	0	0	0
Trees	0	0	0	2553	14	0	0	0	0
Sheets	0	1	0	34	6239	0	1	0	0
Bare Soil	0	1	1	10	5	8726	0	10	0
Bitumen	0	0	0	1	0	121	6844	0	0
Bricks	0	0	0	2	0	0	0	40512	0
Shadows	0	0	0	0	0	0	0	4	2727

Fig. 4. Confusion matrix of the proposed method over PaviaC dataset

accuracy over PaviaC with 4.4%, 5%, 9% and 15% training data respectively. The chart also shows that beyond these training sizes accuracy begins to saturate due to higher training samples.

In order to analyze the proposed model performance for class-wise accuracy, we depict confusion matrix for PaviaU and PaviaC datasets in Figs. 3 and 4 respectively. Highlighted diagonal elements represent accurately classified samples for each class whereas, non-diagonal elements represent error of omission and error of commission values. This quantitatively expresses the amount of agreement between the ground truth class and the predicted HS class.

The accuracy comparison of proposed model with existing state-of-the-art methods Lee et al. [17] and Hamida et al. [18] are shown in Table 5. Table 5 represent overall accuracy for PaviaU and PaviaC datasets respectively. These results are obtained from proposed model architecture after hundred epochs with 4.4% and 5% training size. From the report stated in Table 5, it can be deduced that proposed model is accurately able to classify minute features of the input HS images. It can also be observed that the proposed method performs better or equally well when compared with other state-of-the art methods. More specifically, the proposed method outperforms [17] and [18] by 9.58% and 5.61% on PaviaU dataset with 4.4% training data. Similarly, our model outperforms [17] and [18] by the margin of 5.49% and 6.4% on PaviaU dataset with 5% training data.

Table 5. Performance comparison of the proposed method with existing approaches. All the results are computed with 7×7 spatial neighborhood as input to the network

	Proposed method		Hamida et al. [18]		Lee et al. [17]	
Train (%)	4.4%	5%	4.4%	5%	4.4%	5%
PaviaU	**94.19**	**94.57**	88.58	88.17	84.61	89.08
PaviaC	**98.46**	**98.52**	98.13	98.05	98.17	98.88

Table 6. Kappa coefficient value obtained for proposed method

Dataset	Training size (%)	Kappa coefficient
PaviaU scene	4.4	0.924
	5	0.929
PaviaC scene	4.4	0.978
	5	0.979

We also tabulate different values of kappa coefficient for the proposed model with training size of 4.4% and 5% in Table 6. Kappa coefficient is a discrete multivariate technique, suitable for analyzing remote sensing data due to its discrete and multi-variate nature. It is measure of degree of enhancement by the classifier over pure random allocation of classes. As given in Table 6, the proposed method achieves robust kappa coefficient measures in both PaviaU and PaviaC dataset.

5 Conclusion

Hyperspectral image classification requires design and development of spatio-spectral feature aware CNN network. The proposed 3D CNN based architecture is successfully able to extract detailed spectral features from input HS image using 3D convolutional kernels in combination with localized residual connections. By using residual connections and robust 3D CNN design, we overcome the issue of dimensionality, overfitting of deeper layers and gradient vanishing while back-propagating. The proposed method outperforms recent state-of-the-art approaches for hyperspectral image classification on PaviaU and PaviaC datasets. The model still needs to be tested on low resolution hyperspectral images. In future, model robustness can be improved to effectively classify not only high-resolution image but also low-resolution images.

Acknowledgement. This work was supported by the Science and Engineering Research Board (under the Department of Science and Technology, Govt. of India) project SERB/EEQ/2017/000673.

References

1. Bioucas-Dias, J.M., Plaza, A., Camps-Valls, G., Scheunders, P., Nasrabadi, N., Chanussot, J.: Hyperspectral remote sensing data analysis and future challenges. IEEE Geosci. Remote Sens. Mag. **1**(2), 6–36 (2013)
2. Lacar, F.M., Lewis, M.M., Grierson, I.T.: Use of hyperspectral imagery for mapping grape varieties in the Barossa Valley, South Australia. In: Proceedings of the IEEE International Geoscience and Remote Sensing Symposium, pp. 2875–2877 (2001)
3. Agarwal, A., El-Ghazawi, T., El-Askary, H., Le-Moigne, J.: Efficient hierarchical-PCA dimension reduction for hyperspectral imagery. In: Proceedings of the IEEE International Symposium on Signal Processing and Information Technology, pp. 353–356 (2007)
4. Zhou, Y., Peng, J., Chen, C.L.P.: Dimension reduction using spatial and spectral regularized local discriminant embedding for hyperspectral image classification. IEEE Trans. Geosci. Remote Sens. **53**(2), 1082–1095 (2014)
5. Du, B., Zhang, L., Zhang, L., Chen, T., Wu, K.: A discriminative manifold learning based dimension reduction method for hyperspectral classification. Int. J. Fuzzy Syst. **14**(2), 272–277 (2012)
6. Koonsanit, K., Jaruskulchai, C., Eiumnoh, A.: Band selection for dimension reduction in hyper spectral image using integrated information gain and principal components analysis technique. Int. J. Mach. Learn. Comput. **2**(3), 248–251 (2012)
7. Li, W., Prasad, S., Fowler, J.E., Bruce, L.M.: Locality-preserving dimensionality reduction and classification for hyperspectral image analysis. IEEE Trans. Geosci. Remote Sens. **50**(4), 1185–1198 (2011)
8. Li, Y., Zhang, H., Shen, Q.: Spectral–spatial classification of hyperspectral imagery with 3D convolutional neural network. Remote Sens. **9**(1), 67–88 (2017)
9. Chen, Y., Jiang, H., Li, C., Jia, X., Ghamisi, P.: Deep feature extraction and classification of hyperspectral images based on convolutional neural networks. IEEE Trans. Geosci. Remote Sens. **54**(10), 6232–6251 (2016)
10. Mei, S., Yuan, X., Ji, J., Zhang, Y., Wan, S., Du, Q.: Hyperspectral image spatial super-resolution via 3D full convolutional neural network. Remote Sens. **9**(11), 1139–1161 (2017)

11. Liu, Q., Zhou, F., Hang, R., Yuan, X.: Bidirectional-convolutional LSTM based spectral-spatial feature learning for hyperspectral image classification. Remote Sens. **9**(12), 1330–1348 (2017)
12. Zhu, L., Chen, Y., Ghamisi, P., Benediktsson, J.A.: Generative adversarial networks for hyperspectral image classification. IEEE Trans. Geosci. Remote Sens. **56**(9), 5046–5063 (2018)
13. He, M., Li, B., Chen, H.: Multi-scale 3D deep convolutional neural network for hyperspectral image classification. In: Proceedings of the IEEE International Conference on Image Processing, pp. 3904–3908 (2017)
14. Chen, Y., Li, C., Ghamisi, P., Shi, C., Gu, Y.: Deep fusion of hyperspectral and LiDAR data for thematic classification. In: Proceedings of the IEEE International Geoscience and Remote Sensing Symposium, pp. 3591–3594 (2016)
15. Li, H., Li, C., Zhang, C., Liu, Z., Liu, C.: Hyperspectral image classification with spatial filtering and $l_{2,1}$ norm. Sensors **17**(2), 314–333 (2017)
16. Fauvel, M., Chanussot, J., Benediktsson, J.A.: Kernel principal component analysis for the classification of hyperspectral remote sensing data over urban areas. EURASIP J. Adv. Sig. Process. (1), 783194 (2009)
17. Lee, H., Kwon, H.: Contextual deep CNN based hyperspectral classification. In: Proceedings of the IEEE International Geoscience and Remote Sensing Symposium, pp. 3322–3325 (2016)
18. Hamida, A.B., Benoit, A., Lambert, P., Amar, C.B.: 3-D deep learning approach for remote sensing image classification. IEEE Trans. Geosci. Remote Sens. **56**(8), 4420–4434 (2018)
19. Santara, A., et al.: BASS Net: band-adaptive spectral-spatial feature learning neural network for hyperspectral image classification. IEEE Trans. Geosci. Remote Sens. **55**(9), 5293–5301 (2017)
20. Mou, L., Ghamisi, P., Zhu, X.X.: Deep recurrent neural networks for hyperspectral image classification. IEEE Trans. Geosci. Remote Sens. **55**(7), 3639–3655 (2017)
21. Cao, X., Zhou, F., Xu, L., Meng, D., Xu, Z., Paisley, J.: Hyperspectral image classification with Markov random fields and a convolutional neural network. IEEE Trans. Image Process. **27**(5), 2354–2367 (2018)
22. Song, W., Li, S., Fang, L., Lu, T.: Hyperspectral image classification with deep feature fusion network. IEEE Trans. Geosci. Remote Sens. **56**(6), 3173–3184 (2018)
23. Mandal, M., Shah, M., Meena, P., Vipparthi, S.K.: SSSDet: simple short and shallow network for resource efficient vehicle detection in aerial scenes. In: Proceedings of the IEEE International Conference on Image Processing, pp. 3098–3102 (2019)
24. Mandal, M., Shah, M., Meena, P., Devi, S., Vipparthi, S.K.: AVDNet: a small-sized vehicle detection network for aerial visual data. IEEE Geosci. Remote Sens. Lett. **17**(3), 494–498 (2020)

A Novel Approach for False Positive Reduction in Breast Cancer Detection

Mayuresh Shingan[1(✉)], Meenakshi Pawar[1], and S. Talbar[2]

[1] SVERI, Pandharpur, Maharashtra, India
mayureshshingan@gmail.com
[2] SGGS IE&T, Nanded, Maharashtra, India

Abstract. Breast Cancer is the most prevalent cancer among women across the globe. Objective of the Computer Aided Diagnosis (CAD) in breast cancer analysis is to detect the tumorous mass such that patient could get the proper treatment within time. However, existing algorithm undergoes detection of false positives. Thus, reduction of false positives is one of the challenging tasks to improve the performance of the diagnosis systems. In this paper, we propose a convolution neural network based approach for false positive reduction. We propose residual learning (ResNet) for false positive reduction. Masses segmented using a respective segmentation algorithm are given as an input to the proposed network to classify between true positive (tumorous mass) and false positive (non-tumorous mass). Proposed approach is validated on set of mammography scans collected from the Tata Memorial Cancer Hospital (TMCH). The performance of proposed algorithm is measured using precision, recall and F-score and compared with existing deep networks. Performance analysis shows that proposed approach outperforms other existing deep networks for false positives reduction.

Keywords: False positives · Tumour · Breast cancer · Deep network

1 Introduction

Breast cancer is one of the common cancer type across womens. Also, it became a major cause of death for Indian womens who are suffering from cancer. As per the estimation, nearly, 1,798k womens would suffer from Breast cancer by the year 2020 [27]. This rise in breast cancer incidences causes high mortality. The main reason behind the increase in the breast cancer incidences is the lack of awareness such as screening test in society about the disease [16]. However, cancer detected in early stage can be cured by proper medical treatment and care. Thus, it is essential to identify the breast cancer in early stage to give healthy life to the person suffered from it.

Nowadays, advancement in the medical imaging makes life simpler as it helps Radiologists to identify the disease and to analyze its growth. Among existing medical scans, Mammography is best suited and accepted image modality to

© Springer Nature Singapore Pte Ltd. 2020
N. Nain et al. (Eds.): CVIP 2019, CCIS 1148, pp. 364–372, 2020.
https://doi.org/10.1007/978-981-15-4018-9_33

detect and to analyze the growth of the breast cancer. For preliminary examination, mammography is most reliable as well as cost efficient [20]. Even tough mammograms helps Radiologists while diagnosis of breast cancer it has been observed that diagnosis varies as per the experience of the Radiologists. Also, some times symptoms are overlooked. According to the study [10,28,33], nearly 30% of the visible tumours are overlooked (They are called as false positives) and nearly 20%–30% of biopsies are actual (they are called as true positives). This motivates us to propose a novel approach which will able to reduce the false positives and at the same time will improve the true positives. In this paper, we propose a computer-aided diagnosis system for accurate breast cancer detection. In next Section, we have discussed about the existing approaches proposed for breast cancer detection.

1.1 Literature Survey

A general pipeline of the computer-aided diagnosis system for breast cancer detection comprises of enhancement of input mammogram image followed by feature extraction and classification of tumour into benign or malignant. Among these, enhancement of the mammogram image plays a major role. Researchers [2,12,34,35] proposed contrast enhancement algorithms as well as approaches were proposed for the removal of pectoral muscle [35]. Further, some novel approaches [13,17,26] were proposed for effective segmentation of the region of interest followed by feature extraction and classification.

In computer vision, image segmentation is majorly classified into three types namely: region-based, clustering-based and contour-based [30]. Among these, region-based and contour-based segmentation approaches are commonly used by the researchers. Gorgel et al. have proposed a wavelet transform based approach for segmentation. They propose a seed region growing algorithm and experimented on local dataset. Rouhi et al. [39] proposed region growing approach for segmentation. They make use of artificial neural network for classification task. Breast maas contour segmentation approach was proposed by Berber et al. [3] for breast cancer detection. They achieved 6 false positives per image (FPI). Followed by the successes of region growing algorithm, researchers [38] proposed hybrid level-set algorithm for tumour segmentation which is a combination of level-set and region growing algorithm. The artifacts present in the MIAS database causes the sensitivity of their algorithm to vary from 78 to 100%. The major hurdle in the region based approaches is these algorithms requires seed point to be given manually. Further, various approaches were proposed for breast tumour segmentation using K-means and Fuzzy C-means clustering [20,40]. K-means clustering requires number of clusters to be given before initialization of the algorithm. Also, it uses simple euclidean distance measure to classify the given pixel in one of the cluster. Simple distance measurement fails in complex regions. To overcome these limitations, learning-based segmentation approaches such as self-organizing map [4,21] were proposed to segment input image into number of regions. Pawar et al. [35] make use of self-organizing map for mammogram segmentation.

It is observed that in most of the cases detected tumour regions are not actually a tumour but are dense masses. These falsely detected regions are called as false positives. False positives consumes valuable time of the Radiologists to declare them as a false positive. Some times these false-positives causes unnecessary painful biopsies. To reduce the effect of false-positives (FPs), researchers proposed FPs reduction [19,23,40] approaches for breast cancer detection. Usually, the false-positive reduction algorithm is a post-processing step in computer-aided diagnosis of breast cancer. Before of the Fp reduction, robust feature extraction and classification are the two important building blocks of the CAD systems. Methods based on frequency analysis [9,10,37], morphological analysis [39] and textural analysis [5,8,11,15,25] were proposed for feature extraction. For texture feature extraction, Ojhala *et al.* [29] have proposed local binary pattern (LBP). Due to its efficient and less complex nature, LBP were used by many researchers for various computer vision applications. LBP extracts the local micro-structure information associated with the local region of the image. Variants of the LBP were proposed for different applications [8,11,14,15]. Due to is computational efficiency, it is used for FP reduction and classification task for breast cancer detection from mammograms [1,24]. Existing approaches for FP reduction are based on the hand-crafted feature extraction thus, fails to extract robust features. On the other side, convolution neural network has been utilized in various computer vision task such as for object recognition [18,22,41,42], image enhancement [6,7], style transfer [43], moving object segmentation [31,32]. Inspired from the success of the CNN, in this work, we propose a CNN-based approach for FP reduction. We utilized residual learning (ResNet) [18] to differentiate between the false detected mass (FP) and true tumorous mass (TP).

The rest of the paper is organized as follows: Introduction to the CAD of breast cancer detection is given in Sect. 1. Existing methods proposed for breast cancer detection and FP reduction are discussed in Sect. 2. Section 3 gives the proposed method for FP reduction in mammogram images. The experimental analysis on local database has been carried out in Sect. 4. Lastly, Sect. 5 concludes the proposed approach for FP reduction in breast cancer detection.

2   Proposed Method

In this section, the proposed approach for false positive reduction based on convolution neural network is discussed. We utilized a Tata Memorial Cancer Hospital Breast Cancer cases for the analysis. Initially, mammography scans are pre-processed and segmented using the approach proposed by [35]. Further, segmented masses are feeded to the proposed network for the classification between TP and FP. Overview of the proposed network is shown in Fig. 1. Residual block is a basic building block of the proposed network. We utilized pre-trained weight parameters of the ResNet to extract the features from the segmented masses. Next sub-section gives the architectural details of the proposed network.

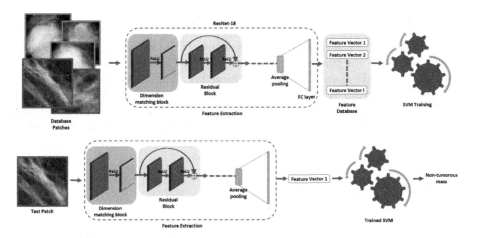

Fig. 1. Proposed approach for false positive reduction.

2.1 Feature Extraction

Vanishing gradient is a core problem in training of deep networks. As the network depth increases this problem becomes more intense which reflects into the decrease in the accuracy. He *et al.* [18] proposed the residual learning approach (ResNet). According to [18], convolution networks can be substantially deeper, more accurate, and efficient to train if they contain shorter connections between the network layers. They call it as a identity mapping. The concept of identity mappings is coarsely analogous to the connections in the visual cortex (as it comprises of feed-forward connections). Thus, it outperforms in different popular vision benchmarks such as object detection, image classification, and image reconstruction. Proposed network comprises of 18 residual blocks to extract the robust features from input segmented mass. Figure 1(b) shows the residual block used in proposed network. Mathematical formulation of the residual block is given by Eqs. 1–2.

$$F_1^k = \left\{ W_1^k | k \in [1, N] \right\} * F + \left\{ B_1^k | k \in [1, N] \right\} \tag{1}$$

$$F_2^k(x) = \begin{cases} F_1^k(x) & if \ F_1^k(x) > 0 \\ 0 & else \end{cases} \tag{2}$$

where, $*$ represents the convolution operation, W_1^k, B_1^k are the filter and biases at $conv1$ respectively, N is the number of filters used at $conv1$, F is the input feature map to the residual block and F_1^k represents the output of the first convolution layer.

F_2^k is the output of the rectified linear unit (ReLU) layer. Similar to Eq. 1, F_3^k is the output of the second convolution layer. Equation 3 shows the functioning of the identity mapping.

$$F_4^k(x) = F_3^k(x) + F^k(x) \tag{3}$$

given that F_3 and F has same number of feature maps.

2.2 False Positive Reduction

As discussed earlier, existing approaches segments the mammography scan however, some of the non tumorous masses are detected as a tumorous mass *i.e.* false positives. This affects to the system performance. Thus, false positive reduction is a most important task in CAD systems. Here, we utilized support vector machine (SVM) classifier to differentiate between true tumorous masses and false tumorous masses. Robust feature extraction using ResNet (as discussed in previous sub-section) followed by the SVM identifies the true tumorous masses and discards the false one.

3 Training Details

In total, we considered, 192 true tumorous masses and 800 non-tumorous masses for the experimentation. Among which 20% of the data is considered for the training and remaining data is considered for performance evaluation of the system. We make use of pre-trained weight parameters of the ResNet [18] to extract features from segmented mammography masses. Further, we trained support vector machine (SVM) on extracted features of 20% data. Training process is been carried on GHz, Intel Xeon i7 processor having 2.59 GHz, NVIDIA GTX 1080 8 GB GPU.

4 Database Information

In this Section, we have discussed about the database we collected from the Tata Memorial Cancer Hospital (TMCH), Mumbai. This database [36] is a study of 90 patients. It consists of 360 full field digital mammograms (FFDMs) having 180 CC views and 180 MLO views from right and left breast. It has 180 malignant and 180 normal breast images verified by the expert Radiologists. It uses biopsy proven breast cancer patients pathological data approved by the Institutional Research Ethics Committee of Tata Memorial Cancer Hospital (TMCH), Mumbai, India. Histopathological Reports are used to mark the tumorous region in these mammogram images. Nearly 35 patients are examined using "Hologic Selenia System" (Scanner1) gives 16-bit. The remaining 55 patients were examined with "GE Medical Senograph System" (Scanner2) providing 8-bit true color mammogram image in DICOM format of 4096×3328 or 2294×1914 pixels each measuring size $50 \times 50 \, \mu m^2$.

These scans are pre-processed and segmented using method proposed in [35] and segmented masses are considered for the analysis. In total, 192 true tumorous masses (Scanner 1–143 and Scanner 2–49) and 823 (Scanner 1–600 and Scanner 2–223) falsely detected tumorous masses are considered for the analysis.

Table 1. Comparison of the proposed approach with existing deep network for false positive reduction.

	Scanner I		Scanner II	
	Recall	Precision	Recall	Precision
AlexNet	0.9563	0.9435	0.9623	0.9512
VGG16	0.962	0.9578	0.9634	0.9565
PM	**0.9823**	**0.981**	**0.9856**	**0.9889**

Algorithm 1. Input: *Image*, Output: *Class Label*

1: Load an input mammography scan.
2: Do pre-processing as given in [35]
3: Segment mammography scan as per the process given in [35].
4: Collect segmented masses from each mammography scan from database and create database of true tumorous masses and falsely detected tumorous masses (patches).
5: Extract features from each collected patch using ResNet-18.
6: Use 20% patches to train the SVM for classification.
7: Load test patches.
8: Extract features from test patch using ResNet-18.
9: Classify these features using trained SVM.
10: Consider output of the SVM as a decision for the input patch *i.e.* true tumorous mass or falsely detected tumorous mass.

---

5   Results and Discussion

To analyze the effectiveness of proposed method for false positive reduction, we utilized Tata Memorial Cancer Hospital (TMCH) mammography scans. This database is consists mammography scans generated from two different scanners. The parameters used for measurement of systems accuracy obtained using proposed approach are Precision (P), Recall (R) are given by the Eqs. 4–5.

$$Precision : P(I_q) = \frac{Total\ number\ of\ correctly\ classified\ images}{Total\ number\ of\ images\ in\ that\ class} \tag{4}$$

$$Recall : R(I_q) = \frac{Total\ number\ of\ correctly\ classified\ images}{Total\ number\ of\ images\ classified\ in\ that\ class} \tag{5}$$

Table 1 gives the results of the proposed method for false positive reduction. Also, it shows the comparison between proposed and existing deep networks for false positive reduction. From Table 1 it is observed that the proposed approach outperforms other existing deep network for false positive reduction.

6   Conclusion

In this paper, we identified the false detection of tumorous mass using existing segmentation algorithms. To reduce the false detection of tumorous masses and to improve the performance of computer aided system for breast cancer detection, we propose a false positive reduction approach. Proposed approach is based on the convolution neural network. Input mammography scan is segmented using existing segmentation algorithm followed by the proposed classification framework reduces the number of false positives. We utilized ResNet-18 for the robust feature extraction followed support vector machine for the classification task. Proposed approach effectively detects the false positives and reverses the decision of the CAD system and improves the performance. We validated our approach using benchmark TMCH database. Performance of the proposed approach is measured using precision and recall parameters. We compared proposed approach with existing deep network. Comparison with existing deep networks shows that proposed approach outperforms them in terms of the precision and recall.

Acknowledgement. The TMCH database used for this work was given by Department of Radiodiagnosis, Tata Memorial Cancer Hospital, Mumbai.

References

1. Abdel-Nasser, M., Rashwan, H.A., Puig, D., Moreno, A.: Analysis of tissue abnormality and breast density in mammographic images using a uniform local directional pattern. Expert Syst. Appl. **42**(24), 9499–9511 (2015)
2. Anand, S., Gayathri, S.: Mammogram image enhancement by two-stage adaptive histogram equalization. Optik **126**(21), 3150–3152 (2015)
3. Berber, T., Alpkocak, A., Balci, P., Dicle, O.: Breast mass contour segmentation algorithm in digital mammograms. Comput. Methods Programs Biomed. **110**(2), 150–159 (2013)
4. Demirhan, A., Güler, İ.: Combining stationary wavelet transform and self-organizing maps for brain MR image segmentation. Eng. Appl. Artif. Intell. **24**(2), 358–367 (2011)
5. Dudhane, A., Shingadkar, G., Sanghavi, P., Jankharia, B., Talbar, S.: Interstitial lung disease classification using feed forward neural networks. In: International Conference on Communication and Signal Processing 2016 (ICCASP 2016). Atlantis Press (2016)
6. Dudhane, A., Murala, S.: C²MSNet: a novel approach for single image haze removal. In: IEEE Winter Conference on Applications of Computer Vision (WACV), pp. 1397–1404. IEEE (2018)
7. Dudhane, A., Singh Aulakh, H., Murala, S.: RI-GAN: an end-to-end network for single image haze removal. In: Proceedings of the IEEE Conference on Computer Vision and Pattern Recognition Workshops (2019)
8. Dudhane, A.A., Talbar, S.N.: Multi-scale directional mask pattern for medical image classification and retrieval. In: Chaudhuri, B.B., Kankanhalli, M.S., Raman, B. (eds.) Proceedings of 2nd International Conference on Computer Vision & Image Processing. AISC, vol. 703, pp. 345–357. Springer, Singapore (2018). https://doi.org/10.1007/978-981-10-7895-8_27
9. Eltoukhy, M.M., Faye, I., Samir, B.B.: A comparison of wavelet and curvelet for breast cancer diagnosis in digital mammogram. Comput. Biol. Med. **40**(4), 384–391 (2010)

10. Eltoukhy, M.M., Faye, I., Samir, B.B.: A statistical based feature extraction method for breast cancer diagnosis in digital mammogram using multiresolution representation. Comput. Biol. Med. **42**(1), 123–128 (2012)

11. Galshetwar, G.M., Patil, P.W., Gonde, A.B., Waghmare, L.M., Maheshwari, R.: Local directional gradient based feature learning for image retrieval. In: IEEE 13th International Conference on Industrial and Information Systems (ICIIS), pp. 113–118. IEEE (2018)

12. Gandhamal, A., Talbar, S., Gajre, S., Hani, A.F.M., Kumar, D.: Local gray level S-curve transformation-a generalized contrast enhancement technique for medical images. Comput. Biol. Med. **83**, 120–133 (2017)

13. Ganesan, K., Acharya, U.R., Chua, K.C., Min, L.C., Abraham, K.T.: Pectoral muscle segmentation: a review. Comput. Methods Programs Biomed. **110**(1), 48–57 (2013)

14. Ghadage, S., Pawar, M.: Integration of local features for brain tumour segmentation. In: IEEE 13th International Conference on Industrial and Information Systems (ICIIS), pp. 173–178. IEEE (2018)

15. Gonde, A.B., Patil, P.W., Galshetwar, G.M., Waghmare, L.M.: Volumetric local directional triplet patterns for biomedical image retrieval. In: Fourth International Conference on Image Information Processing (ICIIP), pp. 1–6. IEEE (2017)

16. Gupta, A., Shridhar, K., Dhillon, P.: A review of breast cancer awareness among women in India: cancer literate or awareness deficit? Eur. J. Cancer **51**(14), 2058–2066 (2015)

17. Hambarde, P., Talbar, S.N., Sable, N., Mahajan, A., Chavan, S.S., Thakur, M.: Radiomics for peripheral zone and intra-prostatic urethra segmentation in MR imaging. Biomed. Signal Process. Control **51**, 19–29 (2019)

18. He, K., Zhang, X., Ren, S., Sun, J.: Deep residual learning for image recognition. In: Proceedings of the IEEE Conference on Computer Vision and Pattern Recognition, pp. 770–778 (2016)

19. Hussain, M.: False-positive reduction in mammography using multiscale spatial Weber law descriptor and support vector machines. Neural Comput. Appl. **25**(1), 83–93 (2013). https://doi.org/10.1007/s00521-013-1450-7

20. Kanadam, K.P., Chereddy, S.R.: Mammogram classification using sparse-ROI: a novel representation to arbitrary shaped masses. Expert Syst. Appl. **57**, 204–213 (2016)

21. Kohonen, T.: The self-organizing map. Proc. IEEE **78**(9), 1464–1480 (1990)

22. Krizhevsky, A., Sutskever, I., Hinton, G.E.: ImageNet classification with deep convolutional neural networks. In: Advances in Neural Information Processing Systems, pp. 1097–1105 (2012)

23. Li, Y., Chen, H., Yang, Y., Cheng, L., Cao, L.: A bilateral analysis scheme for false positive reduction in mammogram mass detection. Comput. Biol. Med. **57**, 84–95 (2015)

24. Liu, X., Zeng, Z.: A new automatic mass detection method for breast cancer with false positive reduction. Neurocomputing **152**, 388–402 (2015)

25. Lladó, X., Oliver, A., Freixenet, J., Martí, R., Martí, J.: A textural approach for mass false positive reduction in mammography. Comput. Med. Imaging Graph. **33**(6), 415–422 (2009)

26. Maitra, I.K., Nag, S., Bandyopadhyay, S.K.: Technique for preprocessing of digital mammogram. Comput. Methods Programs Biomed. **107**(2), 175–188 (2012)

27. Malvia, S., Bagadi, S.A., Dubey, U.S., Saxena, S.: Epidemiology of breast cancer in Indian women. Asia Pac. J. Clin. Oncol. **13**(4), 289–295 (2017)

28. Muramatsu, C., Hara, T., Endo, T., Fujita, H.: Breast mass classification on mammograms using radial local ternary patterns. Comput. Biol. Med. **72**, 43–53 (2016)
29. Ojala, T., Pietikäinen, M., Harwood, D.: A comparative study of texture measures with classification based on featured distributions. Pattern Recognit. **29**(1), 51–59 (1996)
30. Oliver, A., et al.: A review of automatic mass detection and segmentation in mammographic images. Med. Image Anal. **14**(2), 87–110 (2010)
31. Patil, P., Murala, S., Dhall, A., Chaudhary, S.: MsEDNet: multi-scale deep saliency learning for moving object detection. In: IEEE International Conference on Systems, Man, and Cybernetics (SMC), pp. 1670–1675. IEEE (2018)
32. Patil, P.W., Murala, S.: MSFgNet: a novel compact end-to-end deep network for moving object detection. IEEE Trans. Intell. Trans. Syst. **20**(11), 4066–4077 (2018)
33. Pawar, M.M., Talbar, S.N.: Genetic fuzzy system (GFS) based wavelet co-occurrence feature selection in mammogram classification for breast cancer diagnosis. Perspect. Sci. **8**, 247–250 (2016)
34. Pawar, M.M., Talbar, S.N.: Local entropy maximization based image fusion for contrast enhancement of mammogram. J. King Saud Univ. Comput. Inf. Sci. (2018)
35. Pawar, M.M., Talbar, S.N., Dudhane, A.: Local binary patterns descriptor based on sparse curvelet coefficients for false-positive reduction in mammograms. J. Healthc. Eng.**2018** (2018)
36. Pawar, M.M., Talbar, S.N., Dudhane, A.: Tata Memorial Cancer Hospital database (2018). http://eureka.sveri.ac.in/blog/2018/06/17/tmch-database/. Accessed 15 Aug 2019
37. Raghavendra, U., Acharya, U.R., Fujita, H., Gudigar, A., Tan, J.H., Chokkadi, S.: Application of Gabor wavelet and locality sensitive discriminant analysis for automated identification of breast cancer using digitized mammogram images. Appl. Soft Comput. **46**, 151–161 (2016)
38. Rouhi, R., Jafari, M.: Classification of benign and malignant breast tumors based on hybrid level set segmentation. Expert Syst. Appl. **46**, 45–59 (2016)
39. Rouhi, R., Jafari, M., Kasaei, S., Keshavarzian, P.: Benign and malignant breast tumors classification based on region growing and CNN segmentation. Expert Syst. Appl. **42**(3), 990–1002 (2015)
40. Salazar-Licea, L.A., Pedraza-Ortega, J.C., Pastrana-Palma, A., Aceves-Fernandez, M.A.: Location of mammograms ROI's and reduction of false-positive. Comput. Methods Programs Biomed. **143**, 97–111 (2017)
41. Shaha, M., Pawar, M.: Transfer learning for image classification. In: Second International Conference on Electronics, Communication and Aerospace Technology (ICECA), pp. 656–660. IEEE (2018)
42. Simonyan, K., Zisserman, A.: Very deep convolutional networks for large-scale image recognition. arXiv preprint arXiv:1409.1556 (2014)
43. Thengane, V.G., Gawande, M.B., Dudhane, A.A., Gonde, A.B.: Cycle face aging generative adversarial networks. In: IEEE 13th International Conference on Industrial and Information Systems (ICIIS), pp. 125–129. IEEE (2018)

Classification of Effusion and Cartilage Erosion Affects in Osteoarthritis Knee MRI Images Using Deep Learning Model

Pankaj Pratap Singh[1(✉)], Shitala Prasad[2], Anil Kumar Chaudhary[1], Chandan Kumar Patel[1], and Manisha Debnath[1]

[1] Department of Computer Science and Engineering, Central Institute of Technology Kokrajhar, Kokrajhar, BTAD, Assam, India {pankajp.singh,b15cs054,b15cs066,b15cs073}@cit.ac.in
[2] CYSREN, NTU, Singapore, Singapore shitala@ieee.org

Abstract. In today's digital world, the major working culture is shifted to electronic devices such as computers and smartphone causing minimal physical exercise. Due to this reason, at old age people suffer many kinds of health issues such as Osteoarthritis (OA) disease. OA is the most common chronic condition of joints, which is also called as degenerative arthritis or degenerative joint disease. To identify this disease level an automated detection and classification method is required. This also requires expert knowledge person in this area. In this regard some other support is also required which is related to expert person of this area. The current strategy for OA identification includes clinical investigation and medical imaging techniques. In this paper, we detect and classify OA disease in knee from medical images using deep features. As we know, in medical imaging noise has a major role and therefore in this paper we focus on denoising medical images for accurate detection and classification. This paper also focuses on handling huge amount of image data by utilizing some High Performance Computing (HPC). The dataset used in this paper for detection and classification is a keen MRI image. Thus, an integrated discussion of various detection techniques regarding OA is done in a scientific way.

Keywords: Osteoarthritis · Detection and classification · Medical imaging · SSD · LabelImg · Deep learning

1 Introduction

Recent advances in artificial intelligence have led to fully automated work-flows that often exceed human performance. State-of-the-art neural network methods can detect and classify into thousands of categories more accurately and even faster than human prediction. In most of the cases, they have been trained on tens of thousands, or even millions of data samples. Neural networks have also found great success in the field of medical image analysis where data sets are often much smaller and challenging. Computer vision (CV) and patter recognition (PR) have solved many complex and challenging problems like object detection and segmentation [1–4], species and disease

© Springer Nature Singapore Pte Ltd. 2020
N. Nain et al. (Eds.): CVIP 2019, CCIS 1148, pp. 373–383, 2020.
https://doi.org/10.1007/978-981-15-4018-9_34

identification [5, 6], biometrics verification [7], leaf analysis and extraction, and aesthetic transfer [8–10]. The CV techniques and the process of creating images of various human body parts for diagnostic and treatment purposes is called medical imaging [17, 20]. Medical imaging techniques are also used for the visual representation of internal functions of some organs or tissues. This technique is the part of biological imaging which includes various radiological imaging techniques such as X-ray, Magnetic Resonance Imaging (MRI) Computed Tomography (CT), etc. [18, 23]. Osteoarthritis (OA) is one of the most common form of arthritis disease that basically starts from the late 40s onwards and is seen mostly in females, overweight and elderly people. It is a joint disease that mostly affects cartilage. Cartilage is a connective tissue found mainly in joints between the bones. Cartilage is the protective connective tissue whose main function is to connect bones together are help them to move smoothly without rubbing each other. Articular disorders and musculoskeletal diseases are one of the major health problems in recent years and affect especially the aging population. The human knee joint is commonly affected by OA, a degenerative disease that is the primary cause of persistent disability. It can be characterized by progressive degradation of the diarthrodial joint tissue.

Detection and monitoring of knee osteoarthritis can be done by measuring anatomical and biochemical changes which are associated with the tissues such as articular cartilage, meniscus, ligaments, and subchondral bones. An early detection and monitoring of OA is possible by measuring pre-structural and structural changes associated with the tissues such as articular cartilage, meniscus, and subchondral bones. Soft tissues within the diarthrodial joint have been success- fully used to detect and monitor knee OA, changes associated with the subchondral bones are a promising imaging biomarker for assessing knee OA conditions [20]. The main symptoms of OA are joint pain and difficulty in moving the joints. Its evaluation is based on the symptoms, the clinical diagnosis and the radio-graphic assessment techniques like MRI, CT scans. Though there have been various methods proposed, but Kellgren-Lawrence (KL) system is a validated method for classifying individual joints into 5 grades of OA analysis [22]. The different grades of OA diseases [15] is shown in Table 1.

Table 1. Different grades of OA disease.

KL Grades	OA Analysis
(i) Grade 0	(i) No features of OA present
(ii) Grade 1	(ii) Doubtful OA
(iii) Grade 2	(iii) Mild OA
(iv) Grade 3	(iv) Moderate OA
(v) Grade 4	(v) Severe OA

The common X-ray findings of OA include destruction of joint cartilage; joint space is diminished between the adjoining bones and the bone spur formation. MRI scans may be ordered when X-rays do not give clear reason for joint pain or when the

X-ray suggests that other type of joint tissues can be damaged. The current methods used for clinical diagnosis of OA are not accurate enough to efficiently measure the quality of disease. Thus, it is indeed more significant methods which are multifactoral to access the parameters and progression of OA. Detection and classification of OA in knee from medical images is one of the active fields. Since the Convolution Neural Networks (CNN) can detect the objects with more than 90% of accuracy [5], it can use as a fine-tuned neural network to detect a new object class. In respect to above-mentioned issues, this work will utilize deep neural network to calculate deep feature maps for detecting and classifying the OA in medical images. A large database is incorporate for better accuracy in this work. This paper focuses on the machine learning techniques for the classification and detection of osteoarthritis disease.

TensorFlow Object Detection API which is an open source framework for object detection purpose, is used for training and testing a Single-Shot Multibox Detector (SSMD) with Mobilenet-model [14]. The model is tested as pre-trained with a MRI images dataset consisting of OA. The detected OA region is shown in Fig. 1.

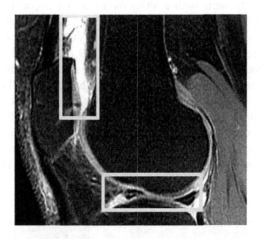

Fig. 1. Sample of detected OA region in knee MRI.

2 Object Detection Approach for Localized Affected Area in MRI Images

In this paper, an object detection problem is attempted using deep learning to localize the actual region of OA in knee MRI images. While detection of OA affected region, the prepared model is to be trained on dataset of OA affected region in MRI. To achieve this, the modern open source based solutions is explored for object detection in images for OA disease. TensorFlow Object Detection API is used for training and testing of Single shot detection (SSD) with Mobilenet model. This model is tested with a dataset consisting of OA knee MRI images [19].

2.1   An Architecture Description with Feature Mapping Approach

The network used in this paper is based on SSD [11, 14, 16]. The block architecture is shown in Fig. 2.

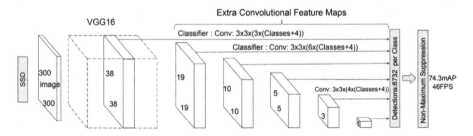

Fig. 2.   SSD architecture.

The SSD normally starts with a VGG model which is converted to a fully connected convolutional network. Thereafter some extra convolutional layers are attached that help to handle bigger objects. The output at the VGG network is a 38×38 feature map (conv4 3). The added layers produce 19×19, 10×10, 5×5, 3×3, 1×1 feature maps. All these feature maps are used for predicting bounding boxes at various scales (later layers responsible for larger objects). Thus the overall idea of SSD. Some of the activations are passed to the sub-network that acts as a classifier and a localizer [14]. Anchors are the collection of boxes overlaid on image at different spatial locations, scales and aspect ratios, act as reference points on ground truth images. A model is trained to make two predictions for each anchor. One is a discrete class and another a continuous offset by which the anchor needs to be shifted to fit the ground truth bounding box [16].

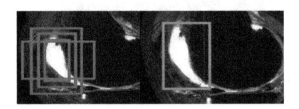

Fig. 3.   Non-maxima suppression

During training, SSD matches ground truth annotations with anchors. A number of anchors are associated with each element of the feature map. If the value of IoU (jaccard distance) for any anchor is greater than 0.5 then considered as a match. Both have been matched on different feature maps. The loss function used is the multi-box classification loss [14]. The classification loss used is the softmax cross entropy.

Non-maxima suppression is used during prediction to filter multiple boxes per object that may be matched as shown in Fig. 3.

2.2 A Workflow of the Model for Detecting OA Affected Region

Image classification can perform some pretty amazing feats, but a large drawback of many image classification applications is that the model can only detect one class per image. In the case of an object detection model, multiple classes can be classified in one image, but the main issue is to target an object inside an image with a bounding box with framing the object [13]. The TensorFlow models github repository has a large variety of pre-trained models for various machine learning tasks, and one excellent resource is their object detection API. The object detection API makes it extremely easy to train our own object detection model for a large variety of different applications.

Whether we need a high speed model to work on live stream high frames per second (fps) applications or high-accuracy desktop models, the tensor-flow object detection API makes it easy to train and export a model [14, 16]. Figure 4 shows the work flow of the model for detection of osteoarthritis affected region in knee MRI. Initially data sets are prepared by selecting the MRI images of osteoarthritis affected knee. Then the prepared data sets are to be labeled into two parts, training set and test set. The labeled data sets are then re-scaled to n × n pixel and read as grayscale. The data sets are divided in such manner like as 303 images for training purpose and 46 images for testing along with respective .xml files. The data sets are then reshaped appropriately for TensorFlow object detection API to build the model [14, 19]. The steps for building a custom object detection model using TensorFlow API are discussed in subsequent steps.

Data Set Collection

MRI OA disease related data sets such as Pascal and COCO are accessed for our work [12, 16], Whereas to train a custom object detection class, it is required to create and label own data sets related to this research work. The data set of osteoarthritis affected knee MRI images are collected from the National Institute of Health (NIH) and also from Invectus Innovation Pvt. Ltd., Noida. In this paper, osteoarthritis affected 300–400 knee MRI images are only used. Then the data set splitted as 303 images for training. The data set contain the two possibilities of OA affected regions namely, effusion and cartilage erosion (concern with the medical expert). Due to the limited amount of collected data, testing data are limited upto 10% of the total MRI images. For the convenience of feature extraction, it is decided to resize all the images to 384 * 384 pixels which help to create the required bounding boxes.

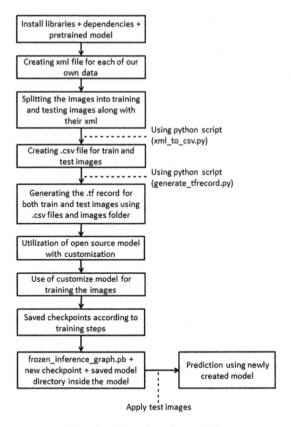

Fig. 4. Object detection model.

Feature Extraction Based on Creating Bounding Boxes

To train the object detection model, bounding box is needed for each image with the image's width, height, and each class with their respective xmin, xmax, ymin and ymax [12, 14]. For this work, the total number of MRI images labeled are 345 which include both training and test MRI images. Bounding box is the frame that captures targeted area where the possibilities of OA affected area is in the MRI image. Figure 5 is shown with a labeled MRI image sample which showing the bounding boxes. These labels are created using LabelImg tool which is an excellent open source software that makes the labeling process easier. Using LabelImg, individual 'xml' labels for the corresponding images are saved, which is converted into a 'csv' extension file for training purpose.

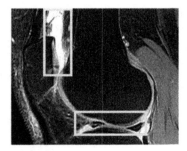

Fig. 5. Sample showing bounding boxes.

Conversion of the Labels into the TFRecord Format

After converting xml files into csv files then it is required to generate a TFRecord file to optimize the data feed.

Model Selection

The TensorFlow API consists of several models which can be used depending on the needs [16, 19]. SSD network works best for a high-speed model that can work on detecting image feed at high fps. Some other object detection networks detect objects by sliding boxes of different sizes across the image and running the classifier many times on different sections of the image [21]. But it can be costly due to consumption of resource. As its name suggests, the SSD network determines all bounding box probabilities in one go. Hence, it is a very faster model and with SSD the speed is gained at the cost of accuracy. SSD as the bounding box framework is used in this paper, and the MobileNet model is used for the neural network architecture. The 'config' file for SSD MobileNet has already configured and depending on the computer, the batch size in the config file is lowered to stop running out of memory.

Retrain the Model with Collected Data with Clip Off Last Layer

Now the entire SSD MobileNet model has been trained on collected data from scratch. This took a lot of time in training. So the easier solution for this is to take a model already trained on a large data set and clip off the last layer, which has the output classes from the trained model, and then it can be replaced with the required classes. In this manner, all the feature detectors which is trained on the previous model will be used and these features will be then used to detect the new classes. Now only the last layer of this mobilenet model is to be retrained, a high-end GPU is not required due to the get rid of training of all layers except the last one. Once the loss is consistently around the value of 1 or starts rising, TensorFlow training is to be stopped. It is simply required to run the train.py file in the object detection API directory for training purpose [19].

Implementation of New Model with TensorFlow

Before starting an experiment with the newly trained model, exported the graphs for inference from TensorFlow [12, 14]. The latest checkpoint (ckpt) from the data directory is used. Various graphs are showing the development of classification loss, localization loss and total loss during the training of the model which is depicted in

Figs. 6 and 7. The validation of the model's performance is checked on the images which are used first time in this model. Some MRI images of different patient's knee MRI are used for validation images. About 10% of the total images are assigned to be validation images. It is to be noted that multiple checkpoints can be tested with a good number of validation images to see which one performs best.

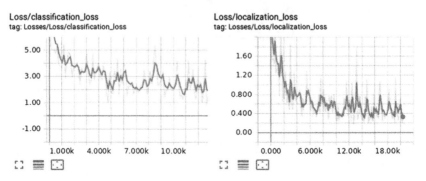

Fig. 6. Graph showing development of classification loss and localization loss during training.

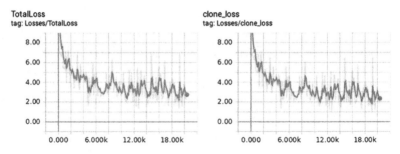

Fig. 7. Graph showing development of total loss during training our model.

Due to the limited amount of data, 100 images is used as per the class (for this work, effusion and cartilage erosion). This much amount of data is not quite enough to get a robust model. The detector is able to detect all the OA affected regions. But it is not able to detect the regions, may be due to absence of OA in knee MRI or due to limited amount of data in some scenario. But, this osteoarthritis OA detection problem explores the API's capabilities in nice manner [19, 21]. This is the main goal to gather all the steps to create such kind of an object detection model which will be capable to detect OA affected regions in knee MRI images.

Object Detection with Customized Model
The pre-trained SSD model (ssd_mobilenet_v1_coco) is used for training medical image data [12]. A provided configuration file (ssd_mobilenet_v1_coco.config) is used as a key basis for the model configuration. The checkpoint file for ssd_mobilenet_v1_coco is used as a starting point. The training is stopped after 12550 time

steps when the mean average precision (mAP) somewhat leveled out. The total loss value is reduced rapidly for this model due to starting from the pre-trained checkpoint file.

The pre-trained model is fed to TensorFlow Object Detection API and tested on the test data from osteoarthritis knee MRI images [19, 21]. In addition, on test data taken from different patient's osteoarthritis knee MRI to see how model's specific factors affect the result.

3   Result and Discussion

This experiment is executed on NVIDIA GPU server. The implementation part is done on TensorFlow framework using python programming language. Dataset contains 349 OA affected knee MRI images which are collected from NIH and Invectus Innovation Pvt. Ltd., Noida. Each MRI images consists of 384×384 pixels. Dataset are divided as training and test set images. The detected results are based on data which reflects the model network is trained. For this detection of OA disease, results are obtained from different OA patient's knee MRIs and the methodology for detection is applied on 349 MRIs from which 303 and 46 MRIs corresponding with .xml files are used for training and testing purpose respectively. Figure 8 shows the input MRI for detecting the presence of OA disease in the detection model.

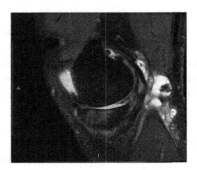

Fig. 8. Input knee MRI for detecting OA.

After the successful detection of OA disease which is shown in Fig. 9. It depicts the result obtained as two type OA disease affected regions, which is the cause of cartilage erosion and effusion.

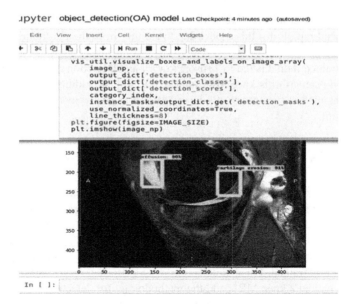

Fig. 9. Output detected as OA due to presence of both cartilage erosion and effusion.

4 Conclusion

This work describes the detection of Osteoarthritis disease area by predicting the OA affected regions with bounding box. OA affected knee MRI images are trained with SSD model using Tensorflow object detection API. This proposed approach is able to detect the actual location of the diseased portion from the osteoarthritis knee MRI images. The SSD pre-trained model gives the better results as compared to the Faster R-CNN model. The experiments utilize GPU server to execute the model for producing the better results in fast manner. After achieving these results, the model depicts the sufficiency and prediction level in cases of OA as well as would also be able to detect the cause of OA presence in a patient's knee. In the continuation of this research work, a large dataset of OA knee MRIs can be used with other two views such as axial and coronal.

Acknowledgement. The authors of this paper is thankful to National Institute of Health and Invectus Innovation Pvt. Ltd., Noida to provide medical datasets for evaluating and analyzing it. Authors are also thankful to CIT kokrajhar for utilizing NVIDIA GPU server to carry over this work.

References

1. Voigtlaender, P., Krause, M., Osep, A., Luiten, J., Sekar, B.B.G., Geiger, A., Leibe, B.: MOTS: multi-object tracking and segmentation. In: Proceedings of the IEEE Conference on Computer Vision and Pattern Recognition, pp. 7942–7951 (2019)
2. Singh, P.P., Garg, R.D.: A hybrid approach for information extraction from high resolution satellite imagery. Int. J. Image Graph. **13**(2), 1340007(1-16) (2013)

3. Singh, P.P., Garg, R.D.: Fixed point ICA based approach for maximizing the non-Gaussianity in remote sensing image classification. J. Indian Soc. Remote Sens. **43**(4), 851–858 (2015)

4. Singh, P.P.: Extraction of image objects in very high resolution satellite images using spectral behaviour in LUT and color space based approach. In: IEEE Technically Sponsored SAI Computing Conference, pp. 414–419. IEEE (2016)

5. Prasad, S., Kumar, P.S., Ghosh, D.: An efficient low vision plant leaf shape identification system for smart phones. Multimedia Tools Appl. **76**(5), 6915–6939 (2017)

6. Prasad, S., Peddoju, S.K., Ghosh, D.: Multi-resolution mobile vision system for plant leaf disease diagnosis. Sig. Image Video Process. **10**(2), 379–388 (2016)

7. Chai, T., Prasad, S., Wang, S.: Boosting palmprint identification with gender information using DeepNet. Future Gen. Comput. Syst. **99**, 41–53 (2019)

8. Prasad, S., Singh, P.P.: Vision system for medicinal plant leaf acquisition and analysis. In: Contractor, D., Telang, A. (eds.) Applications of Cognitive Computing Systems and IBM Watson, pp. 37–45. Springer, Singapore (2017). https://doi.org/10.1007/978-981-10-6418-0_5

9. Prasad, S., Singh, P.P.: Medicinal plant leaf information extraction using deep features. In: IEEE Region 10 Conference TENCON 2017, pp. 2722–2726. IEEE, November 2017

10. Prasad, S., Singh, P.P.: A compact mobile image quality assessment using a simple frequency signature. In: 2018 15th International Conference on Control, Automation, Robotics and Vision (ICARCV), pp. 1692–1697. IEEE, November 2018

11. Krizhevsky, A., Sutskever, I., Hinton, G.E.: ImageNet classification with deep convolutional neural networks. Commun. ACM **60**(6), 84–90 (2017)

12. Francis, J.: How to create your own custom object detection model. https://www.oreilly.com/ideas/object-detection-with-tensorflow. Accessed 17 June 2019

13. Gangawane, P., Kalshetti, P., Jaiswal, A., Rastogi, N.: Object Detection. Final year project report (2016)

14. Gilleman, D.: Multibox Single Shot Detector Overview. http://www.deeplearningessentials.science/singleShotDetector/. Accessed 17 June 2019

15. Kubakaddi, S.K., Ravikumar, K.M.: Measurement of cartilage thickness for early detection of knee osteoarthritis (KOA). In: IEEE Point-of-Care Healthcare Technologies (PHT), pp. 208–211. IEEE, Bangalore (2013)

16. Kurt.: Object Detection Tutorial in Tensor-Flow: Real-Time Object Detection. https://www.edureka.co/blog/tensorflow-object-detection-tutorial/. Accessed 17 June 2019

17. Shapiro, L.M., McWalter, E.J., Son, M.S., Levenston, M., Hargreaves, B.A., Gold, G.E.: Mechanisms of osteoarthritis in the knee: MR imaging appearance **39**(6), 1346–1356 (2014)

18. Mustamo, P.: Object detection in sports: Tensorflow object detection API case study. Thesis Report (2018)

19. Introduction and Use – Tensorflow Object Detection API Tutorial. https://pythonprogramming.net/introduction-use-tensorflow-object-detection-api-tutorial/. Accessed 17 June 2019

20. Gornale, S.S., Patravali, P.U., Manza, R.R.: A survey on exploration and classification of osteoarthritis using image processing techniques. Int. J. Sci. Eng. Res. **7**(6), 334–355 (2016)

21. Daniel, S.: Step by step tensorflow object detection API tutorial-selecting a model. https://medium.com/@WuStangDan/step-by-step-tensorflow-object-detection-api-tutorial-part-1-selecting-a-model-a02b6aabe39e. Accessed 17 June 2019

22. Kohn, M.D., Sassoon, A.A., Fernando, N.D.: Classifications in brief: Kellgren-Lawrence classification of osteoarthritis. Clin. Orthop. Relat. Res. **474**(8), 1886–1893 (2016)

23. Westbrook, C.: MRI at a Glance. 3rd edn., Wiley, Hoboken (2016)

Object Detection

A High Precision and High Recall Face Detector for Equi-Rectangular Images

Ankit Dhiman and Praveen Agrawal[(✉)]

Samsung Research Institute-Bangalore, Bengaluru 560037, Karnataka, India
{ankit.dhiman,praveen.agr}@samsung.com

Abstract. 360° cameras have recently become quite popular as it captures the entire surrounding environment in one frame. The captured image, however, suffers from spherical distortions because of the fish eye lenses used in these cameras. Equi-rectangular (ER) projections are the most popular and convenient 2D projections when it comes to dealing with 360° images. However, these projections suffer from heavy deformations in the pole regions which distorts the parts of the image falling near pole regions, thus increasing the difficulty of face-detection in these regions. In this paper, we address face-detection in 360° images in their ER projection. We discuss how a detector can be trained efficiently on ER images. We also introduce a deep learning pipeline which has high precision and high recall despite the presence of these spherical distortions.

Keywords: 360° cameras · Face detection · Equi-rectangular images

1 Introduction

360° imaging devices capture an entire scene with respect to a viewpoint. These imaging devices are gaining popularity with consumers due to the increase in demand of 360° content on social media platforms. The captured scene is stored as an equirectangular (ER) image which can further be used to generate different modes (like spherical view-mode) for viewing by the user. With this a necessity has come to bring existing tasks in image processing and computer vision for 360° media. One such computer vision task is of detecting a face in the captured scene.

Face detection is an indispensable technique for a multitude of face applications such as face-effects, face recognition, facial expression analysis etc. This is a difficult task because of difference in size, pose, lighting conditions, gender, age, occlusions, accessories and ethnicities of face in the captured image. These problems to some extent have been addressed in the previous literature work which is focused on normal images (not 360° images) captured through traditional imaging devices. For example, Viola Jones famous face-detection algorithm [8], uses cascade of classifiers working on Haar features to detect a face in real-time on

© Springer Nature Singapore Pte Ltd. 2020
N. Nain et al. (Eds.): CVIP 2019, CCIS 1148, pp. 387–397, 2020.
https://doi.org/10.1007/978-981-15-4018-9_35

normal images. Though this works well for simpler cases, it could not achieve great results in complex scenarios like occlusion, low lighting and different human face poses.

Recently deep learning methods, especially architectures using Convolutional Neural Network (CNNs), have illustrated their superior performance in computer vision tasks when compared with orthodox techniques. Zhang et al. [11] proposed a multi-task cascaded CNN based framework for face detection and facial landmarks localization. This method has outperformed several state of the art face-detectors across challenging standard face-datasets. Jiang et al. [3] demonstrate the efficacy of generic object detector Faster R-CNN [7] for face detection task when trained on suitable face-datasets like WIDER [10] and FDDB [2]. Wang et al. solves the problem of occluded faces by proposing an anchor-level attention algorithm over a base net which they call FAN [9]. Detecting small faces in a dataset with faces of varying sizes is another major issue in the task of face detection. Hu et al. [1] solves this problem by implementing a multi-task model exploring the role of scale invariance, image resolution and contextual reasoning in this problem.

This paper is arranged as follows. Section 2 discusses challenges to be solved for detection in equi-rectangular images. Section 3 describes the deep learning model used in the pipeline. Section 4 compares the proposed method and presents results obtained on the 360° ER image data-set.

2   Problems in Current 360° Face Detectors

A 360° image is generated after stitching multiple views from two or more camera lenses. Commercially available 360° devices likes Samsung Gear 360, Ricoh Theta S use two fish-eye lenses to generate a 360° image. More complex 360° cameras use more than two lenses to generate the 360° image. The most common projection to visualize these kind of images is the equi-rectangular (ER) projection. In this section, we discuss the problems specific to face-detection in the ER projection.

- The spherical distortions are more towards the pole region in an ER image. Such kind of distortions make face detection task challenging due to the variations in the shape of a face with respect to its position in the image. As shown in Fig. 1, face near the pole region is heavily distorted. Finding facial features using the conventional face detectors will not work in this scenario.
- ER images have a discontinuity at its two ends. Due to this, a face can be split in an ER image as shown in Fig. 2. These kind of scenarios can go undetected when normal face-detectors are used.

We attempt to solve these problems in this work. The following sections showcase our approach and the results we achieved.

Fig. 1. Examples of scenario where face is highly distorted near the pole region

Fig. 2. An example of split face at ends of an ER image

3   Model Architecture and Training

We design a deep learning architecture with an attempt to solve the problems mentioned in the previous section. Our primary aim is to create a high precision and high recall face detector for 360° images. In order to detect faces with all the distortions as described earlier, using a single model, we decide to use a highly sensitive model that doesn't miss out on any potential face. This is then followed by a network that only runs on the proposals generated by the previous model and eliminates the false positives thereby giving us the best results in terms of both precision and recall. The first part of our face detector which is the high recall network and second part a refine network, both inspired by Tiny YOLO model [5].

3.1   High Recall Model

We use a modified version of Tiny YOLO model [5] for this part. Tiny YOLO is a state of the art, real-time object detection system that can detect over 9000 object categories and is a faster version of the original YOLO (You Only Look

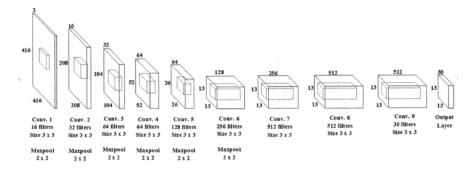

Fig. 3. High recall model based on Tiny yolo v2

Once) model [5]. In our case we only need to detect one class, i.e, face. This means that there are many kernels in this network which are redundant for us and hence, we make the model lighter by pruning it. The pruning strategy is explained in Sect. 3.2. In order to increase the sensitivity of the model towards faces, we keep the confidence threshold for detection pretty low for this model. In addition to this, we also keep a low IoU (Intersection over Union) threshold. Reduced threshold for both these parameters increases the number of false positives but also makes sure that we do not lose any true positive. The architecture of the high recall model based on yolo v2 is shown in Fig. 3.

3.2   Model Pruning

The tiny YOLO model was originally designed with the objective of detecting objects for 9000 categories. This means it has filters learning rich and diverse features. In our problem we only need to detect only one class, which is a face. This means the types of features that need to be learned in our case are far lesser. This opens up the scope for pruning the tiny YOLO model. We use 'mean activations' of the kernels as the criteria for pruning, i.e. for a given layer we take the average of the filter activations from all the training images and remove those kernels which have negative or very low (close to zero) mean activations. The performance metrics during the pruning procedure are shown in Table 1. We observed as we fine tuned the model by removing the kernels playing no role in face detection, our performance metrics improved overall giving us better detection results.

3.3   Refinement Network: High Precision Model

The high recall model architecture, in our previous stage, will have more number of false positives because of model's high sensitivity to classify any potential region in an image as a face. Refinement network (R-net) is a CNN after high recall model for removing false positives and adjusting bounding box's position of a true positive. This model is more tuned to reject false-positives, hence a high precision network.

Table 1. Variation in performance metrics with pruning of high recall network

Parameters pruned	After pruning			After retraining		
	Precision	Recall	IoU	Precision	Recall	IoU
0%	0.760	0.800	0.630	0.760	0.800	0.630
6.8%	0.907	0.895	0.699	0.923	0.932	0.748
30.03%	0.564	0.963	0.659	0.844	0.966	0.728
59.96%	0.937	0.93	0.675	0.918	0.957	0.749
71.19%	0.937	0.935	0.734	0.922	0.953	0.747
73.99%	0.969	0.626	0.597	0.94	0.953	0.747

Given a detection candidate (x, y, w, h) with (x, y) being the coordinates of the top-left corner and (w, h) being the bounding-box width and height, from previous model, the region is cropped (the cropped region is kept a little larger than the original bounding box size) and resized to input size of 256×256 for R-net. This network rejects false candidates and adjusts the bounding-box parameters for better boxes. Impact of this network on overall framework is discussed elaborately in Results section.

The architecture of this R-net is similar to that of high-recall network (Fig. 3). It takes a fixed input of size 256×256. As this CNN architecture only sees a portion of an input image, a different training strategy is used to train this model. Here, training data is generated using three different kinds of process: (1) Negatives: Regions with IoU less than 0.25 with ground-truth faces, (2) Partly Positive: Regions with IoU in between 0.4 and 0.65 with ground-truth faces and (3) Positives: Regions with IoU greater than 0.65 with ground-truth faces. Positives and partly positives are used for bounding box regression. Positives and partly positives are generated from predictions from the previous stage as well as ground-truth faces. Negatives are kept twice the total number of positives and partly positives, and are generated randomly from input image.

3.4 Data Augmentation

The major challenge in detecting a face in ER image is to detect a face near pole, where face-features are heavily distorted as shown in Fig. 1. We aim to develop a face-detector for ER images which is invariant of the distortions arising from it's spatial location in the image. Already, traditional image transformations like flipping, rotating etc. are techniques used for data-augmentation. The motive of our data augmentation technique is to generate more faces near the pole regions, so that the deep learning model learns the distorted features.

For each labelled ER image, we projected it onto a sphere, rotated it in the spherical domain and projected it back to the ER domain to yield a new image with different distortions. The output of this augmentation technique is shown in Fig. 4.

Fig. 4. Left Image: Faces are near the equator region. Right Image: Face near equator is distorted by bringing it to the pole region

3.5 YOLO-v3 Network

Predictions from Tiny YoloV2 model [5] does not make good use of features from the shallower layers. In YOLOV3 [6] network, prediction is done at 3 different scales using similar concept of feature pyramid networks. We use a modified version of Tiny YOLOV3 [6] (modifications made same as in case of v2 by pruning the network with the same strategy) and the architecture is shown in Table 2. Similar, to our previous pipeline,we refine predictions from this network with a high-precision network. We use Tiny YOLOV3 [6] for both high recall and high-precison network. Training data for the refinement network (the high precision network) is generated as discussed in Sect. 3.3. We use the same architecture as shown in Table 2 for the high precision network. The size of input image for the high precision network is 256×256 and that of the high recall network is 832×416.

4   Results

In this section, we analyze the efficacy of the proposed pipeline by benchmarking the performance of our trained model on popular metrics for detection frameworks, namely mean precision, mean recall and mean intersection over union (IoU). We also publish the inference time of our pipeline on NVIDIA Titan X.

4.1 Setup

We train the deep learning framework on in-house generated 360° images dataset. The dataset is discussed more elaborately in Sect. 4.2. We train the high recall model, as discussed in Sect. 3.1, over a pre-trained ImageNet model. For pruning, we followed an approach to prune the deeper layers before pruning the shallower layers in the high recall model, as discussed in Sect. 3.2. Also, for fine-tuning the model, we degrade the learning rate from 0.001 by a factor of 0.1 after 100 epochs. For the quantitative comparison on test data-set, any detection from the high recall network which has IoU score greater than 0.20 is considered positive.

Table 2. Our network inspired by Tiny YOLOV3

Type	Filters	Size
convolutional	16	$3 \times 3/1$
maxpool		$2 \times 2/2$
convolutional	32	$3 \times 3/1$
maxpool		$2 \times 2/2$
convolutional	64	$3 \times 3/1$
maxpool		$2 \times 2/2$
convolutional	128	$3 \times 3/1$
maxpool		$2 \times 2/2$
convolutional	128	$3 \times 3/1$
maxpool		$2 \times 2/2$
convolutional	256	$3 \times 3/1$
maxpool		$2 \times 2/1$
convolutional	512	$3 \times 3/1$
convolutional	128	$1 \times 1/1$
convolutional	256	$3 \times 3/1$
convolutional	18	$1 \times 1/1$
detection		
route (from layer 13)		
convolutional	128	$1 \times 1/1$
upsample		$2\times$
route (from layers 19, 8)		
convolutional	128	$3 \times 3/1$
convolutional	18	$1 \times 1/1$

4.2 Dataset

One of the biggest challenges in addressing the problem of face detection in ER images was the unavailability of a standard dataset for faces in 360° images. We collected and labelled in-house data of around 8,000 360° images. The dataset was made as dynamic as possible by taking the right mix of a wide range of difficulties & variations including illuminations (bright lighting as well as low lighted scenarios), blur, face shape & size, spherical distortions, occlusions, indoor & outdoor scenes, etc. This data is used to train and validate our model for the face-detection task. Further, as discussed in Sect. 3.4, data augmentation enhanced our dataset by 4000 images to give a final set of nearly 12000 360° images.

Our test data-set contains 1500 ER images with good distribution in terms of number of faces per image and also in terms of faces sizes. In Fig. 5(a), we give distribution details of number of faces in an image. In Fig. 5(b), we illustrate distribution of relative size of faces w.r.t the total image size.

394   A. Dhiman and P. Agrawal

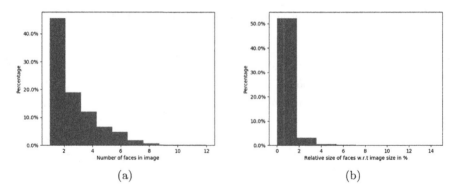

(a)   (b)

Fig. 5. Statistics of 360° test data-set. (a) Percentage Plot of number of faces per image in data-set. (b) Percentage plot of relative face-size w.r.t total image-size

4.3   Model Analysis

The proposed pipe-line is inspired from YOLO [6]. We compare the performance of the proposed pipeline with Full YOLO and Tiny-YOLO models mentioned in [5]. These models are re-trained after modifications for detection on only one class in the final convolutional layer.

Data-Set Augmentation to Distort Faces. As explained in Sect. 3.4, the augmentation technique is used to increase the images with more faces near the pole regions. We trained all the models separately with dataset containing only the originally collected images (without data augmentation) and with the enhanced dataset containing the augmented images too. Table 3 showcases the precision, recall and IoU for Yolo V2, tiny-yolo V2 [4], our high recall model and our high recall plus high precision model. From the results, it is clear that data augmentation technique significantly improves performance of our face-detector pipeline.

Table 3. Mean precision, recall and IoU upon training on dataset with and without data augmentation

Model	Without augmentation			With augmentation		
	Precision	Recall	IoU	Precision	Recall	IoU
YOLO v2	0.755	0.873	0.715	0.760	0.800	0.630
Tiny Yolo v2	0.907	0.895	0.699	0.923	0.932	0.748
Our full model	0.882	0.93	0.675	0.918	0.957	0.749

In Table 4 we present the results when we train our custom yolov3 model discussed in Sect. 3.5. It can be observed that cascading the High Recall further with a high-precision model improves the detector accuracy.

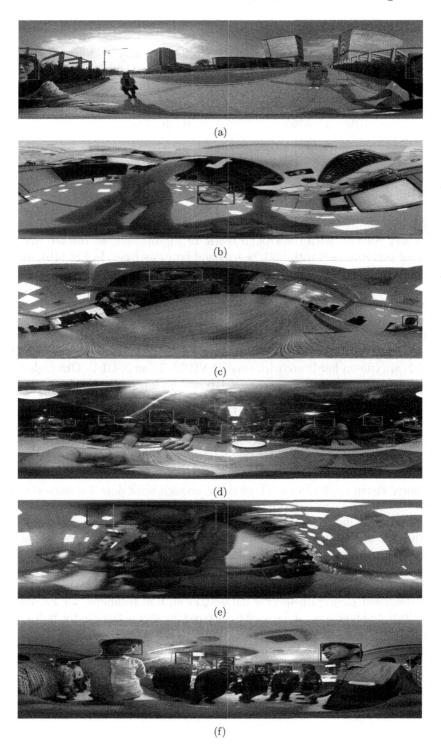

(a)

(b)

(c)

(d)

(e)

(f)

Fig. 6. Results on equi-rectangular images from our face-detection pipeline

Table 4. Mean precision, recall and IoU on Tiny YOLOV3.

Metrics	High-Recall YOLOV3	High-Recall + High-Precision YOLOV3
Precision	0.894	0.919
Recall	0.97	0.955
IOU	0.725	0.729

In Fig. 6, we present images from our test-dataset. It is clear from these results that the detector is invariant to the spherical distortions present in the equirectangular image. The green boxes are the results from our face detector and blue boxes are the ground-truth. As it can be observed, the detected bounding boxes well coincide with the ground truth. In Fig. 6(a)-(f), we can see that our detector performs well with both nearby and far away faces. In Fig. 6(b), inspite of the inherent camera rotation, our detector manages to detect face in the rotated image. In Fig. 6(c) & (e), we show that the detector is able to detect faces in the pole regions too. Our model performs well when compared to state of the art Yolo network. There is an overall improvement in performance metrics with our data-augmentation approach.

The pruning strategy discussed in Sect. 3.2 reduces our model-size by 74%. Models are run on hardware with single NVIDIA Titan X GPU. Our high recall model uses input image of size 832 × 416 and high-precision model uses input patch of size 256 × 256. Our full model runs at under 6 ms on the GPU.

5 Conclusion

Face-detection is an essential computer vision task and has found use in other computer vision applications. There is not enough work done for face-detection in 360° ER images. In this paper, we have proposed a deep learning architecture to detect faces in ER image itself. The proposed face-detection architecture, which is trained on in-house captured 360° ER images data-set, consists of two CNN based models, one which has high recall and the other one which has high precision. Together, our model achieves high precision and high recall rate for face-detection task on the test-set of the data-set discussed in the paper. We have discussed projection-specific data-augmentation technique for ER images and illustrated how this can be used to increase number of faces in the pole region and ultimately improve the overall quality metrics of the detector. Further, the combination is light-weight and can achieve high FPS when run on NVIDIA Titan X GPU.

References

1. Hu, P., Ramanan, D.: Finding tiny faces. In: 2017 IEEE Conference on Computer Vision and Pattern Recognition (CVPR), pp. 1522–1530. IEEE (2017)
2. Jain, V., Learned-Miller, E.: FDDB: a benchmark for face detection in unconstrained settings. University of Massachusetts, Amherst, Technical report UM-CS-2010-009 2(7), 8 (2010)
3. Jiang, H., Learned-Miller, E.: Face detection with the faster R-CNN. In: 2017 12th IEEE International Conference on Automatic Face & Gesture Recognition (FG 2017), pp. 650–657. IEEE (2017)
4. Redmon, J., Divvala, S., Girshick, R., Farhadi, A.: You only look once: unified, real-time object detection. In: Proceedings of the IEEE Conference on Computer Vision and Pattern Recognition, pp. 779–788 (2016)
5. Redmon, J., Farhadi, A.: Yolo9000: better, faster, stronger. arXiv preprint arXiv:1612.08242 (2016)
6. Redmon, J., Farhadi, A.: Yolov3: an incremental improvement. arXiv preprint arXiv:1804.02767 (2018)
7. Ren, S., He, K., Girshick, R., Sun, J.: Faster R-CNN: towards real-time object detection with region proposal networks. IEEE Trans. Pattern Anal. Mach. Intell. **39**(6), 1137–1149 (2017)
8. Viola, P., Jones, M.J.: Robust real-time face detection. Int. J. Comput. Vision **57**(2), 137–154 (2004)
9. Wang, J., Yuan, Y., Yu, G.: Face attention network: an effective face detector for the occluded faces. arXiv preprint arXiv:1711.07246 (2017)
10. Yang, S., Luo, P., Loy, C.C., Tang, X.: Wider face: a face detection benchmark. In: Proceedings of the IEEE Conference on Computer Vision and Pattern Recognition, pp. 5525–5533 (2016)
11. Zhang, K., Zhang, Z., Li, Z., Qiao, Y.: Joint face detection and alignment using multitask cascaded convolutional networks. IEEE Signal Process. Lett. **23**(10), 1499–1503 (2016)

Real-Time Ear Landmark Detection Using Ensemble of Regression Trees

Hitesh Gupta[✉], Srishti Goel, Riya Sharma,
and Raghavendra Kalose Mathsyendranath[✉]

Samsung R&D Institute India Bangalore, Bangalore 560037, Karnataka, India
{hitesh.gupta,srishti.goel,riya.s,
raghava.km}@samsung.com

Abstract. Human face landmark detection algorithms have numerous applications. Current face landmark detection algorithms limit themselves to features around eyes, nose, cheeks and lips. Face landmark detection combined with augmented reality technology has given rise to commercially popular virtual try-on applications. To realize use cases of virtual jewelry try-on like earrings on smartphones, landmark points of human ear is required, but this field is not much explored in the literature. Existing methods are not accurate enough in different face poses and lighting conditions. Proposed method offers solution for ear landmark detection considering the computational requirements of mobility devices, and comprises ear localization followed by ear landmark detection. It adopts Haar cascade based model for ear localization and an Ensemble of Regression Trees for ear landmark detection. The experimental results and comparison with state-of-the-art methods show that the proposed method accurately localizes the ear, provides correct landmark points and is fast enough to run on mobility devices with low memory footprints. Comparison with popular methods shows the novelty points in the proposed approach.

Keywords: Ear landmarks · Jewelry augmentation · Ear localization

1 Introduction

The facial features extraction has numerous applications. Applications such as pose estimation, face recognition, gender detection, expression analysis, age estimation, face morphing is often built upon this. Facial features when combined with augmented reality (*AR*) technologies are creating even more compelling use cases and business models for enterprises e.g. Virtual try on application for virtual-glasses, jewelry, make up, hearing aid.

The Human Ear has a characteristic design with its own unique features. However, little effort has been put to develop accurate ear landmark detection models. Ear landmark detection is very important for use cases such as virtual jewelry and virtual hearing aid. These use cases are dramatically changing the business landscape for enterprises and gaining importance from business leaders. Mobility devices like smartphone nowadays have lower cost and high computational power and can easily

© Springer Nature Singapore Pte Ltd. 2020
N. Nain et al. (Eds.): CVIP 2019, CCIS 1148, pp. 398–409, 2020.
https://doi.org/10.1007/978-981-15-4018-9_36

build these use cases. Enterprises in field of E-commerce, e-tail are keen to tap into potential of these use cases to boost their revenues and profits.

This paper presents a method for ear localization and landmark detection targeting applications on mobility devices like smartphones. Camera based applications need to be fast i.e. real time and run in time order of camera frame rate e.g. 30 fps. In addition, the detection algorithm has to be light weight in terms of memory and power consumption. The proposed method addresses these requirements of mobility devices and is computationally fast and light in terms of memory. From use case perspective, virtual try-on of earing using augmented reality (*AR*) application is targeted, where high degree of accuracy is expected to precisely augment an AR jewelry object.

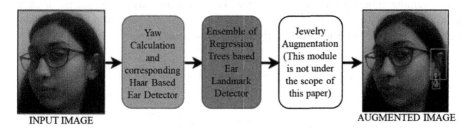

Fig. 1. Flowchart of the proposed method. The red (first) and the blue (second) box are detailed in Figs. 3 and 4 (Color figure online)

The proposed algorithm as described in Fig. 1. uses ear shape and generates an ear landmark detector for accurate jewelry augmentation. The major contributions of the paper are as follows:

Haar-Like Features Based Ear Localization: Haar like features as proposed by Paul Viola and Michael Jones in [1] along with Adaboost and Cascading are used to label the image window as negative or positive w.r.t containing an ear.

Ensemble of Regression Trees for Ear Feature Extraction: Gradient boosting based learning which optimizes over sum of squared error loss is used to train the regressors to estimate the ear landmark positions directly from the pixel intensities themselves. The idea has been taken from facial landmark recognition in [2].

The paper is organized as follows. Section 2 presents the literature survey reviewing popular and state-of-the-art ear landmark detection methods. Section 3 describes the proposed method for ear localization and ear landmark detection. Section 4 describes the results of the proposed landmark detection method and compares it with the state-of-the art solutions. Section 5 mentions the conclusion points and future work.

2   Literature Survey

There has been a lot of discussions amongst the researchers if the ear can be used as a viable biometric or not. Its possibility was first discovered by the French criminologist Bertillon, and after that, many successful studies could only be seen as hints but not evidence due to unavailability of a large ear database. In 2006, the Forensic Ear identification Project (FearID) was initiated by nine institutes from Italy, the UK, and the Netherlands which measured an Equal Error Rate (EER) of 4% which concluded the hypothesis of using ears as biometric.

The ear detection approaches till date have relied on ear properties like its edges or patterns. There has been a lot of methods for ear detection as mentioned by Pfugh and Busch [3]. These methods include skin detection followed by ear template matching, ear contour matching using edge detection, Helix Shape Model, Wavelet and Hough transforms, Adaboost etc., each reporting a good accuracy on different datasets. Out of these, cascaded Adaboost based on Haar-Wavelet features has been used as done by Abaza et al. [4] and Islam et al. [5] for ear localization. Although Islam et al., reported a huge training time for the classifier, Abaza et al. significantly shortened the training phase by using a modified version of Adaboost. This approach was shown to successfully detect 95% of the ears on five different databases.

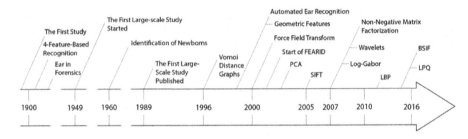

Fig. 2. Selected milestones presented by [6] for the ear recognition approaches through time.

Each ear recognition system consists of a feature extraction method which gives a feature vector for each ear, which in-turn is used for ear recognition. Figure 2 shows the various successful feature extraction methods used in the literature till date. With the increase in the availability of large and complex datasets, the use of neural networks and other statistical approaches has also been successfully explored along with dimensionality reduction techniques like PCA and ICA. Although ear landmark extraction has not been explored in the context of augmentation of jewelry, but it has been used as features for ear recognition in the context of ear biometrics by the following literature. Results of the proposed method has been compared to these methods.

Hansley et al. [7] proposed a solution for ear recognition using a sum-fusion of handcrafted and learned features. They developed a two-stage CNN-based landmark detector, which uses both learned and handcrafted features along with PCA to detect the ear landmarks. The handcrafted features were extracted using a toolbox and includes local binary patterns (LBP), binarised statistical image features (BSIF), local phase quantisation features (LPQ), patterns of oriented edge magnitudes (POEM), HOG, dense scale-invariant feature transform (DSIFT) and Gabor wavelets based features. Since the performance of handcrafted features is known to degrade, they have used the state-of-the-art CNN architecture employed for the face recognition to learn the ear features. Due to availability of only 500 images for training, they have also heavily augmented the data by rotation and scaling. They evaluated their ear recognition results on five datasets: IIT-Delhi, WPUTE, AWE, ITWE and UERC databases.

Zhou et al. [8] uses the complex and unique shape of the ear to identify people. They built a holistic and a patch-based active appearance model (AAM) for ear localization and recognition and both the models were able to align the images of the ears. The results in their paper shows contour fitting over the ear using ear landmarks.

Ravindran [9] in his thesis has used a combination of techniques like Viola Jones Haar-Cascade based ear detector, Active Shape Models (ASM) and Dijkstra's shortest path algorithm to devise a shape model of the ear and mark an accurate contour around the ear using only 2D color images. A set of key landmark points around the ear including the ear anti-helix, outer helix, and the center is extracted using the ASM after the ear detection. The entire ear outer contour is then extracted by tracing out the strongest edge between the adjacent landmarks by Dijkstra's shortest path algorithm.

Many popular applications such as Instagram, Snapchat etc. and commercial applications like CaratLane [10] attempt to augment jewelry on the ear. They create a 3D face mesh and then interpolate the earlobe point for the augmentation of jewelry. The facial characteristics are used in such scenario, which fail to cover the cases when the ear is occluded or cases where the ear shape is irregular.

2.1 Datasets Used

- The University of Notre Dame (**UND**) **HID-J2** ear dataset [11] is used for comparing the results with [9].It is currently the largest 3D ear dataset which consists of 2436 side face 3D scan of 640 X 480 resolution from 415 different persons. It contains images with different lighting conditions but very less pose variations.
- **IIT Delhi** database [12] contains 125 subjects where each has 3 to 6 images taken in grayscale. Images are taken in indoor condition with limited lighting variation. It contains no or occasional occlusions and pose variations.
- **iBug "In The Wild" Ear Database** [8]
    - Collection A consists of 605 images "in-the-wild" collected from Google Images with no specific identity but with annotated 55 ear landmarks.
    - Collection B contains 2058 images with 231 identity-labelled subjects collected from VGG-Face, LFW and Helen database. It contains only bounding box of each ear obtained by HOG and SVM based ear detector.
    - It contains all the naturally clicked images with cluttered backgrounds, occlusions and maximum variation in poses, image size and ear resolution.

3 The Proposed Method

Although Pfugh and Busch [3] has mentioned a lot of techniques for ear detection and ear feature extraction for ear recognition to use it as a viable biometric, but none of the papers have talked about detecting the accurate landmarks on the ear for jewelry augmentation in real-time for mobility devices. This section talks about the same.

Section 3.1 describes the yaw based ear detection model filtering which helps in improving the ear detection rate. Section 3.2 talks about the ear detection approach used and Sect. 3.3 describes the novel approach of using ensemble of regression trees for ear landmark detection. Section 3.4 talks about the possibility of augmenting jewelry on these detected landmarks.

3.1 Preprocessing: Yaw Based Model Filtering

On the input image, first the yaw is calculated using the facial feature points [13] which is further used to select the ear localization model. The proposed approach uses different model for different yaw angles to improve accuracy and to ensure no false detection. Refer Fig. 10 for details. Based on the following yaw angles, the corresponding ear localization model is selected:

[$\sim$- -20], [-20- -10], [-10 – 0], [0 - 10], [10 - 20] and [20 -$\sim$] degrees

Six models have been trained to cover all possible scenarios for ear detection.

3.2 Ear Localization

The proposed approach is based on the concept of Object Detection proposed by Paul Viola and Michael Jones in [1] which describes a machine learning approach for visual object detection. The approach [1] is capable of processing images extremely fast and achieve high detection rates. No novelty is being claimed here as this method has already been much explored in the literature.

The ear detection procedure uses simple "Haar based features" for image classification. Three kind of such features are used, namely two-rectangle, three-rectangle and four-rectangle feature.

The "Integral Image" representation is adopted which allows the features used by the detector to be computed quickly. The images are annotated such that the ears are captured within an accurate bounding box. The Haar based feature extraction then considers adjacent rectangular regions in the box, sums up the pixel intensities in each region quickly using the Integral Image and calculates the difference between these sums. This difference is then used as a feature to categorize subsections of an image.

Adaboost is used to select a small number of critical features from a larger set and train an extremely efficient classifier. As each Haar feature is only a weak classifier, a large number of Haar features are organized into a more complex classifier in a "cascade". This allows background regions of the image to be discarded quickly while spending more computation on promising object-like regions and is used to identify the ear with sufficient accuracy. The cascade is an object specific focus-of-attention mechanism which provides statistical guarantees that the object of interest is unlikely to be in the discarded regions.

Positive (containing an ear) and Negative (not containing an ear) images are used to train the classifier using Boosting and Cascading. During the detection phase, a window of the target size is moved over the input image, and each stage of the cascade classifier labels the region as positive or negative. If the label is negative, classification is complete and the window shifts, else, the region is passed to the next stage for a more confident 'positive' classification.

Figure 3 highlights the flowchart for the training of Haar-Cascade based ear detector.

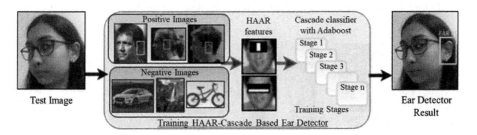

Fig. 3. Training and testing of the Haar-Cascade based ear detector.

3.3 Ear Landmark Detection

Dlib is a machine learning library written in C++. The proposed Ear Landmark Detector uses Dlib's implementation [2]. An ensemble of regression trees is used to estimate the landmark points on the ear directly from a sparse subset of pixels (~ 400) which are selected via a combination of gradient boosting and a prior probability on the distance between pairs of annotated input pixels. Each regression function in the cascade efficiently estimates the ear shape from an initial estimate (mean ear pose) and the intensities of the sparse set of pixels indexed relative to this initial estimate.

$$\hat{S}^{(t+1)} = \hat{S}^{(t)} + r_t\left(I, \hat{S}^{(t)}\right) \tag{1}$$

where the shape vector $S = \left(x_1^T, x_2^T, ..., x_p^T\right)^T \in \mathbb{R}^{2p}$ denotes the coordinates of all the p landmarks in the image I, r_t is the tth regressor in the cascade and $x_i \in \mathbb{R}^2$ are the (x, y) coordinates of the ith ear landmark. In this case, $p = 8$ is chosen to cover all possibilities to get the accurate anchor point for the jewelry rendering task.

It is an iterative process where the initial estimate of the ear shape is used to predict the landmark points which is then used to predict an update vector for the shape and so on, until convergence. This method starts by using:

(1) A training set of 8-point labeled ear landmarks on an image. These images are manually labeled, specifying the specific (x, y)-coordinates of regions surrounding each important ear feature structure with respect to augmenting jewelry.
(2) Priors, the probability on distance between the pairs of input pixels, calculated internally.

So, given the training data, an ensemble of regression trees are trained using Gradient Boosting which optimizes over the sum of squared error loss to estimate the ear feature positions directly from the pixel intensities themselves (i.e., no "feature extraction" is taking place).

The authors, like the other prior arts, have assumed that the estimated shape lie in a linear subspace, which can be easily discovered by finding the principal components of the training shapes, and have used such regressors which produce predictions in the linear subspace defined by the training shapes only.

Figure 4 highlights the flowchart of the complete ear landmark detection process.

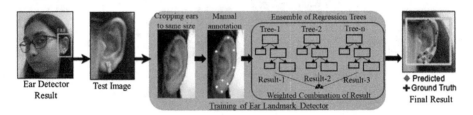

Fig. 4. Training and testing of the ensemble of regression trees based ear landmark detector.

3.4 Augmenting Jewelry in Quasi Real-Time

The 8 points as shown in Fig. 4 "final result" which are predicted using the proposed method of ear landmark detector can be used as anchor points for augmenting jewelry.

4 Results

There is no standard dataset available with 8 annotated points on the ear for training the ear landmark detector for AR jewelry use-case. So, the proposed method was trained on a self-annotated dataset of ∼1000 vectors and evaluated on a dataset of 500 test vectors. Each vector contains an image with at least one face and different levels of yaw. Results are visually compared (due to absence of ground truth) individually by six people, on the aspect of accurate earing augmentation point and then corresponding result is marked as "good" image. The results are benchmarked against Hansley et al. 2017 [7], Zhou et al. 2017 [8] and Ravindran 2014 [9] methods. The final ear landmarks detected from the proposed method as well as from the state-of-the-art solutions are plotted and compared. The results on the standard dataset of IIT-Delhi [12] and iBug "In The Wild" Ear Database [8] and for different yaw angles are plotted as well.

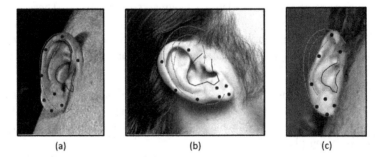

(a) (b) (c)

Fig. 5. Comparison of results of Zhou et al. 2017 [8] (contours) with the proposed method (red points). (Color figure online)

Figure 5 shows that the contour of the ear from the method in [8] is mis-aligned while the proposed method is able to detect the landmarks very accurately.

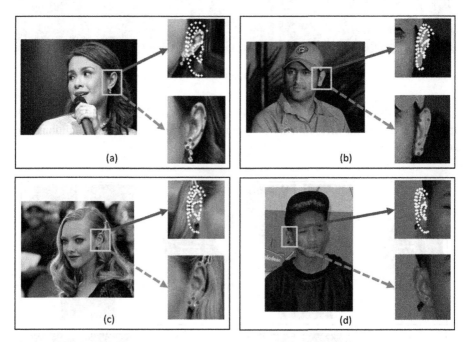

Fig. 6. Comparison of results of Hansley et al. 2017 [7] (red full arrow) with the proposed ear detector (blue dotted arrow) (Color figure online)

Figure 6 shows the comparison between the results of Earnest Hansley et al. 2017 [7] (red full arrow) with the results of the proposed ear landmark detector (blue dotted arrow). The main drawback with this state-of-the-art solution is that it works only for right ear and try to flip the image if left ear is given as input. This methods fails when the ears are asymmetric, while the proposed methods handles this case as it has been extensively ($\sim$600 images) trained on both left and right annotated ears. Figure 6a and b shows the case when a left ear is detected and even after unsuccessful flipping to the right ear, their method tries to anyhow fit the right ear landmarks on the left ear, while the proposed method handles this situation well. Figure 6c shows the case when their method is able to successfully flip the left ear to right and plot the points. Figure 6d shows the case when a right ear is itself given as input. In both Fig. 6c and d, it can be seen that the landmarks alignment on the ear from the proposed method is much accurate than the state-of-the-art method.

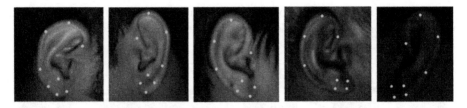

Fig. 7. Results of the proposed approach on IIT-Delhi ear dataset [12]

Figure 7 shows the results of the proposed approach on the IIT-Delhi database [12]. It can be seen that the method handles various ear shapes and lighting conditions very accurately.

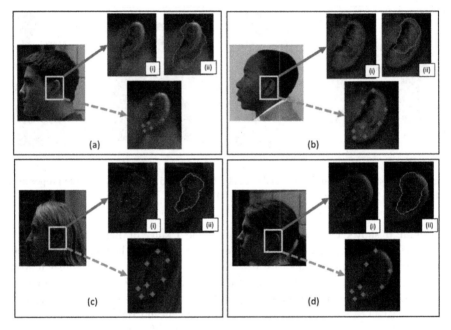

Fig. 8. Comparison of results of Ravindran [9] (red full arrow) with the proposed ear detector (blue dotted arrow) on the UND HID-J2 dataset [11] (Color figure online)

Figure 8 compares the results from the proposed method with the results of Ravindran [9]. The red full arrow from the input images depicts the cases where their method is not able predict the landmark points accurately over the ear (images (i)) and hence end up giving wrong contour around the ear (images (ii)). The blue dotted arrow shows that the proposed method handles these cases very accurately.

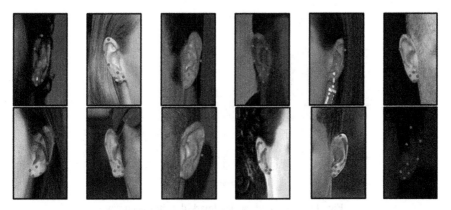

Fig. 9. Results of the proposed approach on iBug "Ears in the Wild" dataset [8]

Figure 9 shows the results of the proposed approach on the iBug "Ears in the Wild" [8] database. This dataset is known to contain the most cluttered backgrounds, occlusions and maximum variation in poses, image size and ear resolution. It can be seen that the proposed method handles these unfavorable conditions also very accurately.

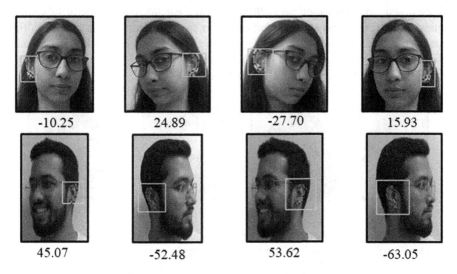

Fig. 10. Images showing detected (red) and ground-truth (blue) landmark points for different yaw angles of face mentioned below each image. (Color figure online)

Figure 10 shows the results of the proposed method on different levels of yaw. The comparison of the predicted landmarks (red) with the ground truth annotations (blue) shows the accuracy of the method.

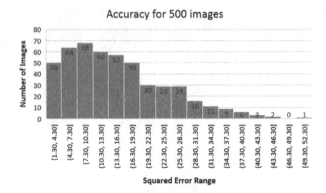

Fig. 11. Proposed ear landmark detector accuracy

Figure 11 also shows the accuracy of the ear landmark detector where the distance between the ground truth pixels and detected landmark pixels has been plotted on the x-axis and number of images on the y-axis. It is to be noted that many images falls on the side of low error only.

Due to absence of any standard dataset with annotated points on the ear, only qualitative results by comparing with other known methods is be presented.

Proposed ear localization takes 90 ms and the ear landmark detector takes 4 ms achieving quasi real-time on Intel platforms. It is calculated by taking average run-time over 500 images. In case of a video feed, with the use of an ear tracker, the accurate results are achieved at the speed of ∼24 frames per second (fps) which validates a smooth jewelry rendering over the ear in camera preview.

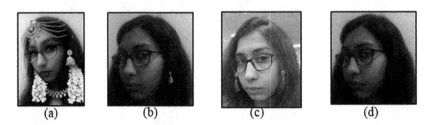

Fig. 12. Comparisons of the current real-life solutions with respect to jewelry augmentation - Snapchat, CaratLane [10] and YouCam Makeup [14] application results.

Figure 12(a–c) shows the performance of the current real-life applications for jewelry augmentation use-case when the ear is occluded with hair. In this case, as shown in Fig. 12a, the widely used application Snapchat as well as the online shopping application of CaratLane in Fig. 12b and makeup application YouCam [14] in Fig. 12c fails to handle occlusion and augments the jewelry on top of the hair. This happens because they interpolate the earlobe point from face feature points. The proposed method in Fig. 12d handles the above drawback accurately since the ear detector does not find an ear and hence no ear landmarks are extracted.

5   Conclusion and Future Work

In this paper, a method which creates a multi model ear detector where the model selection is based on the yaw angle of the face and a regression tree based ear landmark detector which provides accurate 8-point landmark points for jewelry augmentation in quasi real-time has been proposed. All the results shown above proves that the proposed method performs significantly better than the state-of-the-art methods. It accurately predicts the ear landmarks even in poor lighting conditions and pose variations for both the left and right ear of both the genders. It is fast enough to be ported on mobility devices for AR jewelry use-case.

In future, landmark point detection for other body parts like neck, nose, wrist and fingers would be targeted which would aid in the virtual try-on use case for jewelry. We plan to parallelize the landmark point detector for both left and right ear and optimize the proposed method at system level using heterogeneous computing.

References

1. Viola, P., Jones, M.: Rapid object detection using a boosted cascade of simple features. In: IEEE Computer Society Conference on Computer Vision and Pattern Recognition, vol. 1, pp. 511–518 (2001)
2. Kazemi, V., Sullivan, J.: One millisecond face alignment with an ensemble of regression trees. In: IEEE Conference on Computer Vision and Pattern Recognition, vol. 1, pp. 1867–1874 (2014)
3. Pflug, A., Busch, C.: Ear biometrics: a survey of detection, feature extraction and recognition methods. IET Biometrics **1**(2), 114–129 (2012)
4. Abaza, A., Hebert, C., Harrison, M.: Fast learning ear detection for real-time surveillance. In: Fourth IEEE International Conference on Biometrics: Theory Applications and Systems, pp. 1–6 (2010)
5. Islam, S., Davies, R., Bennamoun, M., Mian, A.: Efficient detection and recognition of 3D ears. Int. J. Comput. Vision **95**, 52–73 (2011)
6. Emersic, Z., Struc, V., Peer, P.: Ear recognition: more than a survey. Neurocomputing **255**, 26–39 (2017)
7. Hansley, E.E., Segundo, M.P., Sarkar, S.: Employing fusion of learned and handcrafted features for unconstrained ear recognition. IET Biometrics **7**(3), 215–223 (2017)
8. Zhou, Y., Zaferiou, S.: Deformable models of ears in-the-wild for alignment and recognition. In: 12th IEEE International Conference on Automatic Face & Gesture Recognition (FG 2017), pp 626–633 (2017)
9. Ravindran, S.: Ear Contour Detection and Modeling Using Statistical Shape Models (2014). https://tigerprints.clemson.edu/all_theses/1992
10. CaratLane. https://www.caratlane.com/virtual-try-on/
11. Yan, P., Bowyer, K.W.: Biometric recognition using three-dimensional ear shape. IEEE Trans. Pattern Anal. Mach. Intell. **29**(8), 1297–1308 (2007)
12. Kumar, A., Wu, C.: Automated human identification using ear imaging. Pattern Recogn. **45**(3), 956–968 (2011)
13. FacePlusPlus. https://api-us.faceplusplus.com/facepp/v3/detect
14. YouCam Makeup Application. https://play.google.com/store/apps/details?id=com.cyberlink.youcammakeup&hl=en_IN

Object Recognition

A New Hybrid Architecture
for Real-Time Detection
of Emergency Vehicles

Eshwar Prithvi Jonnadula and Pabitra Mohan Khilar[✉]

Department of Computer Science and Engineering,
National Institute of Technology, Rourkela, India
{714cs1038,pmkhilar}@nitrkl.ac.in

Abstract. VANET is a vital part of wireless networking. Vehicular movement is expanding indefinitely everywhere and is causing terrible problems to daily life. Almost all of the traffic lights now feature a fixed green light sequence and so green light sequence is determined without taking the existence of emergency vehicles into consideration. Consequently emergency vehicles such as ambulances, fire engines, police vehicles etc. are struck in traffic which might cause loss of valuable life and property. In this paper we present a new hybrid architecture for detection of emergency vehicles in real time. This hybrid architecture is based on the mixed features of image processing and machine learning. We also show the percentage decrease in the search space for the processing which results in faster detection of emergency vehicles.

Keywords: VANET · ITS · Emergency vehicles · Image processing · Machine learning

1 Introduction

VANET is the abbreviation of Vehicular Ad-Hoc Network is an integral part of wireless communication. It is a subset of MANET abbreviated as Mobile Ad-Hoc Network. VANET plays a crucial role in smart cities and smart computation. VANET consists of three major components namely Vehicle to Vehicle communication (V2V), Vehicle to Roadside units communication (V2R) and Roadside units to Roadside units communication (R2R) [3]. The primary challenge of VANET is the rapid movement of vehicles. The vehicles interact with others and with the roadside infrastructure units to send and receive relevant information regarding the traffic scenario, emergency information, etc.

In today's world travel and transport has become a standard and non-removable part of life. According to surveys, 40% of people spend 60 min on a trip on average each day. Humans and society are becoming more dependent on transportation every day. This problem of increased vehicles leads to congestion of traffic and increased pollution. In Beijing, China alone 400 thousand

© Springer Nature Singapore Pte Ltd. 2020
N. Nain et al. (Eds.): CVIP 2019, CCIS 1148, pp. 413–422, 2020.
https://doi.org/10.1007/978-981-15-4018-9_37

vehicles are present at the start of the year 2000 and nearly double the number, i.e., 800 thousand vehicles are added to the society at the end of that particular year [18].

The critical feature of traffic management is traffic lights. The traffic light assumes a primary job of smart city and adaptive traffic management. The time interval of green light and continuance of the green light are crucial things which are responsible for adaptive traffic control. In most of the parts of world traffic signals are settled and of a fixed time interval. These fixed traffic signals are however is only suitable for balanced traffic and not suitable for changing traffic conditions. These traffic signals are also programmed in such a way that these do not take the possibility of the presence of emergency vehicles which may come at any instance of time into consideration. In this case, the emergency vehicles get stopped in traffic and might result in loss of life and assets.

The NHTSA - National Highway Traffic Safety Organization collected the vehicle accident information for the USA in the span of the year 1992 to 2011. The data shows that there are about 4500 annual vehicle mishaps which include a lot of emergency vehicles. This report also gives an insight which tells that nearly 300 casualties take place every year despite a good traffic control system [10]. On observing the figures mentioned about the horrific accident, we need to have an intelligent transportation system which needs to take into consideration external factors like weather, traffic blockage and most importantly emergency vehicles as these can save lives [7]. The main difficulty in implementing an above system is the detection of an emergency vehicle which has been detailed addressed in the following section of this paper. This paper presents a new hybrid based method for real-time detection of emergency vehicles and also shows the decrease of search space for the algorithm compared to the previous strategies which are currently being used.

This paper is organized into six sections for a clear understanding of the reader. This paper mainly focuses on ambulances as it is an essential emergency vehicle. So in the future emergency vehicle refers to an ambulance. In the first section, we have explained the basics of VANET and the necessity of identification of emergency vehicles. In the second section, we will be discussing the various works done by the other researchers and authors. In the third section, we will give a basic idea of different types of methods available and useful for the detection of an emergency vehicle. In the fourth section, we see the difficulties involved in the existing practice, and we give a clear explanation of the proposed method for the detection of emergency vehicles. In the fifth section, we show the results of the proposed method discussed in the previous section. In the last and final part, we conclude the paper and give the details of future scope so that it can help other researchers in the right direction.

2   Related Works

For the clear understanding of the problem and how the issue has is solved we first need to study the previous works done by different researchers and authors in this particular field of study. We have done a significant amount of research and gathered the best-related works suitable for our paper for a better understanding of the readers.

- Dangi et al. [4] uses real-time image processing techniques to implement an intelligent traffic controller. Various edge detection algorithms have been used in identifying vehicles.
- Saravanan et al. [15] develops a system that can count the number of vehicles in a particular field of view which is done using an optical sensor. The traffic signal timing is dynamically changed according to the traffic intensity.
- Andronicus et al. [2] use a ripple algorithm to detect an ambulance. This algorithm uses the concept of template matching techniques to compare the input to the existing template.
- Parthasarathi et al. [11] use a new idea of the centroid of lights to detect the ambulance. First, the red and blue colors of the ambulance siren are identified using the segmentation method of image processing. Then the ambulance is recognized if and only if the distance between the centroids of the red and blue part is less than a predefined threshold.
- Sundar et al. [16] uses a basic hardware module called the RFID - Radio Frequency Identification which is installed on every ambulance to identify the presence of the ambulance.
- Placzek et al. [12] makes use of data gathered by GPS - Geographical Position System and pattern matching algorithms to recognize the presence of emergency vehicles.
- Deepa et al. [5] use the basic concepts of computer vision. The emergency vehicles are recognized using the OCR - Optical Character Recognition method.

3   Ambulance Detection

Ambulance detection is the first and primary step in adaptive traffic lights and smart cities. Only when the ambulance is detected, then the signals can change and give preference to the ambulance. There are many methods for detection of ambulances. Here are a few methods mentioned below.

- **RFID-Radio Frequency Identification:** In this method, every ambulance is fitted with an RFID device which has its frequency. When an ambulance passes through a junction, it is recognized by the RFID reader.
- **OCR-Optical Character Recognition:** In this method, we use the fundamental image processing technique of OCR to detect the key works like "AMBULANCE", "EMERGENCY" etc. to recognize the emergency vehicles.
- **Template Matching:** In this method, the ambulance is detected by comparing the vehicles present on the road to a predefined existing template of the ambulance which is obtained by various edge detection methods.

- **Siren Lights Frequency:** In the method, the ambulance is detected by the frequency of the siren light. In many countries, there are fixed standards for the timing, color and frequency of the siren light of an ambulance. These can be exploited for the identification of the ambulance.
- **Centroid of Siren Lights:** In this method, we detect the ambulance if the distance between the centroid of the blue and red light of siren is less than a predefined threshold.

4   Implementation

4.1   Difficulties of Video Processing

The general method for the implementation of the detection of emergency vehicles is video processing. This task involves various methods as mentioned in the previous section. These may always not give the correct results. However, in our case, we need to be very sure as the detection of emergency vehicles is a crucial factor in the smart city and traffic management. So machine learning techniques are used. This process is called activity recognition.

Action recognition task includes the recognizable proof of various actions from video cuts (a succession of 2D outlines) where the action could conceivably be performed all through the whole span of the video [1]. Despite the great achievement of profound learning classification in the picture, advance in models for video classification has been slower. The following are the reasons for the slow and arduous process of action recognition.

1. **Enormous Computational Cost:** A simple two-dimensional convolution for ordering 100 classes has nearly a million variables to learn whereas similar classification when it comes to the three-dimensional structure has nearly 33 million variables to learn. It takes nearly 100 h to train three-dimensional convolution network the standard benchmark dataset of UCF101 [17].
2. **Catching Huge Data:** Action recognition includes catching the spatial-temporal data of the frames of video. Furthermore, this data caught must be made up for the movements and motion of the camera. All the details from very small to the main action needs to be precisely captured to recognize the activity accurately.
3. **Lack of Standard Dataset:** The mainstream and benchmark dataset have been UCF101 and Sports1M for quite a while. To get a perfect structure for the convolution neural network for these datasets can be incredibly costly. However, these datasets only contain a few actions like sports which are not useful for most of the problem in the real world. In our case, we need a dataset of ambulance moving in the streets for the action recognition task. This issue of lack of dataset is added but not completely solved with the kinetic dataset [8].

4.2 Proposed Method

The difficulties of performing a machine learning or deep learning techniques have been mentioned in the above subsection. In our case, we cannot afford such a delay as the detection of an emergency vehicle should be done in real time.

So we propose a two-stage hybrid algorithm which combines the power of machine learning and the simplicity of image processing. This method has adopted due to the lack of video dataset of emergency vehicles on the road. As the dataset is not available, the standard machine learning algorithms are not possible to implement (Fig. 1).

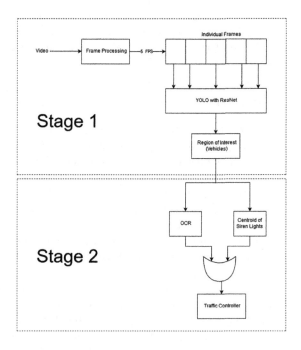

Fig. 1. Architecture of proposed model

Generally, a video consists of 25 to 30 frames per second. Hence the search space for the existing machine learning and deep learning becomes enormous. This results in multiple methods and large computation time.

So we use only five frames per second, i.e. 200 ms for each frame. This process is done to decrease the search space. This process does not result in missing of emergency vehicles as the vehicle cannot escape the camera in a short span of 200 ms.

These five frames are sent to the highly efficient and advanced neural network designed and developed by Google called ResNet which integrated with object detection algorithm called You Only Look Once (YOLO) [14] is used.

This network is used in finding all the vehicles in the five frame. This process is technically called the region of interests.

This region of interests is passed to the modules of image processing in stage 2. Here in our proposed architecture, we use two image processing methods to detect the presence of an emergency vehicle namely Optical Character Recognition and Centroid-based method of the siren bar. The output of these is binary as it only tells if the vehicle is an emergency vehicle or not. In our case, it tells if it an ambulance or a regular vehicle. This output is passed through the binary OR gate. This is done if the any of the image processing modules gives the output as true, i.e. an emergency vehicle is present then the traffic controller will change the traffic lights to give the priority for the emergency vehicle. The OR gate is selected because every time we cannot relay that the image processing modules work correctly in all situations like day or night. So we alert the traffic controller even if any one of the models of image processing says that there is an emergency vehicle present.

4.3   You only Look once

YOLO is state of the art object detection algorithm but requires high-end processing hardware for the training of the dataset. In YOLO, object detection is considered as a regression which takes input as pixels and outputs bounding boxes and probabilities of the object being in different classes. YOLO is completely different from the sliding window method. YOLO analyses the whole pixels in the picture at the time of training so that the prediction of class labels in testing can be much faster as compared to other object detection methods. YOLO makes less number of mistakes in detecting background than a Fast R-CNN because YOLO looks at the entire pixels during the training phase [14].

In YOLO the image is divided into a number of square pieces using grid. Each grid has the duty of detecting the object at its grid center. Each grid gives the output of bounding boxes and the probability of object belonging to different classes. Some of these bounding boxes which have a low Intersection over Union value with the anchor boxes predefined are eliminated. For each output there are five values corresponding to it. X coordinate of the center of bounding box, Y coordinate of the center of bounding box, width of bounding box, height of bounding box and the highest probability of which class the object belongs to.

4.4   ResNet

A Residual Network commonly referred to as ResNet is a type of neural network that uses skip connections to bypass or skip some layers in the neural network. One of the main reason the skip connections are used is the effect of vanishing gradient in a deep neural network. By theses skip connection the activation function value from one layer can be used in the layer which is deep down in the network. This effectively takes into consideration the initial weights of the network even after the vanish gradient problem occurs [6]. These skip connection are responsible for the simplicity of the neural networks by using a less number of layer in training phases. Skip connections reduce the effect of vanishing gradient problem as the data flows through less number of layers bypassing the intermediate layers using these skip connections. So the features are retained even at the end of the network.

5   Results and Analysis

5.1   Results

The above architecture is implemented in Python 3.5. We have taken the pre-trained weights from Darknet [13] used in YOLOv2 network which can identify vehicles on the road. In the first stage of the architecture, the five frames are sent to YOLOv2 with ResNet to get the regions of interest (Figs. 2 and 3).

Fig. 2. Input for YOLO V2 with ResNet   **Fig. 3.** Output with the regions of interest

After Stage 1 we get all objects present in the frame. But we aim to consider the three out of 80 different types of classes of COCO dataset [9] as mentioned below.

1. Car
2. Bus
3. Truck

We crop out this region of interests and send them to stage 2. Here we show the processing of 'bus' object for the understanding of the readers. We send this cropped portion to the OCR module to detect the text 'Ambulance'. Since in some cases the ambulance will be the mirror image we flip the image accordingly so that the character can be recognized easily (Figs. 4 and 5).

Fig. 4. Input for OCR module after flipping   **Fig. 5.** Text recognized by the OCR

The output of the OCR module will be all the text that it recognizes in the image. These can also be garbage texts present in the frame. But we are particularly interested in the word **AMBULANCE**. Once this word is recognized, we send the control information to the traffic controller module.

5.2   Analysis

The primary purpose of this hybrid architecture is to decrease the search space. In this subsection, we do an approximate calculation of search space reduced compared to the simple video processing techniques in the identification of emergency vehicles.

Video Processing

In this method consider we have an HD camera installed at the traffic junction. Generally, the videos have a 25–30 FPS. Let us consider the best case of 25 FPS.
Each frame in single channel = 1280 * 720 pixels
Each frame in all three channels = 1280 * 720 * 3 pixels
Each Second = 1280 * 720 * 3 * 25 pixels
This search space is approximately 70 Million pixels per second.

Proposed Hybrid Architecture

Consider the same camera installed. But in our architecture, we use only 5 FPS.
In Stage 1
Each frame in single channel = 1280 * 720 pixels
Each frame in all three channels = 1280 * 720 *3 pixels
Each Second = 1280 * 720 * 3 * 5 pixels
This search space is approximately 14 Million pixels per second.
In Stage 2
Consider 50% of pixels objects detected.
Let us assume 50% of those objects are cars, buses and trucks. Let the remaining 50% be pedestrians, motorcycles etc.

One module $= 14\,\text{M}*0.5*0.5$

Two modules $= 14\,\text{M}*0.5*0.5*2$

This search space is approximately 7 Million pixels per second.

So the total search space for the hybrid architecture is $14+7 = 21$ Million pixels per second.

As we can see the percentage decrease in search space is nearly 70%.

This comparative analysis is done considering the best case conditions for video processing and average case conditions for the hybrid architecture.

6 Conclusion and Future Scope

6.1 Conclusion

An expanded number of vehicles does not only boosts the response time of emergency vehicles and also hikes the chances of them ending up in accidents.

This paper presents a way to detect the presence of emergency vehicles using a new hybrid algorithm which combines the power of machine learning and the simplicity of image processing. We also saw that the search space of the architecture is decreased.

6.2 Future Scope

The output of both video processing and audio processing can be used together. They can be integrated into a system for a module for intelligent traffic controller. This traffic controller can give priority to the emergency vehicles if detected by the audio and video processing sub modules. This detection of emergency vehicles plays a significant role in smart city development.

References

1. Deep Learning for Videos: A 2018 Guide to Action Recognition. http://blog.qure.ai/notes/deep-learning-for-videos-action-recognition-review
2. Andronicus, F., Maheswaran: Intelligent ambulance detection system. Int. J. Sci. Eng. Technol. Res. (IJSETR) **4**(5), 1462–1466 (2015)
3. Bhoi, S.K., Khilar, P.M.: Vehicular communication: a survey. IET Netw. **3**(3), 204–217 (2013)
4. Dangi, V., Parab, A., Pawar, K., Rathod, S.: Image processing based intelligent traffic controller. Undergrad. Acad. Res. J. (UARJ) **1**(1) (2012)
5. Deepa, Navya, K., Manisha, K., Manu, M., Kshama S: Smart detection of emergency vehicle in traffic. Int. J. Curr. Eng. Sci. Res. (IJCESR) **5**(4), 15–18 (2018)
6. He, K., Zhang, X., Ren, S., Sun, J.: Deep residual learning for image recognition. In: Proceedings of the IEEE Conference on Computer Vision and Pattern Recognition, pp. 770–778 (2016)
7. Jonnadula, E.P., Khilar, P.M.: Comparison of various techniques for emergency vehicle detection using audio processing. Machine Learning, Image Processing, Network Security and Data Sciences, March 2019

8. Kay, W., et al.: The kinetics human action video dataset. arXiv preprint arXiv:1705.06950 (2017)
9. Lin, T.-Y., et al.: Microsoft COCO: common objects in context. In: Fleet, D., Pajdla, T., Schiele, B., Tuytelaars, T. (eds.) ECCV 2014. LNCS, vol. 8693, pp. 740–755. Springer, Cham (2014). https://doi.org/10.1007/978-3-319-10602-1_48
10. Nellore, K., Hancke, G.P.: Traffic management for emergency vehicle priority based on visual sensing. Sensors 16(11), 1892 (2016)
11. Parthasarathi, V., Surya, M., Akshay, B., Siva, K.M., Vasudevan, S.K.: Smart control of traffic signal system using image processing. Indian J. Sci. Technol. 8(16) (2015)
12. Placzek, B., Golosz, J.: The in-town monitoring system for ambulance dispatch centre. arXiv preprint arXiv:1706.03699 (2017)
13. Redmon, J.: Darknet: Open source neural networks in C (2013–2016). http://pjreddie.com/darknet/
14. Redmon, J., Farhadi, A.: YOLO9000: better, faster, stronger. arXiv preprint arXiv:1612.08242 (2016)
15. Saravanan, S.: Implementation of efficient automatic traffic surveillance using digital image processing. In: 2014 IEEE International Conference on Computational Intelligence and Computing Research (ICCIC), pp. 1–4. IEEE (2014)
16. Sundar, R., Hebbar, S., Golla, V.: Implementing intelligent traffic control system for congestion control, ambulance clearance, and stolen vehicle detection. IEEE Sens. J. 15(2), 1109–1113 (2015)
17. Tran, D., Ray, J., Shou, Z., Chang, S.F., Paluri, M.: ConvNet architecture search for spatiotemporal feature learning. arXiv preprint arXiv:1708.05038 (2017)
18. Zhang, J., Wang, F.Y., Wang, K., Lin, W.H., Xu, X., Chen, C.: Data-driven intelligent transportation systems: a survey. IEEE Trans. Intell. Transp. Syst. 12(4), 1624–1639 (2011)

Speed Prediction of Fast Approaching Vehicle Using Moving Camera

Hutesh Kumar Gauttam$^{(\boxtimes)}$ and Ramesh Kumar Mohapatra

National Institute of Technology, Rourkela, India
huteshgauttam@gmail.com, mohapatrark@nitrkl.ac.in

Abstract. Vehicle accidents are increasing day by day as a result of high-speed vehicles on highways so, the speed determination of fast-approaching vehicles is becoming a challenging task with moving camera. Most of the vehicles are driven above the prescribed vehicle speed. On expressways, light motor vehicles are unaware of the speed of the rapid vehicle following to them. So in this paper, an algorithm has been proposed to predict the speed of the fast approaching vehicle by a moving camera to offer better safety. The proposed method comprises of mainly three successive steps, vehicle detection using YOLO (You Only Look Once) algorithm on the video stream, vehicle position tracking over the continuous frame and speed calculation of approaching vehicle using a moving camera. The relative speed is determined using relative distance travelled by vehicle over a number of frames. This proposed algorithm is giving on an average 90% accuracy in speed prediction of approaching vehicles.

Keywords: Speed detection camera · Vehicle detection · Vehicle tracking · Machine learning techniques (YOLO algorithm)

1 Introduction

Advanced Driver Assistance Systems (ADAS) [1] are technologies contributing to increase the awareness of car drivers and automate tasks in their cars. These technologies help the driver in the driving procedure alongside expanding vehicle safety and more generally road safety [2]. ADAS has been widely applied into vehicle detection, vehicle number plate detection, traffic sign recognition, lane detection [3] and lane change assistance on road. Different speed detection instruments are also developed using ADAS technologies [4] to control traffic. One of the most required feature of ADAS nowadays in the traffic scenario to estimate the speed of the heavy and fast vehicles that are approaching nearer from the rear side. As indicated by NCRB (National Crime Records Bureau) [5], almost 33% of over-speeding cases result in fatalities. Unexpectedly, over-speeding is also remaining the single biggest reason for road accident deaths in India with over 36% of all road traffic accident happening exclusively because of this reason [5]. As Fig. 1 demonstrates the increasing road accident statistics [6]

© Springer Nature Singapore Pte Ltd. 2020
N. Nain et al. (Eds.): CVIP 2019, CCIS 1148, pp. 423–431, 2020.
https://doi.org/10.1007/978-981-15-4018-9_38

in India. Road accidents killed 148,000 individuals in 2015 compared to 136,000 in 2011, according to the Accidental Deaths and Suicides in India report released by the NCRB. Road accidents represented 83% of all traffic-related deaths in India and 43% of all accidental deaths in 2015. As of now, ADAS innovation is extensively utilized for different applications generally in video observation frameworks. So video surveillance using ADAS is being used for example tracking moving vehicles and evaluating vehicle speed.

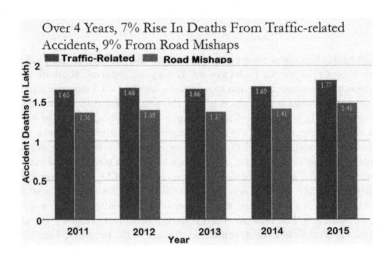

Fig. 1. Increasing road accident statistics related to traffic

The official statistics for road accidents, injuries, and fatalities [7] for the year 2017 is released by the Indian government. As indicated by the latest data, in 2017, a total of 4,64,910 road accidents were accounted in India, claiming 1,47,913 lives and making injuries to 4,70,975 people, which converts into 405 life and 1,290 injuries every day from 1,274 accidents. This additionally implies 16 individuals are killed and another 53 are harmed each hour on Indian roads. Taking into account that these are the formally revealed accidents, there must be a reasonable number that goes unreported all over India. As shown in Fig. 2 the maximum number of accidents for the two-wheeler, accounted for 29% of all fatal road accidents [6] in 2015. To reduce this problem, a new approach has been proposed here for estimating the speed of approaching vehicles while considering the scenario of movement of the camera.

The remainder of this paper is sorted out as pursues. Section 2 exhibits the past work on speed calculation with a static camera. Section 3 presents the YOLO algorithm for real-time object detection. Section 4 depicts the proposed method and Sect. 5 shows experiment results of the different experiment on genuine real-life recordings. Concluding remark is given in Sect. 6.

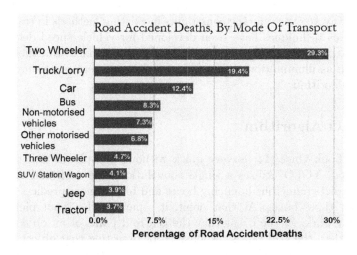

Fig. 2. Road accident statistics in different mode

2   Literature Review

There is a great deal of work has been done on vehicles detection using different image processing algorithms and speed detection of the vehicles with a static camera. Background subtraction algorithm [8] is more often used to extract vehicle in every frame. In background subtraction, background frame and the current video frame is used to detect the vehicle in the current frame. A hybrid algorithm also proposed based on combining an adaptive background subtraction [9] technique with a three-frame differencing algorithm which ratifies the major drawback of using only an adaptive background subtraction algorithm. A detected vehicle needs to be tracked in every frame correctly using some features. Multiple objects tracking, which consists of three successive operations, Object segmentation, Object labeling, and Object centroid extraction [10]. A vehicle coming towards the static camera is tracked in every frame while considering only one lane of the road as ROI (Region of Interest) with starting and end boundary [10]. So, feature extraction is the key point in moving object tracking in the continuous frame. Usually, centroid and histogram of vehicles surrounded by bounding box are considered as the features to observe the position of the vehicle in the continuous frames.

Consideration of area of a vehicle in each frame is also helpful to track the vehicle in continuous frame and speed is calculated with travelled distance by vehicle and time interval as a vehicle enters in ROI and vehicle leaves ROI [11]. Kassem, Kosba, and Youssef [12] also represented a design and analysis of ReVISE including its vehicle identification and speed estimation modules. The recognition module can separate between a vacant road, stationary vehicles, and moving autos dependent on a multi-class SVM approach that utilizes highlights from the RF flag quality. This likewise present two novel speed estimation systems dependent on statistical and curve fitting approaches.

In this way, to overcome the constraints in existing methods in traffic surveillance, various techniques have been developed for vehicle speed determination using image processing while considering the camera is fixed. But the contingent factors such as illumination changes, tree waving, camera noise may affect the output of algorithm.

3 YOLO Algorithm

You Only Look Once [13] is very quick without a doubt in real time object identification. YOLO utilizes a single convolutional network and at the same time, it predicts numerous bouncing boxes and furthermore predicts class probabilities for those boxes. At that point, it separates the input picture into a $S \times S$ framework. In the event that the centroid point of an object falls into a grid cell than the grid cell is responsible for detecting that object. Every cell predicts B bouncing boxes and certainty scores for those boxes. Additionally, in each picture, numerous grid cells don't contain any item. This pushes the confidence scores of those cells towards zero. These certainty scores reflect how sure the model is that the box contains an object and furthermore how precise it supposes the box is that it predicts. Each bounding box comprises of 5 output values that are x and y coordinates of the centroid, width, height of box and confidence. The underlying convolutional layers of the system extract features from the picture while the completely connected layers predict the yield probabilities and coordinates.

4 Proposed Method

The proposed method is used to detect vehicle speed which is approaching towards the moving-camera by following the motion of the vehicle through continuous frames of video. This method mainly consists of 3 steps. Firstly, the video is converted into frames. YOLO object detection algorithm is used for moving vehicle detection in every frame. YOLO gives the centroid of the vehicle in every frame. In the second step, the position of the centroid of the detected vehicle is observed over subsequent frames. In the third step, speed is calculated using a particular distance travelled by the vehicle with the relative velocity of approaching-vehicle and moving-camera and furthermore. This system is helpful for giving awareness about high speed approaching vehicle.

4.1 Pre-processing of Video

Testing videos are recorded using a Logitech camera which is connected to android phone through USB cable. In pre-processing, the frames are extracted from the video. After pre-processing, it gives the total number of frames, the frame rate of the camera. The frame rate of the Logitech camera is 15 frames per second and the frame size of each frame is 1280×720 pixels.

4.2   Vehicle Detection Using YOLO Object Detection

Detection of approaching vehicles precisely towards moving camera in continuous frames of video is a troublesome undertaking. Two primary disadvantages of the background subtraction technique in vehicle detection with the moving camera is observed during this research. First is that when the camera is moving with a speed it is hard to detect moving object in each frame because the background is dynamic. It is changing in every frame ceaselessly.

The second drawback is that when the camera is moving at that time in every frame different types of the roadside objects like waving trees, signboards are detected using this algorithm but the requirement is the only heavy and high-speed vehicles that are approaching towards moving camera. So this methodology is not helpful here to detect the vehicle in each frame.

In this proposed technique, the YOLO Object Detection algorithm is used to detect the approaching vehicle in every frame. It used a trained data set for vehicle detection in every frame [14]. Here Fig. 3 shows the output of the vehicle detection algorithm in which bounding box in drawn on the vehicle and its center denotes the centroid of the vehicle.

Fig. 3. Approaching vehicle detection in frame

4.3   Vehicle Tracking

In this proposed strategy, vehicle tracking is based on tracking the centroid of a detected vehicle in continuously generated frames. This method considers the tracking of a single vehicle in subsequent frames. In tracking of a single vehicle in subsequent frames the coordinate of the centroid in current frame and in next frame belongs to the same vehicle. This method observes the movement of the centroid of the vehicle until the speed is calculated for it.

4.4   Speed Calculation

After tracking the vehicle, the next step is to calculate the speed. These steps explain the methodology for calculating the speed of the approaching vehicle:

Fig. 4. Vehicle entering in ROI

Fig. 5. Vehicle coming outside from ROI

- As the detected vehicle is tracked over a number of frames. So, it is considered only in the scenario when this vehicle is inside the ROI.
- As the vehicle enters in ROI at frame number x as shown in Fig. 4 and comes out at frame number y as shown in Fig. 5. So, total no of the frame is $y-x$. The frame rate of the camera is f. So, time taken to cross the ROI in seconds is

$$t = \frac{y - x}{f} \qquad (1)$$

- To fix the ROI, it is analyzed from a lot of testing video samples that as approaching vehicle reaches towards the moving-camera the motion of the centroid in the continuous frame is not straight. It is in diagonal direction as shown in Figs. 4 and 5.
- Now as the vehicle is going in a diagonal direction so, to define the ROI the starting and end boundaries are parallel to diagonal of frame and almost perpendicular to the motion of the vehicle.
- The starting and end boundaries are at a particular distance with respect to two real points on the road as shown in Fig. 6 as pole1 and pole2 are two real points and these two lines in every frame are passing through two points which are situated apart from a consistent separation.

Fig. 6. ROI boundaries passing through two real point

- As the distance between these two genuine points in actual is D. But the major task is to find the distance travelled by the vehicle when it crosses the ROI with relative motion. This is the Relative Distance travelled in the diagonal direction with the relative velocity of the vehicle with respect to the camera.
- The actual distance travelled by every vehicle while crossing ROI is determined by testing with a vehicle with known speed v and camera speed u and time taken to cross ROI T.

$$Relative\ speed\ of\ vehicle\ and\ camera = v - u \qquad (2)$$

$$time\ to\ cross\ ROI = T \qquad (3)$$

$$Relative\ Distance = (v - u) \times T \qquad (4)$$

- So, this Eq. (3) is the actual distance used in this algorithm to calculate the speed of the vehicle that is approaching towards the camera. So, Relative Predicted Speed (RPS) of the vehicle using Relative Distance from Eq. (4) and time from Eq. (1) is

$$RPS = \frac{Relative\ Distance}{t} \qquad (5)$$

- Relative Predicted Speed shows how fast the vehicle is approaching towards the camera. So, predicted speed (V) of vehicle is

$$Predicted\ Speed\ (V) = RPS\ +\ Camera\ Speed \qquad (6)$$

The mathematical expression of the predicted speed using the Eqs. (4), (5) and (6) is

$$Predicted\ Speed\ (V) = \frac{(v - u) \times T}{t} + u \qquad (7)$$

5   Experiment Result

This algorithm is giving approx 90% accuracy in the speed prediction when it has tested on real-life recorded videos. As the table shows moving camera speed, actual vehicle speed (VS), predicted vehicle speed (PS) using this method and accuracy that shows the comparison between actual speed and predicted speed (Table 1).

$$Accuracy = 100 - \frac{|PS - VS| \times 100}{VS}$$

Table 1. Speed prediction results on real world videos

Serial no.	Camera speed	Vehicle speed	Predicted speed	Accuracy
1	40	60	60.81	98.65
2	40	50	46.73	93.46
3	20	35	35.81	99.9
4	20	40	32	80
5	20	35	28	80

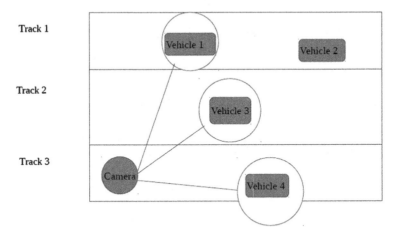

Fig. 7. Multiple vehicle tracking

6   Conclusion

The proposed method represents a speed calculation algorithm for the vehicle when the camera is likewise in movement. For vehicle detection in real time, it uses YOLO algorithm that detects vehicle precisely which is better when contrasted with past image processing techniques. Furthermore, in this technique distance is mapped on the frame as an ROI by calculating it from the real world testing. The proposed method considers the single vehicle since taking care of the situation of multiple vehicles in speed calculation with the moving camera is a complex task. But the ideology mentioned here can help to handle the situation of multiple vehicles as shown in Fig. 7. This contains three different lanes on road and using YOLO all vehicles are detected in every lane, but the significant part is to the selection of vehicles for speed prediction. There is no need to predict the speed of all vehicles in the frame at a time. According to this proposed method, our objective is the front vehicle in every lane who is approaching towards camera because the rear vehicles cannot cross this front vehicle without changing the lane if rear vehicles are faster to this.

References

1. Brookhuis, K.A., De Waard, D., Janssen, W.H.: Behavioural impacts of advanced driver assistance systems-an overview. Eur. J. Transp. Infrastruct. Res. **1**(3), 245–253 (2019)
2. Staubach, M.: Factors correlated with traffic accidents as a basis for evaluating advanced driver assistance systems. Accid. Anal. Prev. **41**(5), 1025–1033 (2009)
3. Geronimo, D., Lopez, A.M., Sappa, A.D., Graf, T.: Survey of pedestrian detection for advanced driver assistance systems. IEEE Trans. Pattern Anal. Mach. Intell. **32**, 1239–1258 (2009)
4. Shaout, A., Colella, D., Awad, S.: Advanced driver assistance systems-past, present and future. In: Seventh International Computer Engineering Conference, pp. 72–82 (2011)
5. https://www.youthkiawaaz.com/2015/12/speed-limits-and-road-accidents-in-india/
6. https://scroll.in/article/826264/three-killed-every-10-minutes-road-accident-deaths-in-india-up-9-in-4-yearsPage
7. https://www.autocarindia.com/industry/road-accidents-in-india-claim-more-than-14-lakh-lives-in2017-410111
8. Ramasamy, B.: A review on vehicle speed detection using image processing. Int. J. Current Eng. Sci. Res. **4**, 23–28 (2017)
9. Ibrahim, O., ElGendy, H., ElShafee, A.M.: Speed detection camera system using image processing techniques on video streams. Int. J. Comput. Electr. Eng. **3**(6), 771–778 (2011)
10. Joshi, S.: Vehicle speed determination using image processing. In: International Workshop on Computational Intelligence (IWCI) (2014)
11. Haque, M.R., Moazzam, M.G., Islam, S., Das, R., Uddin, M.S.: Vehicle speed determination from video streams using image processing. In: International Workshop on Computational Intelligence (IWCI), Dhaka, pp. 252–255 (2016)
12. Kassem, N., Kosba, A.E., Youssef, M.: RF-based vehicle detection and speed estimation. In: IEEE 75th Vehicular Technology Conference (VTC Spring), pp. 1–5 (2012)
13. Redmon, J., Divvala, S., Girshick, R., Farhadi, A.: You only look once: unified, real-time object detection. In: Proceedings of the IEEE Conference on Computer Vision and Pattern Recognition, pp. 779–788 (2016)
14. http://pjreddie.com/yolo/

Improved Performance of Visual Concept Detection in Images Using Bagging Approach with Support Vector Machines

Sanjay M. Patil[1]([✉]) and Kishor K. Bhoyar[2]

[1] Department of Computer Technology,
Yeshwantrao Chavan College of Engineering, Nagpur, India
sannit.99@gmail.com
[2] Department of Information Technology,
Yeshwantrao Chavan College of Engineering, Nagpur, India
kkbhoyar@yahoo.com

Abstract. With rapid advances in imaging devices and internet, millions of images are uploaded on the internet without much information about the image. An efficient method is necessary for detecting the concept of the desired image from this vast collection of images. In this paper, Support Vector Machine (SVM) based architecture is presented to detect concept of a given input image. To enhance the performance of proposed system, a bagging approach is implemented. Color moments, HSV Color Histogram, Grey level co-occurrence matrix, Wavelet Transform and Edge orientation histogram are used for image representation purpose. These low-level feature descriptors are used to train multiple SVM models. The final concept of the query image is obtained by voting from outputs of these multiple models. The proposed system is evaluated on Wang's Corel 10K. Results of proposed system indicate its improved performance over existing systems.

Keywords: Visual concept detection · Bagging · Feature extraction · SVM

1 Introduction

With the recent advances in the imaging devices, telecommunication and mobile computing, an enormous amount of images are being generated by people. Everyday large numbers of images/datasets are uploaded on the internet. Visual concept detection is an effective technique to manage and retrieve these images. The visual content in images and its accompanying metadata are used to detect the concepts [1]. Humans can easily understand the content of images, but search engines have a limited ability to recognize the image or scene content unless the images are tagged. To detect a concept, the low-level features extracted from visual data are mapped to high-level semantics. Huge amount of training data is required for learning a classifier model [2].

Various concept detection techniques exist for different applications. Content Based Image Retrieval (CBIR) systems search or retrieve images from large repositories based on their visual contents. Generally, these systems represent some visual properties of an image in terms of feature vectors. A feature database stores the features of all images in

© Springer Nature Singapore Pte Ltd. 2020
N. Nain et al. (Eds.): CVIP 2019, CCIS 1148, pp. 432–442, 2020.
https://doi.org/10.1007/978-981-15-4018-9_39

the dataset. A query image is searched in the dataset by using some feature matching method. A flaw exists in these systems. If the concepts in dataset are similar e.g. "Sun" and "Fire", they will have similar visual features which leads to misclassification of concepts. This is known as semantic gap between low level image features and high level image semantics. There is a need to bridge this gap by using novel classifiers that result in intelligent concept detection systems.

In this paper, a SVM based framework for concept detection is proposed. First, SVM is trained using different sizes of training dataset and having different combinations of color, texture and shape features. The aim of this system is to detect concept of the query image using the best performing combinations of features, and using a bagging approach. The rest of the paper is organized as follows: Sect. 2 gives a brief overview of previous approaches to concept detection. Section 3 describes various image descriptors used in this paper as well as the summary of the dataset used. It also explains the methodology used for classification. Section 4 presents experimental results and analysis. Section 5 concludes the paper.

2 Related Work

Concept detection of an image is a useful and a very challenging task. Lots of research is being carried out in this area. With the rapid growth in visual information, the need of innovative methods of searching it increases. Researchers have explored various methods of concept detection. Machine learning approaches are widely used for concept detection in images. The two stages in machine learning are training and testing. In training stage, the classifier is trained to learn the various concepts using features of the training images. In testing stage, the trained classifier decides about the concept of test image.

Support vector machine is an effective supervised classifier, especially when the numbers of training images are small [3]. SVM uses kernel mapping and hence can classify both linear and non-linear data. The main advantage of SVM as compared to other classifiers is it finds maximum distance between concepts to achieve optimal class boundaries. It is mainly applied in classification areas such as concept detection, image annotation, objects recognition and text classification [4–9]. An SVM classifier builds a hyperplane for separating images on the basis of a training dataset. An SVM carries out binary classification. However, concept detection system needs multiclass classification which is implemented by training separate SVMs for each individual concept to give the probability value. Finally the outputs of each individual SVM are fused to get the concept of a query image.

Chapelle et al. [5] trained different individual SVMs for different image concepts. HSV histogram feature is used to train the SVM. Images of a particular concept used for training individual SVMs are considered as positive samples. The images of other classes are considered as negative samples. This classification technique is called as "one vs all". For detecting concept of a query image, the features of query image are given as input to all individual SVMs. The concept given by SVM with maximum score is assigned to the query image. Shi et al. [6] first applied k-means algorithm to segment a image and then train 23 SVMs for learning 23 different concepts. SVM are used simultaneously for segmentation and classification [7]. Some of the previous works use multi-level frameworks for classification. In these frameworks, first the individual classifier training is carried out. The decisions of subsequent levels are then fused. Goh et al. [8] classified images using a 3-level approach. Each classifier is trained using different subset of training dataset. A similar framework is presented in Qi and Han [9] where the method for fusing the decisions is different.

Thai et al. [10] developed combined model using ANN and SVM for Roman number image classification. First ANN classifier is trained on input feature vector sub images and then SVM is applied on top of ANN classifier. Tian [11] used HSV histogram globally and mean standard deviation, skewness of color moment features, Gabor wavelet texture features and edge orientation histogram on block wise image partitioned into four blocks using multiple kernels of SVM. Laib et al. [12] used weighted combination of color, texture and points of interest to extract features and then applied SVM classifier to detect the image class for CBIR. Janwe et al. [13] proposed work on multi-class SVM trained on a set of low-level visual features with small size.

The existence of semantic gap [14] between the low-level visual features and the human interpretation of the visual information is one of the major issues in concept detection models. Hence there is a requirement of image concept detection model which can bridge this semantic gap by extracting high-level representations from raw image pixels.

3   Materials and Proposed Methodology

Every concept detection model must extract suitable features and use a learning mechanism to detect the concept of the input image. This section outlines the dataset used in our experiment, the individual features and a novel approach used for classification.

3.1   Dataset Used

We have used Wang's Corel 10K [15]. These datasets has 100 classes with 100 images in each class. The images have size of either 384×256 or 256×384. Corel includes images from various classes like natural scenes, artificial objects, etc. Some of the classes defined are mountain, flower, tree, motorcycle, car, bus, African, light house and gun. Figure 1 shows sample images of these classes.

3.2   Image Descriptors

For efficient concept detection appropriate feature selection and extraction from images is very important. The various low-level image descriptors used in this paper are color, texture and edges.

Color Features
Color feature is one of the most widely used low level features. It is an important measure of human visual perception for differentiating and recognizing visual information. It is independent of the rotation and zoom of image. The color feature used in our experiment is:

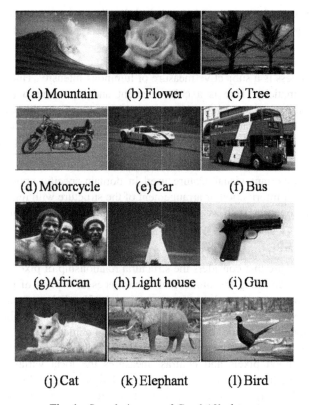

(a) Mountain   (b) Flower   (c) Tree

(d) Motorcycle   (e) Car   (f) Bus

(g)African   (h) Light house   (i) Gun

(j) Cat   (k) Elephant   (l) Bird

Fig. 1. Sample images of Corel 10k dataset

(a) Color Moments: Color moments are measures that can be used to distinguish images based on color. Equations 1, 2 and 3 shows the mean, variance and standard deviation moments for each channel of a color space where i is the index of each channel, I_{ij} is the value of the jth pixel in channel i and N is a total number of image pixels.

$$E_i = \frac{1}{N} \sum_{i=0}^{N} I_{ij} \tag{1}$$

$$\sigma_i = \sqrt{\left(\frac{1}{N} \left(\sum_{j=1}^{N} (I_{ij} - E_i)^2\right)\right)} \tag{2}$$

$$S_i = \sqrt[3]{\left(\frac{1}{N} \sum_{j=1}^{N} (I_{ij} - E_i)^3\right)} \tag{3}$$

(b) Color Histogram: Ahistogram is a graphical representation of the tonal distribution in a digital image. The number of bits used to represent each pixel of an image decides the total number of grey levels in histogram. Color histogram [9] is the simplest and

most common way of expressing the statistical distribution of colors and the essential tone of an image. The color histogram is invariant to translation and rotation of the imaging axis.

Human visual system characterizes a color image by its brightness and chromaticity. Brightness is a subjective measure of luminous intensity. Hue and Saturation define the chromaticity. Hue is a color element and represents a dominant color. Saturation is an expression of the degree to which white light dilutes a pure color. The HSV model is motivated by the human visual system as it better describes a color image than the RGB model [16].

Texture Features

Texture is an important visual feature used in domain-specific applications. It is a repeated pattern of information or arrangement of the structure with regular intervals. It quantifies the properties such as smoothness, coarseness and regularity in an image. The texture feature used in our system is:

(a) Gray Level Co-occurrence Matrices (GLCM): GLCM is a statistical method of characterizing texture that considers the structural relationship of pixels in an image. It is constructed by counting the number of occurrences of a pairs of pixel with same values in an image, at a given displacement and angle. Haralick et al. in [17] proposed statistics such as energy, contrast, correlation and homogeneity for obtaining texture features from the GLCM. The displacement of 2 and 4 angles [0°, 45°, 90°, and 135°] are considered. Correlation measures the probability of occurrence of grey levels among neighborhood pixels and contrast measures the local variations in GLCM. Homogeneity measures how close are the distribution of GLCM elements to the GLCM diagonal. Energy provides the sum of squared elements in the GLCM. The dimension of GLCM feature vector is 24.

(b) Wavelet Transform: The wavelet transform is one of the current popular feature extraction methods used in texture classification. The wavelet transform is able to decorrelate the data and provides orientation sensitive information which is vital in texture analysis. It uses wavelet decomposition to significantly reduce the computational complexity and enhance the classification rate.

Wavelet Transform decomposes the image into a series of high pass and low pass bands and extracts directional details that capture horizontal, vertical and the diagonal activity. Since lower spatial frequencies of an image are more significant for the image's characteristics than higher spatial frequencies, further filtering of the approximation is useful. At each level of decomposition for HL, LH and HH sub-bands, energy, mean and standard deviation can be calculated. Figure 2 shows second level of decomposition of discrete wavelet transform.

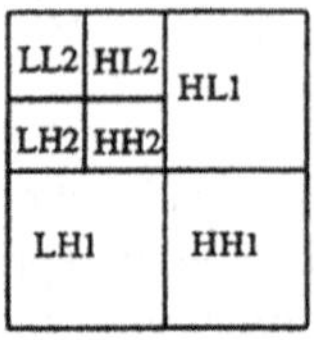

Fig. 2. Second level decomposition of discrete wavelet transform

Shape Features
Shape is an important basic visual features used to describe image content. A commonly used shape feature is edge. Human eyes are known to be sensitive to edge features for image perception.

(a) Edge Orientation Histogram: The edge histogram descriptor consists of the distribution of local edges in the image. The histogram of the directions of the gradients of the edges is build. It is used to find similarity of two images by comparing their edge information. Edge detection is carried out using canny edge detector and the number of edge pixels in five directions i.e. vertical, horizontal, two diagonals and non-directional are counted [18]. Table 1 presents the feature set used in the implementation.

Table 1. Feature set used in the experiment

Feature	Feature description	Dimension
Color Moments	Low order moments (mean and standard deviation)	6
HSV Color Histogram (HSV)	Each of H, S and V channel is quantized to $8 \times 2 \times 2$ bins respectively	32
Gray Level Co-occurrence Matrices (GLCM)	Energy, contrast, correlation and homogeneity	16
Wavelet Transform (WT)	Mean square energy and standard deviation	40
Edge Orientation Histogram (EOH)	Histogram of the 5 directions of the gradients of the edges	5

3.3   Proposed Methodology

Figure 3 shows the block diagram of the proposed concept detection system using SVM. SVMs are used for classification as they are capable of handling large input vector. The proposed model is trained using different combinations of training and testing images as discussed above. In the training process, first low-level features of images are extracted. Color moments, HSV Color Histogram, Grey level co-occurrence matrix, Wavelet Transform and Edge orientation histogram are the features used in our experimentation. Different combinations of features are used to train the SVM classifier. These features are normalized using the standard Min-Max method. Normalization is carried out using data normalization which scales data in the range of [0, 1].

Equation (4) shows the min-max normalization functions [19], where X_{min} and X_{max} are the minimum and is the maximum value of X. The normalized features are now concatenated. The length of the feature vector is 99. These features are used to train the SVM classifier.

$$X' = \frac{X - X_{min}}{X_{max} - X_{min}} \tag{4}$$

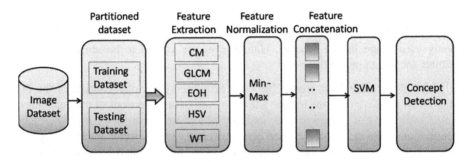

Fig. 3. Block diagram of the proposed SVM system

Concept detection is implemented in two stages, training and testing. The experiments are carried out using three cases of training and testing images. In case 1, first 70 images of each class are used for training and last 30 images for testing. In case 2, last 70 images of each class are used for training and first 30 images for testing, and in case 3, first 60 images of each class are used for training and last 40 images for testing.

Building SVM Classifier
The visual concept detection in images is implemented using LIBSVM-3.23 package and One-Versus-All (OVA) technique for multi-class classification is adopted. The steps for building SVM classifier are as follows:

a. Scaling: conduct simple scaling of the training and test dataset feature vectors.
b. Selection of proper kernel function: e.g. Radial basis function.
c. Parameter tuning: find best parameters C and γ using cross-validation.
d. Training: train the model using best combination of C and γ.
e. Testing: predicting the class of query image.

Parameter Selection
Hsu et al. [20] and Ben-Hur et al. [21] discuss the optimal SVM parameter search for the kernel parameter γ and the SVM penalty factor C. These parameters are generally optimized using grid search and cross validation methods. A coarse-to-fine grid search starts with exponentially growing sequences for γ $(2^{-5}, 2^{-4} \ldots 2^5)$ and C $(2^{-15}, 2^{-14} \ldots 2^3)$. A SVM is trained and validated for each parameter combination. In our experiment the parameters are fine tuned to C = 32 and $\gamma = 0.011$.

3.4   Visual Concept Detection Based on Bagging Approach

In order to test the stability of our proposed system, we have implemented the approach of bagging on Corel 10K dataset using the best performing features of our proposed method. Bagging stands for Bootstrap Aggregation and is an ensemble learning method. The algorithm used for implementation of bagging is as follows:

Algorithm: Bagging approach for Visual concept detection

Input: Image dataset with n images per *m* classes.

Output: Concept of query image.

1. The image feature dataset is divided into two parts for training and testing. 80% of samples of each class are used for training and remaining 20% samples are used for testing.
2. Five separate SVM models (M_1 to M_5) are trained by randomly selecting 60% samples from training dataset with replacement.
3. For each test image, extract features, normalize and concatenate them to form a single feature vector.
4. The test image feature vector is passed through all the five pre-trained SVM models and the output class with maximum voting is considered as the predicted class.

4 Experimental Results and Analysis

In this experiment, LIBSVM-3.23 package with Matlab implementation is used and One-Versus-All method for multi-class classification is adopted. RBF kernel is used. The two parameters, cost (C) and gamma (γ) are fine tuned to C = 32 and $\gamma = 0.011$. The performance of the SVM model is evaluated by training the model using different combination of low-level features of training images on Corel 10K dataset. Three combinations of training and testing dataset are demonstrated. In case 1, first 70% images of each class are used for training and last 30% images for testing. In case 2, last 70% images of each class are used for training and first 30% images for testing, and in case 3, first 60% images of each class are used for training and last 40% images for testing.

The performance of concept detection is evaluated by computing accuracy, precision, recall and F-score. Tables 2, 3 and 4 shows the results for case 1, 2 and 3 respectively using different Feature Combinations (FC Column in Tables 2, 3, and 4) of low-level features on Corel 10K dataset. The number of concepts in this dataset is 100. There is lot of variation in these concepts. The best performance (please refer Table 3, row one) is achieved by using CM, HSV Histogram, WT, GLCM and EOH features (Accuracy = 99.22%, Precision = 0.5974, Recall = 0.6090 and F-score = 0.5963).

Table 2. Results of case 1 on Corel 10k dataset for various feature combinations

FC#	Features					ACC	PR	RC	F-score
	CM	HSV	WT	GLCM	EOH				
1.	√	√	√	√	√	**0.9915**	**0.5738**	**0.5773**	**0.5681**
2.	√	—	—	√	√	0.9880	0.3268	0.4007	0.3600
3.	√	—	√	—	√	0.9894	0.4455	0.4683	0.4398
4.	—	√	√	√	—	0.9908	0.5329	0.5417	0.5301
5.	—	√	—	√	√	0.9891	0.4391	0.4567	0.4350
6.	√	√	√	√	—	0.9912	0.5502	0.5577	0.5467
7.	√	√	√	—	√	0.9911	0.5491	0.5530	0.5429
8.	√	√	√	—	—	0.9904	0.5134	0.5200	0.5082
9.	—	√	√	—	√	0.9908	0.5359	0.5393	0.5282

*— indicates the feature is not used in the experiment.

Table 3. Results of case 2 on Corel 10k dataset for various feature combinations

FC#	Features					ACC	PR	RC	F-score
	CM	HSV	WT	GLCM	EOH				
1.	√	√	√	√	√	**0.9922**	**0.5974**	**0.6090**	**0.5963**
2.	√	—	—	√	√	0.9886	0.3549	0.4313	0.3894
3.	√	—	√	—	√	0.9894	0.4114	0.4700	0.4388
4.	—	√	√	√	—	0.9914	0.5576	0.5720	0.5592
5.	—	√	—	√	√	0.9898	0.4680	0.4897	0.4658
6.	√	√	√	√	—	0.9919	0.5818	0.5943	0.5820
7.	√	√	√	—	√	0.9916	0.5643	0.5790	0.5642
8.	√	√	√	—	—	0.9912	0.5411	0.5577	0.5416
9.	—	√	√	—	√	0.9912	0.5484	0.5610	0.5470

Table 4. Results of case 3 on Corel 10k dataset for various feature combinations

FC#	Features					ACC	PR	RC	F-score
	CM	HSV	WT	GLCM	EOH				
1.	√	√	√	√	√	**0.9910**	**0.5455**	**0.5523**	**0.5410**
2.	√	—	—	√	√	0.9879	0.3283	0.3955	0.3588
3.	√	—	√	—	√	0.9887	0.3799	0.4373	0.4066
4.	—	√	√	√	—	0.9904	0.5092	0.5217	0.5072
5.	—	√	—	√	√	0.9889	0.4289	0.4443	0.4227
6.	√	√	√	√	—	0.9907	0.5237	0.5345	0.5211
7.	√	√	√	—	√	0.9906	0.5183	0.5298	0.5151
8.	√	√	√	—	—	0.9900	0.4854	0.5000	0.4839
9.	—	√	√	—	√	0.9902	0.4975	0.5112	0.4945

4.1 Experimental Results Using Bagging

Table 5 shows the results of bagging tested on Corel 10K dataset with 100 classes. It is observed that the results of bagging are improved as compared to our proposed method without bagging. The accuracy of bagged SVM classifier is better than individual SVM classifiers.

Table 5. Results of bagging on Corel 10k dataset

Model no.	Precision	Recall	F-score	Accuracy (%)
M1	0.6387	0.6295	0.6199	99.20
M2	0.6351	0.6260	0.6194	99.25
M3	0.6321	0.6305	0.6217	99.26
M4	0.6391	0.6355	0.6275	99.27
M5	0.6442	0.6375	0.6312	99.27
Bagging performance using M1 to M5 (Voting)	**0.6442**	**0.6375**	**0.6310**	**99.27**

5 Conclusion

Visual concept detection in still images is a challenging field of research with an aim to reduce the semantic gap. In this paper, a visual concept detection framework based on feature fusion and bagging is proposed and evaluated on benchmark dataset of Corel 10K. It is observed that the accuracy of the SVM classifier remains nearly the same, over different folds (training and testing data). However, with bagging approach there is a considerable improvement in the overall performance of the SVM in terms of all the parameters (i.e. accuracy, precision, recall, and F-score). These experiments show that, for obtaining optimal model performance, machine learning experiments must consider various factors such as different feature combinations, different proportions of training and testing data and the synergy of multiple models with bagging approach.

References

1. Foschi, P.G., Kolippakkam, D., Liu, H., Mandvikar, A.: Feature extraction for image mining. In: Proceedings of International Workshop on Multimedia Information System, pp. 103–1099 (2002)
2. Tang, S., Zheng, Y.-T., Cao, G., Zhang, Y.D., Li, J.T.: Ensemble learning with LDA topic models for visual concept detection. In: Multimedia A Multidisciplinary Approach to Complex Issues, pp. 175–200 (2012)
3. Vapnik, V.: The Nature of Statistical Learning Theory. Springer, New York (1995). https://doi.org/10.1007/978-1-4757-2440-0
4. Jiang, W., Zavesky, E., Chang, S.-F., Loui, A.: Cross-domain learning methods for high-level visual concept classification. In: ICIP, pp. 161–164 (2008)

5. Chapelle, O., Haffner, P., Vapnik, V.N.: Support vector machines for histogram-based image classification. IEEE Trans. Neural Netw. **10**, 1055–1064 (1999)
6. Shi, R., Feng, H., Chua, T.-S., Lee, C.-H.: An adaptive image content representation and segmentation approach to automatic image annotation. In: Enser, P., Kompatsiaris, Y., O'Connor, N.E., Smeaton, A.F., Smeulders, A.W.M. (eds.) CIVR 2004. LNCS, vol. 3115, pp. 545–554. Springer, Heidelberg (2004). https://doi.org/10.1007/978-3-540-27814-6_64
7. Cusano, C., Ciocca, G., Schettin, R.: Image annotation using SVM. In: Proceedings of the Internet Imaging IV, vol. 5304. SPIE (2004)
8. Goh, K.S., Chang, E.Y., Li, B.: Using one-class and two-class SVMs for multiclass image annotation. IEEE Trans. Knowl. Data Eng. **17**(10), 1333–1346 (2005)
9. Qi, X., Han, Y.: Incorporating multiple SVMs for automatic image annotation. Pattern Recognit. **40**(2), 728–741 (2007)
10. Thai, H., Hai, T.S., Thuy, N.T.: Image classification using support vector machine and artificial neural network. Int. J. Inf. Technol. Comput. Sci. **4**(5), 32–38 (2012)
11. Tian, D.: Support vector machine for automatic image annotation. Int. J. Hybrid Inf. Technol. **8**(11), 435–446 (2015)
12. Laib, L., Ait-Aoudia, S.: Efficient approach for content based image retrieval using multiple SVM in CBIR. Comput. Sci. Inf. Technol. (2016)
13. Janwe, N.J., Bhoyar, K.K.: Semantic video concept detection using novel mixed-hybrid-fusion approach for multi-label data. Electron. Lett. Comput. Vis. Image Anal. **16**(3), 14–29 (2017)
14. Smeulders, W.M., Worring, M., Santini, S., Gupta, A., Jain, R.: Content-based image retrieval at the end of the early years. IEEE Trans. Pattern Anal. Mach. Intell. **22**(12), 1349–1380 (2000)
15. Li, J., Wang, J.Z.: Automatic linguistic indexing of pictures by a statistical modeling approach. IEEE Trans. Pattern Anal. Mach. Intell. **25**(9), 1075–1088 (2003)
16. Sangamnerkar, G.V., Bhoyar, K.K.: A neural network color classifier in HSV color space. In: International Conference on Industrial Automation and Computing (2014)
17. Haralick, R.M., Shanmugam, K., Dinstein, I.H.: Textural features for image classification. IEEE Trans. Syst. Man Cybern. **3**(6), 610–621 (1973)
18. Park, D.K., Jeon, Y.S., Won, C.S., Park, S.J.: Efficient use of local edge histogram descriptor. In: Proceedings of ACM International workshop on Standards, Interoperability and Practices, Marina del Rey, California, USA, pp. 52–54 (2000)
19. De Marsico, M., Riccio, D.: A new data normalization function for multibiometric contexts: a case study. In: Campilho, A., Kamel, M. (eds.) ICIAR 2008. LNCS, vol. 5112, pp. 1033–1040. Springer, Heidelberg (2008). https://doi.org/10.1007/978-3-540-69812-8_103
20. Hsu, C.-W., Chang, C.-C., Lin, C.-J.: A practical guide to support vector classification. BJU Int. **101**(1), 1396–1400 (2008)
21. Ben-Hur, A., Weston, J.: A user's guide to support vector machines. In: Carugo, O., Eisenhaber, F. (eds.) Data Mining Techniques for the Life Sciences. Humana Press, Totowa (2011)

FaceID: Verification of Face in Selfie and ID Document

Rahul Paliwal, Shalini Yadav$^{(\boxtimes)}$, and Neeta Nain

Malaviya National Institute of Technology Jaipur, Jaipur, India
{2017pcp5102,2018rcp9169,nnain.cse}@mnit.ac.in

Abstract. Various activities in everyday life require us to verify our identity by demonstrating our ID document containing face images, for example, voter ID, passports, driver licence, to human administrators. However, this procedure is reluctant, unreliable and labor comprehensive. An automatic framework for verifying ID record photographs to live face pictures (selfies) progressively and with high precision is required. Cross-domain biometrics is another requirement, which represents several additional challenges, including harsh illumination conditions, pose variations, noise, among others. In this paper, we propose an algorithm to meet this objective. We first extract faces from ID document and selfie using Multi-task Cascaded Convolutional Networks. To extract prominent features from the data, we apply a VGG face model which is a CNN-based transfer learning approach. Finally, we validate the methods using a novel FaceId-Selfie dataset comprising 600 individuals using cosine distance measure. Results show that 74% accuracy is achieved on FaceId-Selfie dataset.

Keywords: Face recognition · Face verification · Document photo · Selfies · Domain shift · CNN-based transfer learning

1 Introduction

Currently, there are new and interesting challenges for facial recognition applications, for the most part, because of digital life: organizations, for example, financial establishments, are enabling clients to make accounts utilizing the internet, without any requirement to go to a physical branch. In this context, the client ID is required and can be used to check credibility. This procedure is generally performed utilizing a photo of the client related to the ID picture. For this situation, a client can take a "selfie" and an ID photo with a commodity mobile phone and send them for authentication. This is an example of facial recognition systems [1], which have become increasingly popular in commercial applications in the last decade. Typically, such security systems can also be combined with other biometric traits, such as fingerprint, iris, and voice for multimodal biometric authentication.

Identity verification plays a vital role in our every day lives. For instance, access to control, security check, and international border crossing verify us at

N. Nain et al. (Eds.): CVIP 2019, CCIS 1148, pp. 443–454, 2020.
https://doi.org/10.1007/978-981-15-4018-9_40

entry level with our ID Documents. A pragmatic and common way to deal with this issue includes matching a person's real-time face image with the face image found in ID document, for example, migration and passport authorities check the passport photo to verify.

Representatives in general stores verify the client's face and ID Proof to investigate person age when the client is acquiring liquor. Examples of ID archive photograph coordinating can be found in various situations. Be that as it may, it is fundamentally operated by officers manually, which is tedious, expensive, and also inclined to administrator mistakes. A study tells that passport officers in Sydney, Australia, demonstrates that even the well-trained officers implement ineffectively in verifying unknown faces to visa photographs, with a 14% false acceptance rate [2]. Hence, an efficient and automated system for checking of ID document photos to live faces or selfie is required.

Various automated cross-domain matching frameworks are already conveyed at international borders. In 2007, Australia government deployed SmartGate [3] at international borders (Fig. 1) due to the increasing number of visitors. To utilize the SmartGate, passengers are required to submit to a machine their e-Passport chips containing their advanced photographs, which is matched with their face pictures utilizing a camera attached at the SmartGate. After checking a traveler existence by face correlation, the entrance is opened for the passenger. Comparative systems are also introduced in the UK(e-Passportgates) [4], and USA (US Automated Passport Control) [5] etc. In China, such automatic frameworks have been utilized in different areas, including railway junctions, for coordinating Chinese ID cards with live faces [6]. Notwithstanding global fringe control, a few organizations [7,8] are using face recognition answers for ID archive verification for online administration.

An overview of the rest of the paper is as follows: Sect. 2 describes related work, in Sect. 3 we present dataset used in this paper. Section 4 we describe our proposed method for ID-Selfie matching. Section 5 deals with experimental analysis and discussion. The conclusions, and future directions are discussed in Sect. 6.

(a)   (b)   (c)   (d)

Fig. 1. Examples of automated Face-Id matching systems deployed at international borders: (a) SmartGate (Australia) [3], (b) e-Passport gate (UK) [4], (c) Automated Passport Control (US) [5] and (d) ID cards gate (China) [6]

2   Related Work

A conventional face recognition framework includes three basic steps: face detection, feature extraction, and recognition. Face identification techniques typically utilize Haar-like highlights, similar to the ones proposed by Viola and Jones [9], which are efficient and have high recognition rates. A wide range of techniques are commonly applied in the literature for the feature extraction step. For this purpose, there are mainly two different approaches: handcrafted feature extraction and automatic feature learning. Handcrafted descriptors are created by human specialists, requiring great effort and knowledge to develop appropriate features to describe the image characteristics. Examples of these features applied in the face recognition context are Local Binary Patterns (LBP) [10] and Discrete Cosine Transform (DCT) [11]. In the automatic feature learning approach, the more distinctive features are determined directly from data [12]. Examples of features learned from data in the face recognition scenario are the ones extracted using Convolutional Neural Networks (CNN) [13] and Deep Belief Networks (DBN) [14].

The last part of a face recognition framework involves the actual recognition step, in which machine learning techniques can be connected to order the extracted features, aiming at recognizing the users. For this situation, there are two principle use situations: identification and verification.

The issue of cross-domain face recognition represents various difficulties which are not quite the same as generic face recognition. For a generic face recognition system, the significant issues are because of the pose, illumination, and expression (PIE) changes. Instead, in ID-selfie matching, we compare a digital or scanned document picture to a selfie or live face. Assuming that the client is cooperative, in both photographs are obtained under obliged situations, and extensive PIE variations are not present. Rather, (1) due to image compression document photo has poor quality and (2) the immense time difference between the document issue date and the verification date stay essential difficulties. Additionally, since existing face recognition frameworks depend on deep networks, another challenge is the absence of an extensive training dataset (sets of ID photographs and selfies). Regardless of various utilization and related difficulties, there is a lack of research on ID-selfie matching. The following discussion sheds some light on the methods used in the area of cross-domain face verification (Document ID to Selfie Matching) and more.

2.1   ID Document Photo Matching

Starovoitov et al. [15,16] assumes that all face images are frontal appearances without large disparity. Eyes are located using Hough Transforms then face area is cropped and feature maps is computed using gradient maps. For face comparison, the dissimilarity measure rank correlation coefficient is used to compare two maps. The algorithm is same as genric face recognizer but the only difference is it is used for cross domain face matching.

Bourlai et al. [17,18] contemplated ID-selfie matching as a correlation between poor quality face pictures, for example, scanned document photographs, and a digital captured live face pictures. To take out the degradation brought about by scanning, cascading a picture restoration phase before looking at the photographs utilizing a generic face recognizer. Specifically, to classify the degradation type for a given picture they train a classifier, and after that apply filters to reestablish the degraded photos.

Shi et al. [19,20] presents that when classes have very few samples, the problem of under fitting occur and convergence is slow. To resolve this issue, they proposed a method called dynamic weight imprinting (DWI), which update the classifier weights and perform faster convergence. After this, a pair of sibling network is trained to learn face representation.

Folego et al. [21] characterized that the biometric framework is represented with images taken from two different spaces: (1) client "selfie" acquired under real conditions; and (2) a captured image of a driver's license ID. To adapt to the primary difficulties engaged with this cross-domain face verification issue, they investigated various strategies to enhance the images, decreasing the effects of illumination changes and domain-shift, using features from pre-trained Convolutional Neural Network extracted from pre-processed data. These features were then normalized, combined and used as input for different classifiers. Multiple pipeline combinations were tested aiming at verifying the influence on the final result and the main factors that drive the process. It does not use a gallery or create a biometric profile for each from the "selfie" or the ID.

2.2   Heterogeneous Face Recognition

Heterogeneous face recognition (HFR) is an emerging subject that has turned out to be famous in a previous couple of years [22]. It generally performs face recognition amid two distinct domains, containing visible spectrum images (VIS), near infrared images (NIR) [23], thermal infrared images [24], composite representations [25], and so forth. ID-Photo matching can be viewed as an important case of HFR because face matching is performed in cross-domain and require unique approaches. Accordingly, the strategies utilized in HFR could be useful for the ID-photo issue. Most strategies for HFR can be classified into: synthesis-based methods and discriminant feature-based methods. In the Synthesis-based techniques, images are transformed from one methodology into other [25–28]. While in discriminant feature based strategies either manually plan a methodology invariant visual descriptor or collecting a large number of features from the training set so that images from various domains can be mapped into a shared feature space [29–31].

3   Datasets

In this section we briefly describe the dataset which we used in the proposed approach. Some of the example images of the dataset are shown in (Fig. 2). Due to privacy concerns, the private FaceID-Selfie datset is not released publicly.

3.1   CFPW Dataset

CFPW dataset [32] is a public domain dataset, aims to separate the factor of pose variation in terms of extreme poses like profile, where many features are occluded, along with other "in the wild" variations. This datset contains pictures of 500 celebrities (10 frontal and 4 profile images per person). We use this dataset to train our base network. We cropped pictures to 224×224.

3.2   Public IVS Dataset

Zhu et al. [33] released this dataset for experiment on ID-selfie matching systems. The dataset is designed by collecting ID photos and live face photos of Chinese personalities from web. The dataset is a collection of $1,262$ identities and $5,503$ images (one ID photo and 1 to 10 selfies). This dataset is not purely an ID-selfie dataset, as all ID pictures are not taken from original ID documents. Also it contains pictures which have more than one faces. We cleaned it and use 1260 identities.

3.3   Private FaceId-Selfie Dataset

For experiments, we also created a private FaceID-Selfie dataset to evaluate our developed system. ID Documents is sensitive data so we cannot release data in the public domain. To our knowledge, there is no such dataset available in public domain. It is a collection of Indian faces, containing 160 individuals. The dataset contains id card pic and their corresponding selfie images (Indian Face). Each person has only one ID document photo (i.e. Aadhar card, College id card etc.) and 2 to 5 selfies. The age group is 20–30 years old. The images were captured using commodity 20 megapixel smart phone at a resolution of 1080×2340 pixels. The images were captured in day time from $12:00$ noon to $4:00$ PM in a room with regular lighting conditions in a single session.

3.4   Private Profile and Frontal Face

We also, collected 440 persons (in which 300 are child (age group 5–18) faces [34–36]) frontal (2–4 samples) and profile photos (3–5 samples) at different lighting conditions and pose for handling realistic illumination challenges in face recognition. This dataset also comprises of Indian faces. These images were captured using Canon 20 mega pixel camera in daylight ($12:00$ to $4:00$ PM) in open environment without any artificial lights. The subject was made to stand against a white wall at a distance of 3 feet from the hand held camera. The data was collected on different days in three sessions.

Our final FaceId-Selfie dataset contains $14,486$ Images of 2260 identities. We separate $11,727$ images for training and kept 2759 Images for validation. In addition, we also separate 100 ID selfie pairs for testing. These 100 pairs data is not mixed with training and validation for real test result analysis. This is separate data collected in a different session which contains ID card pic and

selfie of a person. A Only Indian Faces data is not used. Dataset contains Public
IVS dataset ID photos and live face photos of Chinese personalities from web
and it also use CFPW dataset which have collection of celebrity face from whole
world. Private FaceId-Selfie dataset and Private Profile and frontal face dataset
is collected on different sessions. ID Photo is usually old in compare of selfie
photo and frontal face data is collected in 3 sessions (gap of 1 week in each
sessions) and light intensity is also different in every session due to different
time.

Fig. 2. Image examples from (a) CFPW, (b) Public IvS, (c) private FaceId-Selfie and
(d) private Frontal-Profile dataset used for experimental analysis.

4   Methodology

Here, we present the workflow shown in (Fig. 3) used to evaluate different
approaches for each step of a ID-Selfie matching system.

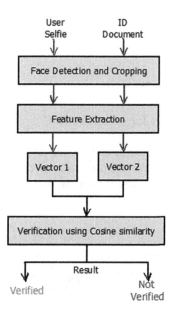

Fig. 3. Workflow of the proposed Faceid-selfie matching methodology.

4.1 Face Detection

As face detection is not our primary task, we adopted Multi-task Cascaded Convolutional Networks (MTCNN) [37] for detecting face. MTCNN is a cascaded
multi-task framework which has three stages: Proposal Network (P-Net), Refine
Network (R-Net) and Output Network (O-Net). P-Net uses non-maximum suppression (NMS) and bounding box regression to refine the output. R-Net further
rejects a large number of false candidates. O-Net gives final output of detected
face. After detecting the face we crop the face image to 224×224 pixels shown
in (Fig. 2(C)).

4.2 Feature Extraction

The output of first phase is a cropped image of size 224×224 which is further
used for feature extraction. Here, We use CNN-based transfer learning approach
to extract complex and prominent features from images. We compute descriptors
using VGG-Face Model [38]. VGG-Face is originally trained on IMDB Dataset
of 2.6 million face images of $2,622$ different identities. It has 16 weight layers and
138 million parameters, which makes it suitable for extracting complex features
from image. We modified and replaced last three layers of VGG-Face according to
our specifications to cater to 2260 classes as shown in (Fig. 4). We add 2260×1
convolution layer, followed by a 2260 Flatten layer, with Softmax activation
function (Eq. 1) at output layer.

$$f(s)_i = \frac{e^{s_i}}{\sum_j^C e^{s_j}} \tag{1}$$

where
$f(s)_i$: softmax function
C: total number of classes
s_j: scores inferred by the net for each class in C

Finally, we trained this modified model on our dataset. Our input layer accepts 224×224 dimension images. We use previous layer of the output layer for representation. It gives 2260×1 dimensions feature vector to compute cosine distance.

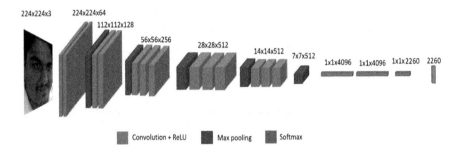

Fig. 4. Proposed modified VGG-Face CNN network.

4.3   Verification

We generate two feature vectors one from ID Face and another from Selfie face and send it for verification. Based on distance between these vector representations, we decide whether both pictures are related to the same person or not. There are two common ways to find the distance of two vectors: Cosine distance and Euclidean distance. Here we use Cosine distance using Eq. 3 for comparing both feature vectors as it we are comparing two sparse vectors in high dimensions. Cosine similarity is only affected by the terms the two vectors have in common, whereas Euclidean distance has a term for every dimension which is non-zero in either vector. Cosine thus has some meaningful semantics for ranking similar vectors, also it handles high dimensional vectors (images) well.

$$\cos(\mathbf{A}, \mathbf{B}) = \frac{\mathbf{A}\mathbf{B}}{\|\mathbf{A}\|\|\mathbf{B}\|} = \frac{\sum_{i=1}^{n} \mathbf{A}_i \mathbf{B}_i}{\sqrt{\sum_{i=1}^{n} (\mathbf{A}_i)^2} \sqrt{\sum_{i=1}^{n} (\mathbf{B}_i)^2}} \tag{2}$$

$$cosine\ distance = 1 - cosine\ similarity \tag{3}$$

5   Experimental Setup

All experiments are performed using Keras (Tenserflow Backend). Since, the FaceID-Selfie dataset is too small, the model trained from scratch overfits heavily and performs very poorly. Similar overfit was observed even when we were trying to train a smaller network from scratch. In comparison, the base model (VGG-Face) pre-trained on IMDB Dataset perform much better even before fine-tuning. This confirms that the features learned by the neural networks are transferable and are used for developing domain specific matchers with small dataset. So we use pretrain VGG-Face model and FineTune it to our requirements and specifications. We also use data augmentation techniques to increase the size of the dataset. We use Batch size 32, learning rate 0.001 and 100 epoch for training, Stochastic gradient descent (SGD) optimizer with momentum 0.9 and categorical cross-entropy loss function as given in Eqs. 4 and 5 respectively.

$$\theta_j = \theta_j - \alpha(\hat{y^i} - y^i)x_j^i \tag{4}$$

where,
x_j^i: training example
y^i: label
α: learning rate

$$CE = -log\left(\frac{e^{s_p}}{\sum_j^C e^{s_j}}\right) \tag{5}$$

where,
S_p: is the CNN score for the positive class.
s_j: scores inferred by the net for each class in C.

The training and testing is performed on single Nvidia Titan V GPU. We Seprate $11,727$ image for training, 2759 images for Validation and 100 pairs of ID selfie data for testing. The testing Data is totally separate from the training and validation data. Cosine distance is used as comparison score for all experiments.

5.1   Experimental Results

Result of verification step is shown in (Fig. 5). We have achieved 74% testing accuracy on our novel dataset. Due to poor quality images in ID document and large time gap between ID card issue date and verification date, there are some false cases. Our model reach 97.62% training accuracy and 72.55% validation accuracy and training loss decrease 7.7231 to 0.0795 and validation loss 7.7223 to 1.7304 as shown in (Fig. 6).

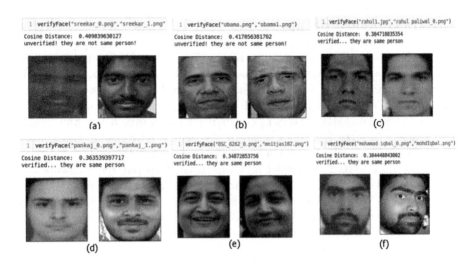

Fig. 5. Face verification results: (a), (b): Unverified faces and (c), (d), (e), (f): Verified faces

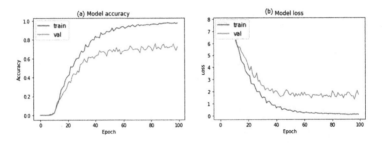

Fig. 6. Proposed model performance (a) Model accuracy (b) Model loss

6   Conclusions and Future Work

In this paper, we explored the face verification issues on a difficult cross-domain setup, looking at images of a "selfie" from a person in real and photos from document IDs. This mechanism can be utilized at numerous practical access control points and situations these days, and has not received enough consideration from academia thus far. To overcome this issue, we propose a new approach, for ID Document to selfie matching. MTCNN model is adopted for extracting face from ID document and selfie. CNN based transfer learning is used for extracting complex features from image. VGG face model is adapted to compute the descriptor, which is compared using Cosine distance measure to check similarity. Experimental results are promising and indicative of further investigation of proposed approach.

The proposed work have following future directions: (i) data collection to increase dataset size for efficient training and better evaluation of the model. (ii)

Authenticity check of ID document of a person and (iii) text retrieval from ID document (i.e. Name, city, date of birth etc.) for better identity comparison of a person.

Acknowledgements. We gratefully acknowledge the support of NVIDIA Corporation with the donation of the Titan Xp GPU used for this research.

References

1. Zhao, W., Chellappa, R., Phillips, P.J., Rosenfeld, A.: Face recognition: a literature survey. ACM Comput. Surv. **35**(4), 399–458 (2003)
2. White, D., Kemp, R.I., Jenkins, R., Matheson, M., Burton, A.M.: Passport officers errors in face matching. PLoS ONE **9**, e103510 (2014)
3. Wikipedia: Australia smartgate, SmartGate (2018). https://en.wikipedia.org/wiki/
4. Wikipedia: ePassport gates (2018). https://en.wikipedia.org/wiki/EPassportgates
5. U.S. Customs and Border Protection: Automated passport control (APC) (2018). https://www.cbp.gov/travel/us-citizens/apc
6. Heng, X.: An Perimeter Security Equipment Co.: What is ID-person matching? (2018). http://www.xjhazj.com/xjhazj/vipdoc/8380983.html
7. Jumio: Netverify ID verification (2018). https://www.jumio.com/trusted-identity/netverify
8. Mitek: Mitek ID verification (2018). https://www.miteksystems.com/mobile-verify
9. Viola, P., Jones, M.: Rapid object detection using a boosted cascade of simple features. In: IEEE Conference on Computer Vision and Pattern Recognition, vol. 1, pp. I-511–I-518 (2001)
10. Ahonen, T., Hadid, A., Pietikainen, M.: Face description with local binary patterns: application to face recognition. IEEE Trans. Pattern Anal. Mach. Intell. **28**(12), 2037–2041 (2006)
11. Hafed, Z.M., Levine, M.D.: Face recognition using the discrete cosine transform. Int. J. Comput. Vis. **43**(3), 167–188 (2001)
12. Chen, D., Cao, X., Wen, F., Sun, J.: Blessing of dimensionality: highdimensional feature and its efficient compression for face verification. In: IEEE Conference on Computer Vision and Pattern Recognition, pp. 3025–3032 (2013)
13. Schroff, F., Kalenichenko, D., Philbin, J.: FaceNet: a unified embedding for face recognition and clustering. In: IEEE Conference on Computer Vision and Pattern Recognition, pp. 815–823 (2015)
14. Huang, G.B., Lee, H., Learned-Miller, E.: Learning hierarchical representations for face verification with convolutional deep belief networks. In: IEEE Conference on Computer Vision and Pattern Recognition, pp. 2518–2525 (2012)
15. Starovoitov, V., Samal, D., Sankur, B.: Matching of faces in camera images and document photographs. In: ICASSP (2000)
16. Starovoitov, V., Samal, D., Briliuk, D.: Three approaches for face recognition. In: International Conference on Pattern Recognition and Image Analysis (2002)
17. Bourlai, T., Ross, A., Jain, A.: On matching digital face images against scanned passport photos. In: IEEE International Conference on Biometrics, Identity and Security (BIDS) (2009)
18. Bourlai, T., Ross, A., Jain, A.K.: Restoring degraded face images: a case study in matching faxed, printed, and scanned photos. IEEE Trans. TIFS **6**, 371–384 (2011)

19. Shi, Y., Jain, A.K.: DocFace: matching ID document photos to selfies. In: BTAS (2018)
20. Shi, Y., Jain, A.K.: Docface+: ID document to selfie* matching. arXiv:1809. 05620v2 (2018)
21. Folego, G., Angeloni, M.A., Stuchi, J.A., Godoy, A., Rocha, A.: Cross-domain face verification: matching ID document and self-portrait photographs. arXiv:1611.05755v1 (2016)
22. Li, S.Z., Jain, A.K. (eds.): Encyclopedia of Biometrics. Springer, Boston (2015). https://doi.org/10.1007/978-1-4899-7488-4
23. Li, S.Z., Chu, R., Liao, S., Zhang, L.: Illumination invariant face recognition using near-infrared images. IEEE Trans. PAMI **87**, 2746–2764 (2007)
24. Choi, J., Hu, S., Young, S.S., Davis, L.S.: Thermal to visible face recognition. In: SPIE DSS-DS107: Biometric Technology for Human Identification IX (2012)
25. Tang, X., Wang, X.: Face photo recognition using sketch. In: ICIP (2002)
26. Wang, X., Tang, X.: Face photo-sketch synthesis and recognition. IEEE Trans. PAMI **31**, 1955–1967 (2008)
27. Liu, Q., Tang, X., Jin, H., Lu, H., Ma, S.: A nonlinear approach for face sketch synthesis and recognition. In: CVPR (2005)
28. Gao, X., Zhong, J., Li, J., Tian, C.: Face sketch synthesis algorithm based on E-HMM and selective ensemble. IEEE Trans. Circuits Syst. Video Technol. **18**, 487–496 (2008)
29. Liao, S., Yi, D., Lei, Z., Qin, R., Li, S.Z.: Heterogeneous face recognition from local structures of normalized appearance. In: Tistarelli, M., Nixon, M.S. (eds.) ICB 2009. LNCS, vol. 5558, pp. 209–218. Springer, Heidelberg (2009). https://doi.org/10.1007/978-3-642-01793-3_22
30. Klare, B., Jain, A.K.: Heterogeneous face recognition: matching NIR to visible light images. In: ICPR. IEEE (2010)
31. Klare, B.F., Jain, A.K.: Heterogeneous face recognition using kernel prototype similarities. IEEE Trans. PAMI **35**, 1410–1422 (2013)
32. Sengupta, S., Cheng, J.C., Castillo, C.D., Patel, V.M., Chellappa, R., Jacobs, D.W.: Frontal to profile face verification in the wild. In: IEEE Conference on Applications of Computer Vision (2016)
33. Zhu, X., et al.: Large-scale bi-sample learning on ID vs. spot face recognition. arXiv:1806.03018 (2018)
34. Deb, D., Nain, N., Jain, A.K.: Longitudinal study of child face recognition. In: International Conference on Biometrics, ICB 2018, Gold Coast, Australia, pp. 225–232 (2018)
35. Chandaliya, P.K., Garg, P., Nain, N.: Retrieval of facial images re-rendered with natural aging effect using child facial image and age. In: 14th International Conference on Signal Image Technology and Internet Based System, Spain, pp. 457–464 (2018)
36. Chandaliya, P.K., Nain, N.: Conditional perceptual adversarial variational autoencoder for age progression and regression on children face. In: 12th IAPR International Conference on Biometrics, Crete, Greece, pp. 200–208 (2019)
37. Zhang, K., Zhang, Z., Li, Z., Qiao, Y.: Joint face detection and alignment using multitask cascaded convolutional networks. IEEE Signal Process. Lett. **23**(10), 1499–1503 (2016)
38. Parkhi, O.M., Vedaldi, A., Zisserman, A.: Deep face recognition. In: British Machine Vision Conference (2015)

Online Handwriting Recognition

A Benchmark Dataset of Online Handwritten Gurmukhi Script Words and Numerals

Harjeet Singh[1]([✉]), R. K. Sharma[2], Rajesh Kumar[2], Karun Verma[2], Ravinder Kumar[2], and Munish Kumar[3]

[1] Chitkara University, Institute of Engineering and Technology, Chitkara University, Punjab, India
`harjeet.singh@chitkara.edu.in, harjeet.research@gmail.com`
[2] Computer Science and Engineering Department, Thapar Institute of Engineering and Technology, Patiala 147004, India
`{rksharma,rakumar,karun.verma,ravinder}@thapar.edu`
[3] Department of Computational Sciences, Maharaja Ranjit Singh Punjab Technical University, Bathinda 151001, India
`munishcse@gmail.com`

Abstract. This paper presents an online handwritten benchmark dataset (OHWR-Gurmukhi) for Gurmukhi script. TIET, Patiala released the unconstrained online handwriting databases, OHWR-GNumerals and OHWR-GScript, which contain isolated strokes samples produced by 190 writers. The OHWR-GNumerals covers 10 stroke classes and OHWR-GScript covers 95 stroke classes to represent the Gurmukhi character set. For data collection, two data sets of Gurmukhi words have been finalized after having a consultation with language experts in order to collect the balanced stroke samples. The preprocessing methods used to prepare these datasets include: size normalization, removing duplicate points, interpolating missing points and re-sampling. The purpose of this benchmark is to create a common platform and make the benchmark dataset publically available for research endeavors in the area of online handwriting recognition. The dataset is available as supplement at https://sites.google.com/view/ohwr-gurmukhi-script/.

Keywords: Online handwriting recognition · Gurmukhi script · Another benchmark dataset

1 Introduction

Online handwriting recognition is an evolving area of pattern recognition. Online handwriting recognition assumes a key role in several human-machine interfaces, including cell phones, smart pads, pen based digital tablets and computers. These devices help us in capturing information with the help of a digital

© Springer Nature Singapore Pte Ltd. 2020
N. Nain et al. (Eds.): CVIP 2019, CCIS 1148, pp. 457–466, 2020.
https://doi.org/10.1007/978-981-15-4018-9_41

pen/stylus. This captured information is stored as co-ordinates (x-, y-values) with progressing time. A sequence of such coordinates is referred as a *stroke* in an online handwriting recognition system. A *stroke* is a sequence of coordinates captured between the two events, namely, pen-down and pen-up with the help of a writing device. A *stroke* is a basic building block in online handwriting recognition systems. A character is formed by combining these strokes and further a word is formed by combining these formed characters. Pen/stylus based interfaces are attractive to users, but the level of handwriting recognition is still disappointing due to the reasons, including and not limited to: variations in handwriting styles, different size of handwriting, and ambiguity in character segmentation. Researchers are therefore challenged to find better handwriting recognition techniques for digital pen-based devices or handheld touch-based smart devices. Moreover, it is also a challenging task to deal with discretized pen trajectory data, arise due to some hardware issues like sensitive surface of a digital device. In contrast, the offline handwriting recognition addresses the problem of recognizing the optical images of handwriting. In other pattern recognition fields, such as speech and optical character recognition, significant progress has been made since large corpora of training and test data are publicly available and public competitions are organized to compare recognition techniques on a fair basis [1,5–8,10,12,13,16]. Khayyat et al. [9] have proposed an effective method of word spotting for Arabic handwritten documents using language models. They have presented experimental results using benchmark CENPARMI dataset. Messaoud et al. [11] have presented their results using benchmarking datasets, namely, DIBCO 2009 and H-DIBCO 2010 and IAM historical dataset. The proposed approach is ranked first when it is applied on DIBCO 2009 dataset and ranked fifth when it is applied on H-DIBCO 2010 dataset. Djeddi et al. [4] have presented a work in ICFHR2016 competition on multi-script writer demographics classification using "QUWI" database. QUWI is a bilingual database which contains writing samples of same individuals in Arabic and English. The competition is aimed at reporting and comparing the latest techniques on these problems under the same experimental settings. Xing and Qiao [17] have proposed a text independent approach to identify the writer for offline handwritten images. In order to extract discriminative features, they employed a deep Convolutional Neural Network (CNN). Experiments are evaluated on two datasets, namely, IAM dataset containing handwritten English text and HWDB dataset containing handwritten Chinese text. Funding agencies have not been interested, so far, in funding a similar effort to collect data for online handwriting recognition and organize competitive tests. We got this opportunity to collect the online handwritten data for Gurmukhi script under the research project "Development of Online Handwriting Recognition system for Indian languages", funded by Technology Development for Indian Languages (TDIL), *DeitY*, MoCIT, Government of India.

Stroke classes (Fig. 1). Zone labels: Upper Zone stroke set, Middle Zone stroke set, Lower Zone stroke set. The following stroke class numbers are shown (each paired with a hand-drawn Gurmukhi stroke symbol):

Upper Zone	Middle Zone						Lower Zone
121	141	156	172	188	204	223	101
122	142	157	173	189	205	224	102
123	143	158	174	190	206	225	103
124	144	159	175	191	207	163	104
125	145	160	176	192	208		
126	146	161	177	193	209		
127	147	162	179	194	210		
128	148	164	180	195	211		
131	149	165	181	196	212		
132	150	166	182	197	214		
133	151	167	183	198	215		
134	152	168	184	200	216		
	153	169	185	201	217		
	154	170	186	202	218		
	155	171	187	203	222		

Fig. 1. Stroke classes for Gurmukhi character set.

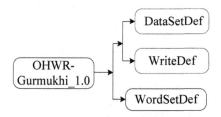

351	352	353	354	355	356	357	358	359	360

Fig. 2. Stroke classes for Gurmukhi numerals.

2 Gurmukhi Script

Gurmukhi is the script used for writing Punjabi language. Punjabi is an Indo-Aryan language spoken by approximately 102 million speakers worldwide. It is the 10th most widely spoken language across the world. The writing structure of Gurmukhi script is cursive and it is written from left to right direction. The strokes for writing characters in Gurmukhi script can be drawn in one of the three horizontal zones, namely, upper, middle and lower zones. The stroke classes (i.e., symbols), used for writing all the Gurmukhi characters (*Consonants, Vowels, and Nasal* [14]) are presented in Fig. 1 and the stroke classes used for writing the Gurmukhi numerals are shown in Fig. 2.

```
OHWR-Gurmukhi_1.0  →  DataSetDef
                   →  WriteDef
                   →  WordSetDef
```

Fig. 3. OHWR-Gurmukhi_1.0 format schema.

```
1  <?xml version="1.0" encoding="UTF-8" standalone="no"?>
2  <OHWRSchema>
3    <dataSetDef>
4      <templateID>GurmukhiVer1</templateID>
5      <language>Punjabi</language>
6      <templateSource>Thapar University</templateSource>
7      <contactName>SMCA-Lab</contactName>
8      <contactEmail>harjeet@thapar.org</contactEmail>
9      <contentDesc>Data consists of text files collected using Tablet PC &amp;
        Desktop</contentDesc>
10   </dataSetDef>
11   <writeDef>
12     <name>Harjeet Singh</name>
13     <uName>usr03</uName>
14     <Password>test</Password>
15     <age>32</age>
16     <gender>Male</gender>
17     <region>Patiala</region>
18     <educationLevel>Post-Graduate</educationLevel>
19     <profession>Office Worker-A</profession>
20     <hand>Right</hand>
21     <deviceType>Tablet PC</deviceType>
22     <index>1024</index>
23     <wordCount>2048</wordCount>
24   </writeDef>
25 </OHWRSchema>
```

Fig. 4. Sub-elements of *dataSetDef* and *writeDef* sections

3   Data Collection and Annotation

The data collection phase is the most critical phase in online handwriting recognition systems. Agrawal et *al.* have discussed the data collection methodology, wherein they have described the data selection and data annotation process [2]. This is important to identify the minimal set of Gurmukhi words that includes all possible stroke classes to form a valid character. This task has been completed with the help of language experts. The experts advised on a word list, consisting of 2,348 words (including Gurmukhi numerals). Further, We divided this list into two sub-list (*i.e.*, 2048 words list and 300 words list). The large words list (*i.e.*, 2048) contains, (i) all the Gurmukhi characters, Vowels, and Nasal symbols, (ii) all the possible combinations of Gurmukhi character with Vowels and Nasal symbols, (iii) Two letters words, (iv) three letter words, (v) two letters words with Vowels and Nasal Symbols, and (vi three letters words with Vowels and Nasal Symbols). We used these lists for data collection. The data has been collected from varied classes of writers, shown in Table 1. The varied classes of writers include, (i) familiar with Gurmukhi script writing, (ii) students studying in schools and colleges at diploma/degree level and (iii) professionals, working in Government offices are selected for data collection from different places in Punjab (a State of India). A total of 190 writers, belonging to different age groups, contributed in data collection. Touch based device, Tablet-PC has been used for capturing the data. Since, writers were not familiar to write words on to the Tablet-PC, therefore we first gave them training about writing on the Tablet-PC then asked to the writer to write in actual. In the data collection process, we did not restrict every writer to write the complete word list in one turn. Our main motive in the data collection was to capture the maximum writing style variations. After collecting data, the annotation of data comes into picture. Annotation means labeling the collected handwritten data in accordance with the designated class-label, depicted in Fig. 1. In this process, the collected data is processed at stroke-level for labeling with the respective stroke-class. A tool has been developed in order to annotate the collected data at stroke-level, depicted in Fig. 5.

Table 1. Writer's information as age group, gender and handedness.

Writers	Age group (10–20)		Age group (20–35)		Age group (35–50)		Total
	Male	Female	Male	Female	Male	Female	
Right handedness	32	23	37	40	25	18	175
Left handedness	3	2	5	2	2	1	15

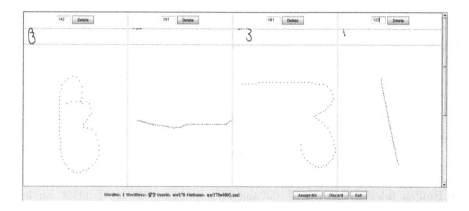

Fig. 5. Data annotation using tool.

4   OHWR Data Format

To facilitate data exchange, a data format, OHWR-Gurmukhi_1.0 has been proposed in this paper, which is unambiguous and easy to comprehend. In addition to this, we have used another format proposed by Belhe et al. [3] for Gurmukhi data collection. Out of 190 writers' data, 140 writers' data is stored in OHWR-Gurmukhi_1.0 format and remaining 50 writers' data is stored in XML based format proposed in [3]. Two signal channels, x and y are used to store the value of x-, y-traces. This information is stored in XML file with respect to stroke sequence number. However, in OHWR-Gurmukhi_1.0 format, there are the provisions for stroke labelling, used at annotation time. Users of this dataset may use

```
1   <?xml version="1.0" encoding="UTF-8" standalone="no"?>
2   <wordSetDef>
3   <wordNo>1</wordNo>
4   <wordDesc> ਉਤੇ </wordDesc>
5   <totalStrokes>4</totalStrokes>
6   <stroke>
7   <strokeId>142</strokeId>
8   <strokeNo>1</strokeNo>
9   <point>
10  <X>267</X>
11  <Y>29</Y>
12  </point>
13  <point>
14  <X>268</X>
15  <Y>29</Y>
16  </point>
17  .
18  .
19  .
20  <point>
21  <X>310</X>
22  <Y>26</Y>
23  </point>
24  </stroke>
25  <stroke>...</stroke>
26  <stroke>...</stroke>
27  <stroke>...</stroke>
28  </wordSetDef>
```

Fig. 6. An example of storing the word sample in the OHWR-Gurmukhi_1.0 format file

her/his own labelling during annotation. Because of the widely differing methods of data collection, a rich annotation is necessary for both the data formats in order to achieve good recognition accuracy.

Our design efforts focused on preparing the format for the dataset. The salient features of this format are:

- Human intelligible without documentation (keywords are explicit English words)
- Easily machine readable
- Compact (few keywords)
- Complete (enough keywords)
- Expandable (one can further enhance it)

5 OHWR-Gurmukhi_1.0 Format Description

To process the collected and annotated handwritten data, it is necessary to store the data in a specific data-format. However, some XML-based data-formats are available according to the scripting language structure. In the present study, we have developed a new data-format, named as OHWR-Gurmukhi_1.0. Using this format (OHWR-Gurmukhi_1.0), we can store, annotate, and access the data at stroke-level easily with minimal number of tags. This format is divided into three main sections, *DataSetDef* section, *WriteDef* section and *WordSetDef* section, as shown in Fig. 3. The *DataSetDef* section provides information about the template used for collecting the handwriting samples, the language used in the template and it traces back to the original data collection template. This section also provides brief description about the template along with the institute name, where the template is created with contact information. The *WriteDef* section provides the details about the writer, i.e., the person actually providing the handwriting sample. The details such as age, gender, education level, region, frequency of writing, right-left handedness of the writer etc. The *WordSetDef* section contains the handwriting information. Handwritten data in *WordSetDef* section is categorized into a tree structure as shown in Fig. 6. Word level data is captured in this format. The elements *wordNo* (information of word number), *wordDesc* (Gurmukhi word given to the writer to write), *totalStrokes* (total number of strokes used to write the word) and *stroke* (stroke details) have been used to present the collected data. Under the stroke element, *strokeID* and *strokeNo* are stored. During annotation, a value is manually assigned to *strokeID* for each captured stroke. The handwriting traces obtained from the touch based capturing devices (*i.e.*, Tablet-PC, smart phones, *etc.*) are also stored under the stroke element. This tree structure maintains a count of the number of elements in each lower level (*i.e.*, word count and stroke count). Further, every captured x-, y-coordinate value is stored in the point element. Thus, point element is the lowest element of this tree structure, which contains the x-, y-coordinates for a stroke. Figure 4 illustrates a sample taken from *usr04*.

Figure 6 illustrates the OHWR-Gurmukhi_1.0 format for storing the Gurmukhi word "ਚੜ". The word "ਚੜ" has been written using four strokes. As there is no restriction on the number of points plotted while writing a stroke, so we store all the points, written against the stroke. This data of unknown number of points to a stroke is called Raw-Data. This Raw-Data is normalized into the feature vector of size 128 (64 x-y coordinate points), after applying the preprocessing steps (*i.e.*, size normalization; removing duplicate points; interpolating missing points; and re-sampling).

In order to validate our benchmark dataset, we performed stroke recognition experiments using the Support Vector Machine (SVM) classifier [14,15]. We first pre-processed the stroke-level data using size-normalization, interpolation, and re-sampling to 64 x- y-coordinates. These coordinates are considered as a feature set. Two zone-wise classifiers were trained to recognize the strokes in their respective zones, namely, Upper and Lower zone. A total of 1,680 stroke samples have been considered to build the Upper zone classifier and 11,502 stroke samples have been considered for Lower zone classifier. Generally, the effectiveness of the classifiers highly depends on the kernel used and kernel parameters such as learning rate (γ) and penalty parameter C. After performing the experimentation with four SVM kernels, we decided to utilize the Radial Basis function (RBF) kernel for training the model. Additionally, the kernel parameters have also been varied considerably to improve the cross-validation accuracy. The maximum cross-validation accuracy of 99.08% for Upper zone and 97.38% for lower zone has been achieved in the experimentation, shown in Table 2.

Table 2. Stroke-level cross-validation accuracy using a SVM-based classifier for two zones.

Zone	Classes	Sample size	Value of k-fold	Cross-validation accuracy (%)
Upper	12	1680	4	98.58
			5	98.83
			6	99.00
			7	99.08
			8	98.91
Lower	81	11502	4	97.11
			5	97.21
			6	97.31
			7	97.31
			8	97.38

6 Web Address of Gurmukhi Database

Online google sites storage has been used to upload the Gurmukhi database. As of now, only raw data of 190 writers has been uploaded in both the formats. To access the database that is publicly available, one can browse the following links:

– https://sites.google.com/view/ohwr-gurmukhi-script/

7 Conclusion

Research in the area of online handwriting recognition systems nowadays is highly affected by the availability of data. In this paper, an attempt to create a benchmark dataset for online handwritten Gurmukhi words and numerals has been made. We believe that the dataset described in this paper has the potential to support future research in this area. We believe that the explanation on the process of creation, collection, and annotation of dataset in this paper will help researchers to better understand the dataset and to utilize the dataset in their research work more effectively. Eventually, we hope that more datasets can be created and shared by different groups.

Acknowledgment. The authors take this opportunity to thank Technology Development for Indian Languages (TDIL) Programme, Department of Information Technology, Government of India for funding this work.

References

1. International Unipen Foundation: The Unipen Project (1994). http://www.unipen.org/home.html
2. Agrawal, M., Bhaskarabhatla, A.S., Madhvanath, S.: Data collection for handwriting corpus creation in Indic scripts. In: International Conference on Speech and Language Technology and Oriental COCOSDA (ICSLT-COCOSDA 2004), New Delhi, India, November 2004. Citeseer (2004)
3. Belhe, S., Chakravarthy, S., Ramakrishnan, A.: XML standard for Indic online handwritten database. In: Proceedings of the International Workshop on Multilingual OCR, p. 19. ACM (2009)
4. Djeddi, C., Al-Maadeed, S., Gattal, A., Siddiqi, I., Ennaji, A., El Abed, H.: ICFHR2016 competition on multi-script writer demographics classification using "QUWI" database. In: 2016 15th International Conference on Frontiers in Handwriting Recognition (ICFHR), pp. 602–606. IEEE (2016)
5. Fisher, W.M.: The DARPA speech recognition research database: specifications and status. In: Proceedings of DARPA Workshop on Speech Recognition, February 1986, pp. 93–99 (1986)
6. Godfrey, J.J., Holliman, E.C., McDaniel, J.: SWITCHBOARD: telephone speech corpus for research and development. In: 1992 IEEE International Conference on Acoustics, Speech, and Signal Processing, ICASSP 1992, vol. 1, pp. 517–520. IEEE (1992)
7. Hemphill, C.T., Godfrey, J.J., Doddington, G.R.: The ATIS spoken language systems pilot corpus. In: Speech and Natural Language: Proceedings of a Workshop Held at Hidden Valley, Pennsylvania, 24–27 June 1990 (1990)
8. Hull, J.J., Fenrich, R.K.: Large database organization for document images. In: Impedovo, S. (ed.) Fundamentals in Handwriting Recognition, pp. 397–414. Springer, Heidelberg (1994). https://doi.org/10.1007/978-3-642-78646-4_24
9. Khayyat, M., Lam, L., Suen, C.Y.: Arabic handwritten word spotting using language models. In: 2012 International Conference on Frontiers in Handwriting Recognition (ICFHR), pp. 43–48. IEEE (2012)
10. Lamel, L.F., Kassel, R.H., Seneff, S.: Speech database development: design and analysis of the acoustic-phonetic corpus. In: Speech Input/Output Assessment and Speech Databases (1989)
11. Messaoud, I.B., Amiri, H., El Abed, H., Märgner, V.: Region based local binarization approach for handwritten ancient documents. In: 2012 International Conference on Frontiers in Handwriting Recognition (ICFHR), pp. 633–638. IEEE (2012)
12. Phillips, I.T., Ha, J., Haralick, R.M., Dori, D.: The implementation methodology for a CD-ROM English document database. In: Proceedings of 2nd International Conference on Document Analysis and Recognition (ICDAR), pp. 484–487. IEEE (1993)
13. Price, P., Fisher, W.M., Bernstein, J., Pallett, D.S.: The DARPA 1000-word resource management database for continuous speech recognition. In: 1988 International Conference on Acoustics, Speech, and Signal Processing, ICASSP 1988, pp. 651–654. IEEE (1988)
14. Singh, H., Sharma, R., Singh, V.: Efficient zone identification approach for the recognition of online handwritten Gurmukhi script. Neural Comput. Appl. **31**, 3957–3968 (2019)
15. Singh, H., Sharma, R., Singh, V.: Recognition of online unconstrained handwritten Gurmukhi characters based on finite state automata. Sādhanā **43**(11), 192 (2018)

16. Wilkinson, R.A., et al.: The first census optical character recognition system conference, vol. 184. US Department of Commerce, National Institute of Standards and Technology (1992)
17. Xing, L., Qiao, Y.: DeepWriter: a multi-stream deep CNN for text-independent writer identification. In: 2016 15th International Conference on Frontiers in Handwriting Recognition (ICFHR), pp. 584–589. IEEE (2016)

Optical Character Recognition

Targeted Optical Character Recognition: Classification Using Capsule Network

Pratik Prajapati[1]([✉]) [ID], Shaival Thakkar[2] [ID], and Ketul Shah[1] [ID]

[1] Dhirubhai Ambani Institute of Information and Communication Technology
(DA-IICT), Gandhinagar 382007, GJ, India
{201711004,201711017}@daiict.ac.in
[2] FactSet Systems India Pvt. Ltd., Hyderabad 500032, Telangana, India
shaivaly.thakkar@factset.com

Abstract. Optical Character Recognition (OCR) is a process of digitizing an image or document containing text in a machine-readable format. In this paper, we are focusing on targeting only the numeric part with a few special characters in the tables. Many firms dealing in financial information would want to parse data from scanned tables and in some cases, they do not focus on the row labels as they might not change a lot. Only focusing on numeric information may also provide language independence to such firms that deal with documents written in a variety of languages. They can have foreign language experts who can just read row labels and have the OCR extract the numeric data. This makes their collection processes fast. We developed a targeted OCR to save time by processing only important characters and it can also overcome erroneous predictions in case of under segmentation of characters. In this paper, we propose a novel approach which segments the document into blocks of text (each line or word into one block) and classifies each block as numeric or non-numeric using a binary CNN. The process of character level segmentation and classification using capsule networks is then applied only to the blocks which are classified as numeric by the binary CNN.

Keywords: Optical Character Recognition · Pre-processing · Segmentation · Feature extraction · Classification · Capsule network

1 Introduction

Optical Character Recognition (OCR) [1] is a system that detects the text from a digital image. The OCR system has to detect the regions with text and segment it character by character and later classify those characters such as letters, numbers, and symbols. The important intermediate steps involved in character recognition are pre-processing, segmentation, feature extraction and classification/recognition. The aim is to build an accurate OCR tool for Numeral and Special Character Data from the business invoices.

© Springer Nature Singapore Pte Ltd. 2020
N. Nain et al. (Eds.): CVIP 2019, CCIS 1148, pp. 469–481, 2020.
https://doi.org/10.1007/978-981-15-4018-9_42

In this paper, we propose a two-step approach which removes textual data before we perform character segmentation so that we can perform more resource-intensive tasks only on areas of interest which are the numeric blocks in our case. We also try to recognize the under segmented characters as a whole, predicting the major content character in the input image after it's segmentation, in cases where character segmentation does not perform well. This will help us retain the complete numeric value in the output, as most of the under segmented blocks have a comma in it which carries no significance in identifying the complete numeric value.

1.1   Problem Statement and Motivation

Problem Statement. Our problem statement is as follows:-

"Given a scanned image, recognize only the numeric regions and give accurate predictions for each character."

Motivation. Firms dealing with the analysis of financial data get a lot of scanned documents from different sources and a variety of markets. Since these documents are scanned, we cannot extract data from these documents directly. So, we want a mechanism to identify such documents and perform OCR on them to convert them to native PDF documents (PDF containing text) which can be utilized later for further processes. But, as the documents are scanned, they have a lot of issues such as noise and shading effects. Some documents have underlines very close to the actual data and these lines can also interfere with the character segmentation. Due to such problems, the OCR output can turn out to be inaccurate. These problems are particularly severe for firms dealing with financial data as OCR serves as the base for many of their consequent operations and a small error can change values from a million to a billion and vice-versa.

2   Literature Survey

A lot of research in the past has been done to compare the performance of various OCR systems. In the past decade researchers have used various approaches including Support Vector Machine (SVM) [2], Hidden Markov Model (HMM) [3], Feed Forward Neural Networks [4], Convolutional Neural Networks [6] and even Transfer Learning [7] models. Recently after Capsule Network [9] was introduced by Hinton et al. on MNIST dataset in 2017, it proved to be a breakthrough in the field of deep learning while addressing few drawbacks in conventional convolutional neural networks.

In past, character recognition was carried out using morphological operations based pre-processing before extracting the features. It then included segmentation and character recognition. Recently, deep learning-based feature extraction approaches are trending in research. The Convolutional Neural Networks (CNN's) [6] are the most powerful networks in performing image processing tasks among the deep learning methods. When CNN is trained on a dataset containing images, a deep feature map representation of the image is constructed. While using CNN for the training phase, Simonyan et al. [8] proposed an improved architecture for image classification. They used pre-trained convolutional layers trained on ImageNet dataset and used transfer learning to train the last few layers of their network on their dataset.

Simon et al. [7] obtained a style representation of an input image and generated results based on the VGGNet [8], which is a CNN that tries to match human performance on a common visual object recognition benchmark task. Compared to other similar methods, the number of parameters of the CNN here is too large, but when compared to Capsule Network that number is smaller. Fortunately, there are a few openly available pre-trained models that can be used for transfer learning (VGG16, VGG19) [7]. Many researchers [8] have suggested that model ensemble is a very powerful technique to increase the accuracy of various machine learning problems.

3   Proposed Approach

The novelty of our approach lies in using block-level segmentation wherein we use binarization and morphological operations to segment lines and words into blocks of text. We then use a binary CNN to classify the blocks into textual and Numeric blocks. Only the numeric blocks are processed further. This helps in eliminating the chances of erroneous recognition in case of alphabets similar to numerals. E.g. B and 8, l and 1, S and 5 etc. We then use capsule network for better accuracy in recognition and also improving recognition for under segmented characters.

3.1   Binary Classification of Segmented Blocks Using CNN [6]

Binary classification between text and numeric regions not only reduces confusion as explained above but also reduces the amount of computation as we only focus on the Numeric regions. For this, we created a dataset of content blocks and divided the blocks into width which is equal to the average width found in most documents. For block-level classification, we need to perform preprocessing steps which include Otsu's binarization [5] and other morphological operations like closing. Also, we remove the lines below the printed characters using hough lines detection and morphological operations. The blocks are then segmented using vertical and horizontal pixel concentration (according to heuristically tuned thresholds). We try to keep the sizes of the blocks same by padding or splitting the blocks as required and get blocks of 240×43 pixels each, which was the initial training sample size while using LeNet-5 CNN architecture. This classification approach gave us near-perfect results for filtering out numeric blocks which were our regions of interest.

The result of our binary classifier which shows only numeric blocks and filters out the text blocks is shown in the below Fig. 1.

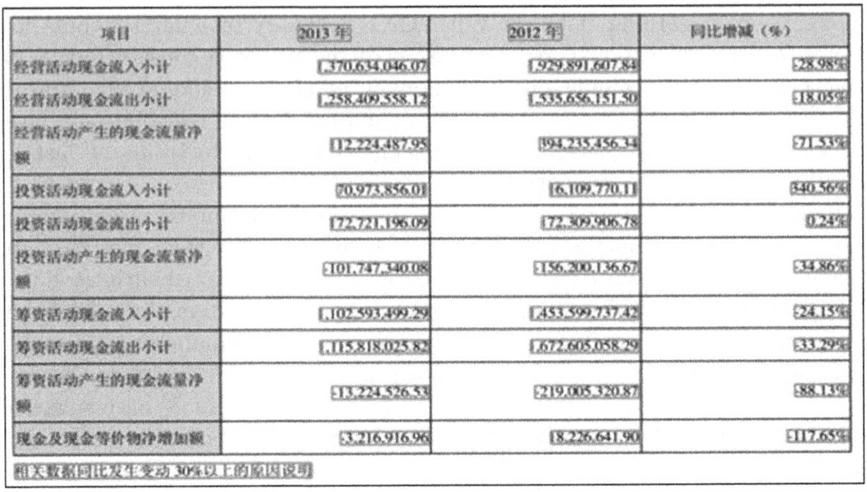

项目	2013 年	2012 年	同比增减（%）
经营活动现金流入小计	1,370,634,046.07	1,929,891,607.84	28.98%
经营活动现金流出小计	1,258,409,558.12	1,535,656,151.50	18.05%
经营活动产生的现金流量净额	112,224,487.95	394,235,456.34	71.53%
投资活动现金流入小计	70,973,856.01	6,109,770.11	340.56%
投资活动现金流出小计	172,721,196.09	172,309,906.78	0.24%
投资活动产生的现金流量净额	-101,747,340.08	-156,200,136.67	34.86%
筹资活动现金流入小计	1,102,593,499.29	1,453,599,737.42	24.15%
筹资活动现金流出小计	1,115,818,025.82	1,672,605,058.29	33.29%
筹资活动产生的现金流量净额	-13,224,526.53	-219,005,320.87	88.13%
现金及现金等价物净增加额	-3,216,916.96	18,226,641.90	117.65%
相关数据同比及年变动 30%以上的原因说明			

Fig. 1. Result of our binary classifier showing only numeric blocks

3.2 Classification of Characters

The character segmentation was done using horizontal pixel concentration for lines and vertical pixel concentration for character segmentation from the lines. We used a statistical approach to split when connected characters would come in the same segment based on one initial round of segmentation to compute average width of characters on a page.

We used three approaches for character classification namely, Convolutional Neural Network (CNN) with LeNet-5 architecture, Transfer Learning with VGG19 architecture pre-trained on Imagenet dataset and retraining the fully connected layers with our dataset. And lastly, a novel approach called "Capsule Network" introduced by Hinton et al. [9] in 2017 which was used on handwritten characters of MNIST dataset. This approach was developed to overcome certain drawbacks of the conventional state of the art convolutional neural networks (CNN). We are using it for classifying printed characters, which would also enhance numeric value recognition by classifying the major region of an undersegmented block of characters. In the end, we also evaluate a model ensemble of Capsule Network and Transfer Learning approach to make more efficient recognition.

The workflow model we have adopted in this paper for OCR is as shown in Fig. 2. Here we take a pdf or a scanned image of the document and perform basic pre-processing to clean the document and obtain segmented blocks which then is passed to our binary classifier to filter out numeric blocks and discard text blocks. We then pass this numeric blocks to get character level segmentation and finally these character images are passed to our ensemble model classifier to predict the classes and finally dump the numeric block values in JSON if required.

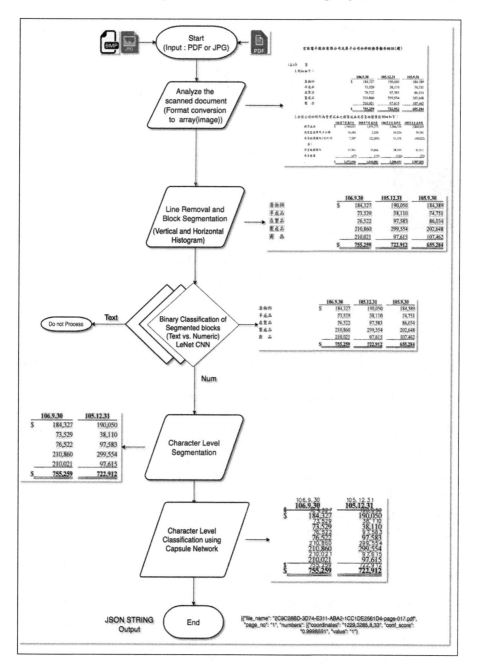

Fig. 2. Basic workflow for our proposed approach

3.3 Capsule Network

The CapsNet [9] Architecture has 2 parts: Encoder and Decoder as shown in Fig. 3 the first 3 layers are an encoder, and the second 3 are decoder: Layer 1. Convolutional layer, Layer 2. PrimaryCaps layer, Layer 3. DigitCaps layer, Layer 4. 1st Fully connected, Layer 5. 2nd Fully connected, Layer 6. 3rd Fully connected.

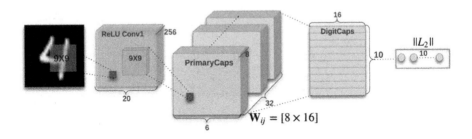

Fig. 3. Basic CapsNet architecture and layers [9]

A Capsule which is a group of neurons that encapsulates all important information in vector form. It uses iterative routing-by-agreement mechanism [9] in which low-level capsules to activate higher-level capsules that represent more complex entities. This Network is inspired by the Human Visual Cortex system. The Human brain deconstructs a hierarchical representation of the image perceived eyes. A Capsule which is a group of neuron encapsulates all important information about the state of the feature they are detecting in vector form. It uses the iterative routing-by-agreement mechanism in which low-level capsules activate higher-level capsules which represent more complex entities with a higher degree of freedom.

Mathematical Explanation
The tensor shapes at each layer in a CapsNet is shown in Table 1.

Table 1. Input & Output tensor shape for various layers in a CapsNet architecture [9]

Layers 1-2-3	Convolutional (1)	PrimaryCaps (2)	DigitCaps (3)
Input	28×28	$20 \times 20 \times 256$	$6 \times 6 \times 8 \times 32$
Output	$20 \times 20 \times 256$	$6 \times 6 \times 8 \times 32$	16×10
Parameters	20992	5308672	1497600

The Convolutional Layer has 256 kernels each of size $9 \times 9 \times 1$. The PrimaryCaps Layer has 32 kernels of $9 \times 9 \times 256$ (8D). The DigitCaps Layer has an

8×16 weight vector (W_{ij}) to help 8D input space to be converted into 16D output capsule layer.

Dynamic Routing by Agreement Between Capsules by Hinton et al. [9] The main backbone of Capsule Network is its novel dynamic routing algorithm which acts between capsules in different layers of the network as shown in Fig. 4 for a lower-level capsule layer (l) and a higher level capsule layer $(l+1)$.

Procedure 1 Routing algorithm.

1: **procedure** ROUTING($\hat{u}_{j|i}, r, l$)
2:    for all capsule i in layer l and capsule j in layer $(l+1)$: $b_{ij} \leftarrow 0$.
3:    **for** r iterations **do**
4:        for all capsule i in layer l: $c_i \leftarrow \texttt{softmax}(b_i)$ ▷ softmax computes Eq. 3
5:        for all capsule j in layer $(l+1)$: $s_j \leftarrow \sum_i c_{ij}\hat{u}_{j|i}$
6:        for all capsule j in layer $(l+1)$: $v_j \leftarrow \texttt{squash}(s_j)$ ▷ squash computes Eq. 1
7:        for all capsule i in layer l and capsule j in layer $(l+1)$: $b_{ij} \leftarrow b_{ij} + \hat{u}_{j|i}.v_j$
    return v_j

Fig. 4. Dynamic routing algorithm in capsule network [9]

The Input prediction vector is given by $u_{j|i} = W_{ij}u_i$. The vector length to probability mapping is done using a non-linear **squashing** function [9] whose graph is shown in below plot in Fig. 5.

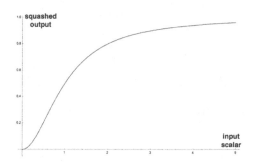

Fig. 5. Squashing non-linearity operation of capsule network [9]

And the vector representation of Output Capsule (v) is given by $v_j = \frac{||s_j||^2}{1+||s_j||^2} \frac{s_j}{||s_j||}$. The length of capsule's output vector represents probability for the presence of entity in input. Here, v_j is vector output for capsule j and s_j is its total input. Where the input vector s_j can be computed as follows:

$$s_j = \sum_i c_{ij}\hat{u_{j|i}}$$

where, c_{ij} is a coupling factor that multiplies output vector from lower-level capsule i and goes as input to a higher level capsule j. c_{ij} is a soft-max function:

$$c_{ij} = \frac{exp(b_{ij})}{\sum_k exp(b_{ik})}.$$

where, b_{ij} are the log prior probability that capsule i should be coupled with capsule j.

Agreement between the current output v_j and the prediction vector $\hat{u}_{j|i}$ from previous capsule i can be given as $a_{ij} = v_j.u_{j|i}$. Lower level capsule will send its input to the higher level capsule that "agrees" with its input. This is the essence of the dynamic routing algorithm.

Margin Loss for Single Digit

$$L_k = T_k max(0, m^+ - ||v_k||)^2 + \lambda(1 - T_k)max(0, ||v_k|| - m^-)^2$$

where $T_k = 1$ if a digit of class k is present, $m^+ = 0.9$ and $m^- = 0.1$. $\lambda = 0.5$.
Figure 6 shows the comparison between a capsule and a neuron.

Capsule vs. Traditional Neuron			
Input from low-level capsule/neuron		vector($\mathbf{u}_i$)	scalar(x_i)
Operation	Affine Transform	$\hat{\mathbf{u}}_{j|i} = \mathbf{W}_{ij}\mathbf{u}_i$	–
	Weighting	$\mathbf{s}_j = \sum_i c_{ij}\hat{\mathbf{u}}_{j|i}$	$a_j = \sum_i w_i x_i + b$
	Sum		
	Nonlinear Activation	$\mathbf{v}_j = \frac{\|\mathbf{s}_j\|^2}{1+\|\mathbf{s}_j\|^2} \frac{\mathbf{s}_j}{\|\mathbf{s}_j\|}$	$h_j = f(a_j)$
Output		vector($\mathbf{v}_j$)	scalar(h_j)

Fig. 6. Comparison of CapsNet and traditional neuron [9]

Dropout: We have used dropout in our capsule network for handling over-fitting and it allows the main network to learn using combinations of many sub-networks. Dropout is a technique proposed to address two issues using a single architecture: reduce over-fitting when the number of training samples is less and to provide a way of approximately combining many different neural network architectures efficiently. Dropout refers to temporarily removing the incoming and outgoing connections of a unit in a Neural Network.

4   Experimental Results

This chapter gives information about the experimental setup, dataset and results.

4.1 Experimental Setup

- Processor: 1xEight-Core Intel Xeon Processor E5-2609 v4 1.70 GHz 20 MB
- RAM: 64 GB RAM
- GPU Support: 1 x Nvidia Geforce Titan Xp Pascal 12 GB Gddr5X

4.2 Dataset

There were two datasets used for this research work.

- We prepared a dataset with 2 classes for text vs. numeral blocks classification. The two classes are Taiwanese Text blocks and Numeric/Special Character blocks. We have cropped blocks after block level segmentation and annotated around 22,000 samples for each of the above two classes. The size of each image is 43 × 240 which is the average size of blocks measured over a wide range of documents (Fig. 7).

Fig. 7. Text and numeric block dataset

- We have also clubbed a dataset for character level classification, consisting of a few samples from "EnglishDataSet" [11] Printed Characters Dataset for numerals but the majority of the data comes from actual financial documents in the Taiwanese language. This clubbed dataset is used for all of the implemented approaches mentioned in this paper (Fig. 8).

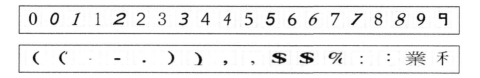

Fig. 8. Sample of numbers and special characters dataset

The Dataset consists of printed characters of 19 classes shown below:

– Numeral from 0–9
– Comma " , "
– Full Stop " . "
– Hyphen " - "
– Opening Bracket " ("
– Closing Bracket ") "
– Dollar " $ "
– Percent " % "
– Colon " : "
– Text "It includes different samples of Taiwanese characters"

We have collected the dataset of printed numerals and special characters [11] in 128×128 and resized them into 32×32 for faster recognition and training. We have 2600 samples of each class, with different fonts and angle and also including a lot of ground truth samples from invoices to get more robust output. Most of the new samples are taken in robust form, with a bit of segmentation error, so that the recognition module becomes prone to bad segmentation too. We are using the same training set for all the implemented approaches.

4.3   Comparison with Other Approaches

We compare three approaches here: LeNet-5 CNN architecture and VGG19 based transfer learning architecture to compare with the capsule network. In the Transfer Learning model, we used VGG19 architecture pre-trained on ImageNet dataset and froze the last layer to introduce our layer and retrained the model on our dataset. We changed input image dimension of our model to 28×28, as we used E-MNIST. The Capsule Network model is similar to the architecture discussed in the literature survey. We finally used a hybrid model ensemble using Capsule Network (17 classes, i.e. except dot and comma) [9] and Transfer Learning (For dot and comma) [7] which gave the best results.

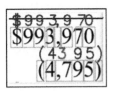

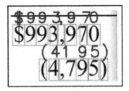
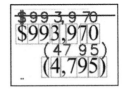

Fig. 9. CNN   **Fig. 10.** Transfer learn   **Fig. 11.** CapsNet

Figures 9, 10 and 11 (6, 7, 8) show the comparison of recognition for under segmented case of ",7". The results of the 3 approaches, CNN [6], Transfer Learning [7] and Capsule Network [9] are shown from left to right. The Capsule Network [9] is able to predict the major character part correctly compared to the other 2 approaches. It can visualize the segment as a whole capsule and also consider the orientation of vectors. So by using Capsule network predictions in case of low confidence predictions of CNN or transfer learnt model, we can achieve better performance.

We have manually checked the accuracy with the ground truth data prepared for 1512 character segments which belong to our 19 classes of interest. So using CNN, we were able to get 1438 correctly identified, and using transfer learning we were able to identify 1498. Using Capsule we were able to identify 1485 but by using a hybrid of Transfer Learning and CapsNet we were able to identify 1503 characters correctly.

Table 2 shows the comparison of the accuracy of all the proposed methods.

Table 2. Accuracy results for various approaches:

Approach	Character level 19 *Classes* (%)
Convolutional neural network	95.12
Transfer learning using VGG19	99.05
Capsule network	98.20
Ensemble of transfer learning & CapsNet	99.45

Qualitative Results for Character Level Classification
See Fig. 12.

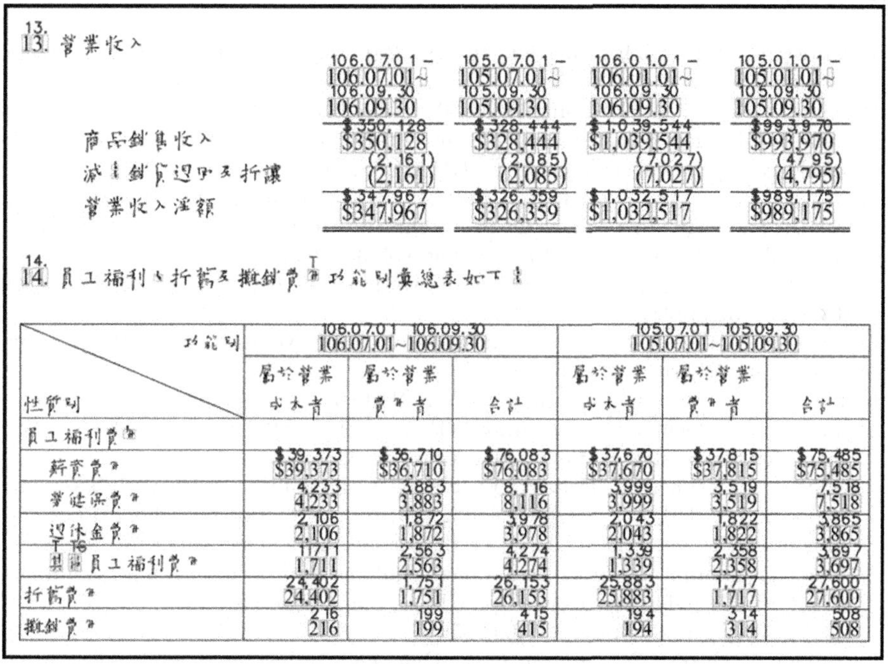

Fig. 12. Capsule network result Taiwanese invoice at FactSet [10]

Timing: Capsule Network [9] took 3.2 s for output on our experimental setup (Sect. 4.1) (Whole process: Segmentation and Recognition).

5 Conclusion

From the results achieved by the method proposed in the paper, it can be concluded that CapsNet (Capsule Network) [9] for recognition proves better in cases of under segmentation and cases where our CNN fails and pairing it with the CNN pre-trained on a larger (and preferably relevant) dataset gets us the best results.

5.1 Future Research Directions

The possible future research directions for the work presented in this paper are that we can make a more robust version which can take any text colour and background colour and provide us with the recognized output. Currently, we are able to detect rotated documents and also widescreen documents and successfully provide recognition output for the same. But, if the document is not properly scanned and it has a different angular shift or orientation other than 90° or 180°, our approach may not give the best results, so we could develop a system in future which can handle more complex problems.

References

1. Christensson, P.: OCR Definition, 16 April 2018. https://techterms.com. Accessed 14 Nov 2018
2. Sharma, S., Sasi, A., Cheeran, A.: An SVM based character recognition system. In: 2017 2nd IEEE International Conference on Recent Trends in Electronics, Information Communication Technology (RTEICT), pp. 1703–1707 (2017)
3. Lu, Z.A., Bazzi, I., Kornai, A., Makhoul, J., Natarajan, P. S., Schwartz, R.: A robust, language-independent OCR system. In: Proceedings of SPIE - The International Society for Optical Engineering, vol. 3584 (2000)
4. Bebis, G., Georgiopoulos, M.: Feed-forward neural networks. IEEE Potentials **13**(4), 27–31 (1994)
5. Otsu, N.: A threshold selection method from gray-level histograms. IEEE Trans. Syst. Man Cybern. **9**, 62–66 (1979)
6. LeCun, Y., Bottou, L., Bengio, Y., Haffner, P.: Gradient-based learning applied to document recognition. Proc. IEEE **86**(11), 2278–2324 (1998). http://citeseerx.ist.psu.edu/viewdoc/summary?doi=10.1.1.42.7665
7. Simon, M., Rodner, E., Denzler, J.: ImageNet pre-trained models with batch normalization. CoRR, vol. abs/1612.01452 (2016). http://arxiv.org/abs/1612.01452
8. Simonyan, K., Zisserman, A.: Very deep convolutional networks for large-scale image recognition. In: 3rd International Conference on Learning Representations, ICLR 2015, Conference Track Proceedings, San Diego, CA, USA, 7–9 May 2015 (2015). http://arxiv.org/abs/1409.1556
9. Sabour, S., Frosst, N., Hinton, G.E.: Dynamic routing between capsules. arXiv e-prints, October 2017

10. FactSet. FactSet Research Systems: Company Name Inc.: Targets and ratings (n.d.). From FactSet database. Accessed 25 May 2018

11. Dataset: Characters from computer fonts with 4 variations (Combinations of italic, bold and normal). http://www.ee.surrey.ac.uk/CVSSP/demos/chars74k/EnglishFnt.tgz

Security and Privacy

An Edge-Based Image Steganography Method Using Modulus-3 Strategy and Comparative Analysis

Santosh Kumar Tripathy[(⊠)] and Rajeev Srivastava

Computing and Vision Lab, Department of Computer Science and Engineering,
Indian Institute of Technology (Banaras Hindu University),
Varanasi 221005, UP, India
{santoshktripathy.rs.cse18,rajeev.cse}@iitbhu.ac.in

Abstract. Steganography or "Covered Writing" is a security tool for hiding secret information inside any media or object of interest. It provides secure communication between the sender and the receiver. The spatial domain image steganography techniques such as the Least Significant Bit (LSB) substitution and edge-based hiding schemes have been vastly exploited in the literature. An improved LSB method using modulus-3 strategy can be found in the literature, which has the advantages of high embedding capacity and resistance against Subtractive Pixel Adjacency Matrix (SPAM) steganalysis. However, one drawback is that it hides data sequentially. The edge-based hiding techniques have the advantage that it can tolerate significant variations in edge pixels and fulfill all the requirements of the Human Vision System (HVS), but it has limited in embedding capacity. The proposed work is a reversible data hiding scheme that takes the advantages of both the edge-based algorithm and the novel modulus-3 hiding scheme. At first, we converted the message into ternary data, and then we applied the novel modulus-3 strategy to hide two ternary bits in the edge pixel of the image thereby taking the advantages of both modulus-3 strategy and edge-based hiding approach. We have implemented four edge detection algorithms like Laplacian, Canny, Sobel, and Prewitt to find four different edge areas from the cover image. For each edge area, we applied the modulus-3 strategy to hide the entire ternary message, and then, we made a comparative analysis based on the performance of four edge detectors, and finally, we concluded.

Keywords: Spatial domain image steganography · LSB · Canny · Laplace · Prewitt · Sobel

1 Introduction

In today's world, securing information from any unauthorized use is a challenging task. Different approaches have been developed to secure data. One such method is steganography. Steganography is an art as well as science which is used to hide secret information inside a media to provide secure communication between the sender and receiver in the presence of third parties. The medium can be image, text, audio, video, etc.; it means steganography is "Covert Communication." The object which is used for

© Springer Nature Singapore Pte Ltd. 2020
N. Nain et al. (Eds.): CVIP 2019, CCIS 1148, pp. 485–494, 2020.
https://doi.org/10.1007/978-981-15-4018-9_43

hiding data is called a "COVER" object, and the object which is formed after hiding information is called as "STEGO" object. Different types of secret data such as text, image, video, audio, etc. can also be used for hiding. As mentioned in [1], basically, there are three various security tools used, such as Cryptography, Steganography, and Digital Watermarking. The purpose of all these methods is the same that is to provide security, but they have different ways of implantation and use. Cryptography means encoding secret information to another format called "Cipher Format/Text" such that it becomes hard to decode by an unauthorized user. That means cryptography shuffles the secret data using a key, and by using this key, only the receiver decrypts the ciphertext to get the original one. On the other hand, steganography hides confidential data inside an object. Unlike cryptography, steganography doesn't open secret information to the third party. Here the third party only sees the stego object where confidential data is hidden, which is very difficult to know whether any secret information exists or not. Digital Watermarking is also used to hide information inside a medium, but it differs from steganography in a sense that watermarking is used for copyright preserving, banknote authentication. So, for covert and invisible communication, we are going to use steganography.

The proposed method is an image steganography approach in which the cover object is an image, and the secret message is text. Image steganography approaches can be broadly classified into spatial-domain, transfer-domain, and spread-spectrum-based approaches. Now a day's researchers exploit machine learning concepts and deep learning approaches [2, 3] to perform steganography. However, our approach is based on the spatial domain technique, and we focus on it. Well, in the spatial domain image steganography technique, the pixels (intensity values) of an image is modified directly to embed secret information. These techniques provide a high embedding capacity compared to other methods.

There are various techniques have been proposed in this domain. Basically, there are many ways by which we can categorize these approaches. On such way is depicted in Fig. 1.

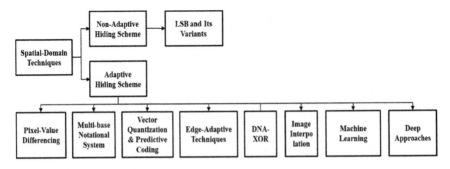

Fig. 1. Types of spatial domain image steganography technique

Adaptive hiding means varying amounts of data hidden into the cover image whereas in Non-Adaptive hiding uniform amount of data hidden. Also, the adaptive hiding scheme considers the region's complexity before embedding data. PVD [4–8], MBNS [9, 10], Histogram-Modification technique [11, 12], Vector quantization [13], Interpolation-based approach [14] and deep learning approaches [2, 3] have been explored to perform adaptive-hiding.

In our proposed method, we utilize both adaptive and non-adaptive hiding scheme by considering an improved LSB and edge-based hiding techniques. In the LSB-based hiding scheme, we manipulate the LSB of the pixel value to hide the secret bit into it. However, in edge-based techniques, we first find the edge of the cover image and usually embed data in this edge area. The main motive of hiding data in the edge area is that we can hide more data in it, and humans cannot readily identify any disturbance due to hiding [10].

1.1 Motivation

There are several aspects which motivates us to do research in steganography. Let us list out some of these issues.

- Increase in data security threats: Recent studies [15–17] shows that over 90% of attacks are based on crypto-mining, over 54% of companies experienced system security threats and it is increasing in numbers. The ransomware attacks and data breach cost a company an average loss of more than $5 million and $3.86 million, respectively. More than millions from revenues spent to handle security threats. So, it is essential to develop a security system that not only restricts attacks but also it should be tough to decode the data. Steganography is one such choice.
- Problems with Cryptography: Cryptography doesn't change the existence of data; instead, it shuffles and encodes it. So, this is why it is open for security threats and costs millions of losses. But steganography just hides secret data inside an object, and by doing so, the appearance of the object is not changing at all which fools the attacker. Again, some steganography algorithms restrict the steganalysis method, which shows it is very reliable and robust against security threats.
- Advantages of Reversibility: Most of the steganography algorithms are reversible, which means we recover the cover object after extracting the secret data from the stego object. So, this reversibility nature of steganography motivates to minimize data transmission and bandwidth cost by hiding multiple data inside an object. Due to reversibility nature, steganography algorithms can be effectively used in the medical domain to secure the patient's information [18].
- Resistance against loss of data: Well, if we delete or do some modification on the encrypted code then original secret message can't be decoded. However, some steganography algorithms resist scale change, translation, or crop the stego image, and we get the original data. So, these are some reasons which indicate that steganography would be a better choice as compared to cryptography.
- Open scope for dual-level data protection: We can take advantage of cryptography as well as steganography for data hiding. Loan [18] utilized the modified LSB substitution and the RC4 encryption to hide patient data in medical images. In the

next section, we review some of the state-of-the-art techniques of LSB and edge-based image steganography.

2   Literature Review and Related Work

A considerable number of studies have made on LSB-based image steganography. The problem with traditional LSB substitution is that it hides data sequentially, cause artifacts upon modifying more LSB bits, and open for steganalysis. Chan [19] found a solution for optimal pixel adjustment for data hiding. Thien [20] proposed a high capacity data hiding scheme using a modulus function that overcomes artifacts caused by the traditional LSB substitution method. Yang [21] estimated the number of k-bit LSBs for data hiding. Meilikainen [22] proposed an LSB-matching algorithm to hide data by making fewer changes to the original image. Li [23] proposed a generalized LSB-matching algorithm and outperformed Meilikainen [22]. Wang utilized a genetic algorithm to perform the LSB-based hiding scheme. The problems with these techniques are that they are open to RS-steganalysis and other statistical steganalysis. Recently, Xu [24] proposes a modified LSB technique using modulus-3 strategy to hide two ternary data in a pixel, and also it shows its resistance against SPAM steganalysis.

Most of the existing LSB-methods don't consider the spatial relationship between image content and the size of the message, which results in poor visual quality and low security. In such a case, Edge-based methods are appropriate. In such cases, the intuition is that edge areas can tolerate more changes as compared to smooth areas. So, more data can be hidden in edge areas as compared to flat areas. LSB, Pixel Value Differencing (PVD) and Difference Expansion (DE) methods have been exploited for edge-based steganography algorithms [5, 25].

Our proposed work is motivated by Xu [24] and Edge-based hiding scheme. The proposed work not only considers the spatial relationship between image content and message size but also takes the advantages of modulus-3 strategy [24]. Let's discuss the modified LSB technique using modulus-3 strategy [24]. First, the message is converted into a ternary. Then, we traverse every pixel sequentially and hide two ternary messages in each pixel at a time by using modulus-3 strategy. Let us describe the embedding and extracting procedure of this method.

2.1   Embedding Algorithm

Initialization: Let the secret data be S which is a ternary message and C be the cover image.

Step-1: Let C_{ij} be the pixel value and convert it to binary string $b_7b_6b_5b_4b_3b_2b_1b_0$.
Step-2: Divide C_{ij} into two sub-segments i.e. $sub\_seg1_{ij} = b_7b_6b_5b_4b_3b_2$ and $sub\_seg2_{ij} = b_1b_0$ by using the following equation.

$$sub_{seg1ij} = \frac{C_{ij}}{2^2} \ and \ sub_{seg2ij} = mod\left(C_{ij}, 2^2\right) \tag{1}$$

Step-3: Embed the first ternary secret number $S(k)$ according to the following equation.

$$Sub1\_stego = \begin{cases} sub\_seg1_{ij}, & if \: mod(sub\_seg1_{ij}, 3) = S(k) \\ sub\_seg1_{ij} + 1, & if \: mod(sub\_seg1_{ij} + 1, 3) = S(k) \\ sub\_seg1_{ij}\_1, & otherwise \end{cases} \quad (2)$$

Step-4: Connect $Sub1\_stego$ generated in Step-3 and sub_seg2_{ij} to construct V_{ij} using the following equation,

$$V_{ij} = Sub1\_stego_{ij} \times 4 + sub\_seg2_{ij} \quad (3)$$

and then embed the second ternary secret number $S(k + 1)$ according to following equation.

$$V_{ij}\_stego = \begin{cases} V_{ij}, & if \: mod(V_{ij}, 3) = S(k+1) \\ V_{ij} + 1, & if \: mod(V_{ij} + 1, 3) = S(k+1) \\ V_{ij}\_1, & otherwise \end{cases} \quad (4)$$

Step-5: For $k = k + 2$ move the location (i, j) to next pixel location and repeat steps 1–4 until all secret messages are embedded.

Step-6: V_{ij}_stego will be the stego image.

Send this stego image and key to the receiver for extraction. The key may be the length of the ternary message.

2.2 Extraction Algorithm

In extraction procedure, each pixel of the stego image is divided into two subparts according to step-2 of embedding algorithm, and the first ternary bit will be obtained by applying the modulus-3 operation to the first subpart of the pixel. The second ternary bit will be obtained by using the modulus-3 strategy to the pixel itself. The same process is repeated for all the pixels. This method handles the underflow and overflow situations very well [24], and here we outline the overall concept.

3 Proposed Method

The modulus-3 strategy [24] has an advantage that it resists SPAM steganalysis [26] but the problem is that it hides the ternary message in pixels sequentially. Also, when we hide data in the edge area, then we cannot easily identify the distortion, and it fulfills the requirements of HVS. It indicates that we can hide more data in the edge area as compared to the smooth region. So, we take advantage of hiding messages in the edge area as well as a modulus-3 strategy [24]. In the data embedding stage, we first covert the secret information into the ternary data, then we find four edge areas from the cover image by using four edge detectors like Canny, Laplacian, Sobel, and Prewitt.

Then, we hide a total ternary message in each of edge area separately by using modulus3 strategy. Finally, we made a comparative analysis of the performance of four edge detectors. We can hide two ternary data in each of the edge pixels. Figure 2 shows the flowchart of the proposed embedding and extraction process. The following algorithm shows the data embedding and data extraction strategy.

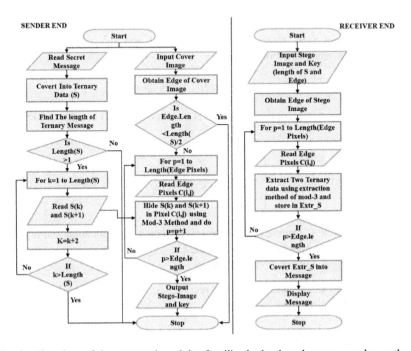

Fig. 2. Flowchart of the proposed modulus-3 utilized edge-based steganography method.

3.1 Embedding Algorithm

Initialization: Let the secret data be S which is a ternary message and C be the cover image.

Step-1: Let C_{ij} be the pixel value and convert it to binary string $b_7b_6b_5b_4b_3b_2b_1b_0$.
Step-2: Divide C_{ij} into two sub-segments i.e. $sub\_seg1_{ij} = b_7b_6b_5b_4b_3b_2$ and $sub\_seg2_{ij} = b_1b_0$ by using the Eq. (1).
Step-3: Embed the first ternary secret number $S(k)$ according to the Eq. (2)
Step-4: Connect $Sub1\_stego$ generated in Step-3 and sub_seg2_{ij} to construct V_{ij} using the following Eq. (3) and then embed the second ternary secret number S $(k + 1)$ according to the Eq. (4).
Step-5: For $k = k + 2$ move the location (i, j) to next pixel location and repeat steps 1–4 until all secret messages are embedded.
Step-6: V_{ij}_stego will be the stego image.

Send this stego image and key to the receiver for extraction. The key may be the length of the ternary message.

3.2   Extraction Algorithm

In extraction procedure, each pixel of the stego image is divided into two subparts according to step-2 of embedding algorithm, and the first ternary bit will be obtained by applying the modulus-3 operation to the first subpart of the pixel. The second ternary bit will be obtained by using the modulus-3 strategy to the pixel itself. The same process is repeated for all the pixels. This method handles the underflow and overflow situations very well [24], and here we outline the overall concept.

4   Experiment and Results

We have taken three publicly available [27] gray images like Lena, Boat, and Baboon as shown in Figs. 3, 5, 7 respectively and random messages of different sizes in bytes like 5393, 7854, 10494 and 13138 bytes. As mentioned in the previous section, we have used four different edge detectors, namely Canny, Laplacian, Sobel, and Prewitt. We performed our experiment on MatLab-2017b. The proposed method retrieves the exact image after extracting the embedded message. After hiding, we made a comparative analysis of the performance of four edge detectors, and Table 1 shows this. We have used PSNR as a performance metric. Figures 4, 6, and 8 show the performance of different edge detectors on test images. By observing Table 1, we can find that the Laplacian edge detector performs better as compared to the other three methods.

Similarly, we have compared our approach with some state-of-the-art techniques [28]. Reference [28] achieved PSNR = 37.66 dB by hiding nearly 15 KB of data on Lena.bmp, whereas our approach achieves more than 40 dB of PSNR. It shows the robustness of our approach.

Table 1. Performance analysis based on PSNRs for different edge detectors

Serial No.	Edge detector	Applying proposed algorithm in	Secret data (Bytes)			
			5393	7854	10494	13138
1	Canny	BOAT.tiff (PSNR (dB))	40.17	38.97	37.97	37.11
	Laplacian		**43.04**	**43.04**	**43.04**	**43.04**
	Sobel		40.43	40.43	40.43	40.43
	Prewitt		40.53	40.53	40.53	40.53
2	Canny	Lena.bmp (PSNR (dB))	40.10	38.59	37.63	36.69
	Laplacian		**44.86**	**44.86**	**44.86**	**44.86**
	Sobel		42.94	42.94	42.94	42.94
	Prewitt		43.05	43.04	43.04	43.04
3	Canny	BABOON.tiff (PSNR (dB))	40.89	39.28	37.97	36.90
	Laplacian		**45.47**	**45.47**	**45.47**	**45.47**
	Sobel		40.41	39.9	39.9	39.9
	Prewitt		40.44	40.00	40.00	40.00

Fig. 3. Lena

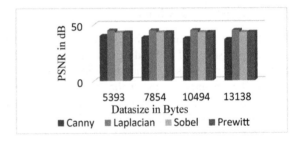

Fig. 4. Performance of edge detectors on Lena

Fig. 5. Boat

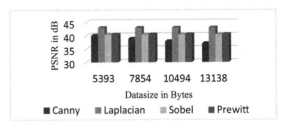

Fig. 6. Performance of edge detectors on Lena

Fig. 7. Baboon

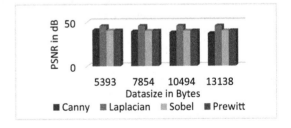

Fig. 8. Performance of edge detectors on Baboon

5   Conclusion

The proposed method takes advantage of both LSB-based modulus-3 strategy and edge-based hiding techniques. For the experiment, we have used three publicly available images and random messages of different sizes. We also made a comparative analysis between different edge detectors like Canny, Laplacian, Sobel, and Prewitt. By observing the Table 1, we can conclude that the proposed data hiding technique using Laplacian edge detection algorithm achieves better performance as compared with other methods. We have compared our result with Tian [28] and k-bit LSB substitution technique and our approach outperform these approaches. One possible drawback of our approach is the formation of salt pepper type noise. So, in our future work, we will try to overcome this situation and also improve performance.

References

1. Subhedar, M.S., Mankar, V.H.: Current status and key issues in image steganography: a survey. Comput. Sci. Rev. **13–14**(C), 95–113 (2014)
2. Yang, J., Liu, K., Kang, X., Wong, E.K., Shi, Y.-Q.: Spatial image steganography based on generative adversarial network, no. 1, pp. 1–7 (2018)
3. Tang, W., Li, B., Tan, S., Barni, M., Huang, J.: CNN-based adversarial embedding for image steganography. IEEE Trans. Inf. Forensics Secur. **14**(8), 2074–2087 (2019)
4. Lee, Y.P., Lee, J.C., Chen, W.K., Chang, K.C., Su, I.J., Chang, C.P.: High-payload image hiding with quality recovery using tri-way pixel-value differencing. Inf. Sci. (Ny) **191**, 214–225 (2012)
5. Yang, C.H., Weng, C.Y., Tso, H.K., Wang, S.J.: A data hiding scheme using the varieties of pixel-value differencing in multimedia images. J. Syst. Softw. **84**(4), 669–678 (2011)
6. Liao, X., Wen, Q.Y., Zhang, J.: A steganographic method for digital images with four-pixel differencing and modified LSB substitution. J. Vis. Commun. Image Represent. **22**(1), 1–8 (2011)
7. Swain, G., Lenka, S.K.: Steganography using two sided, three sided, and four sided side match methods. CSI Trans. ICT **1**, 127–133 (2013). https://doi.org/10.1007/s40012-013-0015-3
8. Hussain, M., Wahid, A., Wahab, A., Ho, A.T.S., Javed, N., Jung, K.: Rightmost digit replacement crossmark. Signal Process. Image Commun. **50**, 44–57 (2017)
9. Geetha, S., Kabilan, V., Chockalingam, S.P., Kamaraj, N.: Varying radix numeral system based adaptive image steganography. Inf. Process. Lett. **111**(16), 792–797 (2011)
10. Hong, W., Chen, T.S., Luo, C.W.: Data embedding using pixel value differencing and diamond encoding with multiple-base notational system. J. Syst. Softw. **85**(5), 1166–1175 (2012)
11. Li, Y.C., Yeh, C.M., Chang, C.C.: Data hiding based on the similarity between neighboring pixels with reversibility. Digit. Signal Process. A Rev. J. **20**(4), 1116–1128 (2010)
12. Zhao, Z., Luo, H., Lu, Z.M., Pan, J.S.: Reversible data hiding based on multilevel histogram modification and sequential recovery. AEU Int. J. Electron. Commun. **65**(10), 814–826 (2011)
13. Ma, X., Pan, Z., Hu, S., Wang, L.: Reversible data hiding scheme for VQ indices based on modified locally adaptive coding and double-layer embedding strategy. J. Vis. Commun. Image Represent. **28**, 60–70 (2015)
14. Tang, M., Hu, J., Song, W., Zeng, S.: Reversible and adaptive image steganographic method. AEU Int. J. Electron. Commun. **69**(12), 1745–1754 (2013)
15. INFOSEC: Security Awareness Statistics. https://resources.infosecinstitute.com/category/enterprise/securityawareness/security-awareness-fundamentals/security-awareness-statistics/#gref. Accessed 13 Aug 2019
16. CSO: Top cybersecurity facts, figures and statistics for 2018. https://www.csoonline.com/article/3153707/top-cybersecurity-facts-figures-and-statistics.html. Accessed 13 Aug 2019
17. VARONIS: 56 Must Know Data Breach Statistics for 2019. https://www.varonis.com/blog/data-breach-statistics. Accessed 13 Aug 2019
18. Loan, N.A., Parah, S.A., Sheikh, J.A., Akhoon, J.A., Bhat, G.M.: Hiding electronic patient record (EPR) in medical images: a high capacity and computationally efficient technique for e-healthcare applications. J. Biomed. Inform. **73**, 125–136 (2017)
19. Chan, C.K., Cheng, L.M.: Hiding data in images by simple LSB substitution. Pattern Recognit. **37**(3), 469–474 (2004)

20. Thien, C.C., Lin, J.C.: A simple and high-hiding capacity method for hiding digit-by-digit data in images based on modulus function. Pattern Recognit. **36**(12), 2875–2881 (2003)
21. Yang, H., Sun, X., Sun, G.: A high-capacity image data hiding scheme using adaptive LSB substitution. Radioengineering **18**(4), 509–516 (2009)
22. Mielikainen, J.: LSB matching revisited. IEEE Signal Process. Lett. **13**(5), 285–287 (2006)
23. Li, X., Yang, B., Cheng, D., Zeng, T.: A generalization of LSB matching. IEEE Signal Process. Lett. **16**(2), 69–72 (2009)
24. Xu, W.L., Chang, C.C., Chen, T.S., Wang, L.M.: An improved least-significant-bit substitution method using the modulo three strategy. Displays **42**, 36–42 (2016)
25. Luo, W., Huang, F., Huang, J.: Edge adaptive image steganography based on LSB matching revisited. IEEE Trans. Inf. Forensics Secur. **5**(2), 201–214 (2010)
26. Pevny, T., et al.: Steganalysis by subtractive pixel adjacency matrix to cite. IEEE Trans. Inf. Forensics Secur. **5**(2), 215–224 (2010)
27. Fabien a. p. petitcolas: Public-Domain Test Images for Homeworks and Projects. https://homepages.cae.wisc.edu/~ece533/images/
28. Tian, J.: Reversible data embedding using a difference expansion. IEEE Trans. Circuits Syst. Video Technol. **13**(8), 890–896 (2003)

Multi-level Threat Analysis in Anomalous Crowd Videos

Arindam Sikdar[✉] and Ananda S. Chowdhury

Department of Electronics and Telecommunication Engineering, Jadavpur University,
Kolkata 700032, India
{arindamsikdar.etce.rs,as.chowdhury}@jadavpuruniversity.in

Abstract. Crowd anomaly detection is a challenging problem in the field of computer vision. An abnormal event in a crowd scene can be labeled as threat in a video. Several existing solutions in this area have marked video frames either normal or abnormal event. Such categorization of frames can be referred as two-class threat labeling problem. However, this notion of two-class threat labeling is not well defined in literature. An event can have multiple aspects as it can be treated as anomalous or non-anomalous based on the situation of occurrence. Based on this argument, we propose a new paradigm of extending this two class threat labeling problem to multi-class labeling. As a solution to this multi-class labeling problem, we cluster frames with low, medium and high threat. We also propose a new feature known as pseudo-entropy for better clustering of threats. Our framework consists of two main components, namely, Earth mover distance (EMD) based anomaly detection system and multi-level threat analysis. As an outcome frame-wise and segment-wise threat representation are also presented to facilitate real time video search for relevant events. Exhaustive internal comparison and statistical analysis over benchmark UCSD and UMN dataset clearly indicates the merit of the proposed framework.

Keywords: Crowd anomaly · Multi-level threat · Frame-wise pattern · Segment-wise pattern · Local motion descriptor

1 Introduction

Automated analysis of crowd behavior has evolved as an important computer vision task towards developing an efficient surveillance system. One such task is crowd anomaly detection where frames in a video are detected as "normal" or "abnormal". Existing methods [1,2] always solved this problem from the two class perspective but such notion is not well defined in literature. Categorizing frames strictly as anomalous or non-anomalous are fundamentally ambiguous at semantic level. For instance, crowd running in a marathon may be regarded as "normal" while the same is not obvious in a pedestrian walkway or an open concert. Also for accurate performance evaluation unbiased ground truth generation is necessary which mostly depends on expert's notion of anomaly. Another way

© Springer Nature Singapore Pte Ltd. 2020
N. Nain et al. (Eds.): CVIP 2019, CCIS 1148, pp. 495–506, 2020.
https://doi.org/10.1007/978-981-15-4018-9_44

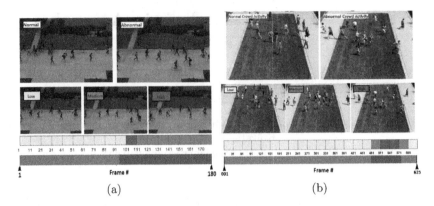

Fig. 1. Results showing demarcation of frames into three different threat levels on UCSD and UMN dataset. Top two rows showing frames separated into to two-class (i.e., normal and abnormal) and three class (i.e., low, medium and high threats). Among the last two bars, the upper color bar represents segment-wise threat indication and the lower bar represents anomalous (red) and non-anomalous (green) portions of the video sequence. (Color figure online)

to analyze crowd anomaly is to solve as a multi-class problem where frames are separated as low, medium and high threat. This is a relaxed approach over two class (anomalous/non-anomalous) problem and can provide useful information even if actual threat is not present in a frame. It can also provide a prior indication of an upcoming threat in a real time video surveillance tasks. We are first to address this problem in such fashion.

In this paper, a multi-level threat analysis is achieved as an extension of our previously proposed anomaly detection system [3]. Using this system, abnormality is determined based on the variation of several motion descriptors generated in a video. The descriptor generation is performed over objects that are appearance wise stable and termed as observers. These descriptors are constructed locally over an object with other spatially neighboring objects (targets) within a frame. The contextual motion variation of these targets relative to observers determines the overall abnormality in a frame. The change in these local descriptors is measured using Earth mover distance (EMD) [4] that contributes to overall frame anomaly. In order to achieve multi-level threat analysis, two features are computed based on the variation of these descriptors. The first one is called frame-EMD and the second one is our proposed entropy like feature (we term as pseudo-entropy). We then cluster and label frames into low, medium and high threat based on these features. Two threat representations are generated based on this clustering, one is frame-wise threat pattern and other is segment-wise threat pattern (see Fig. 1 in this connection). In segment-wise threat pattern, the whole video is fragmented into ordered sequence of several small segments of very short duration. Each such segments are assigned a threat label based on majority voting of labels of frames present in that segment. Frame-wise and

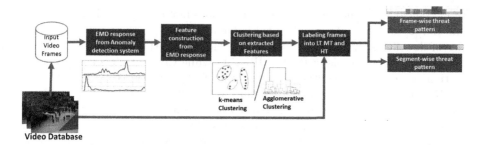

Fig. 2. A schematic flow of our proposed pipeline.

segment-wise threat pattern can be particularly useful to get a multi-level notion of anomaly that can facilitate video search process for other possible events of importance. Though, in principle, it is certainly possible to indicate more than three levels of threat, the practical impact of such an extension would be fairly incremental.

The paper is organized as follows. Section 2 outlines related works. Section 3 presents proposed framework. Experimental results with detailed analysis are described in Sect. 4. Finally, we draw conclusion and future research directions in Sect. 5.

2   Related Work

Several works [1–3,5,6] are found in literature on crowd anomaly detection. Based on the clue types of defining crowd abnormality [7], recent methods can be categorized into two broad approaches, (a) trajectory based methods and (b) motion based methods.

The first category of methods [5,8–10] learn trajectory knowledge from normal scenes and employ it to separates trajectories in abnormal ones. For instance, Wu *et al.* [8] extracted chaotic invariant features from trajectories of lagrangian particles to model abnormal crowd patterns while Basharat *et al.* [9] adopted Gaussian Mixture Model (GMM) to represent the trajectories of normal behavior. Cheng and Hwang [10] followed a different approach of adaptive particle sampling and Kalman filtering to determine abnormal trajectories and localize abnormal events. Zhang *et al.* [5] adopted Bag of Trajectory Graphs (BoTG) feature to model crowd event and classify anomaly using mean shift clustering of trajectories. A recent work by Bera *et al.* [6] employed Bayesian learning over nonlinear pedestrian models for abnormal trajectory analysis. In-spite of having higher level semantics these methods are often computationally expensive, especially over long duration trajectories.

The second category of methods [1,2,11,12] do not involve tracking and mostly use motion or appearance information to separate abnormal events in crowd. Most common works in this context are based on optical flow variations [2,11], where Ali and Shah [11] introduced finite time Lyapunov exponent

(FTLE) field for segmenting regions of anomalous motion pattern. Recently Chen and Lai [2] analyzed abnormal crowd behaviors based on Div-Curl characteristics of flow fields. Kim and Grauman [13] modeled optical flow patterns based on mixture of probabilistic PCA (MPPCA) while Cong et al. [14] represented motion pattern using multiscale histogram of optical flow (HOF). Mehran et al. [15] introduced the notion of dynamics of fluid flow into crowd flow to distinguish abnormal behavior from the normal ones. Some methods like [11,16] portrayed flow as social force models [17] to distinguish normal/abnormal motion patterns. Techniques like [18,19] have utilized hidden Markov models (HMMs) to establish temporal consistency of normal frames over the abnormal ones. In [12], the authors utilized dynamic textures (DTs) to represent crowd behavior. Later Li et al. [20] adopted mixtures of DTs with center-surround saliency detector for detecting anomaly. Recent methods [1,21] have introduced deep models for analyzing crowd behavior analysis and detect anomaly. For example, Hasan et al. [21] used Convolutional Autoencoder to compute abnormality score based on reconstruction error during testing time. Later, [1] improvised the model of Hasan by performing similar task using Generative Adversarial Nets (GANs).

The above mentioned methods solved anomaly detection as two class problem where frames are labeled "normal" or "abnormal" but none have addressed from multi-level threat perspective. Bera et al. [6] were the first to introduce the concept of multi-level threat where they performed an inter-video analysis by categorizing video sequences into low, medium and high threat. Unlike them our solution pipeline performs intra-video threat analysis where we categorize each frame in a video into multiple threats. This is a more informative approach compared to inter-video analysis as we are analyzing from a finer perspective. Moreover, we have performed quantitative evaluation which lacked in the work of Bera [6]. The main contribution of our work are summarized as follows:

1. We propose a new and alternative paradigm for analyzing threat in crowd videos. Here, we solve a multi-class extension of traditional two-class (normal/abnormal) anomaly detection problem where frames are marked as low, medium or high threat. Additionally, we propose frame-wise and segment-wise threat level indication in a video sequence.
2. Secondly, we introduce an entropy like feature (pseudo-entropy) for better clustering of threat and indicate its superior performance over standard variance as a feature.

3   Proposed Framework

Occurrence of abnormality in a frame is determined by overall abrupt change in motion between consecutive frames. The complex motion dynamics of a crowd can be represented using several local motion descriptors [3] generated over the pedestrians [22] in a scene. Some of these pedestrians are appearance wise stable suitable for local motion descriptors. Such pedestrians are termed as observers and rest are targets. Frame level anomaly is computed based on Earth mover distance (EMD) measuring the variation of these local descriptors between consecutive frames. Anomaly in a frame is determined based on aggregate change of all these descriptors. The anomaly of the overall frame termed as frame-EMD. This frame-EMD measure combined with our proposed pseudo-entropy feature clusters frames into different degrees of threat. Thus two major component of our solution pipeline are frame anomaly detection and threat-level determination. A schematic diagram of our solution pipeline is represented in Fig. 2.

3.1 Frame Anomaly Detection

Anomaly in a frame is determined by local motion descriptor generation based on our previous work [3]. The work followed a training less approach for detecting anomaly in crowd scenes. Firstly, each pedestrian (denoted by a bounding box) [22] is associated sequentially in an unsupervised manner based on their appearance consistency within a frame. This association is achieved using an adaptive 3D-DCT [3] based approach. The pedestrians that are appearance wise stable are consistently associated for a reasonable time and are termed as observers. The rest of the pedestrians are termed as targets.

Local motion descriptor is constructed over these observers with neighboring pedestrians as targets. Let the total observers be denoted by β then for i^{th} (where, $i \in \{1, \cdots, \beta\}$) observer the local descriptor is represented as $\{\mathbf{W}_i, \mathbf{F}_i\}_{k=1}^{M}$, where $\mathbf{W}_i \in \mathbb{R}^{1 \times M}$ denotes vector representing weights connecting i^{th} observer with its M spatially neighboring targets and $\mathbf{F}_i \in \mathbb{R}^{4 \times M}$ represents four dimensional feature vector describing motion statistics for each M neighbors. The weight and feature vector are computed over modulated optical flow (MOF) map [3]. MOF is achieved by modulating optical flow map using motion saliency map [23]. Modulation enhances and reduces flow magnitude of certain regions thereby increasing the discriminatory nature of constructed motion descriptors. Weight w_{ij} connecting i^{th} observer with j^{th} neighbor is computed based on distance between histograms of flow vectors of MOF map [3] in their corresponding bounding box regions. The four dimensional feature vector for $\mathbf{F}_i$ represents maximum, minimum, mean and variance of motion energy of MOF map for M neighbors surrounding i^{th} observer. Frame-level abnormality is achieved by measuring dissimilarity of observers over two consecutive frames based on the local descriptor constructed as $\{\mathbf{W}_i, \mathbf{F}_i\}_{k=1}^{M}$. This change is measured using earth mover distance (EMD) [4]. EMD evaluate the distribution between two multi-dimensional distribution in feature space of $\mathbf{F}_i^t$ and $\mathbf{F}_i^{t-1}$ provided with a given ground distance $\mathbf{W}_i^t$ and $\mathbf{W}_i^{t-1}$ for an observer in consecutive frames t and $t-1$. Thus the variation between local descriptor $\{\mathbf{W}_i^t, \mathbf{F}_i^t\}_{k=1}^{M}$ and $\{\mathbf{W}_i^{t-1}, \mathbf{F}_i^{t-1}\}_{k=1}^{M}$ for i^{th} observer in between frame t and $t-1$ can be represented as $EMD(\mathbf{W}_i^t, \mathbf{W}_i^{t-1}, \mathbf{F}_i^t, \mathbf{F}_i^{t-1})$ and the overall frame abnormality can be expressed as:

$$Fa_{raw}^t = \frac{1}{\beta} \sum_{i=1}^{\beta} EMD(\mathbf{W}_i^{t-1}, \mathbf{W}_i^t, \mathbf{F}_i^{t-1}, \mathbf{F}_i^t) \tag{1}$$

where, $EMD(\cdot)$ function compute earth mover distance [4] and Fa_{raw}^t is computed raw frame abnormality value at t^{th} frame instance. Sometimes, erroneous detection results in high Fa_{raw}^t value rendering a normal frame abnormal. Thus we perform a post-filtering operation to smooth out these effects by a weighted moving average filtering. Thus refined frame-EMD response can be expressed as:

$$Fa^t = \frac{1}{2n+1} \sum_{t'=(t-n)}^{t+n} wt \cdot Fa_{raw}^{t'} \tag{2}$$

where, wt is partial weight given by

$$wt = \begin{cases} \frac{1}{n+1} & if \quad t' = n \\ \frac{1}{2(n+1)} & if \quad t' \neq n \end{cases} \tag{3}$$

The filter length is of the form $(2n+1)$, where, $n \in \mathbb{Z}_+$. We have set $n = 3$ according to [3]. Fa^t acts as the first features in determining multi-level threat.

Algorithm 1. Threat analysis in crowd anomaly

Input: Video frames, $k = 3$
Output: $frame^t \leftarrow LT, MT, HT$; $segment^i \leftarrow LT, MT, HT$
1: **for** t=1 **to** T **do** /* T is the total no. of frames */
2:     Compute first feature Fa^t from equation 2
3:     Compute second feature $F_{entropy}^t$ from equation 5
4: **end for**
5: Apply clustering to label each frame
6: Obtain frame labels, $i.e$, $frame^t \leftarrow LT, MT, HT$
7: Construct ns segments of 1 sec duration
8: **for** i=1 **to** ns **do**
9:     Obtain segment labels, $i.e.$, $segment^i \leftarrow LT, MT, HT$ by Majority Voting
10: **end for**

3.2   Threat-Level Determination

Threat labeling can be particularly useful for searching and analyzing long duration videos. We have introduced two ways of threat label determination, (a) Frame-wise threat labeling and, (b) Segment-wise threat labeling. In frame-wise threat labeling, a threat pattern is introduced where each frame is represented by a color depending upon the level of threat. In segment-wise threat labeling, a video segment (consisting of several frames with temporal continuum) is marked with different level of threat. Segment wise threat indication is a coarser form of frame-wise indication. To disintegrate and mark frames in a video into different degrees of threat we solve a 3-class clustering problem where we cluster frames into low threat (LT), medium threat (MT) and high threat (HT). For Clustering of frames two features are constructed, the first feature is final frame anomaly value defined by Eq. 2 and the second feature is entropy like feature constructed from EMD values of observers. Since, entropy works on probability that lies $[0, 1]$ we construct entropy feature over normalized EMD values of all the observers which is computed as:

$$
\begin{aligned}
P(\mathbf{O}_i^{t-1}, \mathbf{O}_i^t) &= \frac{EMD(\mathbf{W}_i^{t-1}, \mathbf{W}_i^t, \mathbf{F}_i^{t-1}, \mathbf{F}_i^t)}{\sum_{i=1}^{\beta} EMD(\mathbf{W}_i^{t-1}, \mathbf{W}_i^t, \mathbf{F}_i^{t-1}, \mathbf{F}_i^t)} \\
&= \frac{1}{\beta} \frac{EMD(\mathbf{W}_i^{t-1}, \mathbf{W}_i^t, \mathbf{F}_i^{t-1}, \mathbf{F}_i^t)}{Fa_{raw}^t}
\end{aligned}
\tag{4}
$$

where, $P(\mathbf{O}_i^{t-1}, \mathbf{O}_i^t) \in [0, 1]$ is normalized EMD value and EMD distance between local motion descriptors of observer i is $EMD(\mathbf{W}_i^{t-1}, \mathbf{W}_i^t, \mathbf{F}_i^{t-1}, \mathbf{F}_i^t)$ given in Eq. 1. Based on the normalized EMD $P(\mathbf{O}_i^{t-1}, \mathbf{O}_i^t)$, the entropy feature can be computed as:

$$
F_{entropy}^t = -\sum_{i=1}^{\beta} P(\mathbf{O}_i^{t-1}, \mathbf{O}_i^t) \cdot \log\left(P(\mathbf{O}_i^{t-1}, \mathbf{O}_i^t)\right)
\tag{5}
$$

where, $F_{entropy}^t$ is entropy like feature as constructed from normalized EMD value of observers at frame t. We term this entropy like feature as pseudo entropy feature as it is computed over normalized EMDs. We analyze different threat pattern using two clustering methods, namely, k-means clustering and hierarchical agglomeration clustering. Once clustering is performed we label each cluster into LT, MT and HT clusters based on the mean of frame-EMD response (Eq. 2). Thus, without loss of generality we can label frames in a cluster based on the criteria as:

$$
\frac{1}{n_L} \sum_{l=1}^{n_L} Fa_l^t < \frac{1}{n_M} \sum_{m=1}^{n_M} Fa_m^t < \frac{1}{n_H} \sum_{h=1}^{n_H} Fa_h^t
\tag{6}
$$

where, Fa_l^t, Fa_m^t and Fa_h^t are frame-EMD responses categorized into LT, MT and HT respectively. Here, n_L, n_M and n_H are total number of frames in each of LT, MT and HT clusters. For labeling video segments, we perform majority voting that fetches the label corresponding to majority of frames in the segment and assigns for that video segment. The overall algorithm of threat labeling of frames and segments is depicted in Algorithm 1. The threat labeling problem can be easily extended to n-class (where, $n > 3$) threat labeling problem but that would be incremental. Moreover, introducing too much threat indication may create more ambiguity rather providing useful information. The overall framework is described in Algorithm 1.

4   Experimental Results

In this section, we first describe the datasets used followed by performance measures. We then tabulate and analyze model performance in details.

4.1 Dataset Description

We use two datasets, namely, UCSD [24] and UMN [25] widely adopted for crowd anomaly detection. **UCSD** [24] dataset consists of two video scenes, namely, *ped1* and *ped2* with 36 and 12 abnormal video sequences. In *ped1* each video sequence is composed of 200 frames recorded at 158×238 resolution, whereas, in *ped2* video sequences are recorded at a resolution of 360×240 with total frames ranging from 150 to 180. **UMN** [25] dataset comprises of three different scenarios, one in indoor and other two in outdoor recorded at 30 fps. The whole dataset consists of 11 video clips that begin with normal crowd activity accompanied by a sudden crowd escape (see Fig. 1(b)). The dataset consists a total of 7740 frames with a resolution of 320×240.

4.2 Performance Measures

We have shown only internal comparisons due to unavailability of ground truth for multi-class threat analysis. The comparisons are achieved based on three widely adopted cluster index as described later. We also perform statistical significance test pairing the experiments of two clustering techniques.

Index for Cluster Analysis:

1. *Average Silhouette width (ASW)*: Silhouette coefficient contrasts the average distance of data points to elements within same cluster with elements in other cluster. Data points with high silhouette value form good clusters and the mean over silhouette value of all the data points is known as Average Silhouette width (ASW) [26]. Higher ASW index is better.

2. *Davies–Bouldin (DB) index:* It is calculated using the following formula:

$$DB = \frac{1}{n} \sum_{i=1}^{n} \max_{j \neq i} \left(\frac{\sigma_i + \sigma_j}{d(c_i, c_j)} \right) \tag{7}$$

where, σ_i and σ_j measures the within cluster scatter for i^{th} and j^{th} clusters; c_i and c_j is the centroid of the cluster i and j, and $d(c_i, c_j)$ distance between the centroid points. Clustering algorithms with low intra-cluster distances and high inter-cluster distance will have low DB value. Clustering with smallest DB index is considered best.

3. *Calinski–Harabasz (CH) index:* This index represents the concept of dense and well separated clusters. The Calinski-Harabasz index is calculate based on ratio between inter-cluster dispersion (BCD) and intra-cluster dispersion (WCD) defined in [27] as:

$$CH(k) = \frac{N - k}{k - 1} \cdot \frac{BCD(k)}{WCD(k)} \tag{8}$$

where, N are total number of data points and k in the number of clusters. Higher CH index is better.

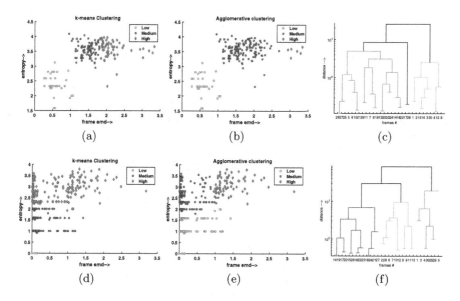

Fig. 3. Performance comparison based on k-means and agglomerative clustering on a sample UCSD and UMN dataset. Here (a) and (b) shown clusters based on k-means and agglomerative clustering while (c) shows dendogram construction based on agglomerative clustering on a UCSD video sequence. (d), (e) and (f) are corresponding representation of (a), (b) and (c) over an UMN video sequence.

Statistical Significance Test: Statistical test measures the similarity between two sets of measurements assuming a null hypothesis that all experiments belong to same distribution. We have perform Kolmogorov-Smirnov (KS) test for measuring statistical significance. Traditionally, a p-value ($\in [0,1]$) is defined as a statistical measure for measuring the rejection of null hypothesis. If the p-value is less than 0.05 then the two experiments are said to be statistically significant at 5% significance level.

4.3 Threat Level Analysis

We perform multi-level threat analysis over different video sequences of UCSD and UMN dataset. In Fig. 3, we have shown the clustering performance on a UCSD and UMN video scene for both k-means and agglomerative clustering. Each data point in the clusters represents a frame in a video sequence. In agglomerative clustering, we form clusters based on bottom up approach, where, frames are iteratively merged into different clusters based on similarity measure. As we are solving a three class labeling problem, the merging process is performed till three clusters are formed and represented using dendogram (see Fig. 3(c) and (f)). Based on the clustering results, a frame-wise and a segment-wise threat pattern is generated where LT, MT and HT threat are indicated using yellow, orange and red colors respectively along with frame-EMD response (see Fig. 4).

For quantitative analysis, we perform cluster analysis for UCSD ped1, UCSD ped2 and UMN dataset separately as shown in Table 1. We have also shown the clustering performance of our proposed $F_{entropy}$ over standard variance as 2nd feature. We have reported the mean and 95% confidence interval (CI) of cluster indexes over all video sequence in a dataset. From our experiments we establish that our proposed feature $F_{entropy}$ have attained better best scores of 0.825, 0.482 and 515.5 over variance as a feature with best scores 0.741, 0.592 and 450.9 in ASW, DB and CH index respectively for UCSD ped1 scenes. The same in also evident in case of UCSD ped2 and UMN dataset. Moreover, agglomerative clustering have performed better in most indexes with an improvement of 4.14% and 15.73% in UCSD ped1; 2.79% and 8.99% in UCSD ped2; and 2.25% and 21.96% in UMN datset in AWS and DB index respectively.

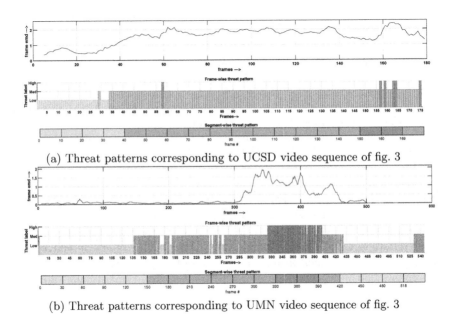

(a) Threat patterns corresponding to UCSD video sequence of fig. 3

(b) Threat patterns corresponding to UMN video sequence of fig. 3

Fig. 4. Top middle and bottom rows of (a) and (b) shows frame-EMD response, frame-wise and segment-wise threat pattern respectively. In segment-wise threat patterns (bottom rows of (a) and (b)) each segment is of 1 s duration.

The KS test is performed over paired indexes of k-means and agglomerative with a null hypothesis that the two experiments are from different distribution. Table 2 reports the p-values of all pair-wise test. It can be seen that all the p-values are less that 5% significance level for all the index except for CH index in UCSD ped2 which is also close to 0.05. The lowest (best) value attained is 6.2×10^{-4} for DB index in UCSD ped1 dataset. Thus the above experiments show that algglomerative clustering is better than k-means for separating frame with various threat of significance.

Table 1. Comparison between k-means and agglomerative clustering

Datasets	2nd Feature	ASW				DB				CH			
		k-means		Agglomerative		k-means		Agglomerative		k-means		Agglomerative	
		Mean	CI	Mean	CI	Mean	CI	Mean	CI	Mean	CI	Mean	CI
UCSD ped1	$F_{entropy}$	0.792	0.0566	0.825	0.0572	0.572	0.0767	0.482	0.0908	515.5	124.1	460.2	114.4
	Variance	0.595	0.0791	0.741	0.0593	0.787	0.0839	0.592	0.0741	302.2	101.9	450.9	95.2
UCSD ped2	$F_{entropy}$	0.608	0.0961	0.625	0.100	0.712	0.176	0.648	0.197	222.6	82.8	178.6	66.7
	Variance	0.571	0.142	0.598	0.0957	0.807	0.187	0.755	0.138	181.5	72.6	209.7	83.8
UMN	$F_{entropy}$	0.798	0.0774	0.816	0.0861	0.515	0.111	0.494	0.119	1276.1	544.8	1476.6	557.9
	Variance	0.667	0.0871	0.768	0.0931	0.674	0.0473	0.526	0.129	881.9	407.5	1351.8	532.9

Table 2. KS test statistics over ASW, DB and CH index pairs of experiments using k-means and agglomerative clustering

Datasets	Cluster index		
	ASW	DB	CH
UCSD ped1	0.0017	0.00062	0.0097
UCSD ped2	0.0026	0.0043	0.056
UMN	0.0037	0.0468	0.014

5 Conclusion

In this paper, we proposed a new paradigm for multi-level threat analysis in anomalous crowd scenes. Here, we extended the notion of two class frame level anomaly detection system to multi-class threat indication. As a result, two threat indication, namely, frame-wise and segment-wise pattern is presented for facilitating video search process. We also propose a new feature based on EMD response of anomaly detection system termed as pseudo entropy. Internal comparison and statistical KS test over UCSD and UMN dataset clearly indicate the effectiveness of the proposed framework. In future, we will be exploring more advanced machine learning techniques such as deep learning.

References

1. Ravanbakhsh, M., Sangineto, E., Nabi, M., Sebe, N.: Training adversarial discriminators for cross-channel abnormal event detection in crowds. In: WACV, pp. 1896–1904 (2019)
2. Chen, X.-H., Lai, J.-H.: Detecting abnormal crowd behaviors based on the div-curl characteristics of flow fields. Pattern Recognit. **88**, 342–355 (2019)
3. Sikdar, A., Chowdhury, A.S.: An adaptive training-less system for anomaly detection in crowd scenes (2019). arXiv:1906.00705
4. Rubner, Y., Tomasi, C.: The earth mover's distance. In: Rubner, Y., Tomasi, C. (eds.) Perceptual Metrics for Image Database Navigation. SECS, vol. 594, pp. 13–28. Springer, Boston (2001). https://doi.org/10.1007/978-1-4757-3343-3_2
5. Zhang, Y., Qin, L., Yao, H., Xu, P., Huang, Q.: Beyond particle flow: bag of trajectory graphs for dense crowd event recognition. In: ICIP, pp. 3572–3576 (2013)

6. Bera, A., Kim, S., Manocha, D.: Realtime anomaly detection using trajectory-level crowd behavior learning. In: CVPR Workshops, pp. 1289–1296 (2016)
7. Sodemann, A.A., Ross, M.P., Borghetti, B.J.: A review of anomaly detection in automated surveillance. IEEE Trans. Syst. Man Cybern. B Cybern. **42**(6), 1257–1272 (2012)
8. Wu, S., Moore, B.E., Shah, M.: Chaotic invariants of Lagrangian particle trajectories for anomaly detection in crowded scenes. In: CVPR, pp. 2054–2060 (2010)
9. Basharat, A., Gritai, A., Shah, M.: Learning object motion patterns for anomaly detection and improved object detection. In: CVPR, pp. 1–8 (2008)
10. Cheng, H.-Y., Hwang, J.-N.: Integrated video object tracking with applications in trajectory-based event detection. J. Vis. Commun. Image Represent. **22**(7), 673–685 (2011)
11. Ali, S., Shah, M.: A Lagrangian particle dynamics approach for crowd flow segmentation and stability analysis. In: CVPR, pp. 1–6 (2007)
12. Mahadevan, V., Li, W., Bhalodia, V., Vasconcelos, N.: Anomaly detection in crowded scenes. In: CVPR, pp. 1975–1981 (2010)
13. Kim, J., Grauman, K.: Observe locally, infer globally: a space-time MRF for detecting abnormal activities with incremental updates. In: CVPR, pp. 2921–2928 (2009)
14. Cong, Y., Yuan, J., Liu, J.: Abnormal event detection in crowded scenes using sparse representation. Pattern Recognit. **46**(7), 1851–1864 (2013)
15. Mehran, R., Moore, B.E., Shah, M.: A streakline representation of flow in crowded scenes. In: Daniilidis, K., Maragos, P., Paragios, N. (eds.) ECCV 2010. LNCS, vol. 6313, pp. 439–452. Springer, Heidelberg (2010). https://doi.org/10.1007/978-3-642-15558-1_32
16. Mehran, R., Oyama, A., Shah, M.: Abnormal crowd behavior detection using social force model. In: CVPR, pp. 935–942 (2009)
17. Helbing, D., Molnár, P.: Social force model for pedestrian dynamics. Phys. Rev. E **51**(5), 4282–4286 (1995)
18. Kratz, L., Nishino, K.: Anomaly detection in extremely crowded scenes using spatio-temporal motion pattern models. In: CVPR, pp. 1446–1453 (2009)
19. Zhang, D., Gatica-Perez, D., Bengio, S., McCowan, I.: Semi-supervised adapted HMMs for unusual event detection. In: CVPR, vol. 1, pp. 611–618 (2005)
20. Li, W., Mahadevan, V., Vasconcelos, N.: Anomaly detection and localization in crowded scenes. IEEE Trans. Pattern Anal. Mach. Intell. **36**(1), 18–32 (2014)
21. Hasan, M., Choi, J., Neumann, J., Roy-Chowdhury, A.K., Davis, L.S.: Learning temporal regularity in video sequences. In: CVPR (2016)
22. Sikdar, A., Chowdhury, A.S.: An ellipse fitted training-less model for pedestrian detection. In: ICAPR, pp. 1–6 (2017)
23. Gangapure, V.N., Nanda, S., Chowdhury, A.S.: Superpixel based causal multi-sensor video fusion. IEEE Trans. Circuits Syst. Video Technol. **28**(6), 1263–1272 (2018)
24. Li, X., Dick, A., Shen, C., van den Hengel, A., Wang, H.: Incremental learning of 3D-DCT compact representations for robust visual tracking. IEEE Trans. Pattern Anal. Mach. Intell. **35**(4), 863–881 (2013)
25. Yuan, Y., Fang, J., Wang, Q.: Online anomaly detection in crowd scenes via structure analysis. IEEE Trans. Cybern. **45**(3), 548–561 (2015)
26. Fox, W.R., Kaufman, L., Rousseeuw, P.J.: Finding groups in data: an introduction to cluster analysis. Appl. Stat. **40**(3), 486 (1991)
27. Calinski, T., Harabasz, J.: A dendrite method for cluster analysis. Commun. Stat. Simul. Comput. **3**(1), 1–27 (1974)

Unsupervised Clustering

Discovering Cricket Stroke Classes in Trimmed Telecast Videos

Arpan Gupta$^{(\boxtimes)}$, Ashish Karel, and M. Sakthi Balan

Department of Computer Science and Engineering,
The LNM Institute of Information Technology, Jaipur, India
{arpan,ashish.karel.y15,sakthi.balan}@lnmiit.ac.in
https://www.lnmiit.ac.in

Abstract. Activity recognition in sports telecast videos is challenging, especially, in outdoor field events, where there is a lot of camera motion. Generally, camera motions like zoom, pan, and tilt introduce noise in the low-level motion features, thereby, effecting the recognition accuracy, but in some cases, such camera motion can have a pattern which can be useful for action recognition in trimmed videos.

In this work, we experimentally discover the types of strokes in a Cricket strokes dataset using direction information of dense optical flow. We use trimmed videos of Cricket strokes taken from Cricket telecast videos of match highlights. The predominant direction of motion is found by summing up the histograms of optical flow directions, taken for significant pixels, over the complete Cricket stroke clip. We show that such a quantized representation of the optical flow direction for the complete video clip can be used for Cricket stroke recognition. Our method uses an unsupervised K-Means clustering of the extracted clip feature vectors and we evaluate our results for 3-cluster K-Means by manually annotating the clusters as *Left* strokes, *Right* strokes and *Ambiguous* strokes. The accuracy for different set of feature vectors was obtained by varying the bin granularity of the histogram and the optical flow magnitude threshold. The best result we obtained, for 3-cluster K-Means and 562 stroke instances, was 87.72%.

Keywords: Cricket stroke · Unsupervised clustering · Farneback optical flow

1 Introduction

There is a lot of scope for technological innovation in the field of sports. Some modern commercial systems, for e.g., HawkEye [1], Snickometer, Hot-spot etc., have helped in improving decision making capability and providing a whole new dimension to the game. These systems work on vision and sensor data for accurate predictions in events like Cricket, Tennis, and Badminton. Apart from decision making, video analytics can be used in sports for identifying playing styles,

© Springer Nature Singapore Pte Ltd. 2020
N. Nain et al. (Eds.): CVIP 2019, CCIS 1148, pp. 509–520, 2020.
https://doi.org/10.1007/978-981-15-4018-9_45

helping the coaches, automated highlight generation [7,17], and for video annotations [6].

Activity recognition in sports videos has been an active area of research in the current decade [25,27]. Researchers have primarily looked into detection and tracking methods [18,29,30] for improved decision making and video content analysis. Finding the position of players and tracking them is the key objective in many sporting events, such as Soccer [13], Basketball [19], and Tennis [26]. The tracking information is then used to detect events based on domain knowledge.

Understanding any sporting event from the raw telecast videos is not a trivial task. Such events involve a set of moving actors, objects of interest and the interactions between them [4]. Modeling these interactions needs some prior knowledge about the rules of the game, which is useful only after one is able to accurately detect the actors, objects and interactions. This task is much easier if the cameras are stationary (such as HawkEye [1]), but is tough if recognition is to be performed in telecast videos. The main source of inherent noise being introduced due to camera motion, such as zooming, panning, and tilting. An action is difficult to recognize with these camera motions, but in a sporting event, such as Cricket, there may be a pattern of camera motion associated with an event of interest.

In this work, we show that in spite of the presence of camera motion in Cricket telecast videos, a simple frame-level motion feature, such as dense optical flow, can be used to detect the direction of stroke play at a coarse level of granularity. We take trimmed Cricket stroke videos from 26 Highlight matches, which has 562 Cricket stroke instances, and cluster the binned angular information of "significant" pixels in a Cricket stroke. The "significant" pixels are defined as the pixels having optical flow magnitude above a certain threshold. We use only the raw RGB frames as input which had a constant frame rate of 25FPS and 360×640 frame size.

Our work tries to solve the problem of Cricket stroke recognition based on the direction of stroke play, without looking at the tasks of detection and tracking of the relevant actors. It can be helpful in automatically annotating a large video data corpus of trimmed Cricket strokes, with the type of stroke being played, thereby, reducing the time and effort of manual annotation. This work is also helpful in generating player analytics, given the players' trimmed video strokes. Since, our work uses only the RGB frames of telecast videos, without the use of any other data modality, therefore, it is widely applicable and does not depend on the availability, synchronization and correctness of other data modalities such as audio or text.

The paper is divided into the following sections. Section 2 gives the literature survey in the field of Cricket activity recognition. Our methodology is described in detail in Sect. 3 followed by the experimentation and results section in Sect. 4. Finally, we conclude and provide ideas for potential future directions in Sect. 5.

2   Literature Survey

There are a few works that have attempted Cricket specific activity recognition. Kolekar in [14] has indexed the Cricket and Soccer video sequences with semantic labels using excitement level in audio features. Similarly, Kolekar et al. in [15,16] have tried to understand the semantics in the video with association rule mining and used it in the task of automated highlight generation. Harikrishna et al. [11] have extended the work by Kolekar et al. [15,17] by "hand-crafting" the features such as Grass Pixel Ratio, Pitch Pixel Ratio etc. which seem less likely to generalize on unseen videos.

Sankar et al. [22] and Sharma et al. [23] have used text commentaries and used dynamic programming based alignment to annotate the segmented video shots. Their method depends on the accuracy of the segmentation step and the availability of text commentary for any given video. Though, it is an effective way of creating an annotated dataset, it does not directly tackle the problem of Cricket activity recognition from videos and bypasses this challenging task, altogether.

Our work uses the directional information of dense optical flow as a distinguishing criteria for discovering Cricket strokes at a coarse level of granularity[1]. It is similar to the histogram of oriented optical flow (HOOF) defined by Chaudhry et al. [5] but our histograms contain only direction information (not weighted by optical flow magnitude) and is summed over the entire clip. In [5], the authors have used Non-Linear Dynamics for histogram sequences in order to classify simple human actions such as walking, running, jumping etc. This approach is useful when human actions have to be recognized and camera motion based noise is negligible. The problem at hand, in our case, does not depend on the temporal dynamics of human actions alone.

Wang et al. [28] used unsupervised clustering of image features for discovery of action classes and do not use motion information while Soomro et al. [24] focus on human action classes by using discriminative clustering of video feature followed by localization of discovered action.

3   Methodology

The strokes played by the batsman in Cricket telecast videos, are the events of interest in Cricket matches. There are a number of cameras that capture this event from various angles for providing better viewing experience. Among these cameras, there are two cameras placed side-by-side, at the stands, in front of the batsman, that capture the bowler run-up, batsman hitting the ball and the direction in which the ball travels. These cameras take the close-up shot and a wide-angle shot, respectively. The close-up view slowly zooms into the pitch area as the bowler is about to bowl the delivery. If the batsman hits/misses the ball and the ball stays near the pitch area, such as in cases when batsman leaves

---

[1] A fine granularity would be if we consider more types of strokes based on direction of stroke play, such as *long-off, long-on, third-man, fine-leg* etc.

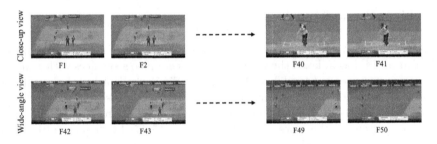

Fig. 1. Sequence of frames in a Cricket stroke (of 2 s duration) from frame 1 (F1) to frame 50 (F50).

the delivery, is bowled out, caught behind or taps the ball gently, then only the close-up shot is sufficient to cover the outcome of the delivery. There is a camera switch from the close-up shot to the wide-angle shot, if the batsman plays the stroke, and the ball goes to some distance in the field. The wide-angle shot follows the direction in which the ball is played and slowly zooms in on the ball. These camera shots, containing either one close-up camera shot or both the close-up and wide-angle shots, constitute a complete Cricket stroke and are considered for manual annotation in Gupta and Balan [9,10]. Figure 1 illustrates a sample Cricket stroke, as described in Gupta and Balan. They use this two camera assumption for manually annotating the Cricket stroke segments in untrimmed Highlight videos. There is a camera switch (CUT transition) after F41 from the close-up view to the wide-angle view and the stroke played is towards the *off* side (*Left*).

3.1   Dataset Description

The Cricket strokes dataset of Gupta and Balan [9] is used in our experiments. This dataset consists of start and end frame number annotations (list of frame number tuples) for all the Cricket strokes in 26 Highlight videos of ICC WT20 2016 tournament. There are a total of 562 annotated clips/strokes having a frame size of 360×640 at 25FPS frame rate. Since, these strokes come from Highlight telecast videos, therefore, they have a greater number of important events (such as sixes, fours, bowled-out etc.) in a shorter time-span, as compared to any full-match telecast video. The dataset had training, validation and testing partitions of 16, 5 and 5 videos, respectively, for the temporal localization task, but for our task of unsupervised clustering, we do not need the partitions.

There are large variations in the types of strokes that are played, and the direction in which the strokes are played. The camera motions involved are zooming, panning and tilting while covering these events. It is to be noted that since, we are using only trimmed Cricket strokes, each stroke video clip has only a single Cricket stroke having one or two video shots (Fig. 2).

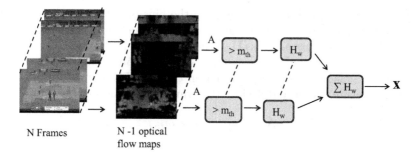

N Frames   N -1 optical
                   flow maps

Fig. 2. Feature extraction pipeline

3.2   Feature Representation

Motion features such as Motion History Images (MHI) and Motion Energy Images (MEI) [2,3] provide a template based representation of foreground motion visualized in 2D. They work well when we have stationary background and are susceptible to camera motion induced noise.

We use sequence of dense optical flow features (Farneback [8]) and get a summed up histogram of the direction information for the complete Cricket stroke. The quantized feature vector represents the direction of prevalent motion in the entire clip.

Let V be the set of trimmed videos, in our case of Cricket strokes. A stroke will be $v_i \in V$, where $1 \leq i \leq |V|$. The number of frames in each stroke clip will be N_i. Traditionally, the optical flow can be computed for any two frames separated by dt temporal interval. The frames follow the brightness constancy assumption for effective results. Taking consecutive frames with $dt = 1$, it is more likely that the brightness constancy assumption holds i.e., the brightness of the pixels remains unchanged, and only their position changes by (dx, dy). Equation 1 denotes the displacement in the two frames using brightness constancy assumption, while Eq. 2 gives the optical flow F, in cartesian coordinate system. The details can be found in [12]. The Eqs. 3 and 4 convert the optical flow displacement values into polar coordinates, denoted by magnitude map M and an angle map A.

$$I(x, y, t) = I(x + dx, y + dy, t + dt) \tag{1}$$

$$F = (dx, dy), \text{ where } dt = 1 \tag{2}$$

$$M = \sqrt{dx^2 + dy^2} \tag{3}$$

$$A = \begin{cases} tan^{-1}\left(\frac{dy}{dx}\right), & \text{if } tan^{-1}\left(\frac{dy}{dx}\right) \geq 0 \\ 2\pi + tan^{-1}\left(\frac{dy}{dx}\right), & \text{if } tan^{-1}\left(\frac{dy}{dx}\right) < 0 \end{cases} \tag{4}$$

We use a global representation of the direction information by taking the binned angle values of the significant pixels in the clip. The significant pixels are the ones having corresponding magnitude values greater than some threshold

(m_{th}), given in Eq. 5[2]. Finally, we sum up the bins over the entire clip, having $N-1$ flow maps from N clip frames. Equation 6 calculates the feature vector $\mathbf{x} \in \mathbb{R}^w$ as the sum of histograms $(\mathbf{H}_w)$ and scaled down by the number of frames in the clip. Since, the angle values are positive, therefore, a larger value in a bin will denote a general direction of motion in the clip. The number of bins is the size of the feature vector and decides the granularity of action categorization.

$$\mathbf{A}_{th} = \arg_A(M \geq m_{th}) \tag{5}$$

$$\mathbf{x} = \frac{1}{N} \sum_{j=1}^{N-1} \mathbf{H}_w(\mathbf{A}_{th}(j, j+1)) \tag{6}$$

3.3 Clustering

The feature vectors are first L2 normalized and then clustered by applying K-Means [21] clustering. Each individual cluster that is formed, represents unique directional information.

4 Experimentation and Results

Extensive set of experiments were carried out on the 562 trimmed videos of the dataset, by varying the number of bins (w), magnitude threshold (m_{th}) and the number of clusters for K-Means. The parameter w defines the vector size of $\mathbf{x}$, and uniformly divides the range of $(0, 2\pi)$ into w bins. Equation 4 ensures that bins have only positive values in range $(0, 2\pi)$. For a finer granularity of clustering, i.e., if there are more number of classes, then more number of bins (higher w value) are needed which may, subsequently, reduce the accuracy.

The parameter m_{th} is the lower bound to the optical flow magnitude or the amount of displacement of the pixels in consecutive frames. This parameter helps in finding the pixels that are eligible for voting towards the creation of the binned histogram. The selected pixels' optical flow direction information is binned and summed up over the entire video. The effect of applying this threshold is two-fold. Firstly, with a large value of m_{th}, less number of pixels get selected, but this does not help in removing the noisy optical flow pixels that may appear at the CUTs. Secondly, a lower value of m_{th} helps in getting more votes, but should not be too low, so as to consider pixels with "jittering" effect or stationary background.

As the first step, we choose the value of $m_{th} = 5$ and $w = 10$ for feature extraction. K-Means clustering was applied to the extracted feature vectors with number of clusters ranging from 2 to 50. A plot of the SSE score (Sum of Squared

---

[2] In the equation, A_{th} is a list of angle values selected from the matrix A, that have a corresponding magnitude value (from M) above a threshold m_{th}. The number of values in A_{th}, for different optical flow maps, can be different.

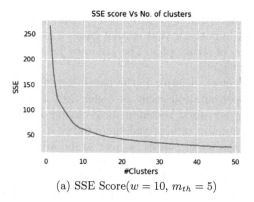

(a) SSE Score($w = 10$, $m_{th} = 5$)

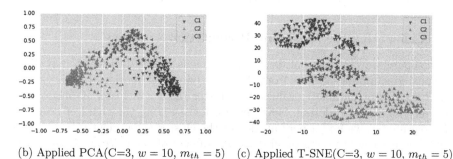

(b) Applied PCA(C=3, $w = 10$, $m_{th} = 5$)   (c) Applied T-SNE(C=3, $w = 10$, $m_{th} = 5$)

Fig. 3. Clusters and SSE scores for different bin sizes.

Error) for different number of clusters is shown in Fig. 3a. The Figs. 3b and c show the K-Means clustering visualization for the above parameters after applying PCA and T-SNE [20], respectively.

A manual examination of the three clusters revealed characteristic camera motion associated with each cluster. Cluster *C1*, in Figs. 3b and c, mainly contained *off-side*[3] played strokes (panning motion towards left), while cluster *C2* contained strokes played towards the *leg-side* (panning motion towards right). These strokes had fast camera motion and covered mostly *lofted* Cricket strokes in their respective directions. The third cluster included strokes with *bowled-out deliveries, straight-drives*, etc. where there was not enough camera motion or the motion was in a different direction.

For evaluation purpose, we independently label all the Cricket strokes as *Left*, *Right* or *Ambiguous*. We manually labeled 177 as *Left*, 224 as *Right* and 161 as *Ambiguous* in a total of 562 samples. We do not consider the motion of the ball, tracking of batsman or the pose of players as a labeling criterion. It is only

---

[3] In Cricket, *leg-side* and *off-side* depend on whether the batsman is right-handed or left-handed, but here we refer to the left of the batsmans' position as *off-side* and right as the *leg-side*.

the camera motion that determines the ground truth annotations. This way, the experiments confirm our hypothesis that regular camera motion can be used for determining the direction of stroke play. The accuracy for $m_{th} = 5$, $w = 10$ and 3-clusters K-Means was 85.59%.

We also study the effect of variation in the parameters for 3-cluster K-Means and find the maximum permutation accuracy (refer Sect. 4.1) by varying w and m_{th}. Figure 4 shows the accuracy heatmap for 3-cluster K-Means. There will be a trade-off with the number of bins and the number of classes (i.e., number of clusters). For more number of classes, a finer granularity of bins is needed. Therefore, $w = 2$ has much lower accuracy for 3-cluster K-Means. Some sample clustering visualization for 3-cluster K-Means are illustrated in Figs. 6 and 7 in the Appendix. The SSE scores, as observed in Fig. 5, show that the number of bins (w) depends on the number of classes chosen for clustering.

4.1 Calculating the Maximum Permutation Accuracy

An unsupervised clustering of data points assigns cluster IDs which may differ at repeated sampling of data from the underlying distribution. In our case as well, for different parameter values, the cluster IDs for 3-cluster K-Means may have any undefined ordering. Nevertheless, for studying the trend of accuracy for a parameter, we find a permutation of the cluster assignment ($C1$, $C2$, $C3$) that when matched to the ground truth assignment of (*Left*, *Right*, *Ambiguous*), has the maximum accuracy. We use this accuracy value in the heatmap (Fig. 4) and infer that accuracy decreases with the increase in m_{th}.

The maximum accuracy value for 3-cluster K-Means was obtained at $w = 6$ and $m_{th} = 11$ with 87.72%.

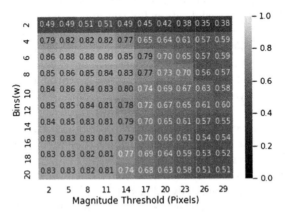

Fig. 4. Accuracy heatmap for different bin sizes and magnitude threshold values.

5   Conclusion and Future Work

In this work, we show that a motion feature, such as direction information of the dense optical flow, can be used for Cricket stroke recognition in telecast videos at a coarse level of granularity. The Cricket telecast videos possess a pattern of camera motion, which does not, necessarily, act as a source of noise for the video. The zooming and panning camera motions, help in recognition of the direction of stroke play. We experimentally show that a quantized histogram of the optical flow directions, taken for significant pixels, can be used for detection of the direction of stroke play. By choosing the number of bins and optical flow magnitude threshold hyper-parameters, we cluster the feature vectors for trimmed stroke videos and study the motion patterns for 3-cluster K-Means clustering. The best accuracy obtained on 562 stroke instances, after manually annotating the strokes as *Left*, *Right* or *Ambiguous*, was 87.72%. Our work can be used, along with the localization of strokes, to automatically extract and label Cricket strokes from raw telecast videos, and, potentially, annotate a large scale Cricket dataset.

In our future work, we plan to use pose-based features and tracking based features of players/object for recognizing the strokes in Cricket dataset.

Appendix: Clustering Visualizations for Some w and m_{th}

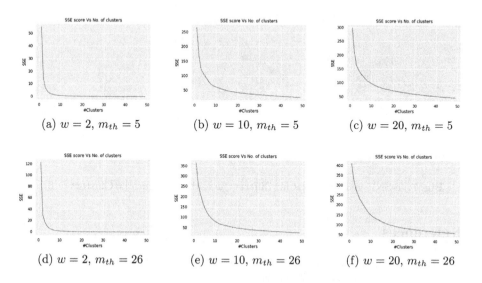

(a) $w = 2$, $m_{th} = 5$ (b) $w = 10$, $m_{th} = 5$ (c) $w = 20$, $m_{th} = 5$

(d) $w = 2$, $m_{th} = 26$ (e) $w = 10$, $m_{th} = 26$ (f) $w = 20$, $m_{th} = 26$

Fig. 5. SSE scores for different w and m_{th} values.

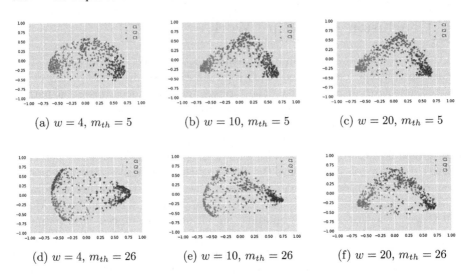

Fig. 6. PCA visualizations for different w and m_{th} values (#Clusters = 3).

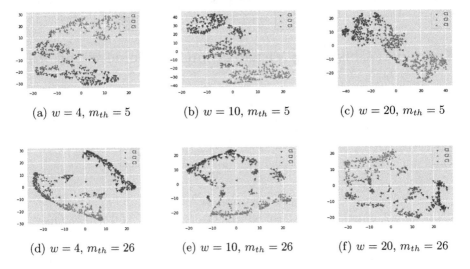

Fig. 7. T-SNE visualizations for different w and m_{th} values (#Clusters = 3).

References

1. Hawk-Eye Innovations hawk-eye in cricket. https://www.hawkeyeinnovations.com/sports/cricket. Accessed 27 Mar 2019
2. Bobick, A.F.: Action recognition using temporal templates. J. Chem. Inf. Model. **53**(9), 1689–1699 (2013). https://doi.org/10.1017/CBO9781107415324.004
3. Bobick, A.F., Davis, J.W.: The recognition of human movement using temporal templates. IEEE Trans. Pattern Anal. Mach. Intell. **23**(3), 257–267 (2001). https://doi.org/10.1109/34.910878

4. Chao, Y., Liu, Y., Liu, X., Zeng, H., Deng, J.: Learning to detect human-object interactions. CoRR abs/1702.05448 (2017). http://arxiv.org/abs/1702.05448
5. Chaudhry, R., Ravichandran, A., Hager, G., Vidal, R.: Histograms of oriented optical flow and Binet-Cauchy kernels on nonlinear dynamical systems for the recognition of human actions. In: 2009 IEEE Computer Society Conference on Computer Vision and Pattern Recognition Workshops, CVPR Workshops 2009, pp. 1932–1939 (2009). https://doi.org/10.1109/CVPRW.2009.5206821
6. De Campos, T.E., et al.: A framework for automatic sports video annotation with anomaly detection and transfer learning. In: Machine Learning and Cognitive Science, Collocated with EUCOGIII (2013)
7. Ekin, A., Tekalp, A.M., Mehrotra, R.: Automatic soccer video analysis and summarization. IEEE Trans. Image Process. 12(7), 796–807 (2003). https://doi.org/10.1109/TIP.2003.812758
8. Farnebäck, G.: Two-frame motion estimation based on polynomial expansion. In: Bigun, J., Gustavsson, T. (eds.) SCIA 2003. LNCS, vol. 2749, pp. 363–370. Springer, Heidelberg (2003). https://doi.org/10.1007/3-540-45103-X_50. http://dl.acm.org/citation.cfm?id=1763974.1764031
9. Gupta, A., Balan, M.S.: Temporal cricket stroke localization from untrimmed highlight videos. In: Proceedings of the Eleventh Indian Conference on Computer Vision, Graphics and Image Processing, ICVGIP 2018. ACM, New York (2018, to appear). https://doi.org/10.1145/3293353.3293415. http://doi.acm.org/10.1145/3293353.3293415
10. Gupta, A., Balan, M.S.: Cricket stroke extraction: towards creation of a large-scale cricket actions dataset. arXiv e-prints arXiv:1901.03107 (2019)
11. Harikrishna, N., Satheesh, S., Sriram, S.D., Easwarakumar, K.S.: Temporal classification of events in cricket videos. In: 2011 National Conference on Communications (NCC), pp. 1–5 (2011). https://doi.org/10.1109/NCC.2011.5734784
12. Horn, B., Schunck, B.: Determining optical flow. Artif. Intell. 17(1–2), 185–203 (1981). https://doi.org/10.1016/0004-3702(93)90173-9
13. Kim, K., Grundmann, M., Shamir, A., Matthews, I., Hodgins, J., Essa, I.: Motion fields to predict play evolution in dynamic sport scenes. In: Proceedings of the IEEE Computer Society Conference on Computer Vision and Pattern Recognition, pp. 840–847 (2010). https://doi.org/10.1109/CVPR.2010.5540128
14. Kolekar, M.H.: Bayesian belief network based broadcast sports video indexing. Multimed. Tools Appl. 54(1), 27–54 (2011). https://doi.org/10.1007/s11042-010-0544-9
15. Kolekar, M.H., Palaniappan, K., Sengupta, S.: Semantic event detection and classification in cricket video sequence. In: 2008 Sixth Indian Conference on Computer Vision, Graphics Image Processing, pp. 382–389 (2008). https://doi.org/10.1109/ICVGIP.2008.102
16. Kolekar, M.H., Sengupta, S.: Event-importance based customized and automatic cricket highlight generation. In: 2006 IEEE International Conference on Multimedia and Expo, pp. 1617–1620 (2006). https://doi.org/10.1109/ICME.2006.262856
17. Kolekar, M.H., Sengupta, S.: Semantic concept mining in cricket videos for automated highlight generation. Multimed. Tools Appl. 47(3), 545–579 (2010). https://doi.org/10.1007/s11042-009-0337-1
18. Liu, J., Carr, P., Collins, R.T., Liu, Y.: Tracking sports players with context-conditioned motion models. In: 2013 IEEE Conference on Computer Vision and Pattern Recognition, pp. 1830–1837 (2013). https://doi.org/10.1109/CVPR.2013.239

19. Lu, W.L., Ting, J., Little, J.J., Murphy, K.P.: Learning to track and identify players from broadcast sports videos. IEEE Trans. Pattern Anal. Mach. Intell. **35**(07), 1704–1716 (2013). https://doi.org/10.1109/TPAMI.2012.242

20. van der Maaten, L., Hinton, G.: Visualizing data using t-SNE. J. Mach. Learn. Res. **9**, 2579–2605 (2008). http://www.jmlr.org/papers/v9/vandermaaten08a.html

21. MacQueen, J.: Some methods for classification and analysis of multivariate observations. In: Proceedings of the Fifth Berkeley Symposium on Mathematical Statistics and Probability, Volume 1: Statistics, pp. 281–297. University of California Press, Berkeley (1967). https://projecteuclid.org/euclid.bsmsp/1200512992

22. Pramod Sankar, K., Pandey, S., Jawahar, C.V.: Text driven temporal segmentation of cricket videos. In: Kalra, P.K., Peleg, S. (eds.) ICVGIP 2006. LNCS, vol. 4338, pp. 433–444. Springer, Heidelberg (2006). https://doi.org/10.1007/11949619_39

23. Sharma, R.A., Sankar, K.P., Jawahar, C.V.: Fine-grain annotation of cricket videos. CoRR abs/1511.07607 (2015). http://arxiv.org/abs/1511.07607

24. Soomro, K., Shah, M.: Unsupervised action discovery and localization in videos. In: The IEEE International Conference on Computer Vision (ICCV) (2017)

25. Soomro, K., Zamir, A.R.: Action recognition in realistic sports videos. In: Moeslund, T.B., Thomas, G., Hilton, A. (eds.) Computer Vision in Sports. ACVPR, pp. 181–208. Springer, Cham (2014). https://doi.org/10.1007/978-3-319-09396-3_9

26. Teachabarikiti, K., Chalidabhongse, T.H., Thammano, A.: Players tracking and ball detection for an automatic tennis video annotation. In: 2010 11th International Conference on Control Automation Robotics Vision, pp. 2461–2494 (2010). https://doi.org/10.1109/ICARCV.2010.5707906

27. Thomas, G., Gade, R., Moeslund, T.B., Carr, P., Hilton, A.: Computer vision for sports: current applications and research topics. Comput. Vis. Image Underst. **159**, 3–18 (2017). https://doi.org/10.1016/j.cviu.2017.04.011. http://www.sciencedirect.com/science/article/pii/S1077314217300711. Computer Vision in Sports

28. Wang, Y., Jiang, H., Drew, M.S., Li, Z.-N., Mori, G.: Unsupervised discovery of action classes. In: 2006 IEEE Computer Society Conference on Computer Vision and Pattern Recognition (CVPR 2006), vol. 2, pp. 1654–1661 (2006). https://doi.org/10.1109/CVPR.2006.321

29. Yao, A., Uebersax, D., Gall, J., Van Gool, L.: Tracking people in broadcast sports. In: Goesele, M., Roth, S., Kuijper, A., Schiele, B., Schindler, K. (eds.) DAGM 2010. LNCS, vol. 6376, pp. 151–161. Springer, Heidelberg (2010). https://doi.org/10.1007/978-3-642-15986-2_16

30. Zhu, G., Xu, C., Huang, Q., Gao, W.: Automatic multi-player detection and tracking in broadcast sports video using support vector machine and particle filter. In: 2006 IEEE International Conference on Multimedia and Expo, pp. 1629–1632 (2006). https://doi.org/10.1109/ICME.2006.262859

Author Index

Printed in the United States
By Bookmasters